Red Dwarfs

David S. Stevenson

Red Dwarfs

Their Geological, Chemical, and Biological Potential for Life

Second Edition

Springer

David S. Stevenson
Sherwood, Nottinghamshire, UK

ISBN 978-3-030-25549-7 ISBN 978-3-030-25550-3 (eBook)
https://doi.org/10.1007/978-3-030-25550-3

This Springer imprint is published by the registered company Springer Nature Switzerland AG
The registered company address is: Gewerbestrasse 11, 6330 Cham, Switzerland

To my adored, beautiful wife, who is a constant sense of inspiration and support.

Preface

In 2011, when I first proposed *Under a Crimson Sun*, few planets had been discovered in orbit around red dwarf stars. Of these, only Gliese 581and Gliese 667C were (modestly) characterized. Moreover, atmospheric modeling was fairly simplistic, and such models applied principally to water planets. Since this time, the number of planetary discoveries has skyrocketed, as have the number of questions surrounding whether the planets of red dwarfs are habitable or not.

Since the publication of *Under a Crimson Sun* in 2013, I have begun to investigate the biological potential of such planets, carving out something of an appropriate niche in astrobiology. I wondered whether one could merge ecological and evolutionary principles with geophysics and astrophysics to provide a clearer picture of the true likelihood of complex biological life emerging on such planets. Without giving too much away, the outcome of these experiments has proved less positive than I initially imagined in 2013. The reasons are relatively straightforward but not entirely obvious from first expectations.

In recent years, the concept of *habitability* has evolved from a simple concept that concerns solely the circumstellar habitable zone to one that embodies a greater range of relevant variables. Therefore, in this book, there is a detailed dissection of stellar and planetary evolution, but with a greater emphasis on the internal workings of the planets themselves. A considerable body of new work has emerged in recent years. This has allowed for a more formal investigation of planetary evolution and what this means for the long-term survival and evolution of life on these worlds. There is an expansion of the discussion of planetary atmospheres and their impact on life from an earlier book I wrote. This book fully explores various atmospheric models and then applies them to address habitability. What still lacks in this area are circulation models for realistic planets with both land and sea. Consequently, I have made some reasonable suggestions how to do this, by adapting these models to incorporate terrestrial observations. Integrating these observations introduces some significant impacts on habitability. These models emerge in Chaps. 5, 7, and 9.

Finally, the role of geophysical processes in determining the geographical landscape is considered. I then apply these principles to derive ecological models, which are based on terrestrial observations. This allows us to make reasonable assumptions as to the true biological potential of exoplanets.

Although this book will always be a "work in progress," it aims to provide two things: first and foremost, it forms a multidisciplinary review of current and ongoing research. Second, it presents plausible outcomes from this research—both from other workers in their fields and from my own. These illustrate the full complexity of the area and perhaps reveal our ongoing ignorance regarding what truly makes a planet habitable for life in general and, more interestingly, complex intelligent life.

Sherwood, Nottinghamshire, UK David S. Stevenson

Acknowledgment

I'd like to thank John Watson for introducing me to the team at Springer and getting my career in book writing going, 6 years ago.

Contents

About the Author

David S. Stevenson was born in Paisley, Scotland, in 1968. He studied Molecular Biology at Glasgow University and obtained a Ph.D. in Genetics from the University of Cambridge. He then studied Astronomy and Planetary Sciences and Geophysics and Geochemistry (S267) at the Open University (OU).

After a stint in academia, Stevenson became a teacher but continued to write science articles for various publications. In July 2007, an article on Type Ia Supernovae, "A Bigger Bang," was published in *Sky & Telescope*, with a second article on supernovae published in the same magazine in October 2011. Three books have followed on supernovae, red dwarf stars, and star clusters (*Extreme Explosions*, *Under a Crimson Sun*, and *The Complex Life of Star Clusters*, respectively). A further article on stellar mergers is at the production stages with *Astronomy* magazine, while a spin-off article from *Extreme Explosions,* called "Supersized Stellar Mystery," was published in *Popular Astronomy* in February 2014.

Despite a background in biology, his interest in astronomy dates from an early age. Many parallels can be drawn between the evolution of life and the evolution of the astronomical bodies life inhabits. These ideas were developed in '*The Nature of Life and its Potential to Survive*' and most recently in 'Granite Skyscrapers'. It is these concepts that are the central focus for this book and a number of accompanying academic peer-reviewed papers. Exploring the crossroads of different scientific disciplines is a key interest and motivation in Stevenson's research and writing.

Chapter 1
The Formation of Stars and Planets at the Bottom of the Main Sequence

Introduction

You cannot consider life on other worlds before first considering their host star. This determines their orbital period, the intensity of the radiation the planet receives and the distribution of wavelengths over which meaningful intensities of radiation are emitted. This chapter is dedicated to the two most abundant classes of stars in the universe: the M- and K-class dwarfs. Although the focus of the chapter is on the former, the latter receive some well-deserved and long overdue discussion and includes a detailed (yet necessarily speculative) account of the lives of K-dwarf stars, certainly an interesting niche for future research.

Scenarios

Red dwarfs, like all stars, begin their lives inside collapsing clouds of gas and dust. Stellar assembly occurs in large star factories that synthesize hundreds or thousands of stars at a time. The nature of the star's life is dictated by the amount of mass it can acquire by birth, its chemical composition and, less obviously, by the interactions the protostar has with its neighbors, both before and immediately after birth. Although it may not be obvious, gravitational cross-talk between protostars and young stars mean that they often interact with one another violently. Perhaps more insidiously, the order in which stars form may grossly affect their fate, with radiation from one star altering the growth of another.

In this first chapter, we bring together this smorgasbord of influences, thereby illustrating the complex origin of the universe's most numerous yet lightweight stars. Consequently, we demonstrate the impact that these forces have on the size, the orbit and the fate of the star's planets.

D. S. Stevenson, *Red Dwarfs*, https://doi.org/10.1007/978-3-030-25550-3_1

The HR Diagram

Stars that fuse hydrogen into helium occupy a special location on the eponymous Hertzsprung–Russell (HR) diagram. Of interest here is the long diagonal band of stars that meanders from the top left to bottom right of the diagram (Fig. 1.1). This is the main sequence, and on it are the 90-odd percent of stars in the universe, which are converting hydrogen to helium in their stellar cores. The point at which a newly formed star is located on the main sequence is called the zero age main sequence, ZAMS for short. Here, the hydrogen fusion becomes sufficient to produce enough energy to support the star against gravity.

The red dwarfs occupy the lower third right-hand corner of the main sequence. They are a diverse bunch of objects that range in mass from a little over half that of the Sun down to around a thirteenth (approximately 0.5–0.075$M_{\odot}$) its size. Proxima Centauri is the nearest star to the Solar System, at 4.1 light years distant. This has a spectral class of M6 and is the outermost object in the Alpha Centauri triple-star system. Despite its proximity, this 0.12 solar mass star is invisible to the naked eye. In comparison to the other two stars that make up this stellar triplet, this star is noticeably dim, with approximately 1/10,000th of Alpha Centauri A's energy output. To get an idea just how faint red dwarf stars truly are, for the *brightest* red dwarf to be visible to the naked eye, the star would have to lie within 1.3 light years of the Sun. In stark contrast, the brightest and very massive O-class stars can be seen over intergalactic distances.

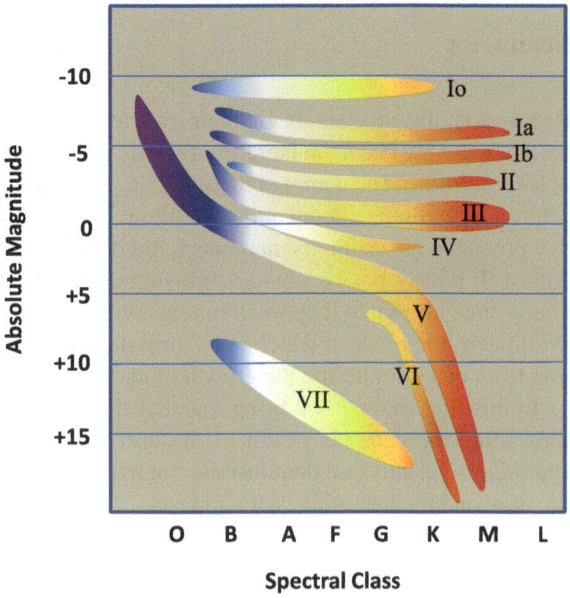

Fig. 1.1 The HR diagram showing luminosity classes I through to VIII. Io—Hypergiants, Ia—Bright Supergiants, Ib—Dimmer Supergiants, II—Bright Giants, III—Red Giants, IV—Sub-Giants, V—Main Sequence, VI—Sub-dwarfs, VII—White Dwarfs. Class L consists of young brown dwarfs and the lowest mass (0.08–0.075 solar mass) red dwarf stars

The red dwarfs belong to spectral class M and the hottest portion of spectral class L. In terms of classification any star on the hydrogen-burning main sequence has the Roman numeral "V" added to its spectral class. Lastly, a second numeral has been added to help signify the strength of certain spectral features—usually the absorption of light by one or more chemical elements. The lower the number the hotter the star is within the spectral class. This gives a star's spectral classification three characters. For example, Barnard's star is a fairly massive red dwarf, with 6/10th the mass of the Sun. This has a classification of M1V. Meanwhile, the cooler and less massive Gliese 581 has a spectral class of M3V, while Proxima Centauri has the classification of M6V, which we mentioned earlier.

Spectral Features of the Coolest Dwarf Stars, M-Dwarfs and Brown Dwarfs

As said above, the vast majority of red dwarfs occupy class M on the main sequence of the HR diagram. However, since around the turn of the millennium things have gotten a little more complex on the nomenclature front, with new stellar and sub-stellar classes having been added to the lower corner of the diagram. The 2MASS program, led by Michael F. Skrutskie, began operation in 1997 and used two highly-automated 1.3-m-wide telescopes. One was located at Mt. Hopkins in Arizona, with the second at CTIO, Chile. This arrangement allowed the program to scan both hemispheres, simultaneously. Each telescope was equipped with a three-channel digital camera, each channel consisting of a 256×256 array of mercury-cadmium-tellurium detectors. These were capable of observing the sky simultaneously at infrared spectral bands, designated J (1.25 microns), H (1.65 microns) and K (2.17 microns). The telescopes were sufficiently powerful to resolve objects with absolute magnitudes as low as 15.8 in the J band and 14.3 in the K band. This allowed the 'scopes to detect a large number of very faint stars and brown dwarfs, as well as a number of interesting other astronomical objects, such as dust-obscured massive stars and active galactic nuclei.

2MASS and other surveys have revealed that a small but sizable proportion of red dwarfs occupy the brightest portion of class L on the HR diagram. This new stellar class had been reserved for the brown dwarf stars. However, scrutiny of these objects by 2MASS and others revealed that while the majority of class L stars were young, and rapidly cooling, brown dwarfs, a smaller number were older hydrogen-burning stars, which often occupied steeply inclined orbits that bisected the galactic disc. Such stars were part of the halo and thus ancient, precluding their recent formation. Consequently, they must be hydrogen-burning entities, or they would long since have cooled into abject darkness. These faint red dwarf stars are in the lowest possible stellar range of 0.08–0.075 solar masses.

Of course, every penny has two sides. The same analysis revealed that some M-class stars (those occupying M6–M8.5) were just young brown dwarfs that had not yet had sufficient time to cool down.

Typical M-class dwarf stars have temperatures that are less than 3900 K, with the lower temperatures encountered commonly around 2200 K, at spectral class M8. However, the 2MASS survey, and other more recent faint object surveys, have marginally extended the region of M-class stars down to a dismal M8.5 (2250–2000 K).

All M-class stars lack lithium in their spectra. Stars of this spectral class have star-wide convection. Such convection ensures that this fragile element is consumed in nuclear reactions in the core, thus removing it from the star. The L-class objects may retain some for a time, hence the L-label of the group, but ultimately this, too, is consumed in the more massive objects as convection drags it from the outside of the star to the hotter interior. Low mass L-class objects, which are brown dwarfs, may never acquire temperatures high enough to combine this element with hydrogen, and thus any L-class object that is young is more likely to be a brown dwarf than a red dwarf.

Beneath the L-dwarfs lie the T-dwarfs. These are all brown dwarfs and have temperatures that extend from around 1200 K down to that of around 700 K. T dwarfs, such as Gliese 229b, are characterized by the presence of methane and water in their spectra. Indeed, methane is diagnostic of temperatures below 1300 K. At these low temperatures, these dwarfs are practically invisible, with the vast majority of their radiation emitted in the infrared band of the electromagnetic spectrum.

Cooler still, a final class of "stellar object" has been added to the stellar menagerie: the Y-class brown dwarf. The spectra of these objects show methane and water vapor, much like the T-class. However, spectra also reveal the presence of the more fragile compound, ammonia. Of this class, only a handful of Y-class dwarfs are known. The exquisitely titled CFBDS0059 was the coolest until recently, with 2MASS 0415-0935 coming a close second. The 15-30 Jupiter-mass CFBDS0059 dwarf was discovered by a consortium of astronomers at the Canada France Hawaii telescope, the Gemini North telescope and the European Southern Observatory in Chile. With a temperature of less than 650 K, or 377 °C, it is utterly invisible to all but the most powerful infrared telescopes, despite the fact it lies only 40 light years away. At 337 °C, its surface is cooler than that of Venus.

Somewhat toastier than CFBDS0059 is 2MASS 0415-0935, with a temperature of around 683 K (411 °C). This "failed star" is a paltry 2 millionth as bright as the Sun, which means that in comparison with the galaxy's brightest object it is less than 5 trillionths as luminous. At these temperatures the concept of a star is unrecognizable. Such objects merge into giant planets, like Jupiter. Indeed, there are countless hot Jupiters that are far hotter than such tepid beasts.

Yet, this is not as low as we can go. In 2011 an object found by WISE—NASA's Wide-field Infrared Survey Explorer was cooler still. WISE 1828+2650 has a temperature that would have British holidaymakers scrambling for their sun cream and clogging the roads to the beach: 25 °C. Although this is only 47 light years away, with an apparent magnitude of only +23.57, it is near the limit of detection—and utterly invisible in the gaze of even the largest optical 'scope. Chilled though WISE 1828 is, it is positively torrid compared to the current record holder, the brown dwarf

(WISE J085510.83-071442.5 [W0855]). This is so far-removed from "stardom" that referring to it as stellar, in any regard, seems disingenuous in the extreme. W0855 is colder than Mars, with a cloud-top temperature of −48 to −13 °C (225–260 K). Spectra taken and analyzed by Jacqueline Faherty (Carnegie Institute of Washington) suggest the 3-10 Jupiter-mass object has water clouds across half of its face and an atmosphere otherwise dominated by methane.

Returning to M-dwarfs, their spectra are exceedingly complex. As the temperature falls below 5000 K, hardy molecules such as carbon monoxide (CO) and hydroxyl (OH) emerge from reactions between carbon, hydrogen and oxygen in the stellar atmosphere and photosphere. At even lower temperatures metal oxides, various organic molecules such as cyanide (CN) and compounds of carbon and hydrogen (CH)) also appear in the stellar spectra. Descending further, below ~2500 K, water (H_2O) appears, and as this and hydrogen react with carbon monoxide, methane is produced. This explains its absence in hotter objects. Therefore, in M-class stars, carbon is found, predominantly, as carbon monoxide rather than methane.

All of this chemical broth blocks a substantial portion of the energy produced in the stellar core. Below 5000 K the opacity of the stellar substance increases rapidly to a maximum near 2500 K. This means it becomes increasingly difficult for light to escape this material. In part, the increase in opacity is due to the condensation of grains and fluff from the cool stellar broth, but it is also because such metals have an increasing abundance of electrons attached to their nuclei. This allows them to absorb more radiation across a broader range of wavelengths. Consequently, stars with a higher opacity are able to sustain nuclear fusion at lower masses than those stars that are more chemically impoverished. Metal-poor Population II stars (subdwarf M stars) could not have masses lower than 0.085 solar masses. Since their formation, stellar death has enriched the soup of gases from which stars form. Therefore, as the abundance of heavy elements has grown over time, the lower limit has come down, until it now lies around 0.075 solar masses.

Although the electrons supplied by metals trap energy within their star, they also ensure that the coolest stars emit a greater than predicted amount of radiation in the infrared rather than at visible wavelengths, making more metal-rich stars redder than their metal poor cousins. On the HR diagram these metal-poor stars appear as the sub-dwarf group (group VI)—a distinct metal-deficient main sequence that lies below, and on the hotter side of, the corresponding band that includes the Sun.

Spectral Features of K-Class Stars

At higher temperatures, absorption lines due to metals and complex organics fade away, while the Balmer absorption line of hydrogen strengthens. At these higher temperatures metal atoms tend to lose electrons more readily and become more strongly ionized. Moreover, the chemical compounds in which they are found at low temperatures are thoroughly vaporized and tend to decompose. K-class stars, thus displaying a blander, but easy to interpret, pattern of chemistry than their cooler

M-class cousins. With less extinction at the visible wavelengths and with higher surface temperatures, more of the energy emitted by K-class stars is in the visible range of the electromagnetic spectrum. Assuming that the atmosphere of a habitable world is similar to ours, the daytime sky of a planet orbiting a K-class star would probably be as blue as ours, or nearly so. Given an atmosphere with a similar composition, sunsets might well be more spectacular, with more red and orange light. However, the daytime sky would be very similar in color to ours. Sunsets might well be more spectacular, with more red and orange light and less blue, but the daytime sky wouldn't be orange.

Star Formation

Stars rarely form in isolation. Most form where large agglomerations of gas and dust accumulate. Such molecular clouds can weigh up to several hundred thousand times the mass of the Sun. Such clouds have far too much angular momentum to fall inwards as a singular mass. Moreover, turbulence and magnetism resist the process of collapse.

Computer models are now able to accurately simulate the broad brushstrokes of star formation. These indicate a ferocious interplay of the various forces that contribute the broad distribution of stellar masses we observe. Some years ago, Matthew R. Bate (University of Exeter) and others modeled the collapse of a modest nebula to create a cluster of stars. Such models resulted in the premature ejection of some protostars, with others regularly exchanging partners. Moreover, many encounters obliterated the protoplanetary discs around these protostars, leaving far smaller discs with radii less than 20 AU.

These results agree with more recent modeling by Christoph Olczak (University of Koln) and colleagues. Here, further modeling has illustrated the effect of changing the density of protostar clusters. In sparse environments, with relatively few stars, only those stars with the greatest mass show regular stellar interactions. Low mass stars tend to be nudged out of the way, leaving their protoplanetary discs intact. However, at high densities even the lowest mass stars are affected, with the majority of their discs truncated or completely obliterated.

Simon F. Portegies Zwart (then at the University of Tokyo) was somewhat of a tour de force in the late 1990s and early 2000s, producing a great number of seminal works that modeled the interactions between stars in clusters of various densities. His work, again based on computer simulations, showed how subtle variations in cluster mass and density could grossly affect the outcome of varying stellar interactions.

For example, consider star formation in the starburst galaxy M82. Within its central region, there are two prominent young star clusters: MGG-9 and MGG-11. In 2003 the University of California's Nate McCrady and colleagues turned their attention to these 10-million-year-old clusters. They mapped the motions of their stars using a simple yet powerful technique, known as velocity dispersion. In essence, the bulk spectrum of the object is taken. If this was a single, static source of radiation, then there would be singular absorption (or emission) lines in the spectrum. However, if the energy is emitted by a variety of sources, their individual motions will broaden each spectral band in a manner determined by the spread of their velocities. The greater the relative motion of the stars in the cluster, the greater the width of each spectral band. This ingenious technique allows you to determine the velocity of the stars (and the cluster mass) without having to resolve the cluster's individual members. MGG-9 had a velocity dispersion of 15.9 km per second, giving it a mass of approximately 1.5 million solar masses. Meanwhile, MGG-11 had a velocity dispersion of 11.4 km per second, giving it a mass of 350,000 Suns.

The second factor to consider is the distribution of mass. A simple measure is the radius in which half the mass of the cluster resides: the half-mass radius. MGG-11 has a half-mass radius of 1.5 light years, while MGG-9 had a half-mass radius of 2.6 light years. Thus, MGG-11 has a little less than one quarter the mass of MGG 9 but is considerably more compact. The interesting point is that despite one cluster having far more mass than the other, the lower mass MGG-11 is the denser of the pair. This has given its members far greater opportunities to collide with one another in the 4 million years or so of the stars' existence. MGG-9 thus hosts lower mass stars, because there are fewer collisions—and is a better place to find planetary systems.

In clusters of the mass and density of MGG-11, stellar collisions will produce huge, short-lived individuals. They also allow frequent gravitational interactions between those stars that do not collide. A key finding of Bate and co-workers' analyses is that their models of star formation accurately reproduced an odd observation—the distribution of stellar masses at the base of the main sequence. Above 0.5 solar masses, stars follow a predictable pattern that has proportionately fewer stars as the stellar mass increases. Known as the Salpeter mass function, this describes (with reasonable accuracy) the distribution of masses of stars that form when a nebula collapses and fragments into a star cluster. Described as an initial mass function, the Salpeter mass function describes fairly accurately the distribution in stellar masses for stars more massive than the Sun.

However, at lower values, in particular below 0.5 solar masses, the distribution flattens out, indicating a deficit of stars below this mass range. Weirder still, below 0.005 solar masses (well into the brown dwarf planet domain) the observed distribution declines precipitously. These variations in the distribution of stellar mass have proved difficult to reconcile with simple models of star formation. Bate and co-workers convincingly demonstrate that stellar interactions are the key to resolving this conundrum. Interactions between low mass protostars and their more massive brethren result in the ejection of many of the smaller ones. This leaves the resulting clusters depleted in the lowest mass objects and proportionately enriched in higher mass stars. It also deprives the smallest protostars of further material from which to grow, thereby truncating the distribution of masses. Observations of the central Orion Nebula cluster support these findings.

The end result is that star clusters may form significant numbers of low mass stars—up to 75% in the Orion Nebula—but that ultimately many of these stars will be expelled, taking with them a diminished mass of material from which planets may subsequently form. However, such interactions likely truncate or ablate the discs of material in which planets would form. Thus, the universe's densest clusters will also likely be the poorest hosts for planets. Less dense clusters—or the stars that are ejected from them—are the likeliest hosts for habitable worlds. The good news is that over time, repeated gravitational encounters between cluster stars leads to the steady dispersal of most of them. Gravitational kneading by the other stars or nebulae also tends to stretch and finally disperse most open clusters. Only the densest stellar clusters last for billions of years.

The other side of the coin may be what happens to those stars that are ejected. An interesting case in point is our nearest neighbor Proxima Centauri. Fabo Feng (University of Hertfordshire) and colleagues have suggested that Proxima Centauri, replete with its potentially habitable planet (Chap. 9) likely wasn't born with its more massive partners Alpha Centauri A and B. Differences in the metal content of Proxima compared with its partners implies that the lightweight Proxima joined Alpha A and B long after the stars formed.

Feng and co-worker's modeling work suggests that Proxima was captured by the Alpha Centauri system after wandering the galaxy for billions of years. This interaction could well have modified the orbit of the potentially habitable Proxima Centauri b and of any planets in the Alpha Centauri system. The 0.12 solar mass Proxima might have begun its formation with enough material to build a bigger star and more planets. However, the interactions that ejected it from its cluster may also have limited its potential to host a decent-sized planetary system.

Thus, when one considers the formation of stars and planets, their synthesis from the dark, cold depths of space must be placed in the context of formation as part of a stellar factory. Periodic smash-ups and ejections will impact the formation of the star—and its planets. The bottom line is that any protostar (and protoplanets) have to live with the hand they are dealt. The universe is not a forgiving place, and while gravity is by far its weakest force, given enough mass to play with, all hell does break loose.

Whether the protostar remains in the cluster or is ejected, it takes with it a declining mass of gas and dust from which it can evolve. A star like the Sun will take around 10 million years to coalesce, while a more massive star may take less than 1 million years. A red dwarf, with less gravitational oomph to play with, will take proportionately longer. Models by Greg Laughlin, Fred Adams and Peter Bodenheim suggested a star with a tenth the mass of the Sun will take up to 2 billion years to settle down to something resembling stellar normality. This is important, as 2 billion years is a long time in terms of planet formation. The long slide to tranquility for these low mass stars will inevitably impact on the formation and early evolution of their worlds, which tend to huddle in tight orbits around their dim suns. During this protracted phase, a multitude of highly luminous stars can come and go, with each generation altering the early evolution of these slowly developing stars.

Grand Tacks, Mass Migrations and Other Planetary Smash-Ups

At this point we should have a clearer understanding of the regular catastrophes that can influence star formation. Do things settle down once the star system has begun to form? Not at all. Further carnage awaits the embryo planets as they coalesce from the firmament.

Investigations of planetary systems reveal a host of peculiar situations. The most obvious is the presence of abundant hot Jupiter planets, which orbit stars such as Peg 51. Jupiter and other giants need material to grow rapidly from before their protoplanetary disc is dispersed by stellar radiation (Fig. 1.2). This requires that they form just beyond the "snowline"—the region where water, ammonia and other volatile gases re-sublimate. However, this is always located outside 3 AU—and likely 5 AU for a Sun-like star. That several Sun-like stars have hot Jupiters is proof that these planets migrated inwards.

Enter computer modeling. Various models have been produced that demonstrate at least three potential scenarios for planetary migration (Fig. 1.3). Each depends on the mass of the planets (or protoplanets) and that of the disc. In Type I migration (Fig. 1.3a), a relatively low-mass planet (one to ten Earth masses) is embedded in a dense disc of material—much of it gaseous. Gravitational and frictional forces between the disc and the planet cause the planet to migrate inwards as its gravitational pull induces density waves in the gas interior to the planet. These waves

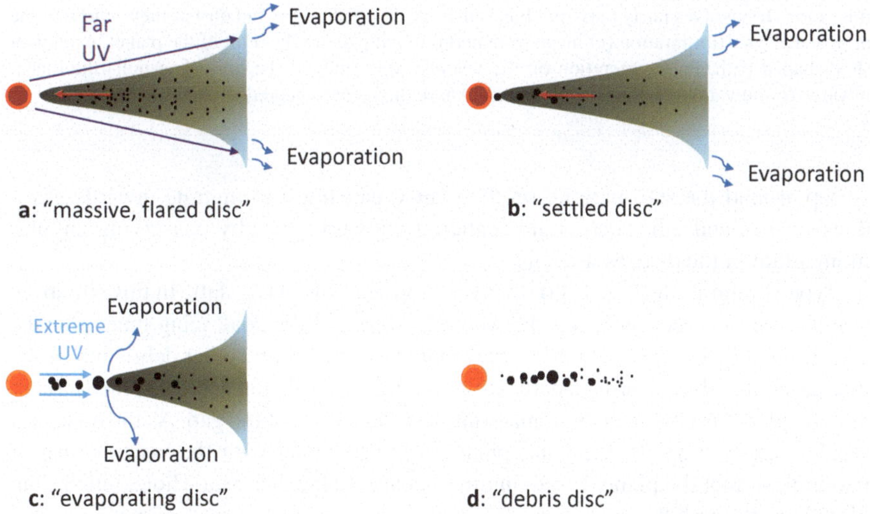

Fig. 1.2 The evolution of a protoplanetary disc. Gas and dust accrete onto the star (red arrow). UV energizes the disc and cause evaporation (wavy arrows). From (**a–d**), the gas is gradually lost, leaving the rocky material behind in a thin disc, which gradually accretes into planets and protoplanets. Base images modified from: https://astrobites.org/2011/03/11/review-article-protoplanetary-disks-and-their-evolution/. Jonathan P. Williams and Lucas A. Cieza, University of Hawaii

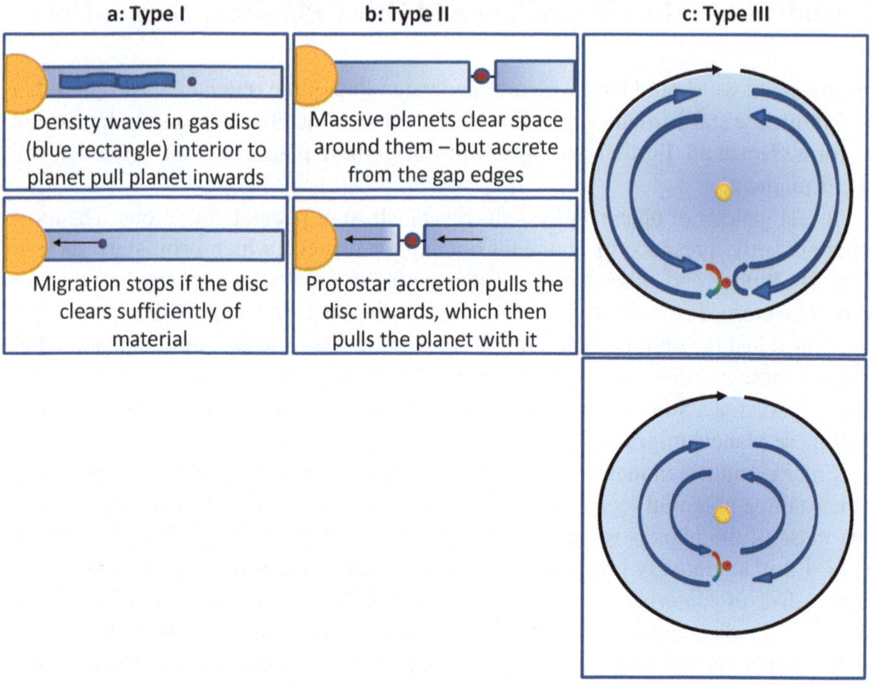

Fig. 1.3 Planetary migration. Planets migrate during development. Type I migration (**a**) involves low (Mars-Earth) -mass planets that interact with gas and dust as they swirl past them. Type II migration (**b**) involves fairly massive objects that are dragged by gas and dust as they swirl into the protostar. Type III migration (**c**) involves material moving along the orbit of the planet in a horseshoe-shaped path. Faster accretion on the leading edge (red) of the horseshoe pulls the planet inwards (or outwards) depending on the side where the greatest accretion occurs

sweep around the star, tugging on the planet, causing it to migrate inwards. As it does so, gas and other debris are scattered outwards, thereby conserving angular momentum in the disc, as a whole.

Type II migration (Fig. 1.3b) involves more massive gas giants. In this situation, gravitational interactions occur between the disc and the coalescing giant planets. This time, the gravitational pull of the planet sweeps gas and other debris out of the disc along its orbit, clearing a path. However, the disc is being accreted onto the star, and the planet is able to accrete material from the edges of the gap. As the disc, as a whole, spirals inwards, the giant planet is pulled along with the viscous disc of material, so that the planet is moving in response to the disc as a whole, rather than ploughing through it.

In Type III migration (Fig. 1.3c), material is moving along the orbit of the planet in a horseshoe pattern. As the planet moves around the star, material in orbits just inside the orbit of the planet move towards the planet. This material moves outwards in front of the planet and then away from the planet, forming the horseshoe. A similar situation occurs behind the orbiting planet, with debris following another horseshoe orbit, away from the planet on the inner edge of the orbital path and towards it on the outer edge of the orbital path—thus completing a large loop around the star, along the planet's orbital path. Under the correct circumstances, the material advancing towards the planet in front of its path picks up enough angular momentum from the planet to accelerate away from it on the outer edge of the orbit. This removes angular momentum from the planet, which causes it to spiral inwards.

From these concepts comes the idea of "pebble-accretion" for terrestrial planets—including super-sized versions known as super-Earths or super-terrans. Dust coagulates under various non-gravitational forces to form pebble-sized lumps. Michiel Lambrecht (Lund University) showed that these pebbles can accrete rapidly onto developing protoplanets and build Earth-sized planets in a million years or so. Here, it is the material that is moving towards the planet on the leading edge that is accreted—potentially at a high rate. Once the Earth-sized planet has reached a respectable size, interactions clear the disc in the path of the planet, creating a gap and ending Type III migration.

Gap clearance further ends the accretion and slows down planetary formation. Ramon Brasser (Tokyo Institute of Technology) has shown that this kind of process can stop Earth and super-Earth mass planets from continuing inwards to their doom. In their simulations, if the disc is hot and dense enough, planetary growth allows them to tug on the disc outside their orbit and pull them back, away from their star. Moreover, the heat generated by rapid accretion of material onto the leading face of the planetesimals can release sufficient energy to alter both the rate of accretion and the pattern of migration. Planetary assembly is a very messy and complex business—something one would expect with such large masses distributed over vast distances.

In a similar regard, observations from the Juno mission suggest something similar to pebble accretion may have facilitated the early growth of Jupiter. Here, within a relatively dense gas disc, rocky material was able to come together without directly colliding. The gas cushioned the motion and helped large volumes of rocky material (perhaps five-Earth's worth) accrete to form a "fluffy" core for the planet—something Juno's investigations imply is still there. Other interpretations are possible, but this is an interesting take on the early accretion of the Sun's planetary system's largest orb.

Our two largest planets, Jupiter and Saturn, also appear to have migrated during their synthesis. Kevin Walsh developed what has become known as "The Grand Tack Model," named after the yachting maneuver, where the boats move into the wind by zigzagging backwards and forwards. In this instance, Jupiter moves in towards the Sun then back outwards. Illustrated in Fig. 1.4, Walsh's model assumes that Jupiter first formed out near 5 AU—at the snowline—around 150,000 years after the Sun condensed.

Pursuing Type-I migration, Jupiter moves inwards scattering planetesimals, asteroids and other space junk outwards. Its inward migration is halted at 1.5 AU by the fast pursuit of the growing body of Saturn. As Saturn draws inwards, it eventually reaches a 2:3 resonance with Jupiter, meaning that for every two orbits Jupiter

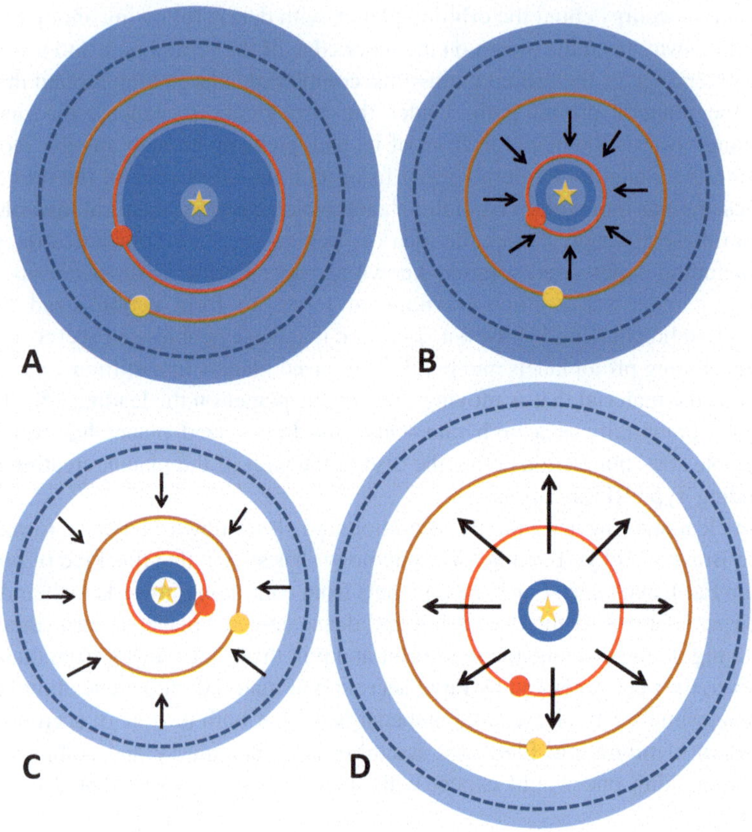

Fig. 1.4 The orbital migration of Jupiter and Saturn. Between 150,000 and 500,000 years after initial formation, Jupiter (red dot) moves inwards from 3.5 to 1.5 AU. Conserving angular momentum, material inwards of this is scattered outwards (**a** to **b**). At **b** Saturn (orange dot) also migrates inwards scattering material as it goes. Eventually it catches up with Jupiter (**c**). Interactions between both cause them to move back out to near where they lie now (**d**). The blue areas represent zones populated by planetesimals. The darker the blue shading, the more massive the debris present

makes around the Sun, Saturn has three. Now, the "Tack" begins. While Jupiter is still "inclined" to move further inwards, consuming the disc from which the terrestrial planets will form, Saturn's arrival constantly accelerates Jupiter until its inward migration becomes a Napoleonic, outwards return march. This pulls Jupiter out to near its current position at 4.75 AU. In the process, the asteroid belt gains denser, silica-rich S-type asteroids in more distal locations, but also on orbits that are more steeply inclined to the ecliptic plane. Meanwhile, carbonaceous C-type asteroids move first inwards, then outwards with Jupiter's contrary movements. This gives rise to two important results: the asteroid belt shows S-type (silicaceous asteroids) in the outer belt—as is observed—while Mars is stunted in size. This leaves Earth and Venus dominating the terrestrial planets—again, as is observed.

Ramon Brasser and co-workers have also carried out some effective reanalysis of Walsh's Grand Tack model. His group has found that moving Jupiter to 1.5 AU means that Earth doesn't form, as Walsh concluded—at least in 95% of the simulations that were run. However, tweaking Jupiter's inward tack to 2 AU rather than 1.5 AU does work and allows Earth to form. What's interesting is that a minor tweak is all that's required to make the Solar System "work." This suggests that the principles laid down in these models are accurate and reflect the processes that happen in the interstellar void. This leads some to conclude that Earth-like habitable worlds will be rare. However, this is a numbers game. To a biologist, the numbers are key. Billions of planetary systems are forming every year. Most of these do not form habitable, Earth-like worlds. However, the population of systems is so high that even with low odds, Earth-like planets in habitable orbits are more than possible.

In principle, all of these mechanisms may operate simultaneously in different parts of the disc, in response to very local differences in the density and property of the disc. Where the discs are turbulent, or have significant variations in density with distance and over time, planets may wander inwards or outwards in what Greg Laughlin (University of California, Santa Cruz) and others refer to as a random walk. In essence, there are many possible outcomes for planets as they condense and perturb one another and the disc from which they are coalescing. This leads to rather complex—but important—outcomes in terms of the kinds of planets that might form around stars, red dwarfs included.

As a case in point, consider TRAPPIST-1. Although we will look at this system in a lot more detail later on it's worth have a quick look at modeling of the planetary system based on an impressive determination of the planetary densities using a variety of measurements.

Simon Grimm (University of Bern) was lead author on a thoroughly international analysis of the TRAPPIST-1 system. The seven TRAPPIST-1 planets showed an unexpectedly wide dispersion in their densities. TRAPPIST-1d has a peculiarly low density, while TRAPPIST-1b and c have higher densities that are still compatible with the presence of thick atmospheres. Meanwhile, the outermost planets—the planets that one might expect to have the lowest densities—are actually rather high. Bottom line: The broad variety of processes that contribute to the final arrangements of planets are so complex and intertwined that crude determinations based on our star system are really rather meaningless. Consequently, attempts to determine the likely habitability of such systems are fraught with danger.

The Late Heavy Bombardment

As an interesting and final coda to this section, we now consider the other part of the Grand Tack—the migration of Jupiter, Saturn, Uranus, and Neptune—and the increasingly controversial impact this convulsion may have had on the early evolution of the Solar System and the origin of life on Earth.

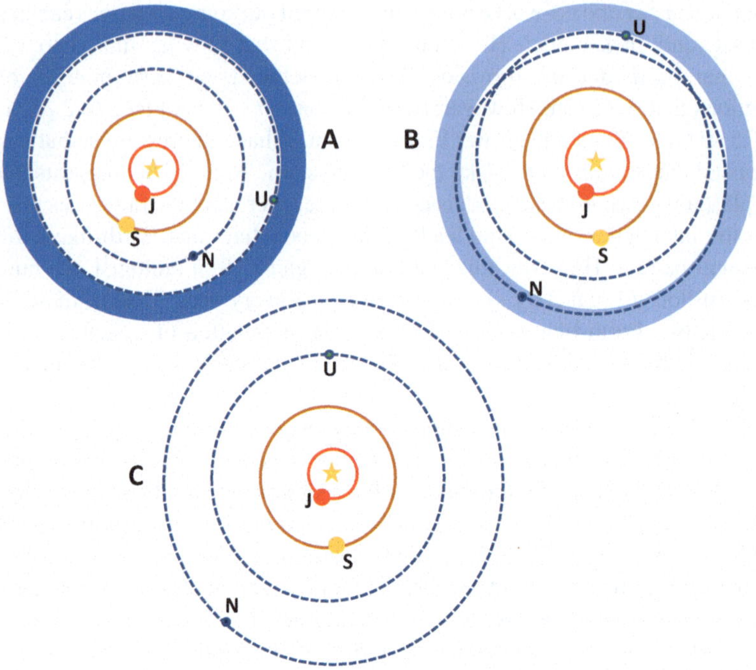

Fig. 1.5 The Late Heavy Bombardment? After formation (**a**) the giant planets lie inwards of a massive disc of debris (blue ring). At **b** slow outwards orbital migration of Saturn (yellow) kicks Uranus (blue) and Neptune (green) out into the disc. Most of the disc is scattered outwards, but much falls in towards the Sun. blasting the inner planets (**c**). The timing of this migration and the accompanying bombardment is disputed

The Apollo missions were widely scattered across the nearside of the Moon. Despite a relatively wide geographical dispersal in their landing sites, all of them found evidence the Moon was pummeled with planetesimal-sized impactors, perhaps a few hundred kilometers across. Dating suggested that these impacts were found in a relatively narrow window around 3.9–4.1 billion years ago. The clustering and the intensity of the impacts led to the notion of a Late Heavy Bombardment (LHB)—late because it happened nearly half a billion years after the formation of the Solar System (Fig. 1.5).

Alessandro Morbideli and co-workers, which include Kevin Walsh and is colloquially known as the "Nice Group" after their location, invoked a late orbital migration to explain the bombardment. As Fig. 1.5 illustrates, the young Solar System is surrounded by a massive disc of icy material, lying outwards of 20 AU. After Jupiter and Saturn approach their final positions, they begin to gravitationally perturb Uranus and Neptune. Each of these worlds lies inwards of this disc, but considerably closer to the Sun than they do now. As Saturn moves into its current orbit at 11 AU from the Sun, Neptune and Uranus are accelerated outwards into this massive debris disc. Scattering most of the material outwards, the Late Heavy Bombardment model suggests a significant proportion went on to fall in towards the Sun. It was this material that bombarded Earth and the Moon.

There are as many implications as there are complications with this hypothesis. For one, the impact debris that was found scattered widely on the Moon may have come solely from one or two massive impacts, the formation of the Imbrium Basin being a prime example. If so, there is no need to invoke a prolonged period of bombardment, merely one or two late impacts in a declining trend from the formation of the Solar System. Secondly, if the LHB is true, then the geological evidence suggests that life started only 50 million years or so after the cessation of the impacts. This isn't long—and the amount of time is important, if we are to understand the complexity or ease of life's origins.

Moreover, the LHB has been invoked to explain the appearance of gold on Earth. Gold is an interesting element—not solely because of its monetary value but also by the question of how it ended up on Earth's surface. Gold is dense and has a strong affinity for iron, which might suggest that during planetary formation most of it (if not all) should end up in Earth's core. However, gold is common (enough), accounting for 1.3 grams per 1000 metric tons of rock in the crust to become part of many fashion items, jewelry, teeth and even additives in some vodkas.

How do we reconcile this element's affinity for our planet's core with its ubiquity on its surface? Imperial College's Matthias Willbold examined rocks in Greenland, which predated the Late Heavy Bombardment and found that the gold we encounter could have been delivered during this catastrophic phase. As an off-shoot of the LHB, this became known as the late veneer hypothesis—a motley addition of material that arrived long after the bulk of the planets had coalesced. It was suggested that a total of 0.5% of the Earths mass—or 20 billion, billion metric tons—was likely added to Earth in this late veneer. If true, all of the otherwise dense, siderophilic (iron-loving) element would be in Earth's deep interior.

However, although this model is superficially appealing, there are problems with it. Notably, the icy material in the outer Solar System would not be expected to be particularly rich in very refractory (high-melting point) materials such as gold. Moreover, discrepancies between the composition of Earth and more refractory chondritic meteorites make it increasingly unlikely that these could have contributed our crust's gold in a terminal Hadean bombardment.

Secondly, the hypothesis ignores geological processes that concentrate gold in the crust, which include magmatism, hydrothermal alteration and metamorphism. Specifically, the late veneer hypothesis makes assumptions about the very limited sample of rocks that were analyzed from Greenland. Tectonic processes are not kind to the hardy bunch of researchers that pursue the earliest epochs of terrestrial history. Very little remains to be analyzed. Consequently, Willbold used tungsten as a proxy for gold, as both elements behave in a chemically similar manner. However, while there certainly are differences in the forms of tungsten (its isotopes) between 4.4-billion-year-old rocks and those that post-date the late heavy bombardment, other explanations are possible. Furthermore, and more fundamentally, assumptions were made about how gold behaves under the extreme conditions of Earth's interior.

If gold does not dissolve readily in the more abundant iron, or it is retained in mantle rocks, then the assumptions about gold's capture by the developing planetary core will be wrong. In this case, there will be no requirement that the gold in Earth's crust will have been added, subsequently, in a late veneer.

Moreover, as experiments improve, we are getting a much better idea of how gold and other metals transform into different materials under extremes of temperature and pressure. Work by Munir Humayun (Florida State University) and others suggests that while most of the planet's stock of gold did go into its core, a significant proportion—perhaps 0.2%—was retained in the 700-km-deep magma ocean that formed as Earth coagulated from interstellar debris. Finally, it's worth noting that while the Grand Tack seems a robust descriptor of how the outer planets came to lie where they are now, the timing of it may well be inconsistent with the idea of a Late Bombardment. Increasingly, the orbital rearrangements, indicative of the Grand Tack, are seen to have happened early in the Solar System's history, negating a need for an increasingly ad hoc—and very belated—planetary migration.

To underscore the problems generated by rare sampling, Matthieu Touboul (University of Maryland) and colleagues examined 2.8-billion-year-old Russian shield rocks. Despite their comparative youth, these rocks displayed the same isotopic patterns for tungsten as the much older rocks from the Greenland shield. This suggested that the patterns in tungsten isotopes seen across the LHB divide have nothing to do with any bombardment, but rather reflected processes that happened when the rocks were formed. And while Touboul agrees with the idea of the late heavy bombardment, he disagrees with its impact on terrestrial chemistry.

The Peculiar Tale of GJ 436b

As a final point, it is worth considering the red dwarf warm Neptune system, Gliese (GJ) 436. GJ436b is a Neptune-mass planet in a highly inclined eccentric orbit around its 0.41 solar mass 5-billion-year-old M2V red dwarf host. The observed orbit is very unusual, being relatively circular yet highly inclined to the rotation axis of the star. This is a peculiar state of affairs, as tidal interactions between the envelope of a red dwarf and its planet would be expected to circularize their orbits over a few hundred million to a billion or so years, but retain the orbit aligned with the spin of the star. The time it takes to do this depends on the mass of the star, its planets and the radius of their orbits.

Vincent Bourrier (Observatoire de l'Université de Genève) and colleagues determined the orbit of GJ436b as it crossed the reference plane of the star—its rotation equator. Although complex, spectral absorption lines were used to determine that the planet's orbital plane was inclined at 80° to the plane of reference, while GJ 436b orbited its star in 2.64 days. The highly inclined orbital is combined with high eccentricity, suggesting that the 5-billion-year-old GJ 436b arrived at its current orbit through a peculiar interaction between planets known as the Kozai-Lidov mechanism

In 1961 Mikhail Lidov identified an interaction between various planetary satellites that explained how a nearly circular orbit could be perturbed so that the orbit of one satellite became eccentric and highly inclined (tilted) to the rotational frame of the planet around which it sailed. In 1962, Yoshihide Kozai identified this same effect when analyzing the orbits of asteroids that were perturbed by Jupiter. In essence planets (or other celestial bodies) in two, near-circular orbits will interact with one another through their mutual gravity. The orbit of the inner satellite exchanges orbital energy and angular momentum with the outer satellite. Gradually, the orbit of the inner satellite becomes eccentric and increasingly tilted—an effect that could result in it rotating in a retrograde manner around its star (orbiting in the opposite sense to the spin of the star).

Bourrier and colleagues suggested that the orbit of GJ 436b was perturbed over the course of 5 billion years. The system itself could be much older, but if so, the interaction that led to the current alignment of the two bodies happened more recently. In Bourrier's model, during the first protracted phase, lasting 4 billion years, an unseen world named GJ 436c induced strong oscillation in the eccentricity of the orbit of GJ 436b. This interaction misaligned GJ 436b's orbital path of with that of the other members of the planetary system. Subsequently, over the next billion years, increasingly large oscillations in eccentricity (the stretch in its orbit) were exchanged for obliquity (the tilt of the orbit). This sent GJ 436b into its current, steeply dipping, but largely circular orbit around the star. Naturally, this model only works if there is another unseen massive planet orbiting further out. Radial velocity surveys should detect this over the next few years, but for now GJ 436b appears to orbit its star alone.

In subsequent work, David Ehrenreich (also at the Observatoire de l'Université de Genève and working with Bourrier) investigated the planet's atmosphere. Earlier, the spectrum of GJ 436b had been thoroughly dissected, leading to an initial, and spurious, claim that methane had been detected. Subsequent analysis refuted this, with a likely explanation put down to misinterpreting some of the complex spectral features of the host dwarf star. Here, in transit, the planet's atmosphere had added its spectrum to that of the underlying star, leading to misinterpretation of the spectrum. However, although detection of methane would have been interesting, something far more unusual was found. Ehrenreich and colleagues identified a large cloud of hydrogen receding from the planet along its orbit, at nearly four times the planet's escape velocity. This cloud is far larger than the planet and partially obscures the host star.

Although such evaporating giants were nothing new, the perplexing factor was that GJ 436b was not particularly close to its host star; certainly not close enough to be irradiated with enough energy to cause evaporation. Spectra revealed that its cloud-top temperature was 712 K (439 °C)—marginally cooler than the surface of Venus. Although toasty by terrestrial standards, this would likely not be energetic enough to allow the planet's atmosphere to evaporate at the observed orbital distance. Yet, GJ 436b is currently losing around 0.1% of its mass per billion years (between 100 million and 1 billion grams per second). The star's radiation is slowly accelerating particles that have boiled out of the planet by the mystery mechanism.

GJ 436b is also hotter than would be expected for a planet with its orbital separation. This suggests it has undergone tidal heating. Again, the current orbital characteristics seem unfavorable. Therefore, astronomers assume that in the last billion years or so, orbital migration flung GJ 436b into its current eclectic orbit. As the planet interacted with the tidal field of its star, its interior was rapidly heated. This led to the expansion and subsequent loss of much of its atmosphere, forming the observed tail.

Although the loss of material is relatively slight today, it's expected that the rate of loss was far higher in the past, soon after the planet attained its current, peculiar orbit. Internal heating drove hydrodynamic escape of a significant mass of atmosphere. Here, much of the atmosphere's mass is sufficiently energized that a significant fraction of its particles has kinetic energies far in excess of that needed to escape the gravitational pull of the planet. In this instance, gas is moving at 100 km per second (compared with the escape velocity of 26 km per second at the cloud top). Interestingly, the situation is reminiscent of Titan. Saturn's large satellite has a massive, yet extended atmosphere. Here, too, it is suspected that tidal-interactions with Saturn led to rapid heating of the atmosphere and its concomitant expansion. However, in this instance, work by Thomas Penz (University of Graz) and colleagues concludes that such early heating was accompanied by a far more vigorous solar wind in the Solar System's infancy. This led to significant loss of Titan's early atmosphere.

Several years ago, Hervé Beust (UJF-Grenoble) made some initial models for the system and the best fits involve the presence of an outer giant with a mass between 0.1 and 50 Jupiters. Best fits place this in an orbit with a 3:1 or 4:1 resonance with GJ436b. As with Jupiter and Saturn, repeated nudging of the inner planet by its outer partner eventually pumped up its eccentricity. This allowed the two worlds to enter a Kozai-Lidov resonance. The process is still in action. GJ 436b's orbit is still marginally eccentric and further tidal interaction with its host star will eventually bring the planet into a more stable configuration. This may take several billion more years, with further losses of the planet's mass.

GJ 436b and others represent the fossilized clues of earlier planetary carnage. In the Beta Pictoris system numerous bodies have been detected that appear to be falling inwards towards β-Pictoris. These FEBs (falling evaporating bodies) are suspected to be sizable icy planetesimals, and the presumption is that they have been sent to their dooms through untoward interactions with planetary mass objects that lurk inside β-Pictoris's large protoplanetary disc. Such interactions are likely to be commonplace simply because stars are born in clusters and are compelled to interact with one another gravitationally. Moreover, once formed changes in the mass and composition of the protoplanetary discs cause further distraction, causing orbital migration—which then leads to final waves of realignment once the planets have coalesced. Planet, and planetary system, formation is thus a protracted—and in some cases ongoing—affair that can last for billions of years, assuming the star lives that long.

Finally, and simply to underscore the chaotic nature of the process, the exit of the star from the main sequence ushers in a final wave of mass migration. Loss of planets into the hot envelope of the giant, coupled to the wholesale loss of stellar mass, leads to expansion of planetary orbits and another wave of gravitational interaction between the system's lesser worlds. Planetary systems are dynamic bodies that slip into a near endless slumber once their star has died.

If one needs a further illustration, consider Mercury. The innermost planet—and seemingly doomed to be vaporized in 6 billion years' time—Mercury might just become embroiled in an altogether different fate: mass migration, collision or expulsion. While Earth and Venus have relatively stable orbits, those of Mars and Mercury are anything but. Of these two smaller bodies, the orbit of Mars is the more secure. Its orbit, while rather eccentric, should remain stable until the Sun expires in a few billion years. Mercury, on the other hand, is far more unstable and prone to migration. Its eccentric orbit processes around the Sun and, like GJ436b, experiences resonances with Jupiter that could well pull it over the edge. This might be sufficient to allow it to collide with another planet.

Over the last three decades the stability of the inner terrestrial planets has been probed by Jacques Laskar and colleagues at the Astronomie et Systèmes Dynamiques in Paris. Rather worryingly, there was a reasonable probability that Mercury's orbit will be stretched and kneaded in such a way that it would collide with the Sun, collide with Venus or be ejected from the Solar System altogether. In all, Laskar's group ran 2501 simulated orbits—each beginning with a slightly different starting position. In these simulations, Mercury came off its current orbit one percent of the time. Now, one percent might not sound like much, but over the course of several billion years that (in astronomical terms) is enough to make a collision worthy of consideration.

In almost all of the simulations, the interactions between the giant planets and the inner Solar System led only to perturbations of Mercury's orbit, which in turn led to its demise. However, in one simulation, after Mercury's orbit was "pumped up," momentum was transferred from the giant planets to the other terrestrial planets. In this simulation, all of the terrestrial planetary orbits destabilized. Thus, 3.34 billion years from now, Earth will likely collide with one of its closest neighbors—Mars or Venus. Our Solar System may be old, but its fate is not yet secure.

The Internal Structure of M- and K-Dwarf Stars

After considering the evolution of planetary systems, we return to the stars themselves. Red dwarf stars are uniquely structured for efficiency and longevity (Fig. 1.6). More massive K-class dwarf stars, and those like the Sun, have distinct layers where energy is transported by radiation and by convection. In the inner portions of stars with masses exceeding roughly a quarter to a third that of the Sun, energy is transported by radiation, the movement of photons of light from electron to electron, within the hot, dense plasma of the stellar interior. Further from the center, where conditions are cooler and less dense, radiation gives way to the roiling motion of gas. Here, the opacity of gases is sufficiently high to prevent the efficient transfer of energy by radiation. Instead, hot gas ascends and cools before slumping back into the interior to heat up once more. This motion—convection—is restricted to the outer layers and is thus only able to mix and homogenize these gases.

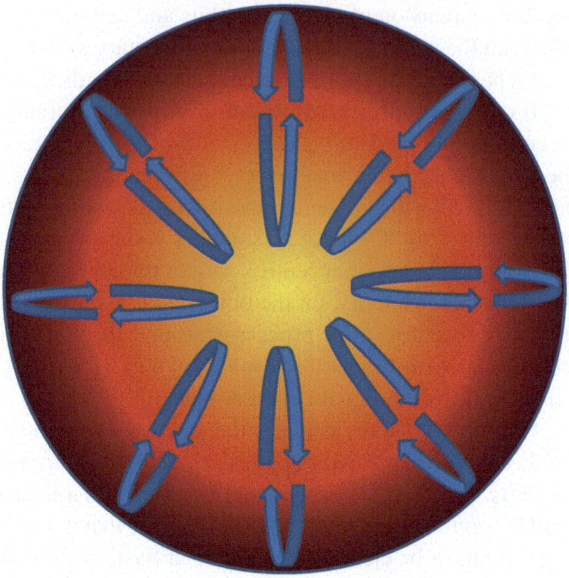

Fig. 1.6 Simplified structure of a red dwarf star. Energy is transported throughout the interior of the star by convection, the roiling motion of gas in the thick, treacly interior of the red dwarf. Energy finally escapes as radiation at the stellar surface, or photosphere, where the density is sufficiently low

A red dwarf, by contrast, with a mass less than a third that of the Sun, has convection throughout its bulk for most of its life. A combination of relatively low temperatures in the envelope and a far greater density throughout ensures that convection is the only efficient process through which energy can be transferred. Thus, fresh hydrogen is continually supplied to the nuclear furnace, while the helium ash is mixed outwards. Stellar evolution is, thus, a steady transformation from an object rich in hydrogen to one rich in helium as nuclear reactions progress. Convection ensures that the bulk of the available stellar fuel is used. Contrast this with a star like the Sun, which is only able to draw on the material within its core, restricting its lifespan. The Sun will consume roughly 10% of its total mass of hydrogen in its main sequence lifetime, then a further 45% or so when it evolves into and through the red giant phase.

In general, as one ascends the main sequence from class L through to A, envelope convection becomes increasingly confined to shallower and shallower outer layers. This process ceases altogether once the mass of the star reaches ~1.5 solar masses. This is around spectral class F5V, for a star with the Sun's composition. Stars that are impoverished in metals will be hotter, and convection will cease at lower masses. At masses in excess of 1.5 Suns, core convection commences. This is driven by a gradual increase in the proportion of energy released through the energy efficient CNO cycle. In the Sun, this process generates less than 10% of the total output, but this increases as the mass of the star is increased. There is some evidence that stars in the range 1.5–1.9 solar masses retain a thin convective skin; and some of these may have a third convective layer, which lies immediately under their photosphere. Such stars reside in the "instability strip" of the HR diagram. Here, such stars physically pulsate. This process is driven by the alternating ionization and recombination of helium in a thin shell, which lies just beneath the photosphere of the star.

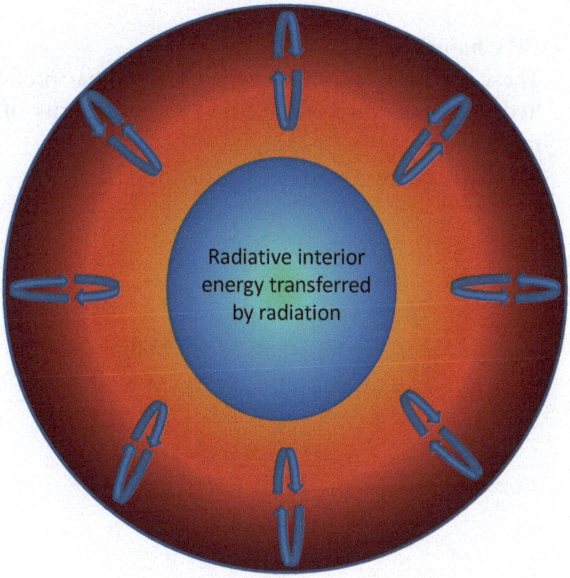

Fig. 1.7 The structure of a K-class orange dwarf star such as Epsilon Eridanae. Convection carries heat only through the outer 50–75% or so of the stellar structure to the photosphere. The proportion diminishes as the mass of the star increases. Radiation carries energy through the heart of the star

At lower masses, K-class dwarf stars are structured much like the Sun (Fig. 1.7). Convection occupies perhaps half the stellar girth, with the volume that convects diminishing as the stellar mass increases.

The Stellar Furnace

The zero-age main sequence (ZAMS) is the point at which the nuclear furnace has ignited and stabilized. i.e., it is releasing just enough energy to halt the further contraction of the star. Core temperatures at ZAMS start at around 2.5 million K for the smallest (0.075 solar mass) dwarfs and rise to approximately 7 million K for stars of 0.5 solar masses (class M0V). Hydrogen normally fuses in the Sun at densities of ~150 grams per cubic cm (or about 150 times the density of tap water). In the smallest stars, where the central temperature is lowest, the densities are far higher (up to 1 million grams per cubic cm). Where the mass is less than ~0.075 solar masses the density becomes so high that the gases become degenerate (at ~10 million grams per cubic cm), and further contraction is halted before the core can become hot enough to ignite bona fide hydrogen fusion through the proton-proton-1, or PP-I-chain. The protostar then ends up as a brown dwarf.

Around the lower mass limit, nuclear reactions may sputter. Alternately, firing up and switching off, ignition releases heat that expands and cools the core to the point at which nuclear reactions shut off once more. Cessation of nuclear burning leads to contraction, re-heating and final reignition in a cycle that continues for the rest of the star's life.

PP Chains

These are a series of branching chains that produce helium directly from hydrogen. Red dwarfs use PP-I for the majority of their lives to release the most energy.

PP-I

$${}_1^1H + {}_1^1H \rightarrow {}_2^2He$$

$${}_2^2He \rightarrow {}_1^2H + e^+ + \upsilon_e$$

This step has a half-life of 1 billion years.

(Overall, this releases 0.42 MeV.)

$$e^+ + e^- \rightarrow 2\gamma \quad (1.022\,\text{eV})$$

$${}_1^2H + {}_1^1H \rightarrow {}_2^3He \quad (5.47\,\text{MeV})$$

$${}_2^3\boldsymbol{He} + {}_2^3\boldsymbol{He} \rightarrow {}_2^4\boldsymbol{He} + 2{}_1^1\boldsymbol{H} \quad (\textbf{Total: 12.86 MeV})$$

Or (PP-II at 14–23 million K)

$${}_2^4He + {}_2^3He \rightarrow {}_4^7Be + \gamma$$

$${}_4^7Be + e \rightarrow {}_3^7Li + \nu_e$$

$$\left(\begin{array}{l}\text{neutrino energies: 90\% as 0.861 MeV / 10\%}\\ \text{as 0.383 MeV}\end{array}\right)$$

$${}_3^7\boldsymbol{Li} + {}_1^1\boldsymbol{H} \rightarrow \boldsymbol{2}{}_2^4\boldsymbol{He}$$

Or (PP-III at temperatures in excess of 23 million K)

$${}_2^4He + {}_2^3He \rightarrow {}_4^7Be + \gamma$$

$${}_4^7Be + {}_1^1H \rightarrow {}_5^8B + \gamma$$

$${}_5^8B \rightarrow {}_4^8Be + e^+ + \nu_e$$

$$\left(\begin{array}{l}\le 14.86\text{ MeV neutrino, with the positron annihilating}\\ \text{to release a further 1.022 MeV as two gamma rays}\end{array}\right)$$

$${}_4^8\boldsymbol{Be} \rightarrow \boldsymbol{2}{}_2^4\boldsymbol{He}$$

More generally, under the "cool" suboptimal temperatures that are found in hearts of red dwarfs, the nuclear reactions that convert hydrogen into helium are not in equilibrium. This means that not all of the PP-I (proton-proton-I) reactions happen with sufficient speed to produce Helium-4 efficiently. Below 8 million K, protons initially combine lazily to make deuterium, but this then fuses much more readily with additional protons to make Helium-3. However, below 8 million K the reaction that converts Helium-3 into Helium-4 occurs over such exceedingly long timescales that the Helium-3 isotope builds up in the star. Only much later in the evolution of the red dwarf does Helium-3 complete the pp-I pathway to produce Helium-4.

Two other PP chains, PP-II and PP-III, occur in the Sun. These reactions only happen when the core temperature exceeds 15 million K. For red dwarfs, with masses less than 0.2 solar masses, this temperature is never reached, at any stage in their evolution and these PP-chains are not utilized to release energy. The most massive red dwarfs (0.2–0.6 solar masses) may burn hydrogen through these chains late on in their main sequence and post-main sequence lives as the abundance of hydrogen in their cores falls away.

Stellar Evolution: An Overview

Stars become progressively dimmer and cooler as we descend the main sequence from the position of the Sun. However, the universe prefers to make stars of low mass. The Salpeter mass function was encountered earlier, and it illustrates how stellar abundance changes with mass. G-class stars, such as the Sun, make up a paltry 7% or so of those in the universe, while cooler class K orange dwarf stars are comprised of perhaps twice this number. The coolest stars in class M (and uppermost of class L) make up the vast majority of the remaining groupings, making up at least 75% of the universe's stellar population.

Despite their abundance, red dwarfs were a bit of a research backwater until Greg Laughlin, Fred Adams and Peter Bodenheimer revisited them in the mid-1990s. Perhaps this was a consequence of their dismal luminosity. Why study such mediocre objects? Whatever the reason, red dwarfs languished in the backwaters of stellar astrophysics for decades. At a rate comparable to the nuclear reactions that power their parsimonious lives, articles emerged steadily but at a geological pace over the years, with roughly two or three per decade from 1960 until the late 1990s.

Greg Laughlin, Fred Adams and Peter Bodenheimer then published a series of research papers that illustrated the full complexity of their lives. Astrophysicists, Laughlin, Adams and Bodenheimer saw this gap in the market, seeing the lack of progress in the field as a fortuitous opportunity for them to revisit the excitement of the 1960s. Here, stellar evolution codes first emerged and were able to unveil the later life stages of stars.

In parallel, researchers Gilles Chabrier and Isabelle Baraffe began work on the physical properties of red dwarfs, determining how their chemistry, luminosity and size varied with mass. Baraffe had previously worked on the structure and evolution of considerably more massive stars, so this was, at the time, a considerable departure for her. However, hers and Chabrier's efforts were fortuitous, as these further advanced considerations of the habitability of red dwarf planets, none of which were known at the time.

With the evolutionary models of Laughlin and colleagues, and the observations of the physical and chemical properties of these stars, subsequent work on habitability could focus on how these peculiar little objects could support living worlds. How would the relatively miniscule energy, liberated by nuclear fusion, impact on neighboring planets? The dim, unconventional fates of nature's least massive stars, and their unique chemistry, mean that red dwarfs have an influence on their retinue of planets that is as unique as their lives. Of these factors, the dominant emission of infrared radiation, and a lifetime orders of magnitudes greater than the central engines of planets, mean that their contribution to habitability is unlike that of any other, more massive star.

Despite languishing in an astrophysical oxbow for most of the last few decades, red dwarfs *did* make a comeback. However, what remains peculiar is that no one appears to have published anything on the evolution of the orange, K-class, stars since 1994. Since the author wrote the first edition of this book, nothing seems to have changed. The 1994 paper (written by F. Fagotto from the Vicolo Osservatorio Padua), which mentions the K-dwarfs, does so as an afterthought. The clear focus of Fagotto's and colleagues' paper is the evolution of much more massive stars. Part of the problem with K-class stars is determining how much mass these stars will lose before they reach the top of the red giant branch. If substantial amounts (more than 25% are lost), then most of these stars will not be massive enough to fuse helium into carbon, and their lives will end prematurely. The best we can do is "guesstimate," based on observations of more massive stars. Mass loss for red dwarfs is largely an assumption. Few measurements have been made of mass loss in stellar winds for any star in class M or K. This means there is something of a mathematical black hole that can readily swallow any calculation of stellar evolution. Until observations plug this proverbial space, evolutionary models are largely guesswork.

Consequently, observations of a special class of evolved stars, known as horizontal-branch giants, can be called on to stand in as models. These are found in star clusters, particularly the old, Population II globular clusters. To varying extents, stars in this region of the HR diagram have lost mass, either on ascent to the tip of the red giant branch or shortly thereafter. This has left them with small but variable masses of hydrogen and a core that is just massive enough to fuse helium. These are not evolved K-class stars, but rather stars that were F- and G-class when they were on the main sequence. Therefore, take these models with a bit of pinch of salt, for now. Although they are reasonable, they are not underpinned with appropriate mathematics. As such, they are façades, rather than fully functional buildings. The lack of modeling is a serious problem, and a rather perplexing one, that in 2019 there is an entire class of stars for which no detailed evolutionary models exist. Given that these objects make up 14% of the total number of stars in the universe, it would seem rather pressing that this gap is plugged.

UV Ceti Stars

Despite their generic faintness, some red dwarfs—including *Proxima Centauri*—belong to a class of variable stars called UV Ceti stars. Strong magnetic fields and a fast rotation stir and heat plasma in their atmospheres, producing powerful flares. These flares brighten the stars by 1–3 magnitudes and generate powerful X-ray and radio bursts. Indeed, the X-ray and radio flares produced by these stars are over one hundred times as energetic as those on the Sun. It seems that these mice can occasionally roar. The precise mechanism by which these magnetic fields arise in the smallest stars is a little unclear, but it must involve some combination of star-wide convection and stellar rotation.

Interestingly, it isn't just the red dwarfs that sustain strong magnetic fields. We already know giant planets such as Jupiter do so, and that these fields generate a lot of radio noise. However, it wasn't clear whether brown dwarf stars would be strong sources of radio emission as well—and some researchers had believed that they would be radio quiet. In the year 2000, *Nature* published an article laying this idea to rest. A young brown dwarf was found emitting bursts of radiation in the X-ray and radio wave bands. These emissions could only come from a hot corona of gas trapped and manipulated by the object's strong magnetic field. So, even though this object had a cool and cloudy atmosphere, the internal magnetic field was still powerful enough to whip up a corona of hot gases above the cloud tops.

Small does not always mean placid. In 2016, John Gizis (University of Delaware) and colleagues published observations a flare from another (young) brown dwarf, 2MASS 0335+23.

In 2013 John Gizis and co-workers found an L-class red dwarf, which bears the name WISEP J190648.47+401106.8. This little star lies 53 light-years from Earth, with an intrinsic luminosity of 0.0002 Suns and a radius of 0.9 Jupiters. With a surface temperature of 2311 K (2038 °C), this is about as cool as a star can get. This 0.8 solar mass dwarf periodically darkened as it rotated every 9 h, indicating that there was a large dark spot on its surface near to its northern pole. Although this could have been a star spot—a region where the magnetic field confined the upward movement of hot gas from the interior—the best fit for the observations was a dark "cloudy" region similar to Neptune's dark spot. It appeared as though this star had a dark, circulating storm, more akin to a planet than a star. What would this cloud be made from? Most likely metal oxides; anything else would evaporate under the conditions experienced.

Such L-class dwarf stars also occasionally host substantial flares. On August 13, 2017, James A. G. Jackman (University of Warwick) and colleagues spotted a massive flare from the L-class dwarf, ULAS J224940.13-011236.9 using the Next Generation Transit Survey (NGTS) telescope. The luminosity of the 9.5-min-long flare was more than 10 times that of the most powerful flare observed from the Sun.

A Life *More* Ordinary: The Evolution of a Red Dwarf Star

This section is almost exclusively derived from the 1990s' modeling work of Adams, Bodenheimer and Laughlin. However, there are a few other additions, notably from Ronald Webbink, who did some modeling of the evolution of these low mass stars in the mid-1970s. We should be a little wary of the numbers. Computer models produce figures that have degrees of precision that may not be warranted. Therefore, if the model says 672 billion years, there may be a substantial range on that figure. However, for the sake of argument, we report the figures that are published in the original academic work.

A red dwarf's uncharismatic lifestyle is its strength. With only microscopic exceptions, all complex life depends on the radiant energy from stars. However, in a little over 100 billion years (depending on the universe's pace of star formation), the only main sequence stars left will be red dwarfs with less than half the Sun's mass. It will be to these dim denizens of the galaxy any future human descendant will be forced to turn to for sustenance. The life span of a star relates inversely and disproportionately to its mass. Weighing in between 0.075 and 0.55–0.6 solar masses (~75 to 550 Jupiter masses), the length of a red dwarf's life is measured more frequently in trillions or hundreds of billions of years, rather than the paltry few million of the most massive stars. While the largest (approximately 0.6 solar mass) red dwarf will last a little over 100 billion years, the meekest red dwarf stars will last 12 trillion years.

Similarly, the process of getting a red dwarf to the main sequence is a protracted affair that only ends hundreds of millions of years after its planets have coalesced. Although a proto-red dwarf will begin glowing within a few million years of commencing its collapse, this energy is derived for the most part from the conversion of gravitational potential energy to heat. During the first few tens of millions of years, additional energy is derived from the fusion of first deuterium to Helium-3, then Lithium-7 to Helium-4, through proton capture. These are meager reserves, but they do allow the proto-dwarf to glow with a light comparable to the early Sun for millions of years.

Only some time later does the core become hot enough to fuse the majority of the star's hydrogen to helium. Collapse then continues until the energy released by nuclear fusion matches that demanded by the inward gravitational pull of the star. This means that for hundreds of millions of years any planet that forms and will lie in its star's habitable zone will be super-heated and likely baked to a crisp. Volatile materials must arrive after the star has largely coalesced and the luminosity is sufficiently reduced. During this phase, violent flaring and substantial coronal mass ejections may plague any young planet. This bombardment may last for 0.5–1 billion years after the dwarf reaches the main sequence. The implications of this protracted irradiation are discussed in Chaps. 5, 7, 8 and 9.

The time taken to perform the transformation from nebula to a 0.1 solar mass star is in itself remarkable—2 billion years, a time equivalent to nearly one thousand generations of the universe's most massive stars. In the time it takes a small red dwarf to stabilize its collapse through the energy generated by the PP-chain, the Sun had nourished Earth for long enough to generate sophisticated biochemistry in millions of different types of bacteria. This in turn led to the formation of our breathable, oxygen-rich atmosphere. Were the Sun a red dwarf, any life dependent on it might barely have gotten going by this point, some 4.6 billion years later, and there certainly wouldn't have been anyone around to write this text or research its contents.

With an interior that convects from core to photosphere, red dwarfs make for efficient furnaces. Stars with masses less than a quarter that of the Sun will leave only 1% of their mass as unused hydrogen. The Sun is a considerably less efficient consumer. By the time it passes, it will have used less than 60% of its total hydrogen reserve. This unspent fuel is largely expelled as a beautiful planetary nebula. Therefore, the story of a red dwarf is one of extreme efficiency, coupled to an intense proclivity for parsimony. Nearly all the fuel is exploitable; there is absolutely no rush to get through it; and the ultimate product, Helium-4, can wait an eternity for its synthesis. A 0.12 solar mass red dwarf, such as Proxima Centauri, will take its time to synthesize Helium-3, and then perpetuity to convert this into Helium-4. There is no rush. In superficial terms, their parsimonious nature makes red dwarfs seem the perfect hosts for the development of complex life. Once life has a foothold on an orbiting world, evolutionary opportunity drifts onwards like a feather, lazily spinning upon the breath of time.

Returning to the star, at birth, a star with a tenth the mass of the Sun will have a surface temperature of around 2200 K—broadly equivalent to burning thermite. Its core temperature hovers at around 2.8 million K, considerably less than that of the Sun. At this stage the 0.1 solar mass red dwarf puts out a paltry 1/3000th that of our star, meaning that to our eyes such a star would be invisible unless it was located within 1.3 light years of Earth.

Subsequently, the evolution of the red dwarf is driven by a delicate step-wise synthesis of Helium-4. Within a cool red dwarf, these reactions run out of kilter with the final manufacturing step lagging far behind the others. That is not to say that it isn't happening; rather, it's just very slow in comparison. The effect of this imbalance is to allow Helium-3 to build up in the red dwarf through a sizable chunk of its life—something that won't happen in the Sun.

As the 0.1 solar mass red dwarf traverses the first 1.4 trillion years of its life, Helium-3 builds up until it occupies almost 10% of the mass of the star. Throughout this period, the little red dwarf slowly expands and heats up slightly (Fig. 1.8). Expansion is a consequence of a slowly warming and an increasingly efficient nuclear furnace, which has more Helium-3 mixed within it. As Helium-3 becomes more abundant, the reactions that create Helium-4 from it can go faster, and the star can generate more energy. For any planet in the star's habitable zone, the effect is slight. The star would appear slightly brighter, but the temperature at the surface of the planet wouldn't change much.

Laughlin and colleagues work suggests that in the core of somewhat more massive red dwarf stars (0.12–0.16 solar masses) the temperature actually falls during the equivalent phase. This may seem counterintuitive, but it's a reflection of the increase in the efficiency of the nuclear furnace. As the amount of Helium-3 increases, the upsurge in energy production is sufficient to expand and cool the core somewhat,

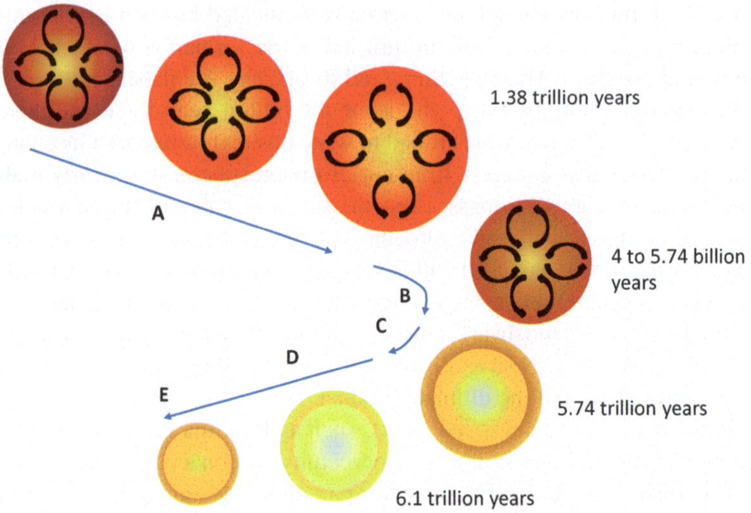

Fig. 1.8 The evolution of a 0.1 solar mass red dwarf. During phase **a** the star steadily coverts its hydrogen to helium, and the increase in energy efficiency causes the star to slowly expand and heat up. During phase **b** a buildup of helium reduces energy production and the star contracts. After 5.7 trillion years (**c**) the star's core stops convecting, and remaining fuel is burned rapidly. Finally, after 6 trillion years (**d**) more than 96% of the hydrogen has been consumed. The star becomes a white dwarf (**e**). Any remaining hydrogen in the star's outer layers is used up before the star begins to cool and fade into invisibility

without jeopardizing energy production. However, once the conversion of hydrogen to Helium-3 becomes too slow to keep pace with its manufacture this phase ends, and the core begins to contract and heat further. At this stage the star has reached the stellar equivalent of middle-age. The increasing efficiency and fervor that was characteristic of the star's youth gives way to the declining slog of maturity. Energy production per unit mass cannot keep pace with the demands of the star as a whole. The core responds by contracting and heating up, to maintain the accelerating rate of energy production.

As the years advance, trillion upon trillion, the amount of Helium-4 steadily increases, further decreasing the efficiency of energy production. The core grows ever hotter, reaching a little over 5 million K at 4.4 trillion years of age. At this stage helium becomes more abundant than hydrogen, and the final struggle to maintain stability gets under way. The red dwarf's surface will now have heated to 2500 K and the star will have brightened to approximately one thousandth that of the current Sun. The little star is still a red dwarf, but it is slowly turning blue. Throughout its main sequence life, convection keeps the composition of the star homogeneous. Convection steals hydrogen from the envelope of the star and drags it into the star's interior. Here, it is steadily replaced with freshly-minted helium that's been synthesized in the core. Once Helium-4 dominates the mass of the dwarf, the amount of energy that the star can generate by fusion reactions decreases. The star is still ticking, but time is finally running out.

As the core collapses, the helium, which is slowly flooding the envelope of the star, makes it increasingly transparent to the radiation generated in its core. This effect is twofold. As hydrogen is turned into helium, there are fewer free electrons to absorb radiation from the core. Helium is far more electronegative than hydrogen; and the stronger affinity of its nucleus for its electrons means that there are fewer (if any) available to absorb radiation at the kinds of temperatures found near the stellar surface. Moreover, as hydrogen is consumed, there is less of it to form organic molecules, such as CH—the simplest compound of carbon and hydrogen. As these compounds are also effective absorbers of radiation, their loss means that the envelope, as a whole, cannot absorb radiation as effectively as before, making it progressively more transparent. Increasingly, radiation simply leaks out of the star without depositing its energy into the envelope. Consequently, the star must contract. Contraction has some unfortunate consequences for the aging star. For one, it means that the star as a whole becomes hotter; and a hotter star has even fewer organic molecules to grab hold of the radiation. With less hydrogen gas in its neutral form even more radiation escapes.

With an increase in the rate of nuclear reactions and a lower opacity, the dwarf brightens further and grows steadily hotter, even as it shrinks in diameter. After 5.74 trillion years, the star has a surface temperature of 3450 K—approaching that of the most massive M-class dwarf stars of today—and a luminosity of approximately 0.5%, or one 200th, that of the present Sun. At this point only 16% of the star's mass is hydrogen, and the end is finally approaching.

After an almighty 5.74 trillion years, helium dominates the composition of the star. The stellar material is now so transparent to radiation that convection is no longer the best means of transporting energy through the star. A direct flow of radiation from the interior in the form of photons is far more effective. With energy now able to escape directly, rather abruptly star-wide convection ceases. The one unique feature of the red dwarf—its whole-star convection—is now over, and the structure comes to resemble that of an old Sun. The red dwarf now has a helium-rich core overlain by a thin layer of hydrogen-rich material. Although convection has ended in the core, it continues in the thin hydrogen-rich shell surrounding it. This stretches out to the star's surface. All the remaining hydrogen in the now stagnant core is rapidly consumed. With this sudden change in structure, the star abruptly contracts a few percent and dims. This is the end of the red dwarf's main sequence and is evident in Fig. 1.9.

After the core runs out of fuel, hydrogen will continue to burn in a shell surrounding it. In turn, a convecting outer hydrogen-rich shell overlies the hydrogen-burning shell. This retains the overall abundance of hydrogen that remained after whole-star convection ended. The helium core then simply contracts and grows hotter as it gains mass from the hydrogen-burning shell, much as it will do in the Sun in 5–6 billion years' time. The hydrogen-burning shell source then burns outwards as the helium core contracts. Hydrogen fusion progressively consumes the remaining fuel reserves until 96% of the former red dwarf has been transformed into Helium-4. During this final protracted phase, the star's envelope contracts and heats rapidly. The Sun will expand during the equivalent phase, but the old red dwarf will not. There is far too much helium in the outer layer for this to happen.

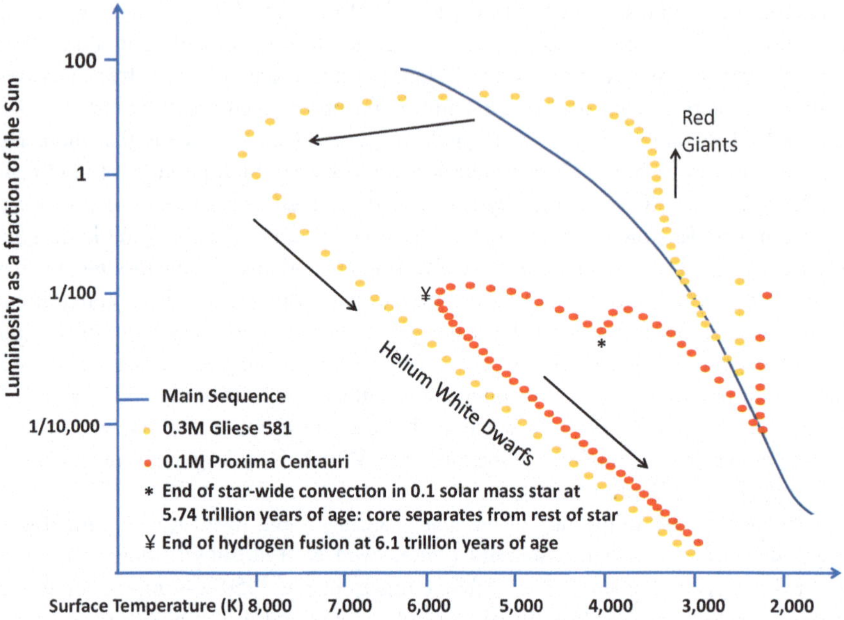

Fig. 1.9 the evolution of 0.1 and 0.3 solar mass red dwarf stars based on the work of Greg Laughlin, Peter Bodenheimer and Fred Adams. The 0.1 solar mass red dwarf (Proxima Centauri) never becomes a red giant, while the more massive Gliese 581 dwarf does. Both red dwarfs ultimately fade away as helium-rich white dwarf stars

Eventually, some 402 billion years after the core has ceased convecting, the star's surface will have heated to roughly 5800 K, making it somewhat hotter than that of the present-day Sun. Although the star is hotter than the Sun, it will only be a little over one hundredth as bright. This is as good as it will get for the 0.1 solar mass red dwarf. The core will have a temperature of 12 million K (again, comparable with that of the present Sun) and, more importantly, will have become fully degenerate. In this state, electrons are confined to particular energy levels, and they take up the strain against further compression. Energy transfer by radiation gives way to conduction, much as it would in a solid material like a metal.

In the degenerate state, electrons in the dense helium plasma resist further compression, and any further contraction that does occur does not increase the temperature. Consequently, although the envelope can contract further, the core is as dense and as hot as it can get, and stellar evolution effectively ceases. The antiquated (former) red dwarf now "turns the corner" on the Hertzsprung–Russell diagram. The dwarf will spend the next 134 billion years contracting, cooling and dimming into invisibility. As it cools and fades from view, the star's hydrogen burning shell will gradually wink out. A colossal 6 trillion 279 billion years after the dwarf reached the main sequence (and eons after the Sun's shriveled remains became as cold and desolate as the space surrounding them) the spent helium dwarf will fade from view forever. The numbers are so vast that once again we have to consider precision. The number of evolutionary steps and the degree of uncertainty will mean that, even in the absence of confounding factors like mass loss through stellar winds, the periods described may be off by tens of billions of years.

The Fate of the Most Massive Red Dwarf Stars

What happens to red dwarfs more massive than the example considered here? For example, two of the stars we will consider later are Gliese 581 and Gliese 667C. These have masses roughly three times greater than the example explored above, and a little under three times the mass of Proxima Centauri. Will their lives differ dramatically from the smaller red dwarf stars we've already discussed? To look at their fate, in general as the mass of the star increases, these more massive stars behave like those we see evolving and dying today. Most significantly is the appearance of a red giant phase.

More massive red dwarfs, including the 0.3 solar mass stars Gliese 581 and 667C, follow broadly similar evolutionary paths as their smaller cousins. However, the main differences are that the stellar core ceases convection at progressively earlier times in the star's history as the stellar mass is ratcheted up. This is important, as it means that proportionately less of the star's hydrogen is used up, with the leftovers locked into the envelope. Furthermore, as the mass increases, the dwarf stars become proportionately hotter once the core ceases convection. The most massive of the aging red dwarfs will turn distinctly blue. A typical 0.2 solar mass dwarf will have a surface temperature above 10,000 K (hotter than Sirius) when it, too, "turns the corner" on the Hertzsprung–Russell diagram.

Laughlin and colleagues concluded that red dwarfs that weighed in excess of roughly 0.16 solar masses underwent expansion once their radiative core appeared. This is in contrast to the 0.1 solar mass star considered, above. Here, expansion occurs throughout the main sequence of the red dwarf, not at the end. These small stars physically shrink once they have consumed their fuel. Stars with twice this mass expand only when the core has exhausted its supply of fuel. Thus, these calculations suggest that 0.16 solar mass stars are the lowest mass red giants that the universe will ever produce.

In the case of a 0.20 solar mass star, Laughlin and colleagues work shows substantial expansion of the envelope. The star expands to more than ten times its original diameter. Although such expansion would only endow these little stars with a girth comparable to the present Sun, it suggests that the universe will flaunt its last wave of red giants in one trillion years, when these stars reach the end of their main sequence lives. Less massive stars would simply heat up and brighten, becoming bluer, before dimming and finally dying.

Why do these more massive stars become giants rather than simply get hotter? The key lies with the amount of hydrogen and helium in their envelopes. The more massive red dwarfs retain more hydrogen in their envelopes and are, therefore, able to trap more of the huge amounts of energy pouring out of the aging stellar cores. A relatively lightweight 0.16 solar mass star constantly mixes hydrogen throughout its bulk, through whole-star convection. Consequently, these small stars are able to increase their luminosity (140-fold) with little expansion.

The bulk of the increased output from the core is accommodated by increasingly vigorous convection. The additional burden of energy is only evident through their higher surface temperature and a very modest expansion; very little additional surface area is needed to radiate the additional energy that the core is liberating. Conversely, a star with 0.25 solar masses cannot dump the extra energy from its core later on in its life; it simply creates too much for the surface to radiate. Instead, an increase in luminosity of a thousand-fold forces the envelope to expand. Only through an increase in surface area can the star keep its interior in equilibrium.

The driving force behind this important change in evolutionary tact is the temperature of the degenerate core. In red dwarfs with less than 0.2 solar masses, the degenerate core maintains a constant temperature even as the hydrogen shell source burns outwards and adds mass to it. The ephemeral neutrino is the source of this consistency. As degeneracy sets in, neutrinos carry energy out of the dense core, thereby allowing it to cool effectively. This means that these small stars never generate so much oomph that their envelopes are overloaded with energy. Contrast this with stars that have more than 0.2 solar masses. These more massive red dwarfs develop progressively hotter degenerate cores as hydrogen fusion in the surrounding shell increases the mass of the core at a rate higher than neutrinos can dump it. This bumps the temperature up, which in turn ups the rate of nuclear reactions. As neutrinos can't keep the star in equilibrium, it must expand to dump the extra energy. Combine this with the higher opacity in the envelope, and more massive red dwarfs are compelled to expand.

Red giants end up in a spiral of expansion because of one final reason. As the temperature falls with increasing size, molecules and grains of matter begin to clog up the stellar surface. This only gets worse as the temperature lowers. Overall the envelope of a star becomes less opaque to radiation as the temperature drops from 5000 to 2000 K before increasing rapidly once more. Therefore, as the temperature initially falls more energy is able to leave the surface of the star by radiation. Above 5000 K, H^+ and H^- ions contribute to how opaque the material is, while below 2000 K metals and complex organic compounds become significant blankets for radiation.

As the red dwarf evolves and its luminosity increases, the star tries to remain in the region of lowest opacity, where energy can most readily escape its surface. However, for the more massive red dwarfs, the temperature of the envelope increases towards 5000 K, and the opacity is eventually forced upwards. This reduces the capacity of the stellar surface to dump energy into space. Since less energy is able to escape, the star has to expand to remain in balance. Hence, both the increase in luminosity and the decrease in the temperature of the photosphere force the more massive dwarfs to become giants.

CNO Cycles

These are series of interwoven cycles that combine hydrogen with carbon to produce helium and carbon, via nitrogen and oxygen intermediates. Only the most abundant reaction is shown here.

$$^{12}_{6}C + ^{1}_{1}H \rightarrow ^{13}_{7}N + \gamma \quad (1.95\text{MeV})$$

$$^{13}_{7}N \rightarrow ^{13}_{6}C + e^{+} + \upsilon_{e} \quad (1.20\text{MeV})(\text{half-life of } 9.965 \text{ min})$$

$$^{13}_{6}C + ^{1}_{1}H \rightarrow ^{14}_{7}N + \gamma \quad (7.54\text{MeV})$$

$$^{14}_{7}N + ^{1}_{1}H \rightarrow ^{15}_{8}O + \gamma \quad (7.35\text{MeV})$$

$$^{15}_{8}O \rightarrow ^{15}_{7}N + e^{+} + \upsilon_{e} \quad (1.73\text{MeV})(\text{half-life of } 122.24\text{s})$$

$$^{15}_{7}N + ^{1}_{1}H \rightarrow ^{12}_{6}C + ^{4}_{2}He \quad (4.96\text{MeV})$$

Final Flings for Mid-Range M-Dwarf Stars

Calculations carried out in the 1970s by Ronald Webbink suggest that stars with a bulk of around 0.2 solar masses may have a few tricks to play before they ultimately fade as white dwarfs. These stars will become red giants, but as their hydrogen envelopes blow away in the stellar breeze, the remaining hydrogen layer will shrink in mass, contract and become degenerate. Hydrogen fusion in these hot hydrogen-burning shells is through the carbon-nitrogen-oxygen (CNO) cycle, which is strongly temperature dependent. This means that minor increases in temperature can cause rapid accelerations in the rate of nuclear reactions. As the remaining hydrogen layer contracts, its temperature strongly increases. This has the potential to lead to explosive burning. Webbink suggested that these ancient red dwarfs may go through a final explosive phase where the remaining hydrogen layer convulses, ejecting much of itself into space in nova-like outbursts. Meanwhile, even more massive red dwarfs can avoid this by maintaining stable hydrogen burning until their remaining reserves are blown away, presumably as diminutive planetary nebulae.

Whatever the ultimate truth, these stars are doomed to a slow fade out. Their hydrogen exhausted, their core will begin to cool, and the remnant will complete its life over tens of billions of years as a slowly fading helium white dwarf.

A Life in the Balance: The Evolution of the Orange K-Dwarfs

Spanning a narrow range of masses, between the parsimonious red dwarfs and the more recognizable home comforts provided by the yellow G-dwarfs lie the orange K-dwarfs. Their masses vary from roughly 0.6 solar masses to 0.75 solar masses, their surface temperatures from 3500 to 4900 K, and it is estimated that they make up about 12% of the total number of main sequence stars in the galaxy. By comparison M-class stars make up perhaps 76.5%; while more massive stars are proportionately rarer. G-class stars make up maybe 7.6%; F-class, 3%; A-class, 0.6%; B-class 0.13%; and the most massive O-class stars a pitiful 0.00003% of the total stellar population. These kinds of figures may well underestimate the true numbers of the faintest stars, and more accurate numbers will await the full data release of Gaia, which is currently mapping the galaxy.

Like their more massive brethren, these stars will be able to burn helium when they eventually leave the main sequence. Moreover, like the red dwarfs, the universe is still too young for any orange K-dwarves to have left the main sequence and begun their journey to their graves. At present, the least massive stars evolving away from the main sequence are metal-poor 0.8 solar mass stars in the galactic halo. These are yellow G8V dwarfs, and their metal-poor structures mean that they are hotter and denser than more metal-rich stars of equivalent mass in the galactic disc. This means that they will evolve somewhat faster, for the reasons we discussed earlier in relation to the highly evolved red dwarfs. It will be approximately another 5 billion years before the most massive galactic K-dwarf leaves the main sequence. Therefore, keep in mind that like the red dwarfs before them, this chapter's evolutionary models are entirely theoretical and will not be confirmed or refuted until the Sun has swollen to become a red giant and every library on Earth has been thoroughly incinerated.

What we do know is that in terms of their future direction, the K-dwarfs will follow the same general route taken by more massive objects such as the Sun. An orange K-dwarf will remain on the main sequence for anywhere between 15 and 100 billion years, with the most massive orange dwarfs leading the shortest lives.

A K-dwarf will take between 50 and 300 million years to contract onto the main sequence, but otherwise follows the same general route as less massive red dwarf stars. On the HR diagram, the contracting protostars of both M- and K-class stars descend almost vertically until they hit the main sequence and their fuel ignites. The path is carved by the released energy from the interior by convection. As long as the star can dump its internal energy through this route, the path remains roughly vertical. More massive stars, such as the Sun, move roughly horizontally towards the main sequence, after an initial decrease in luminosity.

During this early phase, convection keeps up with the increase in energy output, and the protostar dims. However, eventually the increase in energy output causes the star to heat up. Convection cannot dump the energy fast enough. Opacity then falls, as temperature rises above 5000 K; convection then becomes inefficient. In these more massive stars their evolutionary path makes a turn and heads at roughly constant luminosity towards the main sequence. The more massive the star, the earlier the turn to the left occurs, which carries the star towards the main sequence.

Protostars ranging in mass from ~0.3 solar masses up to ~2 solar masses go through a stage called the T-Tauri phase. Here the contracting protostar is racked by violent eruptions, caused by a mixture of debris impacting on the still contracting star and by flaring associated with the star's strong magnetic field. Magnetism is (initially) strong because the young star rotates rapidly and experiences strong convection within its bulk. As the protostar matures, the rate of rotation slows, and the magnetic field declines. This process continues throughout the star's life as its magnetic field brakes against the interstellar medium.

K-dwarfs generate magnetic fields in the same manner as the Sun—the so-called $\alpha\omega$ (alpha-omega) effect. In essence, this complex mechanism relies on two factors: the structure and rotation of the star. In stars with masses in excess of 0.25–0.3 solar masses, the central regions are radiative and rotate rapidly as a single entity. Meanwhile, the outer regions undergo convection and rotate at different rates that depend on the latitude and their distance from the base of the convective shell.

Within the convective layer, the vertical motion of gas is modified by the rotation of the star in a process called the Coriolis effect. (See also Chap. 5.) Although this will be examined in detail in the context of meteorology, the gist of it concerns the conservation of momentum. Consider a packet of hot gas that rises upwards from the base of the convective layer. As it rises, it is now moving faster in an east–west direction than the surrounding gas at greater elevation. This effect increases as the elevation of the packet of gas rises. This causes the gas packet to curve towards the west (if this is the direction of rotation of the star). Conversely, a packet of colder gas that is descending towards the core will be rotating more slowly than the gas into which it descends, relative to the radiative core. Gas then takes a curving trajectory as it descends or ascends, rather than the misleading vertical ascent shown in most diagrams.

Likewise, stars tend to rotate in a non-rigid manner, with the equator rotating faster than the polar regions. Gas moving away from the equator towards the pole is moving faster than the underlying core and curves towards the east. Conversely, gas moving towards the equator is moving more slowly than the underlying core and curves towards the west. The combination of these turning "forces" twists the gas parcels, and this morphs magnetic field lines into various contortions that are manifest as star spots, flares and mass ejections.

In terms of habitability, star spots may reduce the amount of visible light emitted by the star during its more formative years; this could result in large swings in temperature on any planet orbiting it. This is especially true of planets that orbit red dwarfs, where the amount of incident visible radiation is already low, and the spots occupy a proportionately larger area of the star's surface. However, a planet orbiting within the habitable zone of a K-star should expect strong variation in temperature early on if the number of star spots is high, as the overall amount of radiation emitted by the star's surface will be greater than emitted by a smaller red dwarf.

This risk is only apparent under two scenarios. The simplest is early in the star's evolution, when it is rotating quickly. The second scenario is where the star is a member of a tight binary, or where a giant planet orbits close to it. In these situations, the orbital motion of the pair of stars, or of the giant around the system's center of gravity, will tend to maintain vigorous circulation (and rapid rotation) throughout the star's main sequence life. This may be sufficient to maintain active magnetism—something that is apparent in W UMa variables.

Flares, associated with star spots, may release periodic bursts of strong stellar winds. Any planet impinged upon by such outbursts will need the protection of a strong magnetic field if its atmosphere and any life found on the planet needs to be shielded. However, unlike those tidally locked worlds of red dwarfs, the vast majority of planets orbiting within the habitable zone of an orange dwarf should rotate somewhat slowly relative to their stellar host (Chaps. 5, 7 and 9). This is because the stellar habitable zone extends outside the tidal-lock zone for these more massive and more luminous stars. Magnetic fields also heat the outer atmosphere, generating a hot (1–2 million K) corona that is detectable through X-ray emission.

Although a planetary magnetic field confers some protection against the impact of flares and coronal mass ejections, it does not shield against the effects of extreme ultraviolet and X-rays, which are also released in abundance by magnetically active stars. Moreover, although the habitable zone of an orange dwarf extends further beyond its star than that of a red dwarf, it is still comparatively close to its star compared to Earth. This means that a K-class dwarf will expose any potentially habitable planet to several times the intensity of radiation that Earth would experience at a comparable time in its star's evolution.

Evidence came of the intensity and variability of K-dwarf surface magnetism from a study done nearly 20 years ago. John Stauffer (Caltech) and co-workers showed that in the Pleiades K-dwarfs were too blue for their masses. Moreover, they were also about half a magnitude too faint to fit onto the main sequence. Any difference in metallicity was insufficient to explain this spectral discrepancy between the cluster stars. Compared with the older Praesepe cluster, the Pleiades K-dwarfs were oddly dim; yet at longer wavelengths the Praesepe and Pleiades matched one another. This implied that cool, dark star spots covered more than 50% of the surface on the Pleiades K-dwarfs. Meanwhile, stars in the older cluster were less afflicted.

Resolving the conflict came down to understanding how much of the surface was spotty. The surface of Pleiades stars appeared darker than those stars in the Praesepe cluster, yet the associated strong magnetic field also generated periodic flares. These produced bursts of ultraviolet and X-ray emission, which made the Pleiades stars appear bluer than their more mature Praesepe cluster counterparts. Indeed, it now seems apparent that all young K-dwarfs experience strong, localized surface magnetism and exhibit severe acne.

A combination of strong stellar winds and jets reduces the angular momentum of the material that is flowing into the star: this slows the star's overall rate of rotation. This effect diminishes early on, before the star has settled onto the main sequence. Once the star has stabilized, mass loss declines to the region of one trillionth of a solar mass per year. This is a very comfortable level, which wouldn't affect the further evolution of the star. The rate of mass loss is far lower than the rate at which nuclear reactions consume hydrogen. An interesting consideration is the time it takes mass loss to decline to the kinds of levels characteristic of maturity. Brian Wood (U. S. Naval Research Laboratory) and colleagues have shown that magnetically active stars such as the Sun may shed up to 100 times the amount of gas in their stellar winds during the first 700 million years of their lives, as they do now. Other magnetically quiet stars, such as π^1Uma, has far weaker winds, with correspondingly low mass loss. Did Earth (like Mars) lose significant atmospheric mass during the magnetically active youth of the Sun? The loss of volatiles may not necessarily be a bad thing (Chap. 7).

Stellar Evolution of K-Dwarfs

Here, we enter the realm of speculation on the part of the author. The presented models are reasonable extrapolations but remain just that, for now.

The K-dwarfs are very similar in many regards to the more massive but rarer G-dwarfs such as our Sun. It is only their intrinsic faintness that sets them apart. At 0.6 solar masses, the least massive K-dwarf (or most massive M-dwarf) is only 1% as bright as the Sun. At their most massive, the K-dwarfs are approximately a third as bright as the present Sun while on the main sequence. However, it is worth pointing out that this luminosity is only half that of the Sun when it was newborn. Therefore, one of these stars wouldn't look terribly out of place in our Solar System.

After a several tens to a few tens of hundreds of millions of years, the K-dwarf settles onto the main sequence at its zero-age-main-sequence, or ZAMS. After this point, the star will remain stable for the next few tens of billions of years. During this period, the PPI-chain converts hydrogen to helium in a small radiative core that varies in size from 1/5 to 1/4 the total stellar diameter. The smallest K-dwarfs (a little over 0.55 solar masses) have convection in roughly the outer two-thirds of their structure. The thickness of the convective shell decreases quite rapidly with increasing mass. A K0 star, with approximately 0.75–0.8 solar masses, has convection confined to a little less than the outer third of its mass. As with the most massive red dwarf stars, this mode of construction means that the core is isolated from the rest of the star. Like the smaller red dwarfs, which have masses greater than one quarter that of the Sun, this also means that the fuel in the envelope makes little contribution to the life of the star throughout the bulk of the star's life. Thus, when the abundance of hydrogen in the central regions of the core falls below 0.1%, it is unable to generate enough energy to support itself. This initiates a prompt phase of rapid contraction under the influence of gravity.

While the star remains on the main sequence a K-dwarf, like its smaller red dwarf cousins, gradually changes its behavior (Fig 1.10). As the core progressively uses up its supply of hydrogen, the efficiency of hydrogen fusion per gram of stellar mass is lowered. The star compensates by drawing on its immense supply of gravitational potential energy. The stellar core slowly contracts, heating and raising the rate of hydrogen fusion.

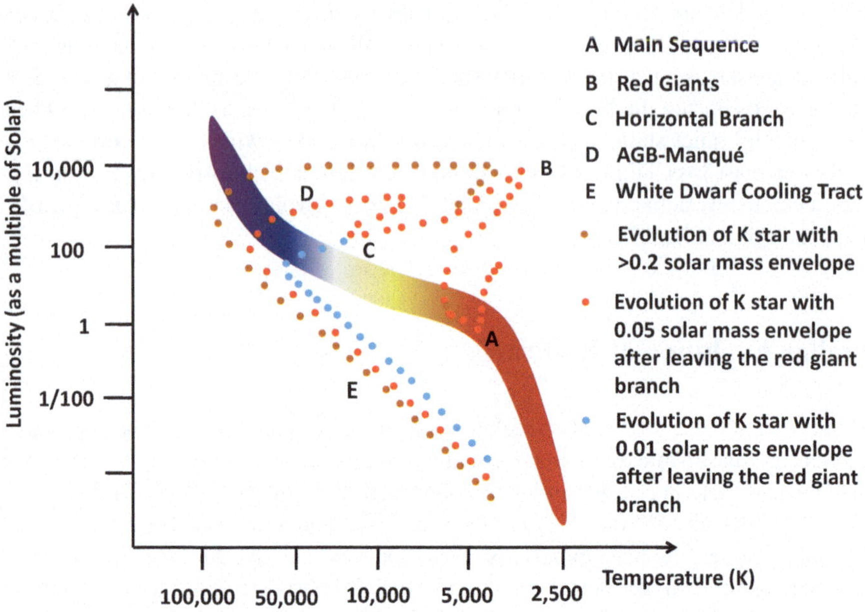

Fig. 1.10 The evolution of K-class stars with differing masses of hydrogen in their envelopes. Those with the lowest envelope masses (●) avoid the return to the red giant branch and simply contract and heat up once helium is exhausted

Indeed, like the M-dwarfs, all stars pull this same trick to compensate for gradually dwindling energy resources. The increase in the rate of energy production in turn causes the star to gradually brighten and become hotter in the same way as the lowest mass M-dwarfs we looked at earlier. On the HR diagram, the star moves upwards to the left, along the main sequence, brightening and becoming steadily bluer with time. This upward trend is terminated when the core finally exhausts the available hydrogen. For the M-dwarfs, this process can take several trillion years and results in the M-dwarf eventually becoming a yellow G-dwarf (see previously). Similarly, a typical K-dwarf may eventually end up as hot as the present-day Sun immediately prior to leaving the main sequence.

Once the core exhausts hydrogen, energy output from the core briefly subsides, and the star is forced to contract and heat up. The star then makes a sharp (but subtle) turn to the left on the HR diagram. This temporary phase ends when the heat released by the contracting core causes the hydrogen lying immediately above it to ignite in a thick shell.

Re-invigorated by this new supply of nuclear energy, the overall stellar energy flow increases while the core contracts, gains mass and grows hotter. Once the shell source is ignited, the star begins expanding in response to the extra energy available, just as it did in the red dwarf. During this phase of core contraction, the volume of the core shrinks from a diameter approximately equivalent to that of Jupiter down to an object approaching the size of Earth (roughly a 100-fold contraction.) while its density rises to about 60 million grams per cubic cm. The star then moves progressively rightwards, across the HR diagram until it reaches class M. It is now a sub-giant.

During its main sequence life, the sluggish proton-proton-I chain liberates energy. The rate of this reaction has a fairly weak dependence on temperature, meaning that there is only a small increase in the rate of nuclear reactions with temperature. However, as the star slowly expands into a sub-giant and then a red giant, hydrogen fusion burns at a much greater rate in the shell through the carbon-nitrogen-oxygen (CNO) reactions. These reactions have a very steep dependence on temperature: rate varies with T^{17} for CNO versus T^4 for the PP-I chain. Therefore, as the temperature rises with core contraction, the rate of nuclear reactions accelerates as well. This is positive feedback: higher core temperature lead to faster nuclear reactions. This, in turn, accelerates the production of helium. More helium in the core makes it heavier, which causes it to contract and heat at an ever-faster rate. This process continues until something gives. Meanwhile, the rapid increase in energy output forces the star to expand and increase its surface area, which allows it to shed the increased energy flow. Lowering temperatures at the surface also increase the opacity of the gas, favoring the expansion of the star as it seeks to release the additional energy that is produced inside.

Over the course of several hundred million to a few billion years, the small K-star takes on the appearance of a red giant. As its surface cools, molecules and grains begin to condense in its outer layers. The increase in radiation pressure generates a strong stellar wind that impacts these molecules and grains. This begins the process that will eventually terminate the evolution of the star—progressive mass loss. If mass loss is excessive, the mass of the star will be whittled down to less than the critical mass needed to ignite helium; somewhat less and the star will be able to begin a new adventure as a helium-burning orb.

The Mass Limit for Helium-Fusion

The orange K-dwarfs and the most massive M-class red dwarf stars are in a precarious position on the HR diagram. If the rate of mass loss exceeds the rate that the core can gain helium through fusion, the core may never reach a critical mass: 0.45 solar masses. Examination of stars that have lost mass in globular clusters reveals that stars with less than 0.45 solar masses never develop hot enough cores to ignite their helium ash. This agrees with theoretical calculations that place the lowermost mass for helium ignition around 0.44 solar masses. A star with 0.45 solar masses will be mostly core, with a thin, low-mass shell of hydrogen-rich material lying on top. K-class stars begin life with such low mass that the loss of mass on the red giant branch might prevent them from forever igniting their helium core. Such failed giants would gradually whittle away, evolving into helium white dwarf stars, much like their smaller red dwarf cousins.

The helium-burning limit is theoretical, but we do know of quite a few select stars with masses around this limit (0.46–0.5 solar masses) that do burn helium. However, for the smallest K-dwarf stars and the largest red dwarf stars, the amount of mass lost during the red giant stage might prevent them from attaining a massive enough core of helium. For the largest single K-stars, this risk is easily avoided, and they will produce a red giant star with a helium core of around 0.5 solar masses. However, at the low end of the range these small K-stars (and most massive red dwarfs) could encounter some difficulty. Therefore, red dwarfs with spectral classes M0 and M1, such as Barnard's Star, may just grow their helium cores sufficiently to ignite them. For example, a star that begins its life with 0.6 solar masses, a loss of 20% of its mass will whittle the star's mass down to 0.48 solar masses—just above the helium burning limit. Such low mass stars may well ignite their helium fuel, but will they evolve into red giants? Figure 1.11 explores their fate.

For the more massive K-dwarfs—those with masses in excess of 0.6 to 0.65 times that of the Sun—the star will ascend the red giant branch. When its luminosity is roughly 2000 times that of the present Sun, its core will hold approximately half a Sun's worth of helium gas in a strongly degenerate state. The interesting thing about stars in the range 0.65–2 solar masses is that regardless of their initial mass, their luminosity will converge on a single figure that is determined by the maximum mass of their helium core, immediately prior to the ignition of helium. In these low mass stars, the luminosity at the top of the red giant branch is relatively fixed. The initial mass is irrelevant, and only the mass of the core prior to helium ignition matters. Since this is fixed at around 0.44 solar masses for all low mass stars, their luminosity doesn't vary between them at the tip of the red giant branch. Stars that are still more massive have helium cores that vary in mass with their total mass on

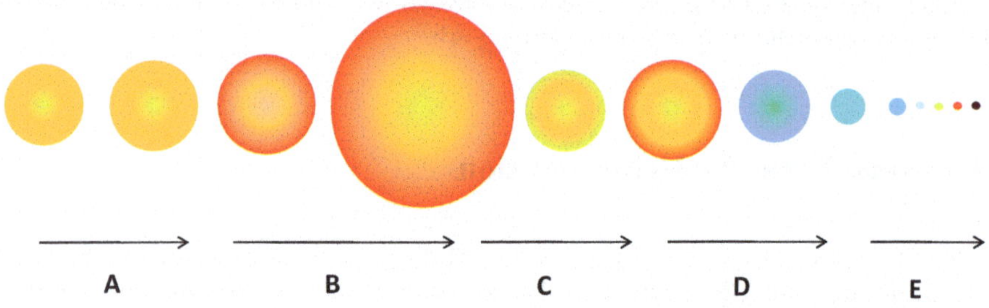

Fig. 1.11 The likely evolution of a 0.60–0.65 solar mass K-dwarf that retains more than 0.52–0.55 solar masses of gas while a red giant. The star leaves the main sequence (**a**) and expands into a red giant (**b**). During **b** and **c** mass is lost from the outer layers in a strong stellar wind, and the star begins to heat up. During **c** the star burns helium for 100–150 million years before the core is exhausted. During **d** the star begins to expand again as its carbon-rich core collapses and another strong wind sets in. This disperses even more of the remaining hydrogen-rich gas. However, where the envelope is thin, most of the radiation from the core escapes directly, and the star enters the AGB-Manquė phase (**d**). During **d** the star meanders lazily up the HR diagram until all the available fuel is spent. Once the thin hydrogen-rich envelope has blown away the star becomes a white dwarf (**e**). There is unlikely to be enough hydrogen left in the outer layer for the star to form a planetary nebula

the main sequence. By the time these stars leave the main sequence, their helium core is already more massive than the 0.45 solar mass cut-off. This means that stars with helium cores that have a mass in excess of this limit *do* vary in luminosity with their total mass.

Our K-class dwarf will now be a red giant star with a diameter more than a hundred times its original value. The surface will have cooled to 2500–3500 K, and it will have approximately 2000 times its original luminosity. Rather abruptly, helium will then ignite. While normal gas will expand with the additional heat from these nuclear reactions, the degenerate helium core will be resistant to such change. Heat will then build up, causing the temperature to skyrocket. As the rate of helium fusion reactions are even more sensitive to changes in temperature than the CNO cycle (T^{40} versus T^{17} in the CNO cycle), there is a catastrophic increase in the rate of reactions. This causes a violent eruption within the core, and the stellar output rises to 100,000 times that of the present Sun. The core shudders and violently expands, while the star remains intact. Expansion lifts degeneracy, cools the core and dampens the pace of nuclear reactions. This allows the star to enter a new normality as a K-class, helium-burning giant.

Such giant stars are orange or yellow in color and have a diameter that is perhaps 30 times that of the Sun, with a luminosity that is around 100 times the Sun's current value. This places them on a strip of stars extending across the middle of the HR diagram. At present low mass, metal-poor stars populate this region, which is known as the horizontal branch. In the future, this zone will be occupied by the lowest mass helium-burning stars, each eking away at the last of its fuel.

Why are helium-burning giants smaller than their predecessors? As the core expands outwards, angular momentum must be conserved, and so the envelope shrinks to balance this change. Moreover, as the core expands it cools down. Not only does this mean the reduction in the rate of helium fusion, but it also cools and largely extinguishes the hydrogen-burning shell that surrounds it. Since hydrogen fusion generates roughly ten times the amount of energy per gram of fuel as helium fusion, the overall energy output from the core falls dramatically once helium fusion commences. In turn, this means that the star has to do less to shed the energy it is generating. Finally, when helium fusion begins and the core is violently disturbed, some helium may enter the envelope and get mixed through it by convection. This lowers the envelope's opacity and allows more energy to escape. The combination of all of these mechanisms is more than sufficient to make the red giant shrink and heat up once more.

Helium fusion lasts roughly 150 million years and is fairly constant for all low mass stars, the Sun included. This is because all stars that have an overall mass less than 2.25 times that of the Sun have essentially the same mass of helium fuel at the point of ignition—approximately 0.44 solar masses. Since the mass of the helium core is roughly the same, the time they spend burning this fuel will also be the same. Once again, stars that are more massive have sufficiently bulky cores to begin helium fusion very soon after they leave the main sequence, and this mass varies with their total mass. However, as these stars have proportionately hotter cores, the time they spend burning this fuel decreases, rather than increases, with increasing mass.

Pulsations

Stars with thinner hydrogen envelopes evolve from the red giant branch across spectral classes A to F (white through yellow-white) along the horizontal branch. Where they settle during helium fusion depends critically on the mass of the remaining hydrogen-rich envelope. If it is greater than 0.1 solar masses these stars will be cooler and lie in spectral classes K to G. Lose more mass and the envelope is thinner and more transparent to radiation. Such stars migrate and settle in spectral classes F to B. Such stars will encounter instability and begin to pulsate.

Pulsations are driven in a similar manner to pulsations in larger, cooler red giants. However, in these hotter stars, helium ionization forces pulsation. Again, it is alternating cycles of ionization and recombination of helium ions with electrons that causes the star to pulsate. At temperatures of around 50,000 K, photons have sufficient energy that they can be absorbed by one of the atoms two electrons. This electron now has sufficient energy to jump off the atom, leaving it ionized. The process absorbs energy, trapping it within the outer layers of the star. Consequently, the star swells up, causing it to brighten. As the star swells, the density of the outer layer declines along with its temperature, which allows the trapped energy to escape. Once this happens, the star is able to relax and contract. The process then repeats as contraction increases the density of the layer, trapping energy once more. Such white giant stars are known as RR Lyrae variables, after the prototype.

Pulsations in RR Lyrae variables typically occur over a day or so with a period that varies with the overall luminosity of the star. Astroseismologists use these pulsations to study the stellar interior; the characteristic relationship between period and luminosity allows these objects to be used to measure stellar distance. RR Lyrae stars and other helium-burning low mass stars form an extended group stretching from roughly midway down the red giant branch towards the white dwarf cooling track at luminosities roughly 100 times that of the Sun. The majority of these stars have relatively thick hydrogen envelopes.

The Likely Fate of K-Dwarf Stars

Stars with Initial Masses in the 0.65–0.75 Solar Mass Range

Once the helium store is exhausted, fusion ends and the core collapses once more. Helium burns briefly in a shell around the core until this, too, is exhausted. Exhaustion occurs because these stars consume helium at a faster rate than hydrogen fusion can replace it. This is particularly true since the hydrogen-burning shell remains inactive while helium is fusing. However, when the helium is exhausted and the core collapses, hydrogen fusion reignites around the inert core. This takes the star back up the red giant branch in a path that is broadly parallel to its first ascension. This time the ascent occurs at slightly higher surface temperatures than before and this route is called the asymptotic giant branch.

As the star ascends, helium fuel is constantly added to the outer core of the red giant. At some point, the mass and density of this layer becomes sufficiently high to ignite. Helium fusion expands and cools the helium shell and the hydrogen-burning layer above it, until it switches off once more. An alternating cycle of hydrogen and then helium fusion then ensues. In the first phase, hydrogen fusion restocks the helium shell. In the second, the helium layer ignites, and hydrogen fusion is extinguished when the helium layer expands and shoves the hydrogen-burning shell outwards.

Every time the hydrogen shell ignites, it generates sufficient energy to re-expand the star. Expansion is interspersed with periods of contraction when helium fusion sets in. However, with every pulse of these fusion shells, the core gains more mass, and the luminosity of the star, as a whole, increases. Each thermal pulse, as these alternating phases are known, lasts for several thousand years until the hydrogen layer is exhausted. However, as every pulse expands the star more and more, the wholesale loss of mass accelerates. A combination of increasing luminosity and decreasing surface gravity seals the fate of the star.

As the luminosity increases, so does the strength of the stellar wind, as radiation pushes on the star's material with ever greater force. Meanwhile, expansion cools the star's outer layers, allowing grains and molecules to condense. These trap more and more radiation and help convert it into bulk gas motion. Moreover, as the star sheds mass, its gravitational pull decreases. This allows more gas and dust to escape the star's clutches. Of course, at a more prosaic level, as the star expands, the stellar surface is increasingly removed from the center of the star's mass. Consequently, how far these dying stars can advance up the asymptotic giant branch will depend upon how much hydrogen it can retain.

Another, subtler problem for red giants is the temperature profile of their envelopes. At temperatures less than 6000 K hydrogen nuclei begin to recombine with free electrons that are found in the plasma. When hydrogen atoms form from ions and free electrons, energy is released, which radiates outwards. This process allows the envelope to relax and contract inwards. As the envelope contracts, the density of the gases rises, and ionization begins once more. This process cycles over periods of months to years, depending on the mass of the star.

Unfortunately, for the dying star, these pulsations serve to further drive mass away from the star. In the end the envelope is whittled down until it has less than 0.02 solar masses of hydrogen remaining. At this stage the envelope is so transparent to radiation that convection is no longer required to transport energy. As with the dying red dwarfs we looked at earlier, this allows the envelope to rapidly shrink like a deflated balloon. The star shrinks to the diameter of a white dwarf and rapidly heats up to several tens of thousands of degrees. This emits copious amounts of ultraviolet radiation, and the dispersing shell of hydrogen fluoresces as a planetary nebula. The remnant star then cools and fades as a white dwarf.

Stars with Initial Masses Between 0.55 and 0.65 Solar Masses

Stars that began life with lower masses are likely to lose the majority of their hydrogen envelopes on the first ascent of the red giant branch. These stars will tend to populate the hotter end of the horizontal branch and retain very thin hydrogen-rich envelopes while they are burning helium. Remember that this is inference. No one has produced models of the evolution of such stars that take into account all of the necessary variables. Therefore, although this discourse is reasonable, it is supposition based on observation of current stars that have lost much of their hydrogen-rich envelope.

Raffaelle Gratton and colleagues (INAF-Osservatorio Astronomico di Padova) showed that the minimum stellar mass required to reach the tip of the AGB was about 0.65 solar masses. Smaller stars, with masses in the range of 0.65–0.54, begin expansion but peel away from the AGB before reaching the tip. Yet smaller stars, less than 0.54 solar masses, never become AGB stars. Rather than become red giants, once helium is exhausted these stars lazily ascend the HR diagram, zigzagging upwards to higher luminosities. These stars are known as "AGB-Manqué" stars. With insufficient hydrogen and helium remaining, these stars will not affect the glory of current planetary nebulae. Instead, they will simply contract and drift onto the white dwarf cooling track once their supply of hydrogen is exhausted or blown off into space. In the process, these stars heat up once more, to temperatures around 80,000 K before shriveling away as white dwarfs.

Subsequently, most of these post-helium-burning stars, and ex-asymptotic giant branch stars, then spend an eternity cooling down. These white dwarf embers may take hundreds of billions of years to shed their significant internal reserves of heat. However, for a precious few there may be sufficient helium in the inert shell that contraction raises its temperature to the point of ignition. For a brief period—perhaps a few centuries—these white dwarfs will enjoy a transitory renaissance. The wave of nuclear fusion inflates their remaining hydrogen and helium-rich layers, driving the star back towards the red giant branch. Vigorous fusion of both fuels generates some peculiar surface chemistry. These born-again giants have surface temperatures of several thousand degrees, glow yellow-white in color and show periodic dips in brightness as carbon soot condenses in the expanding gases. Sakurai's Object did just this in 1996. More than 20 years on, nuclear reactions and vigorous convection have created dense clouds of soot and other matter that obscures its surface.

The fate of these stars is to return to the white dwarf tract, as hot blue, hydrogen and helium-deficient objects.

The Fate of the Lowest Mass K-Dwarfs

Here, we enter what the author would refer to as "authoritative speculation" on his part. Twenty-five years after the last publication considered the fate of the K-dwarfs, nothing further has been written on the evolution of these stars. Therefore, as above,

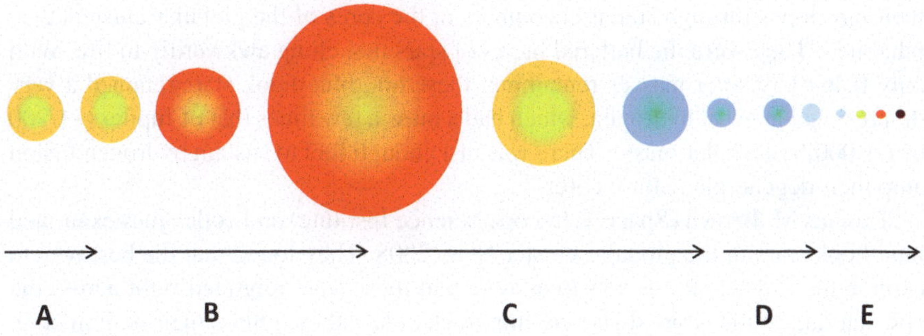

Fig. 1.12 The later evolution of a 0.55 solar mass star (at the M-K dwarf boundary). At the end of the star's main sequence life (**a**) the helium-rich core contracts, and the outer envelope expands. The star becomes a red giant (**b**), perhaps 50–100 times its original diameter. Stellar winds intensify during this phase removing most of the envelope (**c**), leaving a progressively smaller, hotter star that evolves left across the HR diagram and onto the Blue Hook (**d**). For a while helium is then burned to carbon and oxygen (**d**), but so little hydrogen remains that the star slips quietly onto the white dwarf cooling tract (**e**) as soon as the helium fuel is exhausted

what I is presented here is a simple extrapolation of evolutionary models for so-called extreme horizontal branch stars. Bear this in mind when reading. The models seem utterly reasonable, but they are empirical and are not underpinned by thorough mathematical modeling.

However, let's have a little fun and construct a suitable gedanken experiment. Consider a star that has moved onto the red giant branch, but for which mass loss has brought the total stellar mass down to 0.55–0.5 solar masses (Fig. 1.12). If sufficient mass is added to the helium core through hydrogen fusion, so that it breaches the 0.44 solar mass limit, helium will ignite. These stars will still reach the tip of the red giant branch (RGB), but once they ignite helium, they leave this region of the HR diagram for good.

What would such an impoverished star look like? These stars will be neither red, nor giant: instead, they will be hot, extreme horizontal branch stars and most likely very blue in color. Currently, such stars populate spectral class B or maybe O. Remember that the stars that are found in this region of the HR diagram today are not the products of the evolution of small K-dwarfs. The current occupiers are the products of G-class stars that have had most of their envelopes forcefully removed while they were red giants.

If we continue to even lower masses, those stars immediately above the helium-burning limit may traverse the horizontal branch in its entirety until they approach the white dwarf cooling tract. On the HR diagram, one would see the star leave the main sequence and swing upwards onto the base of the red giant branch, before curving along the horizontal branch and down onto the white dwarf cooling tract. These stars would arrive at the so-called blue hook—the extreme end of the horizontal branch at spectral class O. In today's universe, the blue-hook stars we observe are the lucky ones. Formerly, Sun-like stars with comparable masses lost most of their envelopes through stellar encounters in the cores of the globular clusters they inhabited. They were the battered husks of stars that clung awkwardly to life. With only 0.46–0.49 solar masses remaining, these hot, blue-hook stars retained a very thin residual layer of hydrogen, which had a mass a few times that of Jupiter (~1/200 to 1/1,000th of a solar mass). There was just enough fuel to sustain hydrogen fusion atop their degenerate helium core.

Thomas M. Brown (Space Telescope Science Institute) and colleagues examined blue-hook stars in the globular cluster NGC 2808. They found that the best way to explain their observations was to assume that these stars migrated right across the HR diagram to the white dwarf cooling track *before* they ignited their helium core. As these shriveled, remnant stars descended the white dwarf cooling track, hydrogen burning remained just—and only just—sufficient to raise the mass and temperature of the degenerate helium core to the point at which the helium ignited. Ignition was off-center and explosive. As the explosion ripped through the degenerate core, helium burning lifted degeneracy and drove vigorous convection in the super-heated material. Convection sucked some of the remaining hydrogen inwards from the thin, residual envelope. Here, it began to burn with a fury appropriate for such high temperatures. Temporarily, the star begins a phase of vigorous but mixed hydrogen and helium fusion. This brought carbon up to the star's surface, but pulled much of the hydrogen down into the interior.

The structure of these re-born stars is rather complex, even though they only have a mediocre helium core and hydrogen-rich envelope. Their small girth and very steep temperature gradient drive a complex mesh of interacting zones of nuclear fusion, in partly overlapping shells. Remember that these stars are only a few times the diameter of Earth—not much bigger than a white dwarf. Their core boils at 100 million K, while their surface has a temperature of perhaps 30,000 K. This is an enormous temperature gradient for such a small object.

These evolved ~0.46–0.49 solar mass stars spend about 1000 years undergoing the helium flash accompanied by one, then two overlying zones in which hydrogen is burning. Thomas Brown and co-workers' models suggest that early on, during the peak of the flash, all three zones of fusion are mixed together in a very complex mess. However, as the flash dies away, convection, driven by helium burning, retreats as well, and three separate shells emerge.

The innermost convecting zone is associated with continued helium fusion; this is overlain by another convective shell that is strongly enriched in carbon, where hydrogen burns through the CNO cycle. Finally, this shell is capped by a final convective region, where hydrogen burns more leisurely through the *pp*-chain. This outer burning zone extends convection to near the surface of the star a few hundred kilometers further out. These are amazing little stars, and it seems at least plausible that the smallest K-class stars (and perhaps the largest M-class red dwarfs) will end up as these sorts of objects before they become white dwarfs.

Over time, hydrogen-fusion depletes then exhausts the small amount of fuel available to the star. Now, the hydrogen-burning shells and associated zones of convection fade away. The helium-burning zone extends outwards, with a number of minor flashes, and the star finally stabilizes on the hot blue end of the horizontal branch. The luminosity of such blue-hook stars is only marginally greater than that of the present Sun (~1.18 times). However, the surface temperature exceeds 30,000 K. Life for these stars is nearly over.

Like other spent stars, the odd lucky cooling ember, one which had slightly more helium, might undergo a late helium flash, igniting this reserve as it enters the white dwarf track. That aside, its ultimate destiny is simply to chill into a frigid blackness over an interval that will span 300–400 billion years.

Conclusions

We measure the lives of both M- and K-dwarfs in tens of billions to trillions of years. In grossly superficial and anthropocentric terms, this would imply such stars could host habitable planets for periods that are grossly longer than the current age of the universe. Understanding the evolution of these stars is, therefore, critical to our understanding of the long-term habitability of the universe as a whole.

Therefore, it is all the more surprising (and disappointing) that the K-dwarfs have been utterly neglected by the astronomical community. Renewed interest in these stars, in terms of the potential habitability of their planets, may resolve this issue. However, six years after drafting the first version of this book, nothing has changed. Perhaps, someone will consider this neglected spectral class in the next round (or rounds) of grant funding. There are, after all, twice as many K-dwarf stars as G-dwarfs. That many of their worlds will show non-synchronous rotation makes them more likely candidates to host complex biological life than their smaller companions. To understand this aspect, we look to later chapters. The solution is rather prosaic, but seemingly was not considered until 2018. More on this later.

The other side of this chapter is planet formation. Since the first press of this book, rather a lot has happened, both observationally and in terms of modeling. A considerable number of improvements in modeling have resolved some of the issues concerning planet formation. To this end, the thoroughly scrutinized TRAPPIST-1 system will prove to be a testing ground for many of these theories. Already, very sensitive observations are ripping apart the simplistic notions that sprang up regarding the amount of volatile materials thought to be present on each world. Although this chapter only presents an outline of how materials can become mixed and dispersed by aggregating planets, one hopes that it is clear that the situation is, to put it mildly, complex. However, given that we are now verging on observing planets directly, empirical models can replace computer models that are all too often based on outdated preconceptions.

This chapter thus forms a starting point for the further dissection of our notions of habitability. Stellar habitable zones are one thing; stellar evolution another; however, it is the evolution of the host planet that will really set the tone for the remainder of the book. Before we get there, we look at planetary discovery, both in terms of the methodology and in terms of the implications of these discoveries.

References

Planetary Formation and Migration

Benítez-Llambay, P., Masset, F., Koenigsberger, G., & Szulágyi, J. (2015). Planet heating prevents inward migration of planetary cores. *Nature, 520*, 63–65. https://doi.org/10.1038/nature14277.

Boss, A. P. (2004). Evolution of the solar nebula. VI. Mixing and transport of isotopic heterogeneity. *The Astrophysical Journal, 616*(2), 1265–1277.

Boss, A. P. (2006). Rapid formation of super-Earths around M dwarf stars. http://arXiv.org/pdf/astro-ph/0605061v1.pdf.

Boss, A. P.. (2008). Mixing in the solar Nebula: Implications for isotopic heterogeneity and large-scale transport of refractory grains. https://arxiv.org/pdf/0801.1622.pdf.

Bourrier, V., Lovis, C., Beust, H., Ehrenreich, D., Henry, G. W., Astudillo-Defru, N., Allart, R., Bonfils, X., Ségransan, D., Delfosse, X., Cegla, H. M., Wyttenbach, A., Heng, K., Lavie, B., & Pepe, F. (2018). Orbital misalignment of the Neptune-mass exoplanet GJ 436b with the spin of its cool star. *Nature, 553*, 473–480. https://doi.org/10.1038/nature24677.

Brasser, R., Bitsch, B., & Matsumura, S. (2017). Saving super-Earths: Interplay between pebble accretion and type I migration. http://arXiv.org/pdf/1704.01962v1.pdf.

Brasser, R., Matsumura, S., Ida, S., Mojzsis, S. J., & Werner, S. C.. (2016) Analysis of terrestrial planet formation by the Grand Tack model: System architecture and tack location. https://arxiv.org/pdf/1603.01009.pdf.

Brenan, J. M., & McDonough, W. F. (2009). Core formation and metal–silicate fractionation of osmium and iridium from gold. *Nature Geoscience, 2*, 798–801. https://doi.org/10.1038/NGEO658.

Campbell, I. H., & O'Neill, H. C. S. (2012). Evidence against a chondritic Earth. *Nature, 483*, 553–558. https://doi.org/10.1038/nature10901.

Chatterjee, S. & Ford, E. B. (2015). Planetesimal interactions can explain the mysterious period ratios of small near-resonant planets. http://arXiv.org/pdf/arXiv:1406.0521v2.pdf.

Cossou, C., Raymond, S. N., & Pierens, A.. (2017). Convergence zones for Type I migration: an inward shift for multiple planet systems.https://arxiv.org/pdf/1302.2627.pdf.

Day, J. M. D. (2016). Extraordinary world. *Nature, 537*, 320–321.

Ehrenreich, D., Bourrier, V., Wheatley, P. J., des Etangs, A. L., Hébrard, G., Udry, S., Bonfils, X., Delfosse, X., Désert, J.-M., Sing, D. K., & Vidal-Madjar, A. (2015). A giant comet-like cloud of hydrogen escaping the warm Neptune-mass exoplanet GJ 436b. *Nature, 522*, 459–461. https://doi.org/10.1038/nature14501.

Fassett, C. I., & Minton, D. A. (2013). Impact bombardment of the terrestrial planets and the early history of the Solar System. *Nature Geoscience, 6*, 520–524. https://doi.org/10.1038/NGEO1841.

Feng, F. & Jones, H. R. A. (2017). Was Proxima captured by Alpha Centauri A and B? http://arXiv.org/pdf/arXiv:1709.03560.pdf.

Gomes, R., Levison, H. F., Tsiganis, K., & Morbidelli. (2005). Origin of the cataclysmic Late Heavy Bombardment period of the terrestrial planets. *Nature, 435*, 466–469. https://doi.org/10.1038/nature03676.

Johnson, B. C., & Melosh, H. J. (2012). Impact spherules as a record of an ancient heavy bombardment of Earth. *Nature, 485*, 75–77. https://doi.org/10.1038/nature10982. https://www.researchgate.net/publication/224846701_Impact_spherules_as_a_record_of_an_ancient_heavy_bombardment_of_Earth.

Johnson, J. (2017). Compositions of small planets & implications for planetary dynamics. In 229th AAS Meeting, abstract # 413.06.

Kaib, N. A. & Chambers, J. E. (2015). The fragility of the terrestrial planets during a giant planet instability. https://arxiv.org/pdf/1510.08448.pdf.

Kennedy, G. M., Kenyon, S. J., & Bromley, B. C. (2006). Planet formation around low mass stars: the moving snow line and super-Earths. http://arXiv.org/pdf/astro-ph/0609140v1.pdf.

Kley, W. & Nelson, R. P. (2012). Planet-disk interaction and orbital evolution. http://arXiv:1203.1184.pdf.

Knutson, H. A., Madhusudhan, N., Cowan, N. B., Christiansen, J. L., Agol, E., Deming, D., Désert, J. M., Charbonneau, D., Henry, G. W., Homeier, D., Langton, J., Laughlin, G., & Seager, S.. (2011). A Spitzer transmission spectrum for the exoplanet GJ436b, evidence for stellar variability, and constraints on dayside flux variations. http://arXiv.org/pdf/arXiv:1104.2901.pdf.

Proceedings of a workshop held 8–11 2004 in Kaua'i, Hawai'i. A. N. Krot, E. R. D. Scott, and B. Reipurth. San Francisco: Astronomical Society of the Pacific, 883-889. 2.

Labidi, L., Cartigny, J. P., & Moreira, M. (2013). Non-chondritic Sulphur isotope composition of the terrestrial mantle. *Nature, 501*, 208–211. https://doi.org/10.1038/nature12490.

Landin, N. R., Ventura, P., D'Antona, F., Mendes, L. T. S., & Vaz, L. P. R. (2007). Rotational properties of the orion nebula cluster revised. In F. Kupka, I. Roxburgh and K. Chan (Eds.), *Convection in astrophysics, Proceedings of IAU Symposium* #239 held 21–25 August, 2006 in Prague, Czech Republic (pp. 311–313). General Assembly. http://adsabs.harvard.edu/full/2007IAUS.239,311L.

Laskar, J., & Gastineau, M. (2009). Existence of collisional trajectories of Mercury, Mars and Venus with the Earth. *Nature, 459*, 817–819. https://doi.org/10.1038/nature08096.

Levison, H. F., Morbidelli, A., Laerhoven, C. V., Gomes, R., & Tsigani, K. (2007). Origin of the structure of the Kuiper belt during a dynamical instability in the orbits of Uranus and Neptune. http://arXiv.org/pdf/0712.0553v1.pdf.

Makarov, V. V., Berghea, C., & Efroimsky, M. (2013). Dynamical evolution and spin-orbit resonances of potentially habitable exoplanets. The case of GJ 581d. https://arxiv.org/pdf/1208.0814.pdf.

Mann, A. (2018). Bashing holes in the tale of Earth's troubled youth. *Nature, 553*, 393–339.

Nagasawa, M., Ida, S., & Bessho, T. (2008). Formation of hot planets by a combination of planet scattering, tidal circularization, and Kozai mechanism. https://arxiv.org/pdf/0801.1368.pdf.

Nakamoto, T., Hayashi, M. R., Kita, N. T., & Tachibana, S. (2004). Chondrule-forming shock waves in the solar Nebula by X-ray flares. *Chondrites and the Protoplanetary Disk, ASP Conference Series, 341*, 883.

O'Brien, D. P., Izidoro, A., Jacobson, S. A., Raymond, S. N., & Rubie, D. C. (2018). The delivery of water during terrestrial planet formation. https://arxiv.org/pdf/1801.05456.pdf.

Penz, T., Lammer, H., Kuliko, Y. N., & Biernatabd, H. K. (2005). The influence of the solar particle and radiation environment on Titan's atmosphere evolution. *Advances in Space Research, 36*(2), 241–250. https://doi.org/10.1016/j.asr.2005.03.043.

Raymond, S. N., Barnes, R., Veras, D., Armitage, P. J., Gorelick, N., & Greenberg, R. (2009). Planet-planet scattering leads to tightly packed planetary systems. http://arXiv.org/pdf/arXiv:0903.4700v1.pdf.

Roig, F., Nesvorny, D., & DeSouza, S. R. (2016) Jumping Jupiter can explain Mercury's orbit. http://arxiv.org/pdf/1603.02502.pdf.

Walsh, K. J., Morbidelli, A., Raymond, S. N., O'Brien, D. P., & Mandell, A. M. (2011). A low mass for Mars from Jupiter's early gas-driven migration. *Nature, 475*, 206–209. https://doi.org/10.1038/nature10201.

Wertheimer, J. G. & Gregory Laughlin, G. (2006). Are Proxima and Alpha Centauri Gravitationally Bound?. https://arXiv.org/pdf/0607401v1.pdf.

Star Formation

Bate, M. R., Bonnell, I. A., & Bromm, V. (2003a). The formation of a star cluster: predicting the properties of stars and brown dwarfs. *Monthly Notices of the Royal Astronomical Society, 339*, 577–599.

Batygin, K., & Adams, F. C. (2013). Magnetic and gravitational disk–star interactions: an interdependence of pms stellar rotation rates and spin–orbit misalignments. *The Astrophysical*

Journal, 778, 169. https://doi.org/10.1088/0004-637X/778/2/169.

Larson, R. B. (2009). Angular momentum and the formation of stars and black holes. *Reports on Progress in Physics, 73*(2010), 014901 (14pp. https://doi.org/10.1088/0034-4885/73/1/014901.

Malmberg, D., De Angeli, F., Davies, M. B., Church, R. P., Mackey, D., & Wilkinson, M. I. (2007). Close encounters in young stellar clusters: Implications for planetary systems in the solar neighborhood. https://arxiv.org/pdf/astro-ph/0702524.pdf.

Mascareño, S., Rebolo, A. R., González Hernández, J. I., & Esposito, M. (2015). Rotation periods of late-type dwarf stars from time-series high-resolution spectroscopy of chromospheric indicators. http://arXiv.org/pdf/arXiv:1506.08039v1.pdf.

Olczak, C., Pfalzner, S., & Eckart, A. (2009). Stellar interactions in dense and sparse star clusters. http://arXiv.org/pdf/arXiv:0911.0293.pdf.

Portegies Zwart, S. F., Hut, P., McMillan, S. L. W., & Verbunt, F. (1997) Star Cluster Ecology II: Binary evolution with single-star encounters. http://arXiv.org/pdf/astro-ph/9706090.pdf.

Portegies Zwart, S. F., Makino, J., McMillan, S. L. W., & Hut, P. (1999). Star cluster ecology III: Runaway collisions in young compact star clusters. http://arXiv.org/pdf/astro-ph/9812006.pdf.

Stauffer, J. R., Jones, B. F., Backman, D., Hartmann, L. W., Barrado y Navascués, D., Pinsonneault, M. H., Terndrup, D. M., & Muench, A. A. (2003). Why are the K dwarfs in the Pleiades so blue? *The Astronomical Journal, 126*, 833–847.

Brown Dwarfs

Bate, M. R., Bonnell, I. A., & Bromm, V. (2003b). The Formation mechanism and resulting properties of brown dwarfs. *Symposium—International Astronomical Union, 211*, 27–30. https://doi.org/10.1017/S0074180900210206. E. L. Martin (Ed.). https://www.cambridge.org/core/terms.

Kirkpatrick, D., Barman, T. S., Burgasser, A. J., McGovern, M. R., McLean, I. S., Tinney, C. G., & Lowrance, P. J. (2005). Discovery of a Very Young Field L Dwarf, 2MASS J01415823−46335741 J. http://citeseerx.ist.psu.edu/viewdoc/download?doi=10.1.1.337.3284&rep=rep1&type=pdf.

Kirkpatrick, J. D., Cushing, M. C., Gelino, C. R., Beichman, C. A., Tinney, C. G., Faherty, J. K., Schneider, A., & Mace, G. N. (2013) Discovery of the Y1 Dwarf WISE J064723.23−623235.5. http://iopscience.iop.org/article/10.1088/0004-637X/776/2/128/meta.

Leggett, S. K., Cushing, M. C., Saumon, D., Marley, M. S., Roellig, T. L., Warren, S. J., Burningham, B., Jones, H. R. A., Kirkpatrick, J. D., Lodieu, N., Lucas, P. W., Mainzer, A. K., Martin, E. L., McCaughrean, M. J., Pinfield, D. J., Sloan, G. C., Smart, R. L., Tamura, M., & van Cleve, J. (2009). The physical properties of four 600 K T dwarfs. *The Astrophysical Journal, 695*, 1517–1526. http://arXiv.org/pdf/arXiv:0901.4093v1.pdf.

Zuckerman, B. & Song, I. (2013). The minimum jeans mass, brown dwarf companion IMF, and predictions for detection of Y-type dwarfs. http://arXiv.org/pdf/0811.0429v2.pdf.

Stellar Evolution

Adams, F. C., Bodenheimer, P., & Laughlin, G. (2005). M dwarfs: planet formation and long-term evolution. *Astronomische Nachrichten, 326*(10), 913–919. https://doi.org/10.1002/asna.200510440.

Adams, F. C., Laughlin, G., & Graves, G. J. M. (2004). Red dwarfs and the end of the main sequence. *Rev Mex AA (Serie de Conferencias), 22*, 46–49. From Gravitational Collapse: from massive stars to planets, G. García-Segura, G. Tenorio-Tagle, J. Franco, & H. W. Yorke (Eds.).

Boyajian, T. S., von Braun, K., van Belle, G., Farrington, C., Schaefer, G., Jones, J., White, R., McAlister, H. A., ten Brummelaar, T. A., Ridgway, S., Gies, D., Sturmann, L., Sturmann, J., Turner, N. H., Goldfinger, P. J., & Vargas, N. (2013). Stellar diameters and temperatures. III main-sequence A, F, G, and K stars: additional high-precision measurements and empirical relations. *The Astrophysical Journal, 771*, 40. https://doi.org/10.1088/0004-637X/771/1/40.

Brown, T. M., Lanz, T., Sweigart, A. V., Cracraft, H. M. I., & Landsman, W. B. (2012a). Flash mixing on the white dwarf cooling curve: spectroscopic confirmation in NGC 2808. http://citeseerx.ist.psu.edu/viewdoc/download?doi=10.1.1.761.4364&rep=rep1&type=pdf.

Brown, T. M., Sweigart, A. V., Lanz, T., Landsman, W. B., & Hubeny, I. (2012b). Flash mixing on the white dwarf cooling curve: understanding hot horizontal branch anomalies in NGC 2808. http://arXiv:astro-ph/0108040.pdf.

Cassisi, S. (2008). Very low-mass stars: Structural and evolutionary properties. http://arXiv.org/pdf/arXiv:1111.6464.pdf.

Dormanm, B., Rood, R. T., & O'Connell, R. W. (1993). Ultraviolet radiation from evolved stellar populations i. models. *Astrophysical Journal, 419*. https://arxiv.org/pdf/astro-ph/9311022.pdf.

Eisloffel, J., Mohanty, S., & Scholz, A. (2004). Formation and evolution of very low mass stars and brown dwarfs. https://arxiv.org/pdf/astro-ph/0410046.pdf.

Fagotto, F., Bressan, F. A., Bertelli, G., & Choisi, C. (1994). Evolutionary sequences of stellar models with new radiative opacities IV. Z=0.004 and z=0.008. *Astronomy and Astrophysics Supplementary Series, 105*, 29–38.

Gizis, J. E., Dettman, K. G., Burgasser, A. J., Camnasio, S., Alam, M., & Joseph, C. (2015). Kepler monitoring of an L dwarf II. clouds with multiyear lifetimes. https://arxiv.org/pdf/1509.07186v1.pdf.

Gratton, R. G., D'Orazi, V., Bragaglia, A., Carretta, E., & Lucatello, S. (2010). The connection between missing AGB stars and extended horizontal branches (Research Note). http://arxiv.org/pdf/1010.5913v1.pdf.

Li, C., de Grijs, R., Deng, L., & Liu, X. (2013). Blue straggler evolution caught in the act in the Large Magellanic Cloud globular cluster Hodge 11. https://arxiv.org/pdf/1304.4312v1.pdf.

Luhman, K. L. (2012). The formation and early evolution of low-mass stars and brown dwarfs. https://arxiv.org/pdf/1208.5800.pdf.

Rich, M., Sosin, C., Djorgovski, G., Piotto, G., King, I. R., Renzini, A., Phinney, A. E. S., Dorman, B., Liebert, J., & Meylan, G. (1997). Discovery of extended blue horizontal branches in two metal-rich globular clusters. *The Astrophysical Journal, 484*, L25–L28.

van Dokkum, P. G. & Conroy, C. (2010). A substantial population of low mass stars in luminous elliptical galaxies. https://arxiv.org/pdf/1009.5992v1.pdf.

Webbink, R. F. (1975). The evolution of helium white dwarfs in close binaries. *MNRAS, 171*, 555–568.

Wood, B. E., Linsky, J. L., & Güdel, M. (2015). Stellar winds in time. In H. Lammer & M. Khodachenko (Eds.), *Characterizing stellar and exoplanetary environments, astrophysics and space science library* (Vol. 411). Switzerland: Springer International Publishing. https://doi.org/10.1007/978-3-319-09749-72.

Chapter 2
Exoplanetary Discovery

Introduction

To state the obvious, this chapter represents a moving target. Although one can, for example, blithely state that of the end of December 2017, a total of 3605 planets had been confirmed, of which 1414 were Neptune-like; by the time this book goes to press that figure will, most likely, be out of date. Therefore, consider this chapter as an incomplete portrait that has many of its key features presented but not all of the hues and textures indicative of a complete work.

Rather than attempt to detail all of the discoveries that have been made to date, this chapter will focus on the methodologies that have been used to identify alien worlds, then illustrate these techniques with appropriate examples of such planets. Later chapters will then examine several of these worlds in more detail to consider whether or not these planets are habitable—and more poignantly whether such worlds can (or perhaps do) host intelligent life. The backdrop is then the evolution of K- and M- (or early L-) class dwarfs, which we explored in Chap. 1.

A Generation of Discovery

After many false starts extrasolar planets started emerging from astronomical measurements in the 1990s. Although most thought that the paradigm would be a planetary system resembling ours, with its seemingly stable warm star and a gaggle of evenly spaced worlds, reality had some harsh surprises in store. Take the first confirmed exoplanetary system: PSR B1257+12. Even the name sounds uninviting. Spinning wildly, with hundreds of revolutions per millisecond this pulsar does not provide a comfortable abode for its small family of terrestrial worlds.

A millisecond pulsar's first breath is taken in the multi-billion-degree plasma generated in the heart of a collapsing star. As the star falls apart, the core is crushed into a fast-spinning ball of neutrons, iron and elementary particles. This is a

D. S. Stevenson, *Red Dwarfs*, https://doi.org/10.1007/978-3-030-25550-3_2

pulsar—a spinning neutron star. The pulsar generates beams of electromagnetic radiation that sweeps outward from its magnetic poles, scanning surrounding space like the silent beams of a distant lighthouse. However, at the point of stellar detonation this freshly minted neutron star doesn't cavort with the wild abandon that is characteristic of a millisecond pulsar. At this stage the pulsar is positively sedate in its behavior. As the debris of the shattered star clear, the pulsar spins slower than a conventional washing machine, perhaps a few dozen times per second. Subsequently, these pulsars slow and fade from view.

The forerunner of a millisecond pulsar has a different life ahead. Sharing space with a companion star, the intense gravitational field of the neutron star whisks material away from the companion, forming a disc of material around its waist. This accretion disc steadily adds mass to the neutron star. As the pulsar gains mass and momentum, the incoming material spins up the neutron star until it rotates at hundreds of times per second. With increased vigor, the neutron star is reborn as a millisecond pulsar. Its gyrating beams of radiation now rapidly erode what remains of its once vibrant companion.

Pulsar PSR B1257+12 wasn't done yet. Rather than sweep the withering remains of its companion under the cosmic carpet, the considerable gravity of PSR B1257+12 gently nudged what remained into stable orbits. Over the course of the next few million years, the gas-depleted wreckage of the former companion star was molded into a handful of planets. Liberating no detectable radiation of their own, these ghosts of the former companion betrayed their presence through their subtle gravitational interplay with the pulsar host (Fig. 2.1).

Many millennia later, Aleksander Wolszczan and Dale Frail noticed that the flashes of radiation from this millisecond pulsar varied slightly, as if it were gently but repeatedly being pulled in different directions. Rather than a steady stream of blips, slight variations in arrival time meant that something small was, or rather some things were, orbiting it. Judging by their masses these were planets. Not sur-

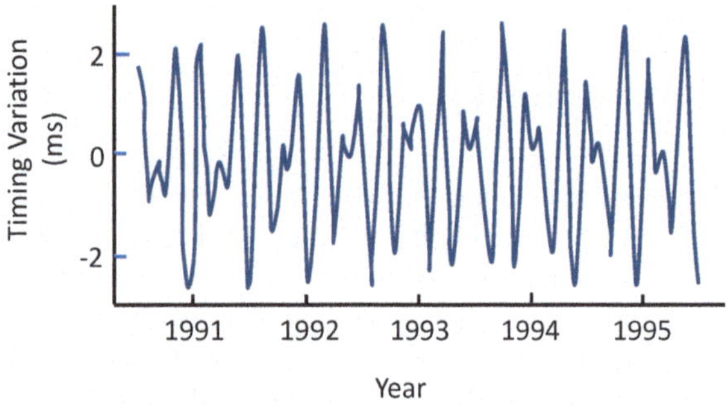

Fig. 2.1 Pulsar timing method. Pulsars emit regular pulses of radiation across the EM spectrum. As an orbiting planet pulls on the pulsar it accelerates the dead star, moving its pulses backwards and forwards in proportion to the planet's mass and orbital distance. In the case of PSR B1257+12, the presence of variations of different magnitude and period told astronomers there were multiple planets in this system

prisingly, given the nature of the star around which these planets were orbiting, many in the astronomy community were more than a tad skeptical. In part this was down to history. A year earlier Andrew G. Lyne had announced the presence of a planet orbiting PSR 1829-10. However, this paper was later retracted. Yet in the case of Wolszczan and Frail's discovery, the data was sound. Planets did orbit this defunct star. Wolszczan and Frail had bagged three firsts: the first confirmed extrasolar planets; the first multi-planet system; and finally, the first super-terrans—planets with marginally more mass than Earth.

Subsequently the method, known as pulsar timing variation, has evolved into a number of spin-off techniques, and although it is a reliable means of detecting mass in orbit around pulsars, the nature of these stars means that such discoveries are unlikely to be habitable.

Pulsars, odd planetary hosts aside, hold one further record: the oldest planetary system known. Orbiting within the dense stellar core of the globular cluster M4 is a pair of dead stars: PSR B1620-26 and its companion white dwarf. With the announcement of this discovery in 2003, this 12.7-billion-year-old system holds a planet with twice the mass of Jupiter in a distant orbit around both corpses. Although this ancient world was probably born in an orbit around the progenitor of the white dwarf, PSR B1260-26b is now locked in orbit around both stars. A distant, peculiar world.

Since the discovery of PSR B1260-26b, a smattering of other pulsar worlds have been discovered, taking the number to eight. In the case of PSR 1829+12, both direct modulation of the pulsar timing and indirect changes to this modulation led to the confirmation of multiple planets in this system. In essence, when a single planet orbits a pulsar the gravitational pull of the planet alters the time at which its beams point towards Earth. When more than one planet is in orbit, the gravitational pull of the other worlds modifies this effect, alternately pulling more or less strongly on the pulsar in different directions, thereby altering the timing in a further predictable manner. To date 1.7% of planet finds have been detected using this method.

Likewise, many variable stars are intrinsically variable; that is, the source of the variability is caused by periodic changes that occur within the star, which cause the star as a whole to pulsate. In F- and A-class main sequence and other dwarf stars, such pulsations occur over periods of hours to a day. These are caused by alternate ionization and recombination of electrons with helium, in a layer just underneath the photosphere (see Chap. 1). Two classes of such stars lend themselves to such investigation. The first are subdwarf B (sdB) stars, which are a collection of hot, low mass and usually helium-fusing dwarf stars. The second group includes main sequence A to F class stars called Delta Scuti variables. As each star has a relatively small diameter, the pulsation period for each is sufficiently short, allowing pulsations to be monitored over relatively short intervals of a week to a month or so. This allows the speedy construction of light curves that can then be easily studied for unexpected variations.

The period is dependent on the overall mass and luminosity of the star, and this allows for ready interpretation of variations caused by orbiting planets. In such objects, planets with short orbital periods will distort the shape of the star and

Table 2.1 The numbers of different planets and planetary systems as of December 2017

Number of planetary candidates	Number of confirmed planets	Number of multiple planet systems	Number of Jupiter-mass planets	Number of Neptune-mass planets	Number of super-terrans	Number of earth and sub-earth planets	Unknowns
4496	3572	2660	1198	1414	53	882	25

Table 2.2 The proportion of the 3605 exo-planets discovered using the different methods described in this chapter as of December 2017

Discovery method	Number of planets	Percentage of total
Transit	2828	77.7
Transit timing (and duration) variation	15	0.42
Eclipse timing variation	9	0.25
Radial velocity	665	18.5
Microlensing	53	1.5
Pulsar timing variation	6	0.17
Pulsation timing modulation	2	0.06
Direct imaging	44	1.2
Orbital brightness modulation (reflection in transit)	6	0.16
Astrometry	1	0.02

Doppler-shift the period of pulsation in fairly predictable ways. Consequently, they are amenable to analysis even where the spectra of the stars are of poor quality. Through this method Roberto Silvetti and colleagues discovered a planet orbiting the sdB star V391 Pegasus in 2007. Several more have been found in the intervening decade. Tables 2.1 and 2.2 summarizes the results so far.

The Radial Velocity Method

In 1988, a few years before a skeptical audience accepted the presence of Wolszczan's and Frail's pulsar planets, three Canadian astronomers had developed a successful method that has been used subsequently to detect hundreds of other worlds. Bruce Campbell, Gordon Walker and Stephenson Yang used a technique known as the radial velocity to look for the gravitational effect of a planet as it orbits its star. The technique involves looking for the subtle toing and froing of the spectrum of a star as a planet moves it gently towards, then away, from an observer, with each orbit (Fig. 2.2).

Campbell, Walker and Yang examined the spectrum of the star Gamma Cephei A. Already known to be part of a binary system, the spectrum of Gamma Cephei A was expected to vary in a predictable manner as this star and its partner orbited their mutual center of gravity. However, the backwards and forwards motion of the brightest star didn't follow the pattern that would be expected if the two stars were alone.

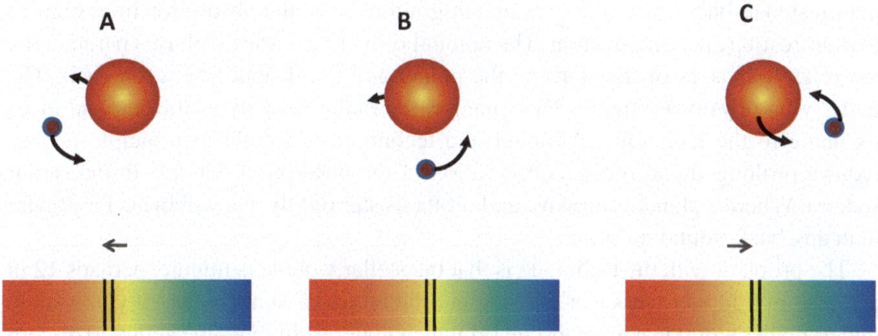

Fig. 2.2 The radial velocity (RV) method of planet detection. Utterly simple in its science yet fiendishly difficult to use, the technique relies on the precise and accurate interpretation of stellar spectra. When a planet orbits a star its gravitational pull yanks the star this way and that. As the planet moves around its orbit the star follows. Both objects swirl around their common center of gravity, much like an Olympic shot-putter and his or her shot. The hidden planet reveals itself by the subtle motion of spectral absorption features to the red- or blue-end of the spectrum. Movement to the blue end occurs as the star moves towards the observer, and to the red end as it moves away

Instead, the spectrum "wobbled" as though a third, Jupiter-mass body was present. Although the technology the team used was cutting-edge in its day, the signal the team was able to detect was only marginally better than the instrumental noise. To add to their woe, the astronomers didn't know the precise mass of the secondary star in the system. Therefore, it was unclear whether the radial velocity measurements were detecting the suspected planet or merely a combination of stellar pulsation in the primary giant and the effect of its binary partner. Thus, despite Anthony Lawton and P. Wright making a further announcement regarding a planet in the same system the following year, confirmation of the planet's existence took more than a decade.

Finally, in 2003, a combination of improvements in the radial velocity technique and measurements of the mass and orbit of both stars confirmed that a 1.59 Jupiter-mass planet circled Gamma Cephei Ab in an eccentric orbit around its giant orange host. Moving between 1.81 and 2.28 AU from Gamma Cephei A, the Jupiter-mass planet takes around 1.4 years to orbit its host giant star. Meanwhile, the final component, red dwarf Gamma Cephei B, had its orbit pinned down to a period of 66–67 years.

Throughout the early 1990s refinements in spectroscopy and computer technology led to the discovery of the first planets around more conventional stellar partners. Much like water escaping through the crack in a dyke, the first discoveries were faltering, hesitant affairs with a relatively long hiatus between each announcement. But by 2000, discoveries were monthly, then weekly, until over 400 worlds were banked. Initially, the majority of these came through the radial velocity technique, used by Campbell and described below. However, as technology has moved forward with increasing pace, more sophisticated techniques have come to the fore.

In most stellar spectra, chemical elements betray their presence with fine absorption features. As a star and any planets orbit their common center of gravity, the pull of any orbiting planet causes the star to wobble in space. This is associated with minute accelerations towards and away from the observer. In turn, this effect is manifested as backwards and forwards migrations of stellar absorption lines relative to their resting spectral location. The amount of wobble varies with two parameters: the relative masses of the star and the planet and the distance between each. The underlying physics is simple. This changes the radial velocity of the star and gives its name to the technique. Although the technique is simple in principle it does require prolonged and precise observation of the absorption features in the stellar spectra. Where a planet is massive and orbits its star tightly, the wobble is far greater than any background noise.

The problem with the technique is that the stellar wobble is minute: perhaps 12 m per orbit of a Jupiter mass world in a tight orbit around a Sun-like star. If you want to find a smaller world in a more distant orbit then your resolution—the ability to discern variability above background noise—must be correspondingly greater. An Earth-mass world pulls with less force than is required to shift its star by a meter. If the orbit of this planet lies within the habitable zone of its star, this weak wobble is extended over 12 months. As radial velocity measurements depend on the precise determination of the location of stellar absorption lines, any interference from atmospheric turbulence or instrumental error can result in spurious identification of planetary signals. As all of the current instruments deployed to use this technique are Earth-bound, our atmosphere presents the biggest problem in the successful isolation of planetary signals.

Furthermore, although a single world orbiting a star produces a nice, clear wobble, additional worlds lying further out produce more subtle resonances that become harder to detect as the number of worlds increases. Not only does each planet generate its own wobble, but by pulling on its planetary neighbors, it also alters theirs. Consequently, the number of measurements is necessarily increased and the sensitivity of these measurements requires greater precision. Unfortunately, this can lead to noise swamping the signal—or spurious signals emerging from the noise. A common method of avoiding this problem is the use of a reference spectrum. Oxygen was proposed some years ago, but it is iodine that was most widely employed until the rise of HARPS (High Accuracy Radial Velocity Planet Searcher at the ESO's 3.6-m telescope at La Silla Observatory in Chile).

In the iodine model, astronomers pass starlight through a vessel containing iodine, and the resulting spectrum is compared to the stellar spectrum. The absorption features of iodine are well characterized, allowing comparison with absorption spectra from the star. This helps identify and remove instrumental noise. Atmospheric interference and thermal motion within the instrument are the principle causes of such interference. Therefore, the use of the reference sources allows some of the instrumental noise to be tracked and its influence reduced.

By the mid-1990s the radial velocity technique was sufficiently well tested and its sensitivity improved so that it could be deployed effectively in the field. In 1995 Michel Mayor and Didier Queloz published the first confirmed extrasolar planet orbiting a conventional Sun-like star: 51 Peg b.

In this context, the word conventional is a somewhat loose term. Only the star is conventional. 51 Peg b is a planet approximating Jupiter in mass, but orbiting its star so tightly that it broils at over 700 °C. As such, Peg b was the first of a growing class of planets dubbed "hot Jupiters."

Over the ensuing years, subsequent discoveries revealed that these hot Jupiters were really rather commonplace and positively mundane. Some wondered if it was our *cold* Jupiter that was unusual. Although the initial wave of discoveries couldn't address this question directly, it was apparent that the modus operandi of the radial velocity technique would bias the pattern of planetary discovery. The radial velocity (RV) technique is, by definition, very self-selecting. Given limited observation time and an (initially) low resolution, the radial velocity technique was always bound to identify the freakish, tightly orbiting, massive worlds. After all, these planets have the greatest impact on stellar motion—and hence stellar spectra. The low resolution of the equipment couldn't identify smaller worlds, as their effect on the motion of their host star was less pronounced.

Over time instrumental precision has improved. In the late 1980s a planet had to swing its star by more than 10 m per second to stand a chance of discovery. The revolution began in 1993 with the installation of Keck's HIRES echelle spectrometer. Conceived by Steven Vogt, this instrument was to capture half the subsequent planet finds, using the RV method.

To give an idea of the sensitivity of HIRES and competing instruments, the captured spectra show variations in the location of absorption bands that are less than 1/1000th of a pixel wide in its CCD array. In 2003 a further upgrade in sensitivity allowed HIRES to detect the stellar wobbles caused by Neptune-mass worlds that orbited within one astronomical unit of their star. HIRES would later go on to capture evidence for three of the four confirmed super-terran planets orbiting Gliese 581 (see Chap. 8).

Although HIRES was impressive, the European Space Observatory's 3.6-m telescope at La Silla in Chile takes the idea of using a reference spectrum to another level altogether. La Silla's instrument is home to HARPS (High Accuracy Radial Velocity Planet Searcher). Beginning operation in 2003, HARPS employed thorium vapor rather than iodine to produce its reference spectrum. In an effort to reduce instrument noise, the whole system was chilled to 0.01 K with liquid helium—a fraction of a degree above absolute zero. This super-chilled spectroscope then sits snugly within a vacuum chamber, which minimizes the effect of thermal noise. The combination of minimal atmospheric disturbance and limited particle motion ensures that HARPS can resolve stellar wobbles down to 30 cm per second—enough to resolve Earth-like worlds in sufficiently tight stellar orbits, or super-terrans in more distant ones. HARPS' resolution is such that the idiosyncrasies of the stars limit its operation, rather than instrumental noise, which is the key limiting factor in other RV systems.

Directed by Peg 51b's discoverer, Michael Mayor, and assisted by Didier Queloz and Stéphane Udry, HARPS has made a number of notable finds, including the first potentially habitable planet, Gliese 581d. This world joined its three (contentious) siblings found with Keck's HIRES. We will dissect the arguments surrounding the Gliese 581 system in Chap. 8, but here we should merely state that while the RV method is still one of the most popular methods for planet hunting, considerable debate surrounds the interpretation of many of its results.

The late 1990s saw detection of the first planets orbiting red dwarfs. Of these, Gliese 876b was the first, and still rare, Jupiter-mass planet orbiting such a star. In 1996 the first multi-planet system orbiting a main sequence star was confirmed using the radial velocity method: Upsilon Andromedae. This system contains three planets, all of which are Jupiter-like. Planets b, c and d were announced in 1996, 1999, and 1999, respectively. With the exception of planet d, all orbit within 1 AU of their host star. Planet d, has an orbit of 2.54 AU, which would place it in our Asteroid Belt, making it at best lukewarm. Consequently, Upsilon Andromedae d became one of the first worlds we'd found that was not roasting in the glare of its host.

Another first for the radial velocity method came in 2001 with the discovery of Iota Draconis b—the first exoplanet found orbiting a red giant. Technically, Iota Draconis is an orange (K-class) giant. Importantly, the survival of Iota Draconis b confirmed that the rise of the red giant phase didn't necessarily mean the annihilation of the entire star system. Unsurprisingly, planets far enough removed from the expanding giant can hang on. Iota Draconis b is a particularly massive planet in an eccentric orbit that carries it on average 1.3 AU from the center of its star. The orbital eccentricity, coupled to its high mass, meant that its orbital signature was easy to separate from the effects of stellar pulsations.

Even more interestingly, in 2007 INAF-Osservatorio Astrofisico di Torino's Roberto Silvotti and colleagues announced the detection of a planetary mass companion to the extreme horizontal branch (EHB) star V391 Peg. Mentioned briefly earlier, Silvotti and colleagues used a method involving stellar pulsation (and the radial velocity method) to detect a 3.2-Jupiter-mass planet orbiting the star V391 Pegasi. The planet orbited at a distance of about 1.7 AU, giving it an orbital period of 3.2 Earth-years. In this regard, subdwarf B (sdB) stars are ideal objects to investigate by these methods. Their compact structure is often combined with pulsations. Such oscillations are extremely stable. Therefore, if a planet pulls on the oscillating star, it causes very small, yet predictable alterations in the arrival times of photons. The pattern reveals the companion, its mass and orbital period, because this effect is proportional to these parameters. In this regard, these stars are as useful as the Delta Scuti variables, mentioned previously.

Since then, a further planet, KIC 7917485 b (aka 2MASS J19574430+4337574 b, WISE J195744.30+433757.3 b or Kepler-1648 b), has been added to the tally for this method, which is known by pulsation-timing variation, bringing the number to a miserly, yet interesting, two. In this instance, Simon Murphy and colleagues monitored the pulsating host A-class star. Like the more diminutive sdB star, V391 Peg, the longer-period pulsations were also modulated by the presence of a super-Jupiter-mass planet, albeit in a much wider, eccentric orbit.

Modeling of V391 Peg suggested that as the star approached the tip of the red giant branch, it expanded with a radius of 0.7 AU. At this stage V391 peg b orbited its swollen star at 1.0 AU. Although this was outside the photosphere of the star, its gravitational pull may have stripped much of the red giant's envelope away, leaving the 30,000 K subdwarf star on the extreme horizontal branch (EHB, Chap. 1).

Although discoveries of sdB star planets form a small proportion of the total, they have addressed a pre-existing problem—how do such stars evolve from considerably more massive stellar objects? Zhanwen Han (Chinese Academy of Sciences) and others carried out mathematical analysis to address this problem. Their work implied a role for sub-stellar companions in the formation of isolated members of this stellar class. As Chap. 1 explored, EHB stars are low mass, helium-burning objects that lie on the blue end of the horizontal branch. Each has a mass between 0.46 and 0.49 times that of the Sun. Such low mass objects can only form if a more massive red giant has shed its outer layers, leaving its hot core behind. Han and others show that in most cases the presence of a stellar companion orbiting within the giant star's envelope is sufficient to explain the loss of this tenuous layer. However, in some instances the companion appears to be far less massive and intrinsically non-luminous. This leaves planets or brown dwarfs as the likely culprits.

V391 Pegasi b clearly orbited outside its former red giant's photosphere, but subsequent discoveries found planets that had clearly been completely immersed by their former giant's skin. In 2011 Stephane Charpinet (Université de Toulouse) and colleagues identified two planetary companions to the extreme horizontal branch star KIC 05807616 at distances of 0.0060 and 0.0076 AU. These 0.44 and 0.65 Jupiter mass planets have orbital periods of 5.7625 and 8.2293 h, respectively, placing each well inside the envelope of the star, during its preceding red giant phase.

In the early 2000s the radial velocity technique had been refined to allow it to detect first Neptune-mass worlds, then super-terrans. In August 2004, the HARPS team discovered a planet orbiting Mu Arae, with a mass of approximately 14 times that of Earth. This world, renamed Dulcinea, completed an orbit of its star every 9.6 Earth days. Why Dulcinea? Dulcinea was the love-interest of the lead character of the novel *El Ingenioso Hidalgo Don Quixote de la Mancha*, and in December 2015, the IAU decided that Mu Arae c should be re-named as such. The system's other, more massive, planet, Mu Arae b, has a much larger orbital period (643.25 Earth days) and a mass one and a half times that of Jupiter. Just to keep the literary theme alive, Mu Arae b is now known as Quijote. Although Mu Arae c most likely had a composition similar to the giants in our Solar System, a rocky composition could make it a particularly massive super-terran—a planet resembling a scaled-up version of Earth.

Throughout the first decade of the millennium, attitudes changed towards red (and later orange) dwarf stars. Their snug habitable zones were narrow, and most if not all of their planets were tidally locked to their star, but their short orbital periods made such planets relatively easy to detect (and of course announce to the public) after short observational periods. Sentiment also switched, with renewed optimism that tidally locked planets could be habitable for quite sophisticated life. Once Gliese 876b was in the proverbial bag, further analysis suggested this star had other, smaller worlds. Attention then focused on similar red dwarf stars, and the hunt was on. K-class dwarfs, very much, took the back seat. The larger orbital periods needed to support habitability around these stars, coupled to their stellar variability, combined with the growing focus on red dwarfs, left these stars rather neglected until the mid-2010s.

Red dwarfs have one big advantage over more massive stars when it comes to planetary searches. Their mass is limited. Therefore, the center of gravity between the star and its orbiting world is located further from the center of the star than it would be with a more massive stellar object. Thus, a smaller planet has correspondingly bigger pulling power around a red dwarf than it would have if it orbited the Sun. Therefore, the radial velocity method is better able to detect such planets if they orbit red dwarfs.

By the end of 2017 18.5% of planetary discoveries were made using the radial velocity (RV) technique. Although this figure is dwarfed by the number of discoveries made using the transit method, the radial velocity method has one advantage to the more prolific method. In order to be detected, planets do not need to have orbits that are aligned with Earth. This means that, with some mathematical tweaking, planets can be detected and their mass determined irrespective of the orientation of the system.[1] The transit method fails on this front—except where there is more than one planet and the gravitational pull of one object on another produces detectable variations in transit time. More on this method shortly!

One of the early successes of the radial velocity method was the detection of the red dwarf planet Gliese 667Cc, often abbreviated to GJ 667C. This 0.31 solar mass M1.5V star is a distant member of a triple-star system that mimics our neighbor, Alpha Centauri. Like the Alpha Centauri system, the core of the Gliese 667 system is a pair of more massive stars (GJ 667AB), in this case of spectral classes K3V and K5V. Each K-dwarf proceeds around their center of gravity in a very eccentric orbit, which takes it anywhere from 5 to 20 AU apart. Gliese 667C orbits the system's center of gravity at a projected distance of approximately 230 AU. Although it is unclear whether either of the two K-class dwarfs have planets, GJ 667C hosts two confirmed super-terrans, the innermost of which is a hot planet in a 7.2-day orbit around the M1.5 star. However, it was the subsequent discovery and confirmation of a 4.5 Earth mass planet, GJ 667Cc, in a 28.1-day orbit that set the community ablaze. This world sat in the red dwarf's habitable zone, making it one of the most significant planetary systems known to date. As such, the 4.5 Earth-mass planet, GJ 667Cc, was the *first* confirmed potentially habitable world (see Chap. 8).

In June 2013 Guillem Anglada-Escudé and co-workers announced that the Gliese 667C system flaunted seven planets—presaging the recent discovery of the multi-planet TRAPPIST-1 system. Critically, this claim included three planets in the star's habitable zone. Anglada-Escudé (Universitat Göttingen) had used a combination of data from the radial velocity measurement and computer modeling to propose a complex planetary system with up to seven members. However, subsequent Bayesian analysis by Farhan Feroz and Mike Hobson (both from the University of Cambridge) suggested that only planet GJ 667Cb and 667Cc likely existed; a further planet in a much larger 91-day orbit remaining possible.

The statistical method the pair employed searches for correlations between pieces of data that may be spurious. Such "red noise" is easily mistaken for a real signal, as it varies in a manner that is consistent with the signal generated by radial velocity variations. Gliese 581d (GJ 581d) was also flaunted in this regard, but its

[1]The mass of the planet is determined through: $M_p = (V_{star} * M_{star})/V_{pl}$, where V_{star} can be found from the Doppler velocity, K. In turn, $K = V_{star} * \sin\theta$, where θ is the inclination of the planet's orbit to the line that is perpendicular to the line-of-sight.

larger diameter and larger orbit (66.8 days) made it a less inviting prospect for life. Indeed, it is still unclear whether planet d exists, despite much wrangling and squabbling. We know three planets with certainty, but conflicting evidence from Bayesian analysis has also placed the existence of planet d in doubt. Both of these systems will be examined later in this book.

Transits, Transit Timing Variations and Transit Duration Variations

Figure 2.3 illustrates two aspects of the transit method. In essence, the transit method is a search for the shadow of a moth in a car headlight—when the car is a few hundred meters away. The transit is a migrating zone where the planet's disc blocks the light from the star. Some more information can be gathered if the planet has an atmosphere. Here, the passage of starlight through the atmosphere allows spectroscopic analysis of its composition.

Although the principle is easy, getting precise enough measurements has been vexing, to say the least. Although various different telescopes began the job in the first decade of this millennium, it took the now defunct Kepler spacecraft to turn the transit method into an industrial process. This has flung thousands of planetary candidates at waiting astronomers.

Although the Hubble Space Telescope had successfully identified the planet HD 209458b using the transit method in 2004, this was a piggy-back discovery. The radial velocity technique had already discovered the planet. Hubble's success was to show a proof-of-principle, identifying the planet through its transit, but also obtain sufficient information to determine part of the composition of the planet's atmosphere. Some of the light passing through the planet's gases was absorbed, yielding precious additional spectral clues that could be subtracted from the stellar glare.

For example, lurking within the data was the spectral signature of sodium vapor. In 2003, the Hubble team identified an enormous comet-like tail of hydrogen-rich debris enveloping the planet. Similar to the cloud enveloping GJ 436b (Chap. 1) but with a different mechanism, HD 209458b orbits its host sun so tightly that its atmosphere is boiling off into space. Later analysis confirmed that the sodium vapor was also present within the extended hydrogen-rich atmosphere at a level corresponding to the planet's stratosphere.

A year later, the Spitzer infrared telescope captured the infrared radiation emitted by two planets: the first of these was (again) HD 209458b, and the second, another hot Jupiter, TrES-1. Jeremy Richardson (Goddard Space Flight Center) observed HD 209458b over a range of 7.5–13.2 μm (in the infrared portion of the spectrum). In both cases, the emission spectrum provided far more detailed information than the limited absorption spectrum captured by Hubble. Detailed spectral emission lines indicated hydrogen and carbon monoxide, but oddly not water vapor, which had been expected. There was an additional strong peak at 7.78 μm, which was

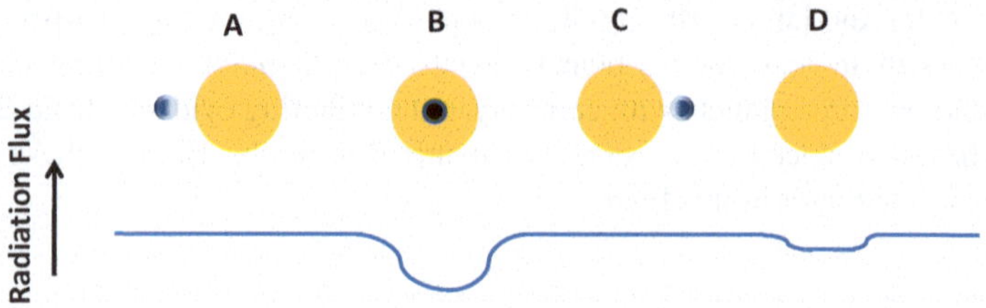

Fig. 2.3 The (exaggerated) effect of a transiting planet on the light curve of its parent star. As the planet moves in front of the star (**b**) the light is partially blocked, causing the star to appear to dim. This is what Kepler is looking for in its field of 150,000 stars. More rarely, a much shallower secondary minimum is detected (**d**) as the illuminated planet moves behind the star and less light is reflected in the direction of the observer. In reality the light curve will be more complex, as the area of the planet illuminated by its star varies throughout its orbit. This variation can also be used to detect planets. CORoT was the first craft to observe this effect with CORoT 1b

unexplained and a further peak near 10 μm attributed to silicate dust—vaporized rock. Later in the same year Travis Barman of the Lowell Observatory and co-workers identified water using a slightly different technique. This claim remained controversial for several years until Mark Swain of NASA's Jet Propulsion Laboratory and colleagues combined Hubble and Spitzer data to confirm its presence. Confirmation followed Carl Grillmair's, Swain's and colleagues breakthrough discovery in December 2008 of carbon dioxide, water and methane around another hot-Jupiter, HD 189733b using Spitzer. These studies opened the gateway on the study of exoplanetary atmospheres and prepared a pathway to investigate habitability.

Meanwhile, in 2006 the SWEEPS program ran for 7 days, using Hubble's Advanced Camera System (ACS). SWEEPS, an acronym for Sagittarius Window Eclipsing Extrasolar Planet Search, monitored 180,000 stars in the direction of the galactic center. Despite only operating for seven days, sixteen planetary candidates were discovered in roughly the same area. These had orbital periods that ranged from 0.6 to 4.2 days. The haul was modest, but an important discovery was planets with orbital periods of less than 1.2 days. These had not previously been detected and were dubbed "ultra-short period planets" (USPPs) by the SWEEPS team.

USPPs were discovered only around low-mass stars, suggesting that larger stars destroyed any planets orbiting too closely or that planets were unable to migrate as far inward around larger stars. Further, recent work with Kepler appears to corroborate this finding. To date, there are no confirmed planets with radii between about 3 and 11 Earths that have an orbital period of less than 1.5 days. Moreover, even when you extend the orbital period to about three days, there is a dearth of larger planets. As a result, many astronomers refer to this region in period-radius volume as the "sub-Jovian desert." The implication is that the orbital region either cannot support the formation of larger planets or gaseous planets are ablated by their host stars. That the majority of ultrashort period planets orbit red dwarfs further suggests that this orbital region is highly destructive for gas giants. Indeed, the only gas giant found in the 1.5- to 3-day orbital range (EPIC 201637175) is thought to be disintegrating.

Important though Hubble and Spitzer's discoveries were, they would always be limited in scope. Both observatories are involved in multiple fields of investigation. What astronomers needed were spacecraft dedicated to planet discovery. This venture would come with the launch of Kepler, CORoT and the initiation of ground-based systems such as WASP in the ensuing years. Of these, Kepler has certainly grabbed the largest share of attention, but all three ventures have revolutionized the field of planetary discovery.

Launched in 2009 after a two-year delay, the Kepler telescope swung into position 1 million km from Earth, in a heliocentric orbit. Here, the Sun and Earth's gravitational pulls balance. Liberated from the erratic pull of Earth and the Moon, the craft's cameras then focused with unique precision on star after star, detecting the subtle variations in starlight caused by transiting worlds.

With a steady gaze, Kepler turned the science of planetary discovery into the banality of working in a cannery—a little harsh, perhaps, but effectively true. With its unblinking 0.95-m-wide eye, Kepler gazed at a small square of sky filled with over 150,000 stars in a field 115°. Every minute of every hour, its sensitive CCD systems captured the subtle variations in stellar brightness caused by the ephemeral moths drawn to their eternal flames.

Indeed, the moth-candle analogy is a poor one. A moth fluttering in a car headlight, when the car is a kilometer away, might be a better descriptor. The variation in stellar luminosity caused by a planet transiting a Sun-like star is miniscule even when the planet is as large as Jupiter. Scale that to see the effect of an Earth-sized world, and you get the picture. This is high-precision science. The variation in stellar luminosity caused by an Earth-like world transiting a Sun-like star is as low as 80 parts in a million—barely above background noise. Thus, confirmation of such diminutive worlds relies on data from multiple transits. Super-terrans—large terrestrial or water-dominated worlds,—or their larger Jupiter-like relatives will clearly impose a greater variation while transiting a Jupiter-sized red dwarf. But are such large planets commonplace around red dwarfs, or are we instead looking for smaller, and hence more recalcitrant, worlds?

After less than 1 year of operation, Kepler had spied and confirmed the dimming effect of over 400 planets. Four hundred is a somewhat embarrassing figure—matching the number found by all other techniques in the preceding 15 years. That said, another 400 or so planets likely skulked in the data from other stars, awaiting further confirmation. Marvelous stuff, indeed. For a long time, planetary astronomy was the preserve of theory or of the study of Sol's worlds. With Kepler, routine discovery has allowed a lot more science to be dissected and refined and true discovery to be made.

Although Kepler grabs much of the media focus on planet discovery, WASP (Wide-Angle Search for Planets) and France's CORoT (from the somewhat forced acronym COnvection Rotation et Transits planétaires) have turned up a slew of interesting worlds. WASP (or more accurately, super-WASP), is a UK-based detection program made up of the Isaac Newton Group, the IAC and six universities from the United Kingdom. This international collaboration uses telescopes at the Roque de los Muchachos Observatory at La Palma and the South African Astronomical

Observatory. Utilizing a series of wide-angle lens cameras, and with observatories in two hemispheres, a whopping 500 square degrees of sky is covered, extending down to 15th magnitude objects. This allows for a very detailed and broad sweep of the heavens for transiting exoplanets.

Notably, WASP's discoveries include a series of super-heated planets in tight orbits around their host sun. WASP detected the hottest known exoplanet using the transit method. WASP-33b is a giant planet in an eccentric orbit around its F-class main sequence star. A combination of a tight orbit—at less than 7% that of Mercury-Sun distance—and a relatively hot central star mean its cloud tops sizzle at over 3000 °C. More bizarre still is WASP-12b. Initially crowned as the hottest exoplanet, with cloud top temperatures in excess of 2200 °C, WASP-12b orbits its star so closely that tidal forces distort it into an egg-shape. The combination of extreme heating and nose-touching proximity to its star means that this egg-shaped world is being steadily torn apart.

Although several evaporating exo-Jupiters were known (e.g., HD 209458b), WASP-12b was a different kind of beast. Whereas the massive HD 209458b would probably survive in some form for the duration of its star's main sequence life, WASP-12b was being shredded at such a rate that it probably wouldn't last much more than 10 million years. To observe the final fling of this planet's short remaining life might have been serendipity in the extreme, or perhaps more likely WASP-12b had suffered some form of interaction with another unseen world, which pulled it into an unstable orbit close to its host Sun. Whatever the true reason, WASP12-b hasn't long left. Within another 10 million years, all that will remain of the once giant planet will be its spectral signature in the gases that make up the star's corona. Ashes to ashes.

At the other extreme, WASP also spied a number of red and brown dwarfs, bagging the coolest known objects in the universe (Chap. 1).

Alongside WASP, CORoT diligently carried out transit sweeps from 2006 until 2012. In essence, COToT paved the way for the later, heavily delayed but ultimately more successful, Kepler mission. Not only did CORoT illustrate that the transit method was an efficient means of planet hunting, it also refined the extent of stellar variability in Sun-like and lower mass stars. This important function constrains a considerable fraction of the background noise, which might otherwise interfere with transit searches. Stellar pulsations or star spots, rotating in and out of view, can readily mimic the periodic dimming associated with transits. In this regard CORoT demonstrated that there was a greater than expected level of pulsation in many Sun-like stars, information that would then limit the ability of Kepler to carry out its task.

In 2010, CORoT unveiled CORoT-9b. This transiting planet was the first known that enjoyed a temperate orbit around its host star. The relatively low mass of its orange, K-class host star meant that the surface temperature of CORoT-9b lay somewhere between –20 and 160 °C. This is despite CORoT-9b orbiting at a distance equivalent to Mercury around the Sun. At its time of discovery, the orbital separation of 0.36 AU was by far the largest separation of any exoplanet from its star that the transit method had identified.

By 2011, CORoT had bagged around 600 planetary candidates, including the first found to show a secondary minimum in the light curve of the parent star. When CORoT-1b moved behind the parent star, the additional light reflected from its surface or clouds is lost behind the parent star. This causes a small dip in the luminosity of the system, as it is observed from Earth. Capturing a secondary minimum is quite a feat, given the minuscule amount of light reflected from the orbiting planet, and proved the principle illustrated in Fig. 2.9.

At the other temperature extreme lies HD 14902b. The N2K Consortium identified this world in 2005 using radial velocity data. The consortium was a two-year-long collaborative venture of Chilean, Japanese and U. S. astronomers to identify extrasolar planets using the radial velocity technique. The project concentrated on 2000 nearby bright main sequence stars that were not already the focus of investigation in the astronomical community.

N2K observations ran at Keck, Subaru, and Magellan, plus an automated 'scope at the Fairborn Observatory. The expectation was that around 60 planets would be found, but in the end, only seven were confirmed. Of these HD 14902b was rather unusual. Each transit decreased the light from its host star by 0.003 magnitudes. This doesn't sound much, but it was enough to allow the amateur astronomer Ron Bissinger, to confirm it. Moreover, it suggested that HD 14902b was a large planet, orbiting close to its star. What's more, the combination of data from various sources allowed a lot of detail to be gleamed concerning this odd world. The concomitant identification by an amateur astronomer using a backyard 'scope, equipped with its own CCD array, opened the door of planet discovery to the amateur astronomy community in general—a significant moment.

Despite being yet another hot Jupiter, HD 14902b stands out. It is extremely dense; in that regard, the analogy to a gaseous Jupiter is misleading. More than three-quarters of its 114 Earth-mass bulk is rock and metal—a whopping 80–110 Earth-masses. The presumption is that this odd world is the torched remains of something even bigger. Bombarded by intense radiation from its host star, much of its volatile material has boiled off into space, leaving the rocky core, flanked by a streamlined gaseous layer. HD 14902b (and CORoT-7b) is now regarded as chthonian worlds—a term coined from the Greek term for underworld. These blistering planets are often massive, but the loss of volatile materials leaves them grossly enriched in much more refractive, dense material. They would not make comfortable abodes for life (Fig. 2.4).

Two variants of the technique have allowed Kepler to dissect multi-planet systems: Transit Timing Variation (TTV) and Transit Duration Variation (TDV). These are illustrated in Figs. 2.5 and 2.6. In the former method, the gravitational pull of one planet tugs on another transiting world. This alters its orbital motion so that transits occur earlier or later than expected. A little bit of math allows the masses of the planets to be determined from the effects of their mutual gravitational pulls on their transit times. Critically, the presence of additional planets can be determined irrespective of whether they are visible in transits or orbit out of alignment with our line of sight. This method was successfully employed by WASP to infer the presence of planet WASP-3c after close scrutiny of the transit times of the transiting world,

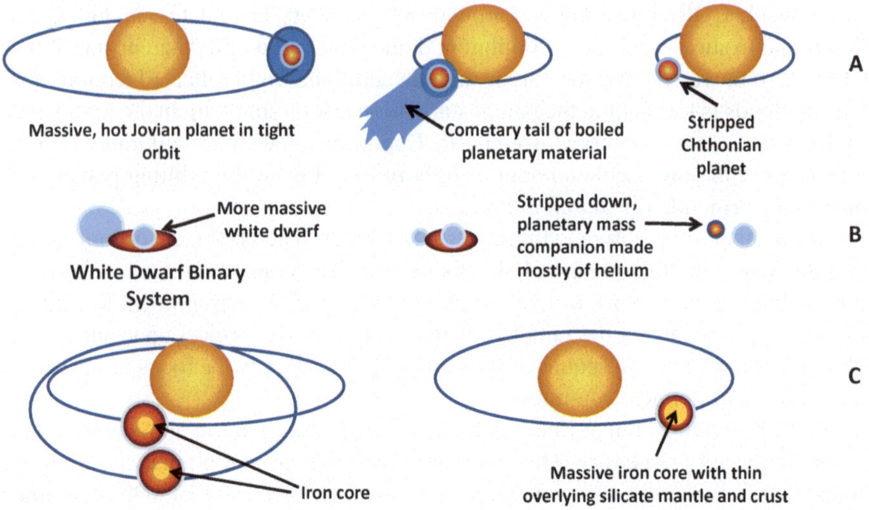

Fig. 2.4 Formation of three exotic worlds: (**a**) Chthonian planet; (**b**) a helium planet and (**c**) a "cannonball." In **a**, a massive hot Jupiter is irradiated and stripped of gases by its star, leaving the massive, rocky core. WASP-7b may be an example. In **b**, two white dwarf stars orbit a common center of gravity and the smaller but more massive one strips material from its helium-rich companion. This leaves a helium-rich, planet-mass companion in a fairly tight orbit around the white dwarf star. In **c**, two differentiated planets collide early on in their history. The iron-rich cores merge, and much of the overlying rocky mantle is blasted off into space. The resulting planet is much like Mercury but could, in principle, be much more massive

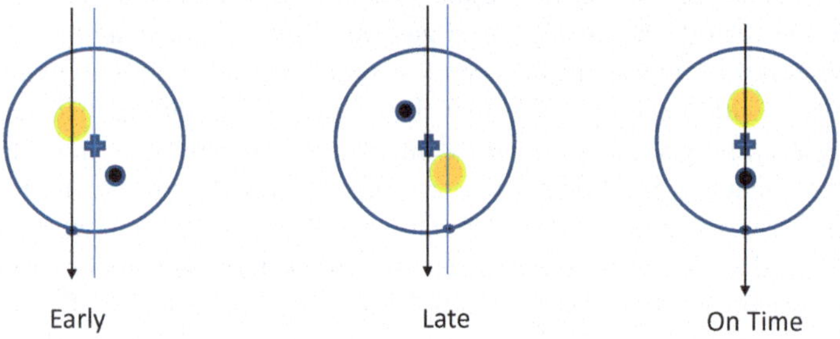

Fig. 2.5 Transit timing variation. If more than one planet orbits the star, the time it takes the known transiting planet to pass in front of the star varies as the unseen planet pulls it back and forth in its orbit. If the star moves leftward, relative to the planet, the transit is early, and vice versa. The cross marks the barycenter, the smaller black circle, the known planet, and the larger black circle the unknown planet. The star is the yellow-orange circle. The black arrow is the line of sight to Earth

WASP-3b. WASP-3c does not transit its star, but its presence is required to produce the observed changes in the transit times of WASP-3b.

Unsurprisingly, Kepler has identified considerably more planets using the TTV method. As Table 2.1 showed, while the number of discoveries of worlds using TTV is dwarfed by direct transit detection, TTV has bagged a small but growing family of planets. These account for 0.42% of the total—around six planets that would have

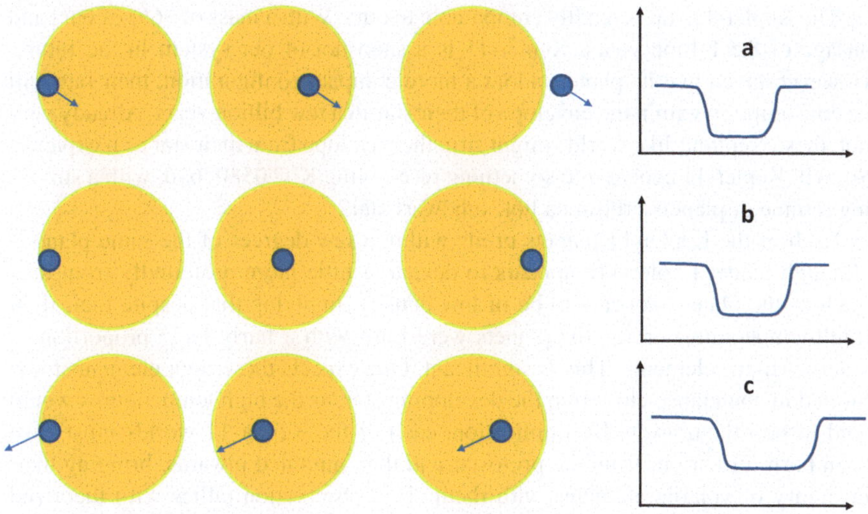

Fig. 2.6 Transit duration variation (TDV). Although TTV measures how quickly a transit starts relative to an expectation, TDV measures how long the transit lasts once it's begun. Gravitational pulls of an unseen planet on a transiting world may accelerate it so that the transit is faster (**a**) or slower (**c**) than expected (**b**). This allows the mass and orbit of both planets to be determined

otherwise gone undetected. Of these, the Kepler-9 system was the first. Kepler-9b is a Saturn-mass world in a 19.2-day orbit, while its TTV-detected (similarly massed) sibling orbits much further away, in a 38.9-day orbit. To illustrate the method and its limitations, the transiting planet Kepler-19b shows TTV with an amplitude of 5 min, over a period of about 300 days. This indicated the presence of a second planet, Kepler-19c, with a period that is a near-rational multiple of the period of the transiting planet, i.e., it is orbital resonance with its unseen partner. Likewise, the Kepler-9 system's two planets are in a near 2:1 orbital resonance, where the inner planet completes two orbits of its star for every one of its more distant partner.

The TTV method allows the maximum mass of the secondary planet to be determined, but unless the orbits of both worlds are fairly close to one another, little else can be gleaned from the data. Kepler-36 and Kepler-88 are exceptions, as the planets in each of these systems orbit close enough to one another to accurately determine their masses.

In 2011, Kepler used this approach to determine the properties of the very tight clutch of worlds orbiting the star Kepler 11. This tight nest of relatively low-mass planets has five worlds moving within the orbit of Mercury. The sixth planet lies somewhat further out, nestling in a region that would place it between the orbits of Mercury and Venus. Planets Kepler-11b and c orbit with a 5:4 orbital resonance, implying that they share a strong gravitational interaction. Aside from the very dense orbital arrangement—the densest known at the time—transit timing variation was able to provide precise parameters for the orbital separation and angular size of each world. The TTV method then allowed the determination of the mass of each world and hence (with radii from transits) their densities.

The Kepler-11 star is readily comparable to ours. With a mass of 96% of ours and an age of 8.5 billion years, Kepler-11 is a snapshot of our system in the future. However, given that its planets adopt a more compact configuration, their fate will be consumption within the envelope of their star in a few billion years. Already very hot, these Neptune-like worlds might strip the envelope from their star as it expands. So, will Kepler-11 evolve into something resembling KIC 05807616, with a surviving retinue of planets orbiting a hot, subdwarf star?

Each of the Kepler-11 planets orbits within a few degrees of the same plane—although planet Kepler-11e appears to deviate a little more profoundly from this. Each of the planets appears to be of low density, implying that despite their tight (high-temperature) orbits, the planets were born with a fairly large proportion of volatile (light) elements. This is significant. One expects these light elements to be limited in abundance close into the developing star, as the high temperatures would tend to boil them away. By implication, each of the Kepler-11 worlds must have been born further out from the protostar and then migrated inwards, bringing their inventory of volatile elements with them. This observation tallies with theorized events in our Solar System—*and,* critically, it means that planets that form around red dwarfs can still be volatile-rich, despite their torrid beginnings. Volatile elements are essential for life; thus the Kepler-11 system constrains some of the parameters needed for the future habitability of exoplanets.

In 2013, the fourth of Kepler's reaction motors failed, and it seemed likely that the mission would be over. The reaction wheels ensure that Kepler remains stable and pointing in the correct plane. With four motors broken only thrusters could keep the craft orientated correctly. However, the Kepler engineers had an ingenious fix. Using the solar panels as sails, the K2 mission used the pressure of sunlight to stabilize the spacecraft, thus allowing it to continue to do science. Since the mission was revitalized, hundreds of new exoplanet discoveries have been made. While it had been expected Kepler's mission would end in 2015, the revitalized K2 mission only ended in October 2018, when the craft's propellant was exhausted. Quite a feat, for a seemingly disabled craft, but TESS now takes its place.

The transit technique's principle advantage over the radial velocity technique is its ability to refine planetary diameter and resolve planetary atmospherics. However, transits only occur when the orbital plane of the planet is aligned with our line of sight. Clearly, this is a limitation, as simple geometrical arguments imply that the majority of stellar orbits will fall outside those that transit the star. Compared to Kepler 11, most planetary systems are dispersed. With longer orbital periods, the TTV method is more difficult to apply. The gravitational influence of each planet on one another is proportionately less significant and the length of the transit too drawn out. Thus, the impact of any additional planet on transit time is more readily lost in instrumental noise. Moreover, transits are limited by the relative diameter of the star, its planet, and the distance between them. Furthermore, where stellar variability is great, the effects of star spots and flares have the potential to obscure the dimming effect of transiting planets. With an increasing focus on red dwarfs, these limitations are more to the fore. Red dwarfs, more than any other type of star, are prone to the youthful problems of spots and flares. The detection of transiting worlds is consequently more complex. A transit is a useful thing, but it isn't perfect.

Yet, astronomers have adeptly found yet another route around the problem of determining planetary characteristics even where the planet's orbit doesn't carry it across the face of its star. In 2012 a novel technique was published in the journal *Nature* that circumvented this limitation. The star system Tau Boötis was known to host a planet. τ Boötis b was one of the first identified using the radial velocity method. However, there was considerable uncertainty as to the nature of the planet's orbit and hence its mass. Moreover, without available transit data, no orbital information or planetary atmospherics could be directly determined for τ Boötis b. To get around this Matteo Brogi (Leiden University) used the Cryogenic Infrared Echelle Spectrograph (CRIRES) at the Nasmyth A, focus of the Very Large Telescope UT1. This unique infrared spectrograph is located at the European Southern Observatory (ESO) on Cerro Paranal, Chiledata.

Scanning the system over many days as the hot Jupiter swept rapidly along its orbit, the CRIRES spectrograph detected the clear and varying signature of carbon monoxide. This noxious gas wasn't present in the atmosphere of the star. It had to be in the atmosphere of τ Boötis b. The thorough baking the dayside of the planet received from its host star meant that the planetary carbon monoxide readily emitted a potent spectral signature.

As the tidally locked planet swung around its star, the carbon monoxide hot spot appeared and disappeared, as first it was blocked by the star and later as the view of it was occluded by the planet itself. Using the periodic signal of the carbon monoxide hot spot, Brogi and co-workers showed that the 5.95-Jupiter-mass planet orbited its star at an angle of 44.5°, carrying it high above the line of sight needed for a transit to be observed from Earth. Not only did this ingenious method bypass the problem of a lack of a direct transit, it also opened the door on the discovery of planets using direct infrared emission. The data had such high resolution that Brogi and colleagues were able to show that the carbon monoxide was concentrated in the planet's lower atmosphere. The presumption was that this noxious gas was destroyed at higher altitudes by ultraviolet emission from τ Boötis. These observations contrast with those of other hot Jupiter worlds, where emission from carbon monoxide has been detected at much greater altitudes. It suggests that there is less intense irradiation of these planetary atmospheres by ultraviolet light from their parent stars, which in turn allows carbon monoxide to persist at these altitudes. Ultraviolet intensity varies with stellar activity—and this varies inversely with stellar age for singleton stars. Older stars generate less intense ultraviolet (and X-ray) emission. Therefore, we expect that these hot Jupiters orbit older stars.

Comparison with 51 Pegasus b was illuminating. In the following year, 2013 Brogi and colleagues used the same instrument to detect the molecular signature of carbon monoxide and water vapor in the daytime atmosphere of 51 Peg b. Back in 1977, Ronald G. Prinn and Stephen S. Barshay (MIT) used the presence of carbon monoxide in Jupiter's atmosphere to infer the extent of convection from the deep interior. In a hydrogen-rich planet, carbon monoxide is unstable at lower temperatures and readily reacts with (and is chemically-reduced by) hydrogen and water vapor. The end result is methane. The detection of carbon monoxide at high levels implies vigorous convection from deeper and hotter levels (around 1100 °C), where carbon monoxide is stable. Therefore, the detection of carbon monoxide in the atmosphere of both Tau Boötes b and 51 Peg b implies that while its atmosphere is relatively hot (around 700 °C for 51 Peg b and approximately 1400 °C for Tau Boötes b), it is still otherwise convecting. The spectra of other hot Jupiters, such as HD 209458b, suggest that their atmospheres don't. Convection appears to be capped by a temperature inversion.

Given that Tau Boötes b is more intensely irradiated than HD 209458b, why does the latter planet have a temperature inversion but not the former? One suggestion is that planets orbiting young, magnetically active stars are more strongly irradiated with ultraviolet light. This destroys carbon monoxide in the upper atmosphere. Moreover, these hot Jupiters are less likely to have thermal inversions because the intense ultraviolet radiation destroys the absorbing chemicals that are required to form a temperature inversion.

More recently, in 2017 astronomers scrutinized transit time variations in the TRAPPIST-1 system. Here, Simon Grimm (University of Bern) and an array of international collaborators used computer modeling to dissect the transit times of the seven terrestrial planets. This allowed them to determine the densities of each of the planets with considerable precision. Without giving too much away, their results showed surprising variations in the densities of the planets that shred many of the simple models of atmospheric evolution that had emerged following the discovery of this planetary system. Lending strong support to mass planetary migration in such compact systems, the TRAPPIST-1 planets show very complex patterns in density, which are certainly not readily explainable Chaps. 1 and 9.

Finally, transits allow one further derivative that has been exploited in planetary discovery: Transit Duration Variation (TDV, Fig. 2.6). This method is very similar to TTV in its modus operandi but exploits the effect of secondary masses in altering transit duration. If the observed transiting planet has a large satellite orbiting it, if it has another planet pulling on it in an eccentric orbit, or even where relativistic effects are altering its orbital path, the time the planet takes to complete a transit will vary. Like TTV, TDV will vary in a predictable manner that is dependent on the mass of the object pulling on the observed transiting world.

Although TDV has had a very limited haul of worlds, it does provide important clues to the nature of planets that have already been detected. In particular "Tatooine-like" circumbinary planets are worlds in orbit around two stars. Here, the mutual orbit of the stars around their center of gravity can grossly perturb the orbit of the planet. However, in this instance the TTV is insubstantial.

Consider the different scenario where the planet orbits one star but not both in a strict sense. Here, transit duration will vary because the planet is orbiting one star, but the star is orbiting the center of gravity (the barycenter) of the system. At some points in the planetary orbit, the planet will be moving in the same direction as its star, lengthening the transit time. This occurs near secondary eclipse when the primary star (and its planet) eclipses the secondary star. At others, the planet will be moving in the opposite direction to its star around the barycenter. Here, because the star and its planet are moving in the opposite directions, the transits are short in duration. Such variations in transit duration resulted in the discovery of Kepler-16b, where the third observed transit across the primary star occurred 8.8 days early, compared to the expected time based on the first two transits. While such transit duration variations were unknown before the turn of the millennium, Jon Jenkins (SETI institute) successfully predicted the observed effect as early as 1996.

Kepler-16b was the first planet found in orbit around a binary pair. This 0.33 Jupiter-mass planet orbits the barycenter of the system in the same plane as each low-mass star in an eccentric orbit, with a period of 229 days. Consequently, the planet transits both stars as they orbit one another. The two stars (an M- and K-class dwarf) have masses of 0.2 and 0.69 times that of the Sun, respectively, and share a 41-day orbit around the system's barycenter. Given these properties, it is expected that the planet formed in its circumbinary orbit. Although this Saturn-mass world lies in its star's habitable zone, the density of the planet and its high mass certainly preclude habitability.

Soon after the confirmation of Kepler-16b came the discovery of the Kepler-47 system. This was the first with multiple planets in orbit around two stars. This young binary pair consists of a star with roughly the mass of the Sun (1.04 solar masses), paired to a 0.36 solar-mass red dwarf, in a tight orbit lasting 7.45 days. The inner and outer planets of Kepler-47 system have radii 3.0 and 4.6 times that of Earth, with masses 8.4 and 23.7 times that of Earth, respectively. This implies that the inner planet is fairly rocky and a likely a super-terran, while the outer planet is a bit more mysterious. Its high density suggests that it is primarily rocky—perhaps an aquaplanet with a modest atmosphere, one that is considerably less dense than that of Neptune. The inner planet, Kepler-47b, has an orbital period of 49.5 days in an orbit a little less than 0.3 AU. The outer planet's orbital period is 303.2 days with an orbit of 0.99 AU, placing it in the stellar habitable zone. Between each is a third planet for which the orbital parameters are less clear, but likely has an orbit of 0.5–0.7 AU and a mass of 23.1-Earth masses. In each case, the mass has been determined using transit time variation.

While a perturbing second star most commonly causes TDV, in principle this method can also be used to detect other unseen planets. Application of TDV to such systems will depend on the sensitivity of the instrumentation that is used and the algorithms employed to dissect the obtained data. However, one will expect to see a number of multiplanet systems found, where some of the planetary orbits produce transit duration variations as well as transit timing variations.

Photometric Detection of Extrasolar Planets

An interesting offshoot of the transit method employs the converse effect to the transit—the reflection of "sunlight" from the orbiting planet. Going by various names, the photometric effect essentially looks for planetary phases as the planet orbits the star. Irrespective of a transit, the orbiting world will exhibit phases as its orbit carries it around its parent star. This was illustrated in Fig. 2.3. As the planet moves away from the transit position, light from its star will illuminate a crescent. This illuminated area will gradually increase as the planet moves along its orbit, reaching a maximum just before it is eclipsed by its sun.

Clearly, though such phase differences are detectable—and amenable to planets that don't transit their star—there is a limit to the difference in light that is detectable. Once the orbital plane is at 90° to our line of sight, then there will be no variations in the amount of light that is reflected, and the observed "phase" will be equivalent to the "quarter Moon"—with half the observable disc illuminated.

It is likely that this method may ultimately score the largest numbers of planet-finding hits, but it is confounded by two problems: surface emissivity and albedo. In the former, if the planet is very hot it will, of course, emit its own visible radiation. Such emitted light will interfere with the detection of phases. Secondly, if the albedo is very low, then little light will be reflected and will not be detectable by Kepler or other telescopes.

Kepler-70b and 70c were the first planets found using this technique. Kepler-76b followed soon thereafter. The host star, Kepler-70 (also known as 2MASS J19452546+4105339), is another low mass, helium-burning, hot sub-dwarf b (sdB) star. As with V391b Peg, the two (very hot) molten planets likely played a hand in dispersing the former giant's envelope. These worlds likely began life as gas (or ice) giants, but lost their gaseous envelopes during the plunge through their star's swollen envelope. Interestingly, Kepler-70c passes only 240,000 km from Kepler-70b during its closest approach, making this the closest recorded approach between planets currently known. Given that each world has a temperature approximating the surface of the Sun, there are likely to be considerable tidal forces acting against one another during such close approaches. Imagine large molten tsunamis surging across the landscape every time the planets pass one another. Moreover, at temperatures of thousands of degrees, elements such as vanadium oxide or titanium oxide could combine with carbon monoxide to form a cloudy, torrid atmosphere that would be utterly alien to us.

In 2015 astronomers had another first with 51 Peg b. Here, an international team of astronomers, led by Jorge Martins from the Instituto de Astrofísica e Ciências do Espaço (IA) and the Universidade do Porto, detected this planet using HARPS. Nothing new here, it might seem, given this was how the planet was found in the first place in 1995. However, in this instance radial velocity information from HAPRS was used to dissect the radiation emitted from the system and thereby isolate the planet's spectrum from that of its star. With the HARPS data, astronomers looked for variations in the radiation that was emitted from the system that would correspond to phases. The results indicated that 51 Peg b is a hot, puffed up and highly reflective planet that orbits its star at 80° to our line of sight. This corroborates the results of Mattheo Brogi and colleagues, who used the emission spectrum of carbon monoxide to produce the same results.

To date, this method has not detected a significant number of planets, with only 0.17% of the total number of extrasolar worlds found this way. TESS may add to this small trove. Most importantly, as the 51 Peg b story illustrates, various techniques can be used to confirm and extend the results from other studies. Kepler-76b is another case in point. Although it was initially identified through its transit, its presence was confirmed using four other methods: through its reflected light; through variations in the shape of the star due to gravitational tugs from the planet; using the radial velocity method; and finally, through its secondary eclipse. Here, as the planet passed behind its host star, the total amount of light emitted by the system declined. Kepler-76b thus forms one of the most heavily probed and verified systems to date. Although this is impressive in its own right, Kepler-76b was also detected using a fifth and rather interesting relativistic effect that we shall return to later in this chapter.

Eclipse-Timing Variation

Here, we encounter a twist of fate. Jon Jenkins and colleagues had used the CM Draconis system to model the photometric system, while another method successfully detected a planet in orbit around this red dwarf binary. The CM Draconis system consists of two equally matched eclipsing M4.5V red dwarf stars. Hans Deeg (Instituto de Astrofísica de Canarias) and co-workers (one of whom worked with Jon Jenkins in 1996), used simple light curve modeling to detect the presence of a third body. In this approach, the planet orbits both stars (a so-called P-type arrangement) and perturbs the timing of the eclipse of one star by its partner. This variant on the transit timing variation method revealed the presence of a Jupiter-mass world. However, the outcome of this analysis was unclear as there were two, equally matched, best fit models. In the first a planet with a few times Jupiter's mass orbited relatively close around the system's center of gravity (its barycenter). Though not a close companion, this model had the massive planet in an 18.5 year-long orbit. The alternative was a somewhat smaller 1.5-Jupiter mass object with a period of hundreds to thousands of years. The data was consistent with a distant and very low mass red dwarf, therefore, but considerably more data is needed to constrain the properties of a system that also includes a very distant white dwarf, GJ 630.1B. In 2011 Kepler-16b became the first planet to be found using this method. At the time of writing the eclipse-timing variation (ETV) method had bagged around 0.25% of the total number of planets thus far identified (Fig. 2.7).

Micro-Lensing and Other Relativistic Methods

Einstein's general theory of relativity states that matter bends space, and the curvature of space alters the path of any incumbent light. Therefore, when a massive object passes in front of a star, the gravitational field of the object focuses the star's light. Much like a magnifying glass, the lensing object causes the star to brighten and then dim again in a predictable manner, dictated by the curvature of space (Fig. 2.8). The extent of the change in brightness tells you the mass of the lens and its distance from the star.

During the 1980s there was debate among cosmologists as to the nature of the mysterious dark matter that dominated space yet couldn't be seen. One group suggested that dark matter was comprised of fundamental particles created in the Big Bang. These were WIMPs, or weakly interacting massive particles. Another camp insisted dark matter was nothing more than dim objects made of conventional matter, such as

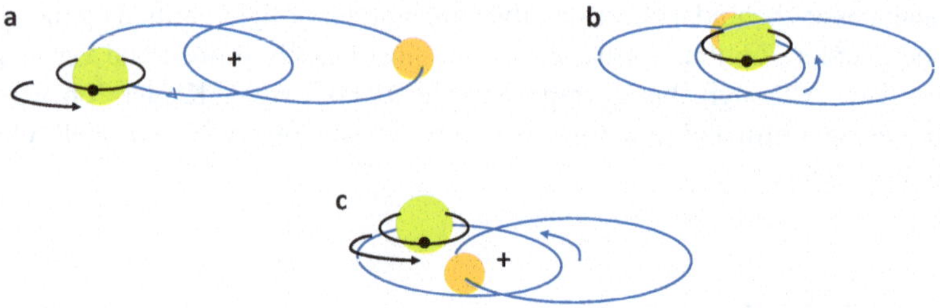

Fig. 2.7 Eclipse timing variation. A variation of TDV, where the pull of the planet on one star in a binary system alters the time at which one star eclipses its partner. In **a**, the planet transits the star where neither star is eclipsed by the other and the transit and eclipse times are "normal." In (**b**) the planet will be moving in the same direction as its star, lengthening the transit time but shortening the eclipse time. While in **c**, the planet is moving in the opposite sense to its star around the barycenter (marked by a cross). Here, eclipses are longer but transits are shorter

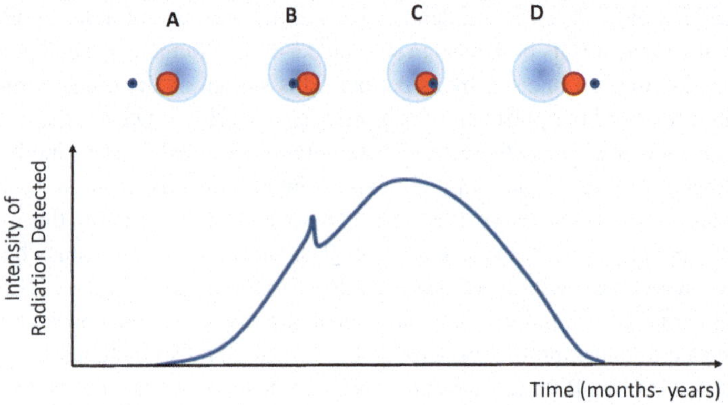

Fig. 2.8 Microlensing. As the foreground red dwarf moves in front of the distant star, its gravitational field focuses the light from it, causing it to brighten in a predictable fashion, the effect varying with the mass of the red dwarf. If a planet orbits the red dwarf, the additional mass causes a further, brief magnification (**b**) on top of that caused by the red dwarf. This boost is transitory in duration, as the time taken for the planet to orbit the host star is considerably shorter than the overall lensing time

brown or red dwarfs, neutron stars and black holes. These were MACHOs, or MAssive Compact Halo Objects—the language further exemplifying many of the rather bizarre attitudes to male sexuality in this decade.

During the early 1990s, Polish astronomers Andrzej Udalski, Marcin Kubiak, Michał Szymański, and the late Bohdan Paczyński from Warsaw established OGLE (the Optical Gravitational Lensing Experiment), based at the University of Warsaw. Most of the observations were made at the Las Campanas Observatory in Chile. OGLE had two principle successes: the first, in November 2002, was the detection of the first planet through the transit method: OGLE-TR56b. This hot-Jupiter held the record for the shortest orbital period until WASP-12b was spied in 2008.

Its second first came in 2004. OGLE and New Zealand's Microlensing Observations in Astrophysics (MOA) identified a planet orbiting a distant star, taking the prize as the first planet detected using the microlensing method. OGLE-2003-BLG-235L b, a 1.5-Jupiter-mass world, was found indirectly. Rather than spy the gravitational lens of the planet, a background G-class star was lensed by a foreground dim M-class dwarf. There is nothing very exciting here, given the abundance of red dwarfs. However, if the M-class star was the lens on its own, the pattern of brightening and dimming would have been monotonic—a simple rise and fall lasting a few days to weeks. That was not the case. Instead, the light flickered in a manner that betrayed the presence of a planet-mass object in orbit around the lensing red dwarf. This additional orbiting mass led to a secondary peak in brightness, which lasted a few hours.

In 2006 OGLE announced it had bagged the most distant exoplanet. The 5.5-Earth-mass planet, glamorously named OGLE-2005-BLG-390Lb, orbits its red dwarf host at a distance of 2.6 AU. The whole system lies a whopping 21,500 light years away, towards the center of the Milky Way. The wide separation of the super-terran planet from its dim red dwarf star also implied that this was the coldest exoplanet known at that time. The recently announced Barnard b makes an interesting modern counterpart.

In 2018 Xinyu Dai and Eduardo Guerras, a postdoctoral researcher and professor from the (University of Oklahoma) took the microlensing method to its logical conclusion: smashing OGLE's distance record by detecting planets outside our galaxy. The pair examined electromagnetic radiation from the distant quasar RX J1131–1231, which lies 3.8 billion light years from the Milky Way. Using the Chandra X-ray observatory and powerful computer analysis, Dai and Guerras identified a multitude of minute blips in the intensity of the radiation from the quasar. This data implied that there were around 2000 unbound planets per main sequence star in this distant galaxy. These small worlds had masses between one hundredth that of Earth (equivalent to the Moon) right up to the mass of Jupiter.

To put Dai's and Guerras' results in a more parochial perspective, back in 2012 astronomers, led by Louis Strigari (KIPAC), carried out a similar analysis of free-floating (or rogue) planets in the Milky Way. Strigari and colleagues extrapolated an estimate based on the small number of such orbs that had been detected thus far, either through direct imaging or by gravitational microlensing. At this time 24 planet-mass objects had been detected by microlensing—14 bound to their parent stars, with 10 apparently free-floating. This gave a very vague estimate of anything from one to 100,000 free-floating planets in our galaxy for every star. The upper estimate put planet formation theories in the dock. Where would all these planets have come from?

In 2016 a further study in *Nature* brought these rather flamboyant claims back down to Earth. Przemek Mróz (Warsaw University Observatory and a collaborator on OGLE) updated the analysis of microlensing events from OGLE and effectively discredited the idea that most of these microlensing events were attributable to unbound, Jupiter-mass planets. Instead, most of these planets were bound to stars. Rather than a massive swarm of such planets, Mróz found that there were likely only 0.25 Jupiter-mass rogue planets per star in the Milky Way. This figure made for not only more comfortable reading but also put the numbers solely back into the theoretical models we had for planet formation.

That said, the analysis done by Mróz and co-workers did leave the door open on a substantial population of Earth-mass worlds that were either distant to their stars or free-floating. Mróz and colleagues admit that the data remains poor, and to date the precise figure remains unclear. Therefore, while 2000 might seem unnervingly high, it is well within the range that is at least suggested for our (and presumably all) spiral galaxies. The number of events has grown to 53 at the time of writing, but the overall conclusion remains the same.

To put this in context, consider the Orion Nebula. Within this famous, dusty shroud lie a variety of free-floating, young massive planets that Hubble spied near the turn of the millennium. S Ori 70 was the smallest. Based on its temperature and presumed age of 5–7 million years, this object appeared to have only five times the mass of Jupiter. Quite whether this was a true planet kicked loose from its parent star, a prematurely terminated brown dwarf or a true planet formed entirely in situ remains unclear.

Further work on microlensing was carried out by PLANET (Probing Lensing Anomalies NETwork) between 2005 and 2015. The PLANET network is a unique collaboration of small (1- to 2-m) optical telescopes that operate from bases in Australia, Chile, Denmark, South Africa, the United Kingdom and the United States. Initially scattered across various lines of longitude in the southern hemisphere, this network allowed it nearly continuous observation of the night sky. The inclusion of similar 'scopes in the UK and the United States, after 2007, extended the network's area of view to much of the northern hemisphere. PLANET scrutinized thousands of stars, scattered across the hub of the galaxy, searching for signs of lensing events. By 2013 PLANET had detected 40 such planetary events, three of which correspond to previously confirmed finds by Kepler.

Finally, in 2016, microlensing detected the most distant discernable star in the universe. Nicknamed Icarus, MACS J1149 is a blue supergiant lying some 9.9 billion light years away.

Although microlensing is a challenging, if serendipitous, method of planetary detection, it gives little away other than the mass of the planet. What about distance? This is far harder to determine unless you have a luminosity for the object. As these lenses are dark, this is clearly not applicable. However, if you have two telescopes observing the microlensing event, you can in principle determine the distance through geometric parallax. Such microlensing events typically last weeks as the lensing object passes between us and a background star.

In 2015, Jennifer Yee (Harvard-Smithsonian Center for Astrophysics) proposed using Spitzer and ground-based OGLE in just this manner. In the same year, these two systems successfully collaborated in this manner. Both devices succeeded in making near-simultaneous observations of a gravitational lens event from their locations, 207 million km apart. The combination of both systems allowed the mass of the lensing system and its distance to be determined. In this case a Jupiter-mass planet was identified some 13,000 light years from Earth. Most microlensing events last perhaps 40 days. However, this one was considerably longer—150 days. This allowed Spitzer and OGLE to collaborate with OGLE, alerting Spitzer 3 days into its observations.

Since this first run, Spitzer has been used to detect twenty-two other events in collaboration with OGLE or other ground-based 'scopes. None of these discoveries were new. Rather, they were used as a testbed for future observations and to determine the approximate scale of microlensing events that would be detectable from Earth. From this data astronomers can examine the population structure of stars and planets in the direction of the Milky Way's bulge.

Kepler was better known as a transit factory, but it was ideally suited to identify planets through microlensing, as the number of stars in its field of vision is substantial. Between April 22 and July 2016 Kepler's K2 Campaign 9 mission scanned part of the Kepler field for microlensing events. However, with the primary means to stabilize the craft gone, there were significant issues associated with this arm of the K2 mission. However, towards the close of 2017 Kepler, Spitzer and ground-based MOA (Microlensing Observations in Astrophysics) announced the simultaneous detection of a microlensing event. MOA is a Japan-New Zealand collaboration based at the University of Canterbury, Mt John Observatory in New Zealand. MOA is seeking out a variety of dark objects through their gravitational effects on light. This includes dark matter, extrasolar planets, as well as an examination of stellar atmospheres. Led by Jennifer Yee this system allowed astronomers to determine the distance to the lensing object through geometric parallax—as well as calculate its mass. Since the lens was observed from three separate locations, subtle differences in the timing of the microlensing event allowed the team to determine that the object was a very low mass red dwarf, located about 2200 light-years away. With a mass immediately above the hydrogen-burning limit (around 0.077 solar masses), MOA-2016-BLG-290 is one of the lightest stars known.

Taking these and other discoveries into account, at the time of writing a total of 53 planets have been found (in the Milky Way) through microlensing events. This is equivalent to 1.5% of the total found so far. Future missions (such as WFIRST) will expand this number in collaboration with the increasingly antiquated Spitzer and now defunct Kepler craft.

Time for BEER

Shude Mao and the late Bohdan Paczynski proposed using microlensing as a means of finding planets a couple of decades ago. This was elaborated and refined by the versatile researcher Abraham Loeb (Harvard) in 1991, alongside fellow astrophysicist Andy Gould. The problem with microlensing is its lack of reproducibility. Lensing events are generally one-offs—chance alignments of stars and planets, with the latter some considerable distance from the light source. However, planets in distant orbits from their host stars are detectable—something that would otherwise take decades of observation to detect radial velocity variation or an even rarer transit. Typically, this would be 1–10 AU, except for those free-floating worlds that will never be encountered again.

Yet, with microlensing, these benefits are grossly outweighed by the negatives. Therefore, it came to Abraham Loeb to develop a further relativistic method of planetary detection—one that could produce reproducible results irrespective of the orbital parameters or chance alignment. Once more, a known planetary system was probed as a proof-of-principle and the results were astounding.

What if the planet does not transit a background star, or otherwise move close enough to one in our line of sight to produce a microlensing event? Can such a world be detected? Abraham Loeb (Harvard-Smithsonian) and Scott Gaudi (now at Ohio State University) suggested a further mechanism, whereby planets might be detectable through their relativistic effects on their parent star's radiation. Relativistic beaming is a minute variation in the intensity and frequency of a star's radiation caused by the motion of the planet around the star. As the planet moves, the density of photons we detect varies as photons bunch up when they move through the warped spacetime continuum around the planet.

This method can determine the orbital eccentricity and the minimum mass of the planet in a manner akin to the radial velocity method. However, unlike the radial velocity method, relativistic beaming does not require an accurate spectrum of a star (Fig. 2.9). Consequently, it can be used more readily to find planets around fast-rotating stars and more distant stars that are otherwise out of reach to the RV method. Clearly, the relativistic effects will be trivial in nature and thus will be far easier to detect where the mass of the planet increases and the distance to the host star decreases. This means that relativistic beaming will preferentially find hot Jupiter and hot Neptune-like planets. Although this is a clear limitation, relativistic beaming can still detect planets where their orbits are not aligned with our line of sight. That said, although relativistic effects are greatest where the orbit is aligned with our line of sight, these will still be detectable, irrespective of the orientation of the orbit.

The method is likely to produce a fairly limited haul of planets, but it does allow the mass to be determined without the need for repeated measurements. In 2013, lead author Simchon Faigler of Tel Aviv University announced the detection of Kepler-76 with this method. Significantly, while Kepler was the instrument used to identify this world, transits played no part in its detection. As we mentioned earlier, Kepler-76 has served as a very useful testbed for a number of methods that have all been employed to detect this hot Jupiter world. Once the new planet was identified through its relativistic effects, follow-up radial velocity observations were gathered by two teams. One led by David Latham was based at the TRES spectrograph at Whipple Observatory in Arizona and other by Lev Tal-Or (Tel Aviv University) using the SOPHIE spectrograph at the Haute-Provence Observatory in France.

Finally, although initial data failed to identify transits, a closer look at the Kepler data revealed that Kepler-76b did transit its Sun-like star. This provided additional confirmation of various measurements relating to the initial discovery. As Kepler-76b was found through these relativistic effects, it earned the planet the slightly dubious and frivolous title in the media of "Einstein's Planet"—something that might be referred to as "clickbait" in today's digital universe.

Although relativistic beaming is useful in its own right, the radiation we detect from the system can be probed in other ways. The full method comes under the irritating acronym BEER—B̲eaming E̲llipsoidal E̲mission/R̲eflection. (One sincerely hopes that the person responsible for this forced acronym doesn't feel

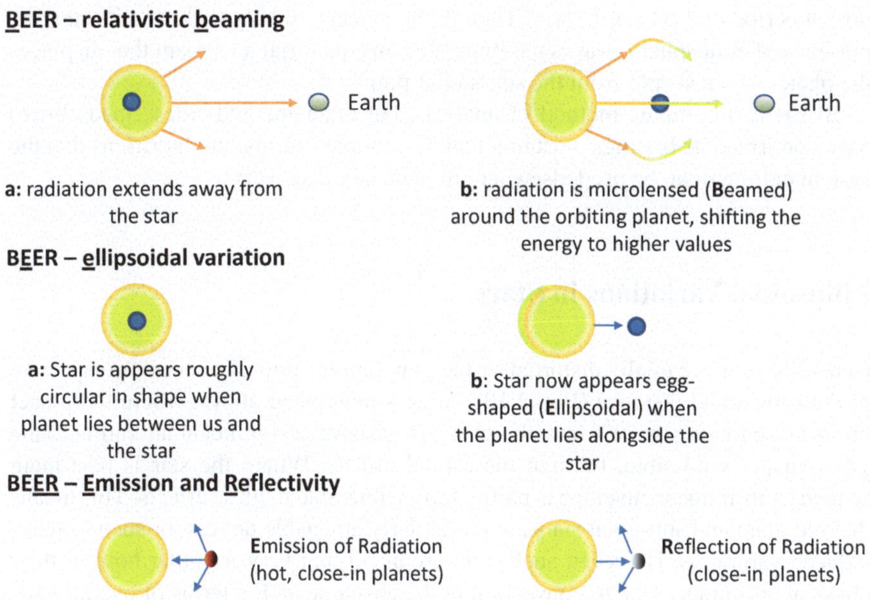

Fig. 2.9 Relativistic beaming and BEER. As a planet moves around the orbit of the star, not only can the planet microlens the radiation from the star, but its motion Doppler shifts the radiation to higher frequencies (top). This is subtle but detectable. In BEER (bottom), a dire acronym for BEaming, Ellipsoidal, and Reflection/emission modulation, microlensing is combined with variations caused by the shape of the star and reflection of light onto the planet

proud of his or her achievement…)[2] The ellipsoidal component of the acronym reflects the tidal effects massive planets have on the envelope of the star, where the orbital separation is small. As the planet orbits, the star's envelope is distorted into an egg shape by its gravitational pull. This also produces detectable variations in the manner in which the star releases radiation compared with emission from a perfect sphere. Measurements of the stellar radiation over one orbital period allows the shape of the star to be determined and hence the mass of the orbiting planet.

Finally, hot Jupiter or other planets will invariably be tidally locked to their star. The region facing the star is known as the sub-stellar point, which naturally becomes very hot if the orbital separation is small. The planet will then emit its own infrared (and possibly) visible radiation. Indeed, Kepler-76b is so close to its star that it has a temperature of around 2000 °C at its cloud tops. Redistribution of this energy by Kelvin and Rossby Waves allows the rotation period of the planet to be determined as well as obtain some information about atmospheric motion. This is the final ER part of the BEER acronym. With these various parameters to hand astronomers determined that Kepler-76b has a mass of 1.74–2.26 times that of Jupiter and has an

[2] Popular Science writer Mary Roach would say BEER is an example of "PLEASE"—Pretty Lame Excuse for an Acronym, Scientists and Experimenters. One concurs…

orbital period of 1.54 Earth days. Though the planet is tidally locked to its star, the presence of atmospheric waves has generated an equatorial jet stream that displaces the planet's hottest spot from the sub-stellar point.

BEER is a complex method of analysis, but other methods (described above) have confirmed its findings, meaning that the complex computer algorithms that the system employs can be used elsewhere in planetary discovery.

Ellipsoidal Variations in Stars

Generally stars are tidally distorted by the gravitational pull of partners, irrespective of their mass and distance (Fig. 2.10). As was mentioned above, where the planet orbits close to the star, or where the planet is massive, a significant alteration to the star's shape is possible, through these tidal effects. Where the star is post-main sequence, its flimsier envelope is particularly vulnerable to these effects. This means that red giant and sub-giant stars are particularly amenable targets for such gravity-induced variations. However, such stars are also often very irregular both in their physical appearance due to convection in the envelope and in terms of overall variability due to pulsations and other factors. Take the nearby red giant π1 Gruis. Claudia Paladini at ESO used the PIONIER instrument on ESO's Very Large Telescope to observe this star, which lies 530 light years from Earth.

The team found that the surface of this red giant has a fairly limited number of very large granules—hot, ascending plumes of gas, each about 120 million km across. This is about a quarter of the star's diameter, or slightly wider than the

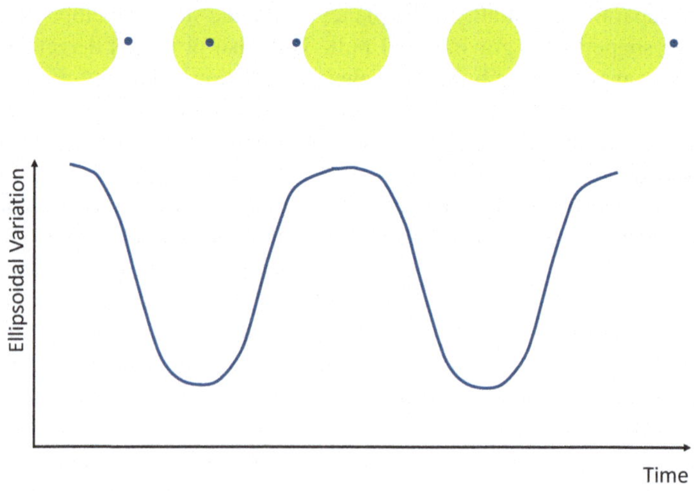

Fig. 2.10 Ellipsoidal variation planetary detection. As part of BEER, regular ellipsoidal changes in the shape of the star (yellow) are caused by the gravitational pull of the planet (blue-black). This changes the visible surface area of the star and the manner in which energy is delivered—if the envelope convects. Such variation changes the luminosity, and this is detectable with sensitive photodetectors

orbital separation of Venus from our Sun. A further complication for measurements of ellipsoidal variation is that the photosphere of many giant stars is obscured by dust, which hinders observations—and varies over periods of days to years, making measurements of ellipsoidal variation complex, to say the least. Consequently, if planets are to be found through ellipsoidal variation, these other interfering factors will need to be controlled—which is far from easy.

Now You See It, Now You Don't: Direct Observation of Fomalhaut b and Other Worlds

How wonderful it would be to see an extraterrestrial world. Before Hubble raised its eyelid above Earth's atmosphere, and long before the Cold War opened an avenue of exploration using adaptive optics, all astronomers could do was daydream that one day they might catch a glimpse of an alien planet, orbiting serenely around some distant star. Earth's turbulent atmosphere distorted the signal of any such world into a murky haze of scattered inferences. Catching a glimpse of an alien world was impossible.

Then along came Hubble, its eye poised above the cloud tops, followed by an array of superlative ground-based optical systems that were able to compensate for the minutia of atmospheric motion.

It had been unknown for some considerable time that several nearby A-class (white) stars had discs of dusty material orbiting them. The warped edge-on shape of the debris disc orbiting Beta Pictoris was such that it required the presence of at least one Jupiter-mass planet to mold it. However, until the middle of the last decade, nobody had thought to look for a planet. The resolution needed was beyond that thought possible. However, Fomalhaut provided a new opportunity. Its disc was nearly face-on and was there to be anything sizable hidden within it, perturbations in the motion of material might reveal the presence of the massive, orbiting body.

Indeed, after 8 years of scrutiny by the Hubble Space Telescope, Thayne Currie (formerly of the Goddard Flight Center) imaged one "blob" moving 115 AU from the central star in what appeared to be a predictable orbit. The blob, apparent in visible light, seemed to follow an orbital path, albeit one cutting through the debris disc. The pattern of infrared radiation detected by Hubble suggested a temperature of a few hundred Kelvin, which was in line with expectations for a giant planet that was a few tens of millions of years old.

Three years prior to this 2008 discovery, Paul Kalas and James Graham had inferred the presence of such a planet from the pattern of motion of a more distant dusty ring. However, its direct observation by Currie's team would be groundbreaking—if true (Fig. 2.11).

Exciting though the discovery was, subsequent observations that were made by Spitzer in 2010 failed to identify the planet at different near-infrared wavelengths. Was the planet merely a specter; a mirage caused by the distant stellar flame scattered through the dusty disc? Attention turned to the Atacama Large Millimeter

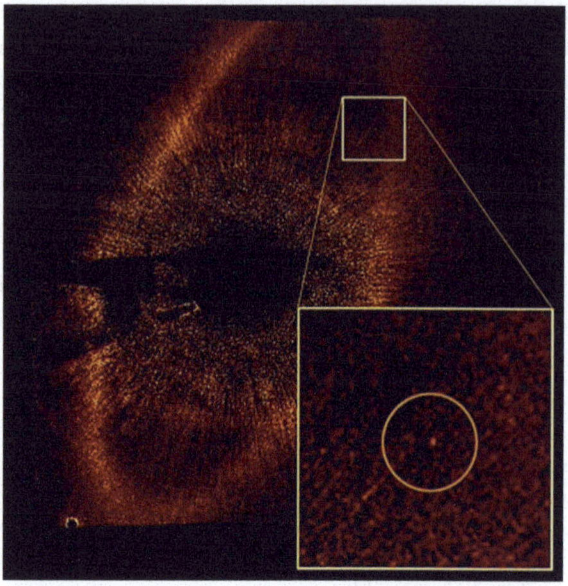

Fig. 2.11 Direct imaging of planets. This is possible with adaptive optics and with the use of a coronograph to block the light from the star. Planets can be detected if they are large enough, either using reflected light or secondary infrared radiation emitted by them. This is Fomalhaut b, with its long orbital period of 872 years. (Image courtesy of Paul Kalas/UC Berkeley, NASA, ESA)

Array (ALMA). This had resolved the edges of rings within the Fomalhaut dust disc. The location of these tight edges supported, in principle, the existence of Fomalhaut b and a more distant planet, Fomalhaut c. That said, the masses implied for each world differ significantly from those implied by the visible-light images. Hubble's images taken in visible light implied a mass similar to Jupiter for Fomalhaut b. Meanwhile, ALMA data suggested that the mass was somewhere between that of Mars and a few times that of Earth. Clearly, something was amiss that required further rounds of infrared and visible observation.

Finally, in 2012 Currie announced the results of further observations using Hubble. The said planet does indeed follow a path aligned with the debris disc. The mass of the world appeared to be around 1.92 Jupiter masses. Fomalhaut b is now known as Dagon—the name of a Semitic deity that is half-man, half-fish, in keeping with its host constellation, Pisces. Confirmation took several years in coming,

Dagon is certainly an interesting object. Although it was initially thought to emit its own light—thermal energy inherent in its youth—it is now thought that this light is scattered starlight. This implies that this Jupiter-mass object is surrounded by an extensive mass of debris, which might either be accreting planetesimals or rings. Dagon has a very extended orbit that is highly elliptical. Calculations by Paul Kalas and others suggest that each orbit around the A-class star takes 1700 years to complete, with the orbital distance varying between 7.4 and 44 billion km (50–300 AU) from the parent star.

Before we depart Fomalhaut, we'll take a look at this peculiar 440-million-year-old star system. Although these A- and K-class stars Fomalhaut A and B were known to be a widely dispersed binary, lying some 0.7 light years apart, it had been thought that they were merely a lonesome pair. However, after an initial investigation by Willum

Luyten in 1979, Eric Mamajek announced nearly four decades later that a third star, a 0.2 solar mass red dwarf, Fomalhaut C, was also moving through the sky in the same manner as the other two gravitationally bound stars. Moreover, it shared a chemical composition with the other two stars, suggesting that it was part of a very widely dispersed triple-star system.

The Fomalhaut C star lies some 2.5 light years from Fomalhaut A and 3.2 light years from the K-class Fomalhaut B. Indeed, this star system may be the widest stellar partnership known, spanning fully eleven full Moons in the night sky. Oddly, while Fomalhaut A and C both have their own debris discs, Fomalhaut B has none. Fomalhaut C's disc is a comet-infested icy mass lying some 10–40 AU from the star, making it loosely analogous to the Kuiper Belt. What explanation accounts for the absence of a similar disc around one of the three stars?

For some time, Fomalhaut A was the most massive star known to host a planetary system. Although Fomalhaut A is still the hottest star known to have planets, Kappa Andromedae, which has 2.5 solar masses, may have just topped it. More recently two giant stars, with masses as high as 5–7 times that of the Sun, hold the current record. The 8th magnitude star HD 13189 is a K-class giant with a radius 45–50 times that of the Sun. This star probably has a mass of 3–7 solar masses and has a massive planet (or low mass brown dwarf) in a relatively distant and elliptical orbit that takes 472 days to complete. Aside from these, at present truly massive stars appear bereft of planetary-mass companions. However, brown dwarfs may be more common at least for intermediate mass stars (those with 1.5–7 solar masses).

Concurrent with the initial discovery of Fomalhaut b was the announcement of the existence of four giant planets orbiting a lower mass star, HR 8799. This object is a 30-million-year-old F-class star with 1.5 times the Sun's mass and just under 5 times its luminosity. Direct imaging of each planet was accomplished by blocking the light from the central star coronagraph. The discovery of the first three of the four giant planets was announced by Christian Marois and his team (University of Montreal) in November 2008. Images were taken using the Keck II and Gemini telescopes in Hawaii. These observations revealed that the outermost planet swung around its star within a very dense, dusty disc of material. Between this outer disc and a less substantial inner one lay the other two worlds. The fourth planet was observed at higher resolution a year later.

The system resembles a scaled-up version of the outer Solar System. Planets e, d, c and b lie at 2–3 times the orbital radii of Jupiter, Saturn, Uranus and Neptune. With a more luminous central star, each receives approximately the same radiation flux as the equivalent worlds in our planetary system. However, here the similarity ends. All four planets are veritable giants. Each would put Jupiter in its place were they to move into orbits around Sol. Each world has a mass exceeding that of Jupiter and approaching the deuterium-fusion limit of 13 Jupiter-masses. At this point these planets might be reclassified as brown dwarfs.

It would seem scaling up the mass of the star also scales up the mass of the planets around it. As Chap. 1 discussed, more massive stars will usually form in more substantial protoplanetary discs. The only exception to this rule is where stellar collisions breed more massive stars—and fresh discs of material around them. Such collisions are only likely in dense star clusters where the crossing-time of the cluster is short and stars can collide while the cluster is young. In essence, a bigger disc of material is needed to build a bigger star. But a more substantial disc of material will also feed the growth of larger planets. This is an important relationship once we get down to the lowest mass stars, red dwarfs. In these systems, massive planets are rare, and the preference appears to be for synthesizing Earth-mass to super-terran planets.

Pipping both optical 'scope discoveries was the identification by Daniel Lafeniére of 1RXS J160929.1−210524 using the infrared eye of Gemini in Hawaii. Although this discovery was announced in September of the same year—2 months ahead of the Fomalhaut b discovery—it took nearly two further years of observation before the distant, 8-Jupiter-mass planet was confirmed, orbiting its star at a distance of 330 AU.

A rather more unusual planetary find was the discovery as part of the 2MASS survey in 2004. The 2M1207 system consists of two objects that orbit their common center of gravity at a separation of 41 AU, or 41 times the Sun-Earth distance. The 3.3-Jupiter-mass world orbits a 25-Jupiter-mass brown dwarf, and both are embedded in a dusty disc that is still being accreted by both objects. Like many protostars, 2MASS 2M1207 appears to direct some of this accreted material into broad, bipolar outflows. Lying 172 light years away, with a temperature of approximately 1000 °C, the planet is visible only in the infrared. However, if you were sufficiently close to it, you would probably see it glow a dull red, through clouds of dust and organic molecules. At present 1.2% of planets have been found through direct observation, but this should grow as optical systems improve.

Polarimetry and Astrometry

Starlight is generally not polarized. This is because stars emit radiation in all directions from atoms, free electrons and molecules that are in random motion. However, in a few instances the cosmos can create a situation where the light we detect is preferentially scattered when it passes through gases or past dust. In that regard if we can detect light that has passed through the atmosphere of a planet it will have become polarized by molecules within it. Although polarization produces a tiny effect on the overall light emitted by a star system, in principle it is easy to detect, as its effect is readily discriminated from the generic starlight from the system.

Moreover, polarization produces patterns that are determined by the chemical composition of the material that generates it. This allows the chemistry of the atmosphere to be determined with reasonable precision. Significantly, such polarized light is then not affected by the instability of Earth's turbulent atmosphere, allowing planets to be detected using ground-based instrumentation. The down side is relatively obvious: planets that have very dark surfaces—or which lack atmospheres—will not be detectable through this method.

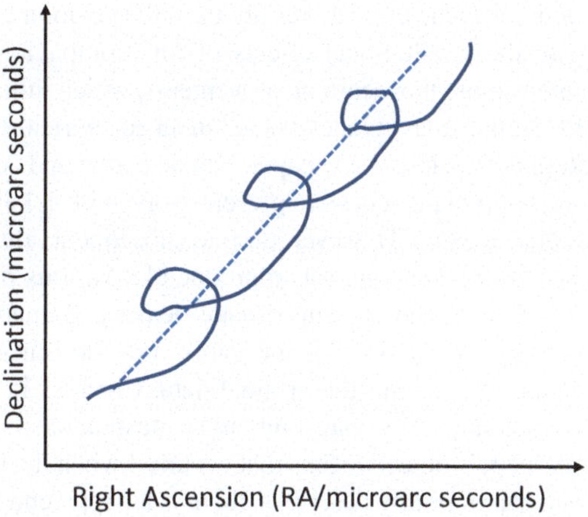

Fig. 2.12 Astrometry. If one is able to precisely plot a star's path across the sky (dashed line) it may be possible to detect a spiraling pattern, caused by the pull of an otherwise invisible planet (solid blue line). In 2002, the Hubble Space Telescope succeeded in using astrometry to identify a previously discovered planet around the star Gliese 876. This had previously been found by the RV method

At present this method hasn't scored any successes in terms of discovery, but it could be coupled to other searches to investigate the atmospheres of planets found by Kepler or other instrumentation. For example, in 2008 polarimetry was used to investigate the hot Jupiter HD 189733 b, which had been discovered 3 years earlier.

Astrometry (Fig. 2.12) is the oldest method that's been trialed to find planets. This consists of precisely measuring a star's position in the sky, then observing how that position changes over time. In the 1700s William Herschel claimed that an unseen companion was altering the position of the star 70 Ophiuchus. In 1855 William Stephen Jacob carried out the first astrometric calculation for this stellar body. Neither his, nor others that were carried out in the twentieth century, were able to identify evidence of this unseen companion.

More generally, during the nineteenth century astrometric measurements were done by hand on written notes, leaving considerable imprecision in the data that was available. However, towards the end of the nineteenth century the introduction of photographic plates opened the possibility that precise measurements could be made and produce tangible results. If a star has a planet, then the gravitational influence of the planet will cause the star itself to move in a tiny circular or elliptical orbit. As the star moves across the sky the orbiting planet will cause the star to wobble as the star moves relative to the Sun. One of the earliest such targets was Barnard's Star. Located six light years distant, this star's trajectory through the plane of the Milky Way causes significant proper motion relative to the Sun. If a planet is present, it will produce a readily detectable oscillation in the path of this nearby star.

Years earlier, Peter Van de Kamp's (Swathmore College) had claimed that Barnard's star had a Jupiter-mass companion. Both the proximity to Earth and the low mass of the red dwarf (0.6 solar masses) would have made astrometry a feasible method to discover such a planet. Van de Kamp dissected the proper motion of Barnard's Star as it tracked across the sky from 1938 until the 1960s. Van de Kamp claimed that there were micrometer-scale variations in the position of Barnard's Star that were consistent with the gravitational effects of this companion. Several people were involved in determining these minute movements, which initially led credence to the idea that this 1.6-Jupiter-mass planet existed in an eccentric orbit some 4.4 AU from the red dwarf. Refined in a later 1969 paper, Van de Kamp and colleagues went on to suggest that not one but two planets were present—one with 1.1 Jupiter masses and the other with 0.8 Jupiter masses. However, despite subsequent attempts by other astronomers, none of these measurements was reproducible. Various papers refuted the existence of these worlds over the ensuing decade, leaving Barnard's Star bereft of any companions. Subsequent analysis of the data—and the times at which it had been collected—seemed to explain the original data. Changes in the astrometric field were noted in various stars. This was found to be attributable to modifications and corrections in the telescope's objective lens that occurred when the telescope was upgraded. Thankfully, as Chap. 8 briefly describes, there is a happy end to the Barnard's Star tale.

Astrometry was back in the dock, having failed to fulfill its promise a second time. Events repeated once more when, in 1996, George Gatewood claimed multiple planets orbited another, nearby red dwarf, Lalande 21185. Once more, these results were irreproducible, and astrometry was shunned once more. Recall that by this time the radial velocity method was churning out planets at the rate of a few per year, and there was great expectation that astronomers had opened the floodgate on exoplanetary discovery. Indeed, they had—and astrometry was firmly off the party list.

Indeed, if we are particularly generous the only "hit" for this method (prior to 2010) was 8 years earlier when the Hubble Space Telescope succeeded in using astrometry to characterize the previously identified planet Gliese 876b. However, despite a further failure in 2009, with the supposed detection of a planet orbiting VB10 (VB10 b), astrometry finally nailed a hit with the spectroscopic system HD 176051. Matthew W. Muterspaugh (Tennessee State University) and colleagues identified six spectroscopic binary systems for which there was evidence of planetary-mass companions. Of the two most likely systems, HD 176051 was definitively shown to possess a 1.5 Jupiter-mass planet. However, although the existence of the planet is secure, which star (or stars) in the binary the planet orbits remains unclear. For now astrometry's minor victory has rescued the technique from derision.

Despite the antiquity of the method, to date only 0.03% of all planetary finds have been made with this method. However, the launch of Gaia in 2013 should change matters, as its precision optics dissects the positions of one billion astronomical objects—roughly 1% of our galaxy's stars. Consequently, it is expected to find tens of thousands of planets via astrometry, of which 1000–1500 are expected to orbit red dwarfs. During its mission, Gaia will examine each of these myriad objects around 70 times, during the first 5 years of operation. While Gaia has a defined five-year mission, this could be extended for a further 4 years, until 2022. Like previous astrometry successes, the majority of finds will be hot-Jupiter planets, since these will have the greatest effect on stellar proper motion.

Conclusions

The last two decades have seen a revolution in our understanding of extraterrestrial worlds. This chapter summarizes the methods and some of those discoveries, thereby illustrating the diversity of planets and methods employed to find them.

Prior to the discovery of the planets orbiting PSR 1257B + 12, planetary science was restricted to our Solar System—and to our wildest imaginings. Now, we have thousands of confirmed examples of foreign systems of alien worlds. These have been found by a multitude of discrete and increasingly fanciful methods that now includes the oldest: astrometry. Gaia and TESS promise to widen our horizons even further, bringing forth a vast and representative zoo of exoplanets in the next 5 years.

References

Adams, E. R., Jackson, B., & End, M. (2016). Ultra short period planets in K2 SuPerPiG results for campaigns 0-5. *The Astronomical Journal, 152*, 13. https://arxiv.org/pdf/1603.06488v2.pdf.

Agol, E. & Fabrycky, D. C. (2017). Transit timing and duration variations for the discovery and characterization of exoplanets. https://arxiv.org/pdf/1706.09849.pdf.

Agol, E., Steffen, J., Sari, R., & Clarkson, W. (2005). On detecting terrestrial planets with timing of giant planet transits. *MNRAS, 359*, 567. https://arxiv.org/pdf/astro-ph/0412032.pdf.

Agol, E., & Steffen, J. H. (2007). A limit on the presence of Earth-mass planets around a Sun-like star. *Monthly Notices of the Royal Astronomical Society, 374*(3), 941–948. https://doi.org/10.1111/j.1365-2966.2006.11213.x.

Anglada, G., Amado, P. J., Ortiz, J. L., Gomez, J. F., Macas, E., Alberdi, A., Osorio, M., Gomez, J. L., de Gregorio-Monsalvo, I., Perez-Torres, M. A., Anglada-Escude, G., Berdiñas, Z. M., Jenkins, J. S., Jimenez-Serra, I., Lara, L. M., Lopez-Gonzalez, M. J., Lopez-Puertas, M., Morales, N., Ribas, I., Richards, A. M. S., Rodreguez-Lopez, C., & Rodriguez, E. (2017). ALMA discovery of dust belts around Proxima Centauri. *The Astrophysical Journal, 850*, L6. https://arxiv.org/pdf/1711.00578.pdf.

Anglada-Escudé, G., Rojas-Ayala, B., Boss, A. P., Weinberger, A. J., & Lloyd, J. P. (2012). GJ 1214b revised: Improved trigonometric parallax, stellar parameters, orbital solution, and bulk properties for the super-Earth GJ 1214b. https://arxiv.org/pdf/1210.8087.pdf.

Anglada-Escudé, G., Amado, P. J., Barnes, J., Berdiñas, Z. M., Paul Butler, R., Coleman, G. A. L., de la Cueva, I., Dreizler, S., Endl, M., Giesers, B., Jeffers, S. V., Jenkins, J. S., Jones, H. R. A., Kiraga, M., Kürster, M., López-González, M. J., Marvin, C. J., Morales, N., Morin, J., Nelson, R. P., Ortiz, J. L., Ofir, A., Paardekooper, S.-J., Reiners, A., Rodríguez, E., Rodríguez-López, C., Sarmiento, L. F., Strachan, J. P., Tsapras, Y., Tuomi, M., & Zechmeister, M. (2016). A terrestrial planet candidate in a temperate orbit around Proxima Centauri. *Nature, 536*, 437–440. https://doi.org/10.1038/nature19106.

Anglada-Escudé, G., Arriagada, P., Vogt, S. S., Rivera, E. J., Butler, R. P., Crane, J. D., Shectman, S. A., Thompson, I. B., Minniti, D., Haghighipour, N., Carter, B. D., Tinney, C. G., Wittenmyer, R. A., Bailey, J. A., O'Toole, S. J., Jones, H. R. A., & Jenkins, J. S. (2012). A planetary system around the nearby M dwarf GJ 667C with at least one super-earth in its habitable zone. *The Astrophysical Journal, 751*, L16. https://arxiv.org/pdf/1202.0446.pdf.

Anglada-Escudé, G., Tuomi, M., Gerlach, E., Barnes, R., Heller, R., Jenkins, J. S., Wende, S., Vogt, S. S., Butler, R. P., Reiners, A., & Jones, H. R. A. (2013). A dynamically-packed planetary system around Gliese 667C with three planets in the habitable zone. *Astronomy & Astrophysics, 556*, A126. https://www.eso.org/public/archives/releases/sciencepapers/eso1328/eso1328a.pdf.

Aurière, M., Konstantinova-Antova, R., Espagne, O., Petit, P., Roudier, T., Charbonnel, C., Donati, J.-F., & Wade, G. A. (2013). Pollux: A stable weak dipolar magnetic field but no planet? https://arxiv.org/pdf/1310.6907v1.pdf.

Ballard, S., Fabrycky, D., Fressin, F., Charbonneau, D., Desert, J.-M., Torres, G., Marcy, G., Burke, C. J., Isaacson, H., Henze, C., Steffen, J. H., Ciardi, D. R., Howell, S. B., Cochran, W. D., Endl, M., Bryson, S. T., Rowe, J. F., Holman, M. J., Lissauer, J. J., Jenkins, J. M., Still, M., Ford, E. B., Christiansen, J. L., Middour, C. K., Haas, M. R., Li, J., Hall, J. R., McCauliff, S., Batalha, N. M., Koch, D. G., & Borucki, W. J. (2011). The Kepler-19 system: A transiting 2.2 R⊕ planet and a second planet detected via transit timing variations. *The Astrophysical Journal, 743*, 20. https://arxiv.org/pdf/1109.1561v1.pdf.

Bear, E. & Soker, N. (2010). Connecting planets around horizontal branch stars with known exoplanets. https://arXiv.org/pdf/1003.4884v2.pdf.

Berdyugina, S. V., Berdyugin, A. V., Fluri, D. M., & Piirola, V. (2007). First detection of polarized scattered light from an exoplanetary atmosphere. https://arxiv.org/pdf/0712.0193.pdf.

Bonfils, X., Delfosse, X., Udry, S., Forveille, T., Mayor, M., Perrier, C., Bouchy, F., Gillon, M., Lovis, C., Pepe, F., Queloz, D., Santos, N. C., Ségransan, D., & Bertaux, J.-L. (2011). The HARPS search for southern extra-solar planets? XXXI. The M-dwarf sample. *Astronomy & Astrophysics, 605*, 7. https://arxiv.org/pdf/1111.5019v2.pdf.

Brogi, M., Snellen, I. A. G., de Kok, R. J., Albrecht, S., Birkby, J. L., & de Mooij, E. J. W. (2013). Detection of molecular absorption in the dayside of exoplanet 51 Pegasi b? https://arxiv.org/pdf/1302.6242.pdf.

Brogi, M., Snellen, I. A. G., de Kok, R. J., Albrecht, S., Birkby, J., & de Mooij, E. J. W. (2012). The signature of orbital motion from the dayside of the planet τ Boötis b. *Nature, 486*, 502–504. https://arxiv.org/ftp/arxiv/papers/1206/1206.6109.pdf.

Buchhave, L. A. & Latham, D. W. (2015). The metallicities of stars with and without transiting planets. https://arxiv.org/pdf/1507.03557v1.pdf.

Buchhave, L. A., Latham, D. W., Johansen, A., Bizzarro, M., Torres, G., Rowe, J. F., Batalha, N. M., Borucki, W. J., Brugamyer, E., Caldwell, C., Bryson, S. T., Ciardi, D. R., Cochran, W. D., End, M., Esquerdo, G. A., Ford, E. B., Geary, J. C., Gilliland, R. L., Hansen, T., Isaacson, H., Laird, J. B., Lucas, P. W., Marcy, G. W., Morse, J. A., Robertson, P., Shporer, A., Stefanik, R. P., Still, M., & Quinn, S. N. (2012). An abundance of small exoplanets around stars with a wide range of metallicities. *Nature, 486*, 375–377. https://doi.org/10.1038/nature11121.

Charpinet, S., Fontaine, G., Brassard, P., Green, E. M., Van Grootel, V., Randall, S. K., Silvotti, R., Baran, A. S., Østensen, R. H., Kawaler, S. D., & Telting, J. H. (2011). A compact system of small planets around a former red-giant star. *Nature, 480*, 496–499. https://doi.org/10.1038/nature10631.

Chauvin, G., Lagrange, A.-M., Dumas, C., Zuckerman, B., Mouillet, D., Song, I., Beuzit, J.-L., & Lowrance, P. (2004). A giant planet candidate near a young brown dwarf. *Astronomy and Astrophysics, 425*(2), L29–L32. https://doi.org/10.1051/0004-6361:200400056. https://arxiv.org/abs/astro-ph/0409323.

Crossfield, I. J. M., Petigura, E., Schlieder, J. E., Howard, A. W., Fulton, B. J., Aller, K. M., Ciardi, D. R., Lepine, S., Barclay, T., de Pater, I., de Kleer, K., Quintana, E. V., Christiansen, J. L., Schlafly, E., Kaltenegger, L., Crepp, J. R., Henning, T., Obermeier, C., Deacon, N., Weiss, L. M., Isaacson, H. T., Hansen, B. M. S., Liu, M. C., Greene, T., Howell, S. B., Barman, T., & Mordasini, C. (2015). K2-3: A nearby M star with three transiting super-earths discovered by K2. https://arXiv:1501.03798.pdf.

Currie, T., Debes, J., Rodigas, T. J., Burrows, A., Itoh, Y., Fukagawa, M., Kenyon, S., Kuchner, M., & Matsumura, S. (2012). Direct imaging confirmation and characterization of a dust-enshrouded candidate exoplanet orbiting fomalhaut. *The Astrophysical Journal, 760*, L32. https://doi.org/10.1088/2041-8205/760/2/L32. https://arxiv.org/abs/1210.6620.

Delfosse, X., Bonfils, X., Forveille, T., Udry, S., Mayor, M., Bouchy, F., Gillon, M., Lovis, C., Neves, V., Pepe, F., Perrier, C., Queloz, D., Santos, N. C., & Ségransan, D. (2012). The HARPS search for southern extra-solar planets. XXXV. Super-Earths around the M-dwarf neighbors Gl 433 and Gl 667C. https://arXiv:1202.2467v1.pdf.

Díaz, M. R., Jenkins, J. S., Tuomi, M., Paul Butler, R., Soto, M. G., Teske, J. K., Feng, F., Shectman, S. A., Arriagada, P., Crane, J. D., Thompson, I. B. & Vogt, S. S. (2018). The test case of HD26965: Difficulties disentangling weak Doppler signals from stellar activity. https://arxiv.org/pdf/1801.03970.pdf.

Doyle, L. R., Carter, J. A., Fabrycky, D. C., Slawson, R. W., Howell, S. B., Winn, J. N., Orosz, J. A., Prsa, A., Welsh, W. F., Quinn, S. N., Latham, D., Torres, G., Buchhave, L. A., Marcy, G. W., Fortney, J. J., Shporer, A., Ford, E. B., Lissauer, J. J., Ragozzine, D., Rucker, M., Batalha, N., Jenkins, J. M., Borucki, W. J., Koch, D., Middour, C. K., Hall, J. R., McCauliff, S., Fanelli, M. N., Quintana, E. V., Holman, M. J., Caldwell, D. A., Still, M., Stefanik, R. P., Brown, W. R., Esquerdo, G. A., Tang, S., Furesz, G., Geary, J. C., Berlind, P., Calkins, M. L., Short, D. R., Steffen, J. H., Sasselov, D., Dunham, E. W., Cochran, W. D., Boss, A., Haas, M. R., Buzasi, D., & Fischer, D. (2011). Kepler-16: A transiting circumbinary planet. *Science, 333*, 1602. https://doi.org/10.1126/science.1210923. https://arxiv.org/ftp/arxiv/papers/1109/1109.3432.pdf.

Eibe, M. T., Cuesta, L., Ullán, A., Pérez-Verde, A. & Navas, J. (2012). Analysis of variations in transit time and transit duration in WASP-3. Evidence of secular perturbations reconsidered. https://www.researchgate.net/publication/258680903_Analysis_of_variations_in_transit_time_and_transit_duration_in_WASP-3_Evidence_of_secular_perturbations_reconsidered.

Faigler, S., Tal-Or, L., Mazeh, T., Latham, D. W., & Buchhave, L. A. (2013). BEER analysis of Kepler and CoRoT light curves: I. Discovery of Kepler-76b: A hot Jupiter with evidence for superrotation. Available from: https://arxiv.org/pdf/1304.6841v3.pdf

Feroz, F., & Hobson, M. P. (2013). Bayesian analysis of radial velocity data of GJ667C with correlated noise: Evidence for only 2 planets. *Monthly Notices of the Royal Astronomical Society, 437*, 3540–3549. https://arxiv.org/pdf/1307.6984v2.pdf.

Ford, E. B., & Scott Gaudi, B. (2006). Observational constraints on Trojans of transiting extrasolar planets. *The Astrophysical Journal, 652*, L137. https://arxiv.org/pdf/astro-ph/0609298.pdf.

Gillon, M., Triaud, A. H. M. J., Demory, B.-O., Jehin, E., Agol, E., Deck, K. M., Lederer, S. M., de Wit, J., Burdanov, A., Ingalls, J. G., Bolmont, E., Leconte, J., Raymond, S. N., Selsis, F., Turbet, M., Barkaoui, K., Burgasser, A., Burleigh, M. R., Carey, S. J., Chaushev, A., Copperwheat, C. M., Delrez, L., Fernandes, C. S., Holdsworth, D. L., Kotze, E. J., Van Grootel, V., Almleaky, Y., Benkhaldoun, Z., Magain, P., & Queloz, D. (2017). Seven temperate terrestrial planets around the nearby ultracool dwarf star TRAPPIST-1. *Nature, 542*, 456–460. https://doi.org/10.1038/nature21360.

Gizis, J. E., Shipman, H. L., & Harvin, J. A. (2005). First ultraviolet spectrum of a brown dwarf: Evidence for H_2 fluorescence and accretion. *Astrophysical Journal, 630*(1), L89–L91. https://doi.org/10.1086/462414. https://arxiv.org/pdf/astro-ph/0507429.pdf.

Gregory, P. C. (2011). Bayesian re-analysis of the Gliese 581 exoplanet system. *Monthly Notices of the Royal Astronomical Society, 415*, 2523–2545. https://arxiv.org/pdf/1101.0800.pdf.

Grillmair, C. J., Burrows, A., Charbonneau, D., Lee, A., Stauffer, J., Meadows, V., van Cleve, J., von Braun, K., & Levine, D. (2008). Strong water absorption in the dayside emission spectrum of the planet HD189733b. *Nature, 456*, 767–769. https://doi.org/10.1038/nature07574. https://arxiv.org/ftp/arxiv/papers/0901/0901.4774.pdf.

Grimm, S. L., Demory, B.-O., Gillon, M., Dorn, C., Agol, E., Burdanov, A., Delrez, L., Sestovic, M., Triaud, A. H. M. J., Turbet, M., Bolmont, É., Caldas, A., de Wit, J., Jehin, E., Leconte, J., Raymond, S. N., Van Grootel, V., Burgasser, A. J., Carey, S., Fabrycky, D., Heng, K., Hernandez, D. M., Ingalls, J. G., Lederer, S., Selsis, F., & Queloz, D. (2018). The nature of the TRAPPIST-1 exoplanets. *Astronomy & Astrophysics, 613*, A68. https://arxiv.org/pdf/1802.01377.pdf.

Guinan, E. F. & Engle, S. G. (2009). The "Living with a Red Dwarf" program. https://arxiv.org/pdf/0901.1860v1.pdf.

Haghighipour, N., Capen, S., & Hinse, T. C. (2013). Detection of Earth-mass and super-Earth Trojan planets using transit timing variation method. *Celestial Mechanics and Dynamical Astronomy, 117*, 75. CrossRefADSGoogle Scholar.

Haghighipour, N., Capen, S., & Hinse, T. C. (2013). On the detection of (habitable) super-Earths around low-mass stars using Kepler and transit timing variation method. *Celestial Mechanics and Dynamical Astronomy, 111*, 267. https://arxiv.org/pdf/1307.6806.pdf.

Han, Z., Podsiadlowski, P., Maxted, P. F. L., & Marsh, T. R. (2003). The origin of subdwarf B stars—II. *Monthly Notices of the Royal Astronomical Society, 341*, 669–691.

Heller, R., Hippke, M., Placek, B., Angerhausen, D., & Agol, E. (2016). Predictable patterns in planetary transit timing variations and transit duration variations due to exomoons. *Astronomy & Astrophysics, 591*, A67. https://arxiv.org/pdf/1604.05094.pdf.

Holman, M. J., Fabrycky, D. C., Ragozzine, D., Ford, E. B., Steffen, J. H., Welsh, W. F., Lissauer, J. J., Latham, D. W., & Marcy, G. W. (2010). Kepler-9: A system of multiple planets transiting a Sun-like star, confirmed by timing variations. *Science, 330*, 51–54.

Holman, M. J., & Murray, N. W. (2004). The use of transit timing to detect extrasolar planets with masses as small as Earth. *Astrophysics, 307*, 1288–1291. file:///F:/astro%20writing%20stuff/Red%20Dwarf%20book/UACS%202%20Papers/Chapter%202%20References%20Discovery/Original%20TTV%20paper.pdf.

Irwin, J., Charbonneau, D., Nutzman, P., & Falco, E. (2008). The MEarth project: Searching for transiting habitable super-Earths around nearby M-dwarfs. *Proceedings of the International Astronomical Union, 4*, 37–43. https://arxiv.org/pdf/0807.1316v1.pdf.

Janson, M., Carson, J., Lafreniere, D., Spiegel, D., Bent, J., & Wong, P. (2012). Infrared non-detection of Fomalhaut b—Implications for the planet interpretation. *Astrophysical Journal, 747*, 116. https://doi.org/10.1088/0004-637x/747/2/116. arXiv:1201.4388. https://arxiv.org/abs/1201.4388.

Johnson, J. et al. (2017). Compositions of small planets & implications for planetary dynamics. In 229th AAS Meeting, abstract # 413.06.

Justham, S., Podsiadlowski, P., & Han, Z. (2010). On the formation of single and binary helium-rich sdO stars. *Monthly Notices of the Royal Astronomical Society, 410*, 984–993. https://arxiv.org/pdf/1008.1584.pdf.

Kalas, P., Graham, J. R., Chiang, E., Fitzgerald, M. P., Clampin, M., Kite, E. S., Stapelfeldt, K., Marois, C., & Krist, J. (2008). Optical images of an exosolar planet 25 light-years from earth. *Science, 322*(5906), 1345–1348. https://doi.org/10.1126/science.1166609.

Kalas, P., Graham, J. R., & Clampin, M. (2005). A planetary system as the origin of structure in Fomalhaut's dust belt. *Nature, 435*(7045), 1067–1070. https://doi.org/10.1038/nature03601. arXiv:astro-ph/0506574. https://arxiv.org/abs/astro-ph/0506574.

Kalas, P., Graham, J. R., Fitzgerald, M. P., & Clampin, M. (2013). STIS coronagraphic imaging of Fomalhaut: Main belt structure and the orbit of Fomalhaut b. *The Astrophysical Journal., 775*(1). https://doi.org/10.1088/0004-637X/775/1/56.

Kane, S. R., Hill, M. L., Kasting, J. F., Kopparapu, R. K., Quintana, E. V., Barclay, T., Batalha, N. M., Borucki, W. J., Ciardi, D. R., Haghighipour, N., Hinkel, N. R., Kaltenegger, L., Selsis, F., & Torres, G. (2016). A catalog of Kepler habitable zone exoplanet candidates. *The Astrophysical Journal, 830*, 1. https://arxiv.org/pdf/1608.00620.pdf.

Kipping, D. M. (2009). Transit timing effects due to an exomoon. https://arxiv.org/pdf/0810.2243.pdf.

Knutson, H. A., Benneke, B., Deming, D., & Homeier, D. (2014). A featureless transmission spectrum for the Neptune-mass exoplanet GJ 436b. *Nature, 505*, 66–68. https://doi.org/10.1038/nature12887.

Knutson, H. A., Charbonneau, D., Allen, L. E., Fortney, J. J., Agol, E., Cowan, N. B., Showman, A. P., Cooper, C. S., & Thomas Megeath, S. (2007). A map of the day-night contrast of the extrasolar planet HD 189733b. *Nature, 447*, 183–186. https://doi.org/10.1038/nature05782.

Kopparapu, R. K., Ramirez, R., Kasting, J. F., Eymet, V., Robinson, T. D., Mahadevan, S., Terrien, R. C., Domagal-Goldman, S., Meadows, V., & Deshpande, R. (2013b). Habitable zones around main-sequence stars: New estimates. *The Astrophysical Journal, 765*(2), 16. https://doi.org/10.1088/0004-637X/765/2/131. https://arxiv.org/pdf/1301.6674.pdf.

Kurster, M., Hatzes, A. P., Cochran, W. O., Pulliam, C. E., Oenner, K., & Oobereiner, S. (1994). A radial velocity search for extra-solar planets using an iodine gas absorption cell at the CAT + CES. *The Messenger, 76*, 51–55. https://www.eso.org/sci/publications/messenger/archive/no.76-jun94/messenger-no76-51-55.pdf.

Lazorenko, P. F., Sahlmann, J., Ségransan, D., Figueira, P., Lovis, C., Martin, E., Mayor, M., Pepe, F., Queloz, D., Rodler, F., Santos, N., & Udry, S. (2010). Astrometric search for a planet around VB 10. *Astronomy & Astrophysics, 527*, A25. https://arxiv.org/pdf/1011.2338.pdf.

Lissauer, J. J., Dawson, R. I., & Tremaine, S. (2014). Advances in exoplanet science from Kepler. *Nature, 513*, 336–344. https://doi.org/10.1038/nature13781.

Lissauer, J. J., Marcy, G. W., Bryson, S. T., Rowe, J. F., Jontof-Hutter, D., Agol, E., Borucki, W. J., Carter, J. A., Ford, E. B., Gilliland, R. L., Kolbl, R., Star, K. M., Steffen, J. H., & Torres, G. (2014). Validation of Kepler's multiple planet candidates. II. Refined statistical framework and descriptions of systems of special interest. *The Astrophysical Journal, 784*(1), 21. https://doi.org/10.1088/0004-637X/784/1/44.

Lithwick, Y., Xie, J., & Wu, Y. (2012). Extracting planet mass and eccentricity from TTV data. *The Astrophysical Journal, 761*, 122. https://arxiv.org/pdf/1207.4192.pdf.

Loeb, A., & Scott Gaudi, B. (2003). Periodic flux variability of stars due to the reflex Doppler effect induced by planetary companions. *The Astrophysical Journal, 588*, L117–L120. https://arxiv.org/pdf/astro-ph/0303212v2.pdf.

Lopez, S., & Jenkins, J. S. (2012). The effects of viewing angle on the mass distribution of exoplanets. *The Astrophysical Journal, 756*, 177. http://iopscience.iop.org/article/10.1088/0004-637X/756/2/177/pdf.

Malhotra, R. (1993). Three-body effects in the PSR 1257+12 planetary system. *Astrophysical Journal, Part 1, 407*(1), 266–275. ISSN 0004-637X.

Mamajek, E. (2005). A moving cluster distance to the exoplanet 2M1207b in the TW hydrae association. *The Astrophysical Journal., 634*(2), 1385–1394. https://doi.org/10.1086/468181. https://arxiv.org/abs/astro-ph/0506574.

Marcy, G. W., Paul Butler, R., Vogt, S. S., Fischer, D., & Lissauer, J. J. (1998). A planetary companion to the nearby M4 dwarf, Gliese 876. *The Astrophysical Journal, 505*, 2. http://iopscience.iop.org/article/10.1086/311623/pdf.

Martin, D. V. (2016). Circumbinary planets II—When transits come and go. *Monthly Notices of the Royal Astronomical Society, 465*, 3235–3253. https://arxiv.org/pdf/1611.00526v1.pdf.

Martins, J. H. C., Santos, N. C., Figueira, P., Faria, J. P., Montalto, M., Boisse, I., Ehrenreich, D., Lovis, C., Mayor, M., Melo, C., Pepe, F., Sousa, S. G., Udry, S., & Cunha, D. (2015). Evidence for a spectroscopic direct detection of reflected light from 51 Pegasi b. *Astronomy & Astrophysics, 576*, A134. https://doi.org/10.1051/0004-6361/201425298. https://www.researchgate.net/publication/275342970_Evidence_for_a_spectroscopic_direct_detection_of_reflected_light_from_51_Peg_b.

McQuillan, A., Mazeh, T., & Aigrain, S. (2013). Stellar rotation periods of the Kepler objects of interest: A dearth of close-in planets around fast rotators. *The Astrophysical Journal, 775*, L11. https://arxiv.org/pdf/1308.1845.pdf.

Mohanty, S., Jayawardhana, R., Huelamo, N., & Mamajek, E. (2007). The planetary mass companion 2MASS 1207-3932B: Temperature, mass, and evidence for an edge-on disk. *Astrophysical Journal, 657*(2), 1064–1091. https://doi.org/10.1086/510877.

Mróz, P., Udalski, A., Skowron, J., Poleski, R., Kozłowski, S., Szymański, M. K., Soszyński, I., Wyrzykowski, Ł., Pietrukowicz, P., Ulaczyk, K., Skowron, D., & Pawlak, M. (2017). No large population of unbound or wide-orbit Jupiter-mass planets. *Nature, 548*, 183–186. https://doi.org/10.1038/nature23276.

Murphy, S. J., & Bedding, T. R. (2016). A planet in an 840-d orbit around a Kepler main-sequence A star found from phase modulation of its pulsations. *The Astrophysical Journal, 827*, L17.

Muterspaugh, M. W., Lane, B. F., Kulkarni, S. R., Konacki, M., Burke, B. F., Colavita, M. M., Shao, M., Hartkopf, W. I., Boss, A. P., & Williamson, M. (2010). The phases differential astrometry data archive. v. candidate substellar companions to binary systems. *The Astronomical Journal, 140*, 6. https://arxiv.org/pdf/1010.4048.pdf.

Naydenkin, K. E. & Kaparulin, D. S. (2017). Correlations between planetary transit timing variations, transit duration variations and brightness fluctuations due to exomoons. https://arxiv.org/pdf/1704.00202.pdf.

Neuhäuser, R., Mugrauer, M., Fukagawa, M., Torres, G., & Schmidt, T. (2006). Direct detection of exoplanet host star companion Cep B and revised masses for both stars and the sub-stellar object. *Astronomy & Astrophysics, 462*, 777–780. https://arxiv.org/pdf/astro-ph/0611427.pdf.

Nichols, J. D. (2011). Magnetosphere-ionosphere coupling at Jupiter-like exoplanets with internal plasma sources: implications for detectability of auroral radio emissions. https://arxiv.org/pdf/1102.2737v1.pdf.

Orosz, J. A., Welsh, W. F., Carter, J. A., Fabrycky, D. C., Cochran, W. D., Endl, M., Ford, E. B., Haghighipour, N., MacQueen, P. J., Mazeh, T., Sanchis-Ojeda, R., Short, D. R., Torres, G., Agol, E., Buchhave, L. A., Doyle, L. R., Isaacson, H., Lissauer, J. J., Marcy, G. W., Shporer, A., Windmiller, G., Barclay, T., Boss, A. P., Clarke, B. D., Fortney, J., Geary, J. C., Holman, M. J., Huber, D., Jenkins, J. M., Kinemuchi, K., Kruse, E., Ragozzine, D., Sasselov, D., Still, M., Tenenbaum, P., Uddin, K., Winn, J. N., Koch, D. G., & Borucki, W. J. (2012). Kepler-47: A transiting circumbinary multi-planet system. *Science, 337*, 1511. https://doi.org/10.1126/science.1228380. https://arxiv.org/pdf/1208.5489.pdf.

Oshagh, M., Heller, R., & Dreizler, S. (2016). How eclipse time variations, eclipse duration variations, and radial velocities can reveal S-type planets in close eclipsing binaries. https://arxiv.org/pdf/1610.04047.pdf.

Pepe, F., Ehrenreich, D., & Meyer, M. R. (2014). Instrumentation for the detection and characterization of exoplanets. *Nature, 513*, 358–366. https://doi.org/10.1038/nature13784.

Pravdo, S. H. & Shaklan, S. B. (2009). An ultracool star's candidate planet. https://arxiv.org/ftp/arxiv/papers/0906/0906.0544.pdf.

Prinn, R. G., & Barshay, S. S. (1977). Carbon monoxide on Jupiter and implications for atmospheric convection. *Science, 198*(4321), 1031–1034.

Richardson, J., Deming, D., Horning, K., Seager, S., & Harrington, J. (2007). A spectrum of an extrasolar planet. *Nature, 445*, 892–895. https://doi.org/10.1038/nature05636.

Rivera, E. J., Gregory, L., Paul Butler, R., Vogt, S. S., Haghighipour, N., & Meschiari, S. (2010). The Lick-Carnegie exoplanet survey: A Uranus-mass fourth planet for GJ 876 in an Extrasolar Laplace Configuration. *The Astrophysical Journal, 719*, 1. http://iopscience.iop.org/article/10.1088/0004-637X/719/1/890/pdf.

Robertson, P. & Mahadevan, S. (2014). Disentangling planets and stellar activity for Gliese 667C. https://arxiv.org/pdf/1409.0021.pdf.

Rogers, L. A. (2015). Most 1.6 Earth-radius planets are not rocky. *The Astrophysical Journal, 801*(41), 1–13. https://doi.org/10.1088/0004-637X/801/1/41.

Rowe, J. F., Bryson, S. T., Marcy, G. W., Lissauer, J. J., Jontof-Hutter, D., Mullally, F., Gilliland, R. L., Issacson, H., Ford, E., Howell, S. B., Borucki, W. J., Haas, M., Huber, D., Steffen, J. H., Thompson, S. E., Quintana, E., Barclay, T., Still, M., Fortney, J., Gautier, T. N., III, Hunter, R., Caldwell, D. A., Ciardi, D. R., Devore, E., Cochran, W., Jenkins, J., Agol, E., Carter, J. A., & Geary, J. (2014). Validation of Kepler's multiple planet candidates. III. Light curve analysis and announcement of hundreds of new multi-planet systems. *The Astrophysical Journal, 784*(1), 45. https://doi.org/10.1088/0004-637X/784/1/45.

Sahu, K. C., Casertano, S., Valenti, J., Bond, H. E., Brown, T. M., Ed Smith, T., Clarkson, W., Minniti, D., Zoccali, M., Livio, M., Renzini, A., Rich, R. M., Panagia, N., Lubow, S., Brown, T., & Piskunov, N. (2007). Planets in the galactic bulge: Results from the SWEEPS Project. https://arxiv.org/pdf/0711.4059v1.pdf.

Schneider, A. C., Windsor, J., Cushing, M. C., Kirkpatrick, J. D., & Wright, E. L. (2016). WISEA J114724.10-204021.3: A free-floating planetary mass member of the TW Hya association. https://arxiv.org/pdf/1603.07985v1.pdf.

Schneider, J., Lainey, V., & Cabrera, J. (2015). A next step in exoplanetology: Exomoons. *International Journal of Astrobiology, 14*(2), 191–199. https://doi.org/10.1017/S1473550414000299.

Schuh, S., Kruspe, R., Lutz, R., & Silvotti, R. (2009). Time-resolved spectroscopy of the planet-hosting sdB pulsator V391 Pegasi. https://arxiv.org/pdf/0901.4885v1.pdf.

Seager, S., Kuchner, M., Hier-Majumder, C. A., & Militzer, B. (2007). Mass-radius relationships for solid exoplanets. https://arxiv.org/pdf/0707.2895v1.pdf.

Shields, A. L., Barnes, R., Agol, E., Charnay, B., Bitz, C., & Meadows, V. S. (2016). The effect of orbital configuration on the possible climates and habitability of Kepler-62f. *Astrobiology, 16*(6), 443–464. https://doi.org/10.1089/ast.2015.1353.

Sigurdsson, S., Richer, H. B., Hansen, B. M., Stairs, I. H., & Thorsett, S. E. (2003). A young white dwarf companion to pulsar B1620-26: Evidence for early planet formation. *Science, 301*, 193–196. https://doi.org/10.1126/science.1086326.

Silvotti, R., Østensen, R. H., Telting, J. H., & Lovis, C. (2014). Planetary companions to sdB stars. https://arxiv.org/pdf/1406.4711v1.pdf.

Silvotti, R., Schuh, S., Janulis, R., Solheim, J.-E., Bernabei, S., Østensen, R., Oswalt, T. D., Bruni, I., Gualandi, R., Bonanno, A., Vauclair, G., Reed, M., Chen, C.-W., Leibowitz, E., Paparo, M., Baran, A., Charpinet, S., Dolez, N., Kawaler, S., Kurtz, D., Moskalik, P., Riddle, R., & Zola, S. (2007). A giant planet orbiting the "extreme horizontal branch" star V391 Pegasi. *Nature, 449*, 189–191. https://doi.org/10.1038/nature06143.

Simon, A., Szatmary, K., & Szabo, G. M. (2007). Determination of the size, mass, and density of "exomoons" from photometric transit timing variations. *Astronomy & Astrophysics, 470*, 727–731. https://arxiv.org/pdf/0705.1046.pdf.

Skowron, J., Udalski, A., Poleski, R., Kozlowski, S., Szymánski, M. K., Wyrzykowski, L., Ulaczyk, K., Pietrukowicz, P., Pietrzynski, G., Soszyński, I. (The OGLE Collaboration) and Abe, F., Bennett, D. P., Bhattacharya, A., Bond, I. A., Freeman, M., Fukui, A., Hirao, Y., Itow, Y., Koshimoto, N., Ling, C. H., Masuda, K., Matsubara, Y., Muraki, Y., Nagakane, M., Ohnishi, K., Rattenbury, N., Saito, T. O., Sullivan, D. J., Sumi, T., Suzuki, D., Tristram, P. J., Yonehara, A. (The MOA Collaboration) and Dominik, M., Jørgensen, U. G., Bozza, V., Harpsøe, K., Hundertmark, M., Skottfelt, J. (The MiNDSTEp Collaboration) (2015). MOA 2011-BLG-028Lb: A Neptune-mass microlensing planet in the galactic bulge. The Astrophysical Journal 820:4. https://arxiv.org/pdf/1512.03422v1.pdf.

Steffen, J. H., & Agol, E. (2005). An analysis of the transit times of TrES-1b. *Monthly Notices of the Royal Astronomical Society, 364*(1), L96–L100. https://doi.org/10.1111/j.1745-3933.2005.00113.x. https://arxiv.org/pdf/astro-ph/0509656.pdf.

Strigari, L. E., Barnabè, M., Marshall, P. J., & Blandford, R. D. (2012). Nomads of the galaxy. *Monthly Notices of the Royal Astronomical Society, 423*, 1856–1865. https://arxiv.org/pdf/1201.2687v1.pdf.

Sumi, T., Kamiya, K., Udalski, A., Bennett, D. P., Bond, I. A., Abe, F., Botzler, C. S., Fukui, A., Furusawa, K., Hearnshaw, J. B., Itow, Y., Kilmartin, P. M., Korpela, A., Lin, W., Ling, C. H., Masuda, K., Matsubara, Y., Miyake, N., Motomura, M., Muraki, Y., Nagaya, M., Nakamura, S., Ohnishi, K., Okumura, T., Perrott, Y. C., Rattenbury, N., Saito, T., Sako, T., Sullivan, D. J., Sweatman, W. L., Tristram, P. J., Yock, P. C. M., Szymanski, M. K., Kubiak, M., Pietrzýnski, G., Poleski, R., Soszýnski, I., Wyrzykowski, L., & Ulaczyk, K. (2011). Unbound or distant planetary mass population detected by gravitational microlensing. *Nature, 473*, 349–352. https://arxiv.org/pdf/1105.3544.pdf.

Swain, M. R., Vasisht, G., & Tinetti, G. (2008). The presence of methane in the atmosphere of an extrasolar planet. *Nature, 452*, 329–331. https://doi.org/10.1038/nature06823.

Swift, J. J., Johnson, J. A., Morton, T. D., Crepp, J. R., Montet, B. T., Fabrycky, D. C., & Muirhead, P. S. (2013). Characterizing the cool KOIs. IV. Kepler-32 as a prototype for the formation of compact planetary systems throughout the galaxy. *The Astrophysical Journal, 764*(14pp), 105. https://doi.org/10.1088/0004-637X/764/1/105. https://arxiv.org/pdf/1301.0023.pdf.

Thompson, S. E., Coughlin, J. L., Hoffman, K., Mullally, F., Christiansen, J. L., Burke, C. J., Bryson, S., Batalha, N., Haas, M. R., Catanzarite, J., Rowe, J. F., Barentsen, G., Caldwell, D. A., Clarke, B. D., Jenkins, J. M., Li, J., Latham, D. W., Lissauer, J. J., Mathur, S., Morris, R. L., Seader, S. E., Smith, J. C., Klaus, T. C., Twicken, J. D., Van Cleve, J. E., Wohler, B., Akeson, R., Ciardi, D. R., Cochran, W. D., Henze, C. E., Howell, S. B., Huber, D., Prša, A., Ramírez, S. V., Morton, T. D., Barclay, T., Campbell, J. R., Chaplin, W. J., Charbonneau, D., Christensen-Dalsgaard, J., Dotson, J. L., Doyle, L., Dunham, E. W., Dupree, A. K., Ford, E. B., Geary, J. C., Girouard, F. R., Isaacson, H., Kjeldsen, H., Quintana, E. V., Ragozzine, D., Shporer, A., Aguirre, V. S., Steffen, J. H., Still, M., Tenenbaum, P., Welsh, W. F., Wolfgang,

A., Zamudio, K. A., Koch, D. G., & Borucki, W. J. (2017). Planetary candidates observed by Kepler. VIII. A fully automated catalog with measured completeness and reliability based on data release 25. *The Astrophysical Journal, 235*, 38. https://arxiv.org/pdf/1202.0446.pdf.

Thorsett, S. E., Arzoumanian, Z., Camilo, F., & Lyne, A. G. (1999). The triple pulsar system PSR B1620-26 in M4. *The Astronomical Journal, 523*, 763–770.

Traub, W. A. (2011). Terrestrial, habitable-zone exoplanet frequency from Kepler. *The Astrophysical Journal, 745*, 20. https://arxiv.org/pdf/1109.4682v1.pdf.

Tuomi, M. (2011). Bayesian re-analysis of the radial velocities of Gliese 581: Evidence in favour of only four planetary companions. *Astronomy & Astrophysics, 528*, L5. https://arxiv.org/pdf/1102.3314.pdf.

Tuomi, M., Anglada-Escudé, G., Gerlach, E., Jones, H. R. A., Reiners, A., Rivera, E. J., Vogt, S. S., & Butler, R. P. (2012). Habitable-zone super-Earth candidate in a six-planet system around the K2.5V star HD 40307. *Astronomy & Astrophysics, 549*, A48. https://arxiv.org/pdf/1211.1617v1.pdf.

Tuomi, M., Pinfield, D., & Jones, H. R. A. (2011). Application of Bayesian model inadequacy criterion for multiple data sets to radial velocity models of exoplanet systems. *Astronomy & Astrophysics, 532*, A116. https://arxiv.org/pdf/1106.5981.pdf.

Walkowicz, L. M., Batalha, N. M., Jenkins, J. M., Rowe, J. F., Cochran, W. D., Fressin, F., Torres, G., Buchhave, L. A., Sasselov, D. D., Borucki, W. J., Koch, D. G., Basri, G., Brown, T. M., Caldwell, D. A., Charbonneau, D., Dunham, E. W., Gautier, T. N., 3rd, Geary, J. C., Gilliland, R. L., Haas, M. R., Howell, S. B., Ciardi, D. R., Endl, M., Fischer, D., Fürész, G., Hartman, J. D., Isaacson, H., Johnson, J. A., MacQueen, P. J., Moorhead, A. V., Morehead, R. C., & Orosz, J. A. Kepler-9: A system of multiple planets transiting a sun-like star, confirmed by timing variations. *Science, 330*(6000), 51–54. https://doi.org/10.1126/science.1195778.

Welsh, W. F., Orosz, J. A., Carter, J. A., Fabrycky, D. C., Ford, E. B., Lissauer, J. J., Prsˇa, A., Quinn, S. N., Ragozzine, D., Short, D. R., Torres, G., Joshua, N. W., Doyle, L. R., Barclay, T., Batalha, N., Bloemen, S., Brugamyer, E., Buchhave, L. A., Caldwell, C., Caldwell, D. A., Christiansen, J. L., Ciardi, D. R., Cochran, W. D., Endl, M., Fortney, J. J., Gautier, T. N., III, Gilliland, R. L., Haas, M. R., Hall, J. R., Holman, M. J., Howard, A. W., Howell, S. B., Isaacson, H., Jenkins, J. M., Klaus, T. C., Latham, D. W., Li, J., Marcy, G. W., Mazeh, T., Quintana, E. V., Robertson, P., Shporer, A., Steffen, J. H., GurWindmiller, D. G. K., & Borucki, W. J. (2012). Transiting circumbinary planets Kepler-34 b and Kepler-35 b. *Nature, 481*, 475–479. https://doi.org/10.1038/nature10768.

Welsh, W. F., Orosz, J. A., Carter, J. A., Fabrycky, D. C., & the Kepler Team. (2013). Recent Kepler results on circumbinary planets. *Proceedings of the International Astronomical Union, 8*, 125–132. https://arxiv.org/pdf/1308.6328v1.pdf.

Welsh, W. F., Orosz, J. A., Short, D. R., Cochran, W. D., Endl, M., Brugamyer, E., Haghighipour, N., Buchhave, L. A., Doyle, L. R., Fabrycky, D. C., Hinse, T. C., Kane, S., Kostov, V., Mazeh, T., Mills, S. M., Mueller, T. W. A., Quarles, B., Quinn, S. N., Ragozzine, D., & Shporer, A. (2014). KIC 9632895: The 10th Kepler transiting circumbinary planet. *The Astrophysical Journal, 809*, 26. https://arxiv.org/pdf/1409.1605v1.pdf.

Whelan, E. T., Ray, T. P., Randich, S., Bacciotti, F., Jayawardhana, R., Testi, L., Natta, A., & Mohanty, S. (2007). Discovery of a bipolar outflow from 2MASSW J1207334-393254 a 24 M_{jup} brown dwarf. *The Astrophysical Journal, 659*, L45–L48. https://arxiv.org/pdf/astro-ph/0703112.pdf.

Whelan, E. T., Ray, T. P., Randich, S., Bacciotti, F., Jayawardhana, R., Testi, L., Natta, A., Mohanty, S., et al. (2007). Discovery of a bipolar outflow from 2MASSW J1207334-393254, a 24 M-Jup brown dwarf. *The Astrophysical Journal., 659*(1), L45–L48. https://doi.org/10.1086/516734. https://www.researchgate.net/publication/231143257_Discovery_of_a_bipolar_outflow_from_2MASSW_J1207334-393254_A_24_M-jup_brown_dwarf.

Winn, J. N., Sanchis-Ojeda, R., Rogers, L., Petigura, E. A., Howard, A. W., Isaacson, H., Marcy, G. W., Schlaufman, K. C., Cargile, P., & Hebb, L. (2017). Absence of a metallicity effect for ultra-short-period planets. *The Astronomical Journal, 154*, 60. https://arxiv.org/pdf/1704.00203v2.pdf.

Wittenmyer, R. A., Mikko, T., Butler, R. P., Jones, H. R. A., Anglada-Escudé, G., Jonathan, H., Tinney, C. G., Marshall, J. P., Carter, B. D., Bailey, J., Salter, G. S., O'Toole, S. J., Wright, D., Crane, J. D., Schectman, S. A., Arriagada, P., Thompson, I., Minniti, D., Jenkins, J. S., & Diaz, M. (2014). GJ 832c: A super-earth in the habitable zone. *The Astrophysical Journal, 791*, 114. https://arxiv.org/pdf/1406.5587.pdf.

Wolszczan, A., & Frail, D. (1992). A planetary system around a millisecond pulsar PSR1257+12. *Nature, 355*, 145–147.

Wordsworth, R. D., Forget, F., Selsis, F., Millour, E., Charnay, B., & Madeleine, J.-B. (2011). Gliese 581d is the first discovered terrestrial-mass exoplanet in the habitable zone. *The Astrophysical Journal, 733*, L48. https://arxiv.org/pdf/1105.1031.pdf.

Zhu, W., Huang, C. X., Udalski, A., Soares-Furtado, M., Poleski, R., Skowron, J., Mroz, P., Szymanski, M. K., Soszynski, I., Pietrukowicz, P., Koz Lowski, S., Ulaczyk, K., & Pawlak, M. (2017). Extracting microlensing signals from K2 campaign 9. *Publications of the Astronomical Society of the Pacific, 129*, 980. https://arxiv.org/pdf/1704.08692v2.pdf.

Chapter 3
Planetary Tectonism

Introduction

In this chapter we explore the processes that sculpt the long-term evolution of a planet's surface. These are integral to developing a deeper understanding of what controls the long-term habitability of planets.

We take it for granted that the surface of Earth is in constant motion. The release of energy from Earth's interior is facilitated by mantle-wide convection, which in turn is manifest as the constant shuffling of the portions of Earth's surface known as plates.

Although plate tectonics is the principle means through which our planet's interior sheds heat, it remains unclear whether this process has operated since our planet's inception—and if it hasn't, how did the early (and presumably hotter) early Earth dispense with heat from its formation and from radioactive decay? In this chapter we examine how plate tectonics operates now, whether another system operated early on, and what will happen to this dynamic system as the planet cools. Moreover, we will investigate how mantle-wide convection may operate on larger planets and the impact that this might have on planetary magnetism. This chapter forms a bridge to Chap. 5, where we link plate tectonics and other processes to the long-term habitability of the planet.

The Early Earth as a Model for Exoplanet Tectonism

What Granites and Granitoids Tell Us About the Early Earth

Determining the processes that operated in the earliest crust—and the solid portion of the mantle known as the lithosphere—is impossible to do directly. The oldest crust that remains has an age of approximately 4 billion years, leaving only the sparsest clues from the first 600 million years of our planet's history. When we

D. S. Stevenson, *Red Dwarfs*, https://doi.org/10.1007/978-3-030-25550-3_3

reach the Archaean era (3.9–2.45 billion years ago), the only remnants are found in the cores of the modern continents. Indeed, it isn't until you reach the Proterozoic era that continental crust becomes abundant.

Oceanic crust, by contrast, is never older than 180 million years on Earth. Older crust is rapidly subducted to its grave at the base of the deep mantle. Slivers of oceanic crust known as ophiolites preserve both some of the history of these lost early ocean basins and reveal approximately when modern-day subduction began. What is telling is a consummate lack of ophiolites prior to the mid-Archaean. Clearly either all trace of our modern style of subduction was destroyed, or something else operated in its place.

What Archaean (and latest Hadean era) rocks we can see contain odd assemblages of dense green rock and granite-like rocks (granitoids) known as greenstone belts. Granitoid is a generic term for silica-rich rocks, which are more than 63% silicon dioxide. True granites have more than 68% silicon dioxide, with the silica-poor granitoids generally known as granodiorites. Such granitoid rocks are poor in iron and magnesium and rich in feldspar and quartz, with quartz becoming increasingly abundant as the percentage of silicon dioxide increases. The combination of *fe*ldspar and *si*lica gives these rocks another generic name: *felsic*. *Ma*gnesium and iron (*f*errous) rich rocks are commonly referred to as *mafic*. Where relevant, we will use these terms from now on.

Compared to modern-day granites, these Archaean and late Hadean granitoid magmas show some chemical peculiarities that are largely absent from modern granites. The majority of Archaean granitoids are known as tonalite-trondhjemite-granodiorite, or TTG, for short. TTGs are richer in sodium but impoverished, in terms of potassium, than most modern-day granites. The precise chemistry suggests that these rocks were derived from the melting of dense garnet-containing rocks known as eclogite.

In modern-day subduction zones, eclogite forms when basalt is subducted to depths in excess of 52 km on Earth. Here, the increasing pressure causes forms of calcium- and magnesium-rich silicates known as pyroxene and plagioclase feldspar to undergo a chemical rearrangement. The end result is the formation of the red semi-precious mineral, garnet. In current terrestrial subduction, this mineral forms at depths where temperatures are far below those needed to melt it. However, in the Hadean and Archaean it appears this was not the case: TTGs appear to have formed from the direct (partial) melting of eclogite. Perhaps this is unsurprising, given that the mantle was hotter; however, at what temperature the eclogite melted is the contentious issue.

For eclogite to form and melt, basaltic rocks must be buried to suitable depths and temperatures. Did this happen in a subduction setting (as eclogite currently forms) or was another process at work? Whatever it was, it happened abundantly, so that large volumes of sodium-rich TTGs were formed. In order to understand the likeliest setting, we have to consider how hot the early crust and lithosphere was. In this context, the crust refers to the solidified scum produced by melting of the mantle.

Melting of mantle or any other rock is usually incomplete, because temperatures never become hot enough to melt the whole rock. Moreover, as temperatures rise some minerals will melt first. Since melts are usually less dense than the leftover minerals, these separate from them, buoyantly rising upwards. Finally, melting often happens in the presence of solvents, of which water and carbon dioxide are the most abundant on Earth. Some minerals are more soluble in these than others. A combination of both of these processes leads to partial melting, where those with the lowest melting points melt, leaving a residue of denser crystals. Terrestrial ocean crust is basalt, the product of 10–25% partial melting of dense upper mantle peridotite. Higher degrees of partial melting produce a denser magma, with far more of the magnesium-rich mineral olivine. This is known as komatiite. Both rocks may have formed early ocean crust, and both will form eclogite if they are buried to suitable depths by subduction or other processes.

The question is how did this crust become buried to such great depth that it melts? In modern subduction, basaltic oceanic crust has already been thoroughly altered (metamorphosed) by the time it enters subduction zones. This process happens the moment that the hot basaltic magma interacts with seawater. Water is absorbed into the crystal structure of the minerals olivine and pyroxene. The former mineral morphs into serpentine, with the latter mineral forming amphibole or micas. The resulting rock isn't basalt anymore; rather it is a less dense, water-rich rock known as amphibolite. It is this amphibolite that will convert into eclogite as it descends into the mantle.

In modern subduction settings, the hydrated crustal rock sinks into the mantle. Here, as temperatures rise, water is driven out of the minerals serpentine, amphibole and any mica. This water rises into the overlying mantle layer, known as the mantle wedge. Here, water depresses the melting point of the mantle, and large volumes of basaltic magma are produced. The chemistry is different from the kind of basalt that formed along the mid-ocean ridges, as it is heavily contaminated with sodium, chlorine, potassium and other water-soluble minerals such as calcium. Moreover in this hot, wet environment, it seems within this basaltic brew, rivulets or streamers of silica-rich fluids ascend. These alter the chemistry of the mantle and begin the process of manufacturing large volumes of granitoid (granite and granodiorite) magma. Thus, modern granitoids are formed from a combination of partial melting of the overlying mantle and fractionation of the resulting (largely) basaltic magma. This occurs in relatively shallow portions of the mantle, in the presence of water, *above* the region where eclogite forms (Fig. 3.1).

By contrast, the Archaean granitoids, which are sodium-rich, formed from the direct melting of eclogite. Therefore, these granitoids must have originated in a layer of oceanic crust that had become thickened and buried to great depth. This tells us that Archaean granites and modern ones form in distinctly different ways.

Clues from Other Worlds

There appears to have been a progressive evolution of the manner through which Earth's interior shed heat. At first the mantle was so hot that abundant volcanic activity simply piled lava on top of lava (Fig. 3.2). This system, where crustal rock is processed largely through vertical movement, is known as pipe or vertical tectonics.

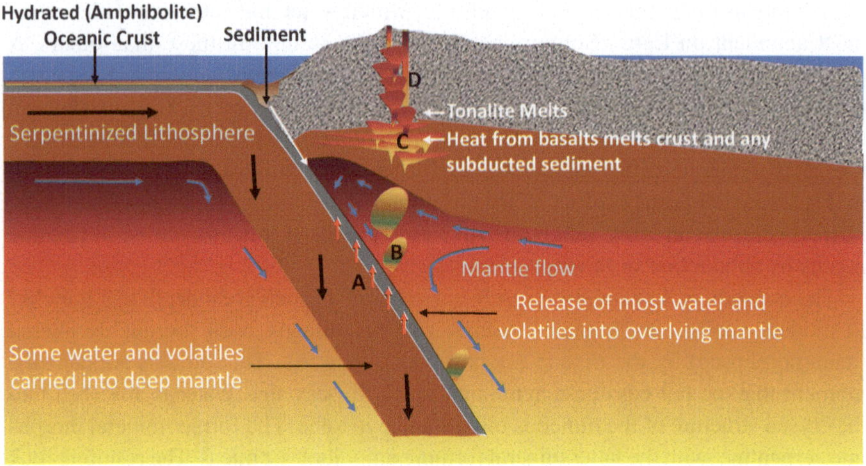

Fig. 3.1 Modern subduction. In **a** subducting oceanic crust releases saline fluids from the dehydration of the crust and at greater depth the underlying lithosphere. In **b** these fluids partially melt the overlying hot asthenosphere, making basaltic magmas enriched in sodium and potassium. These rise to the base of the crust, **c**, where most accumulate. Melting of the crust here and fractionation of the basaltic magmas produces tonalites (**d**), which continue to differentiate into granites. Melting of the oceanic crust may occur when fluids from the underlying lithosphere are stewed out of the serpentinite lower crust and mantle lithosphere

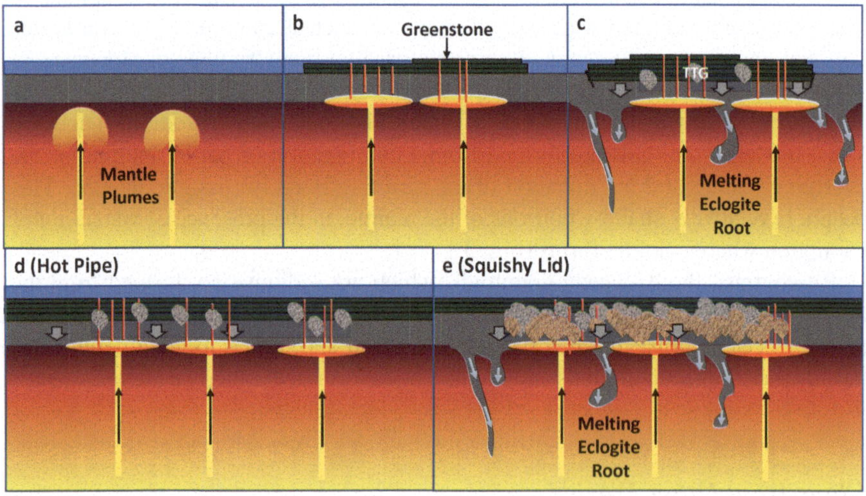

Fig. 3.2 Hadean hot-pipe and squishy-lid tectonics. In the Hadean the mantle lost heat through plumes and conduction (**a**). Extensive melting produced thick crust and undersea plateau (**b**). Over time, these became so thick that their bases melted and collapsed into the underlying mantle (**c**). In the Io-style "hot-pipe" version, the crust is cold but there is extensive volcanism (**d**). In the squishy-lid version, which is thought to dominate Venus, **e**, the crust is hotter, and most of the magmatism is underground (plutonic). Only the squishy-lid model reproduces the observed proportion of different TTG magmas seen on Earth

Although the crust may spread under the weight of material on top, most of the excess weight is accommodated by a sinking motion into the underlying mantle. In this Hadean scenario, once the crust had descended 52 km or so, the komatiites, basalts and any (limited) water-altered derivatives that comprised it would transform to eclogite. Now much denser than the mantle beneath it, these sank further and faster before partially melting to produce the TTGs. Being much less dense than the surrounding basalt-komatiite mix, they rise buoyantly towards the surface, eventually forming the first (micro) continental crust.

The basic Hadean scenario is called the hot-pipe, or heat-pipe, model and is an excellent fit for the cycling of crust on Io—and likely elsewhere in the universe where the upper mantle is particularly hot. Here there is so much energy in the mantle that extensive melting produces basaltic magmas, which largely pile up on top of one another in an otherwise cold crust.

However, though this works for Io, there are problems with the model when applied to the formation of early granitoids. The chemistry and relative abundance of intrusive (plutonic) and extrusive rocks is wrong.

In 2017 a variant of this hot pipe scheme was proposed by Antoine Rozel (ETH Zurich) and colleagues, known as "squidgy-lid tectonics." In this variant of the model it is suggested that rather than an Io-style hot-pipe scenario, the earliest TTGs were formed with a similar volcanism rate at the surface to that seen today (about 40% of the generated magma). Instead, the authors of this study propose a "squishy-lid" model, which is more akin to the volcanism of Venus than that of Io. In the squishy-lid model, a thicker crust is hotter throughout and partly melts at its base. Rather than the products erupting at the surface, most are intruded and differentiate underground. This gives rise to a great thickness of TTGs, formed under a range of pressures and temperatures at different depths. In turn this leads to subtle variations in their chemistry and crystal composition—precisely in line with the observations.

Of course, in reality the whole surface of Earth wasn't broiling under a sea of volcanism. Instead, volcanic activity would be concentrated above mantle plumes. Here, columns of particularly hot and lower density material were rising towards the surface. The mantle was hottest and, therefore, most likely to melt. Thus, while the squidgy-lid model gives the general picture of continental crust formation, it does not place it in its geographical context. To reiterate, this is most likely to occur in localized regions of the upper mantle, where plumes have reached the top and where the temperatures at the base of the crust are naturally highest.

The Hadean-Archean Transition

Over time the underlying mantle cooled somewhat and was able to support thicker crust underlain with a more rigid lithosphere. Remember that the term lithosphere refers to the uppermost rigid mantle and the overlying crust. Imagine the planet cooling down. Initially, when the mantle is very hot, extensive melting produces a thick, mafic crust atop a thin frozen layer of mantle. Most of the lithosphere *is* the crust. However, as the mantle cools, more and more of it solidifies, forming an increasingly thick lithosphere over which the crust forms a proportionately thinner veneer. Now more rigid, this lithosphere is less able to deform than it was in the hotter Hadean and is now better able to support the overlying crust—and, critically, can transmit stresses horizontally without deforming.

Figure 3.3 illustrates how the original vertical, or pipe, tectonics evolved into a subduction-assisted process. Once again the process begins with a mantle plume that ascends from great depth. As it approaches the surface it melts to produce a mixture of komatiites and magnesium-rich basalts. Once again the outcome is a deep plateau similar to modern-day Iceland. Partial melting of the foundering eclogite base also produces TTGs, as it did in the Hadean era.

The difference between the Hadean and Archaean is in what happens next. As the crust sinks into the mantle enough of the stress is accommodated horizontally to weaken the surrounding oceanic crust. Moreover, the thick TTG crust, overlain by abundant dense komatiites, becomes soft enough to spread horizontally. This is too buoyant to subduct but can progressively load the neighboring oceanic crust until it founders under the additional weight. The outcome is a micro-continent flanked on all sides by subducting oceanic crust. This crust will also morph to eclogite as it descends into the mantle. As the Archaean mantle is still hot enough to melt this, more TTG is produced, and the continents can grow outwards over the subsiding oceanic crust.

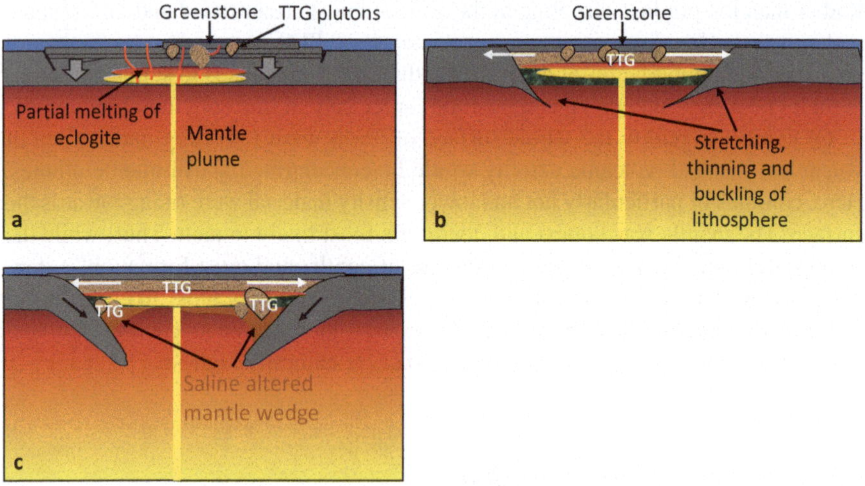

Fig. 3.3 Archaean tectonics and the formation of continents. In **a**, a strong lithosphere supports the microcontinents forming over plumes. A thick layer of TTG granitoids (textured fill) develops from partial melting of its eclogite base (red). The TTGs are overlain by greenstone (basalt and komatiite—green rectangles). The combined weight causes these to spread outwards (**b**). The flanking oceanic lithosphere buckles and begins to subduct around its edges. Fluids from subducting lithosphere rise into the overlying mantle (purple), causing a further wave of TTG formation around the edges of the micro-continent. This causes the microcontinent to grow rapidly outwards (**c**)

Moreover, as with modern subduction, as the eclogite and serpentine-rich crust sinks into the asthenosphere, heating and pressure release abundant fluids into the overlying wedge of mantle. This chemically alters (metasomatizes) the hot, plastic rock and sweats even more TTG from it. Although this is not modern-day plate tectonics, it has many of its hallmarks. Importantly, most measurements suggest that at least 50% of the total continental crust was synthesized during this era, indicating that this mechanism was particularly efficient at extracting granite from the mantle.

Finally, around the margins of these diminutive but growing continents subduction may be asymmetric, meaning that it proceeds at different rates around each edge of the micro-continent. This will cause these micro-continents to migrate across the surface of the globe until they collide with one another. The end result is a series of collisions that form the first contorted continents by 3.5 billion years ago. The Archaean world would have been a peculiar one to our eyes. Iceland-sized micro-continents likely dotted the surface of the globe, forming over hot plumes that then jostled around, leaving trails of islands behind them. Each micro-continent would have had low ranges of mountains along their leading edges as TTGs intruded the crust and raised it upwards.

When did this system of tectonics morph into the one we observe today? Evidence of the transition comes from work focused on the North American craton. Carol Frost (University of Wyoming) and colleagues show that there was a gradual shift in the types of magmatism that were building the continents towards the end of the Archaean. TTGs were most common prior to 3 billion years ago. However, shortly thereafter granitoids with a composition similar to those produced by modern subduction began to overtake the TTGs in abundance.

From the looks of it, the modern granites appeared to form from the melting of pre-existing, early Archaean granitoids. This places their formation in much the same context as modern-day settings with their synthesis occurring above subduction zones. Therefore, we conclude that Earth's granite crust evolved over time, from rocks produced from the direct melting of eclogite, through to ones produced in the mantle wedge—and finally to granites synthesized from the partial melting of pre-existing crustal rocks. These crustal rocks include subduction-eroded continental crust and sedimentary material, eroded from their surfaces. This has important consequences for the habitability of any telluric planet. Figure 3.4 shows changes in the ways in which continental crust forms and changes its composition. The crust becomes thicker and less buoyant. Over time, the crust can emerge from the ocean depths. This is explored in Chaps. 5, 7, 8 and 9.

There's a fundamentally important point to make at this stage—the impact of global tectonics on habitability. Although it is apparent that plate tectonics plays a fundamental role in driving the climate-stabilizing carbonate-silicate cycle on the modern Earth, this was not operational in the Hadean and possibly little of the Archaean. Therefore, irrespective of any theoretical underpinnings, plate tectonics is not essential for the generation and long-term maintenance of habitability. Earth managed to produce living organisms and maintain them for hundreds of millions of years before plate tectonics began. What is missing from this argument is the nature of the surface of Earth, or indeed any planet. As we'll see shortly, the modern

Earth is a very poor analog for its earlier state—and it's this that should form the foundations of our considerations of what makes a planet habitable.

During the geological transition to the modern regime, the planet continued to make its granite heartlands through the melting of eclogite—the metamorphosed remains of oceanic crust. However, from the latest Archaean and Proterozoic eras onwards, the lithosphere was then thick and rigid enough to undergo the kinds of horizontal and vertical movements that characterize modern plate tectonics. It is here, from 3.2 billion years ago, that ophiolites emerge in the geological record. By the time we enter the Proterozoic, the mantle had cooled below the point at which eclogite melted easily. Now, most of the magmas, which were produced by subduction, came from the release of water into the overlying mantle wedge. At this point the planet stopped making abundant TTGs and swapped to more conventional granodiorites and granites. This process continues to dominate today and will do so until plate tectonics ceases at some point in the next few billion years. During this transition, ophiolites emerge, signaling continent-continent collisions and the trapping of slivers of oceanic crust as subduction is crushed out.

Bear in mind, the depth of the oceans has likely decreased since the Archaean Chap. 5. This means that over time, the abundance of many sedimentary rocks would also alter as weathering and erosion became more abundant.

In keeping with the idea that tectonism has been an evolving system on Earth (and likely in all terrestrial planets), Craig O'Neill (Macquarie University) and colleagues have modeled the effect of declining heat production in terrestrial planets. Their models suggest that all planets begin with an internal set up similar to Io (Fig. 3.5). Here, planets have such vigorous internal heating that the mantle sheds heat preferentially through heat pipes. This gradually evolves into a plate tectonic regime as the mantle becomes more viscous and stresses are conducted more fully

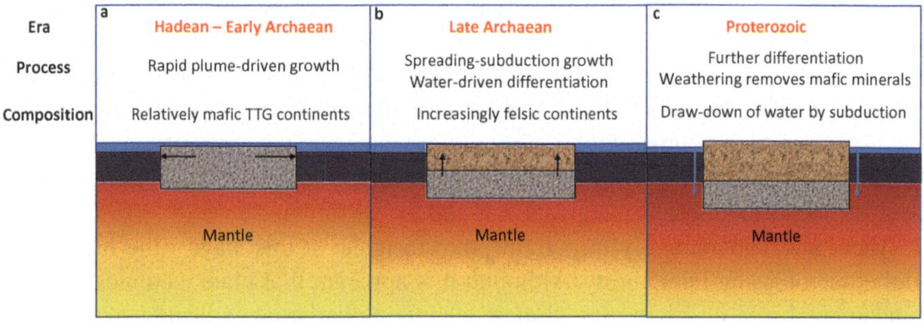

Fig. 3.4 Integrated model for the emergence of continents on Earth-like planets. When the planet is young and hot, continental crust forms rapidly, but this is dense, and the oceans are deep (**a**). Most is submerged. As the planet cools, water influx in early subduction drives the formation of less dense, felsic crust. This begins to emerge from the oceans (**b**). Finally, in **c**, subduction is drawing water back into the mantle, while weathering removes iron and magnesium from exposed crust, lowering its density further. Continued erosion and recycling of continental crust also has this effect

to the thickening lithosphere. For small terrestrial planets, such as Earth, this system can operate for several billion years—perhaps as long as 10–20 billion—until the lithosphere becomes so thick and strong that mantle convection is insufficient to fracture it. Plate tectonics ceases, and the planet evolves into a state of episodic tectonism (see below), until this also fails.

Other planets will fail to operate or maintain plate tectonics if the mantle becomes too viscous or the lithosphere too thick and rigid. Initially, these planets have a convecting mantle and operate regional tectonism (perhaps) like that on Venus. As the mantle cools and the lithosphere thickens (to more than 100 km thick) mantle convection is unable to rupture the lithosphere, and volcanism is confined to regions where the lithosphere delaminates.

On Earth, the principal changes in the pattern of tectonism were driven by the secular cooling of its interior. Granitoid rocks, thus, form a chemical record of these changes that are related to the aging of our world. As our exploration of other worlds continues, we are learning how our planet's behavior fits into the broader zoo of tectonism that we observe elsewhere. Cooling allowed the lithosphere to become rigid enough to support long-range, horizontal movement.

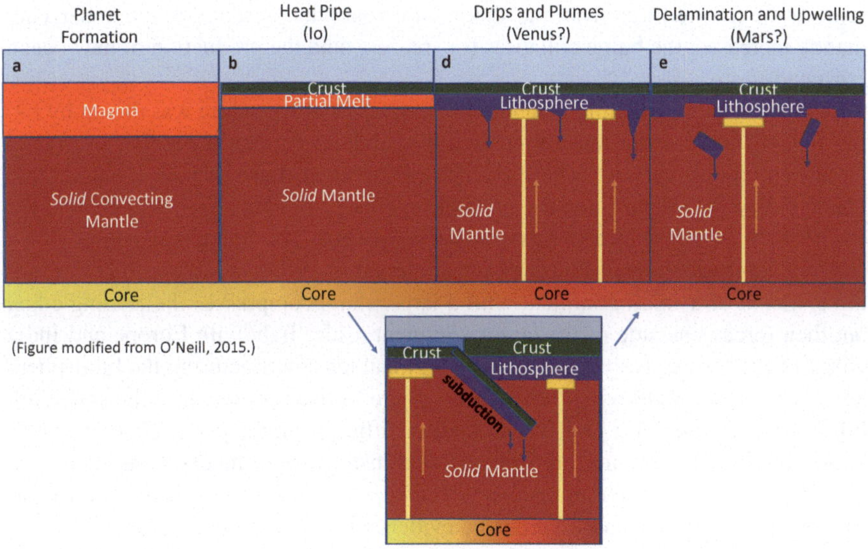

Fig. 3.5 Possible stages in the evolution of telluric planets. After solidification of a magma ocean (**a**), planets may retain a region of partial melt for sometime. This is a hot stagnant lid regime (**b**). If there is sufficient water or other volatiles in this layer and on the surface, planets may evolve to states with plate tectonics (**c**). Alternatively, planets may retain a stable, stagnant lid, with regional tectonism (**d**) then continue to cool to form first a thick Mars-like stagnant lid (**e**) with thick lithosphere, then frigid, final, Moon-like state

Modern Tectonism on Earth

Driving Forces

Modern plate tectonics is driven by subduction, which in turn is driven by density differences between the rigid lithosphere and the underlying hot and soft asthenosphere. Common misconceptions that are frequently reiterated in books and websites are that the mantle convects, with hotter, lower density material rising under ridges, spreading outwards and sinking back into the mantle at subduction zones. Although the principle of this motion is correct, it implies it is the rising of hot material that drives the process. Rather, it is sinking of colder, denser material at subduction zones that pulls the plates along and down (where appropriate) into the mantle. Hotter, low density material rises (passively) upwards, wherever the weight is relieved.

If subduction pulls on the crust, the crust undergoes a brittle fracturing nearer the surface, and a steadier, ductile flow at greater, hotter depths. As the crust is stretched (and with it the underlying lithosphere) the underlying mantle bulges upwards, and the process of continental rifting (or thinning of oceanic lithosphere) occurs. Then the mantle can partly melt, and a new ocean ridge (and basin) is formed. If you remove the tensile force from the subduction zone, the pressure on the underlying mantle increases; the bulge collapses (or freezes), and the production of new ocean crust ends.

The proof of the pudding is in the tasting. If crustal spreading was driven by rising hot mantle and not tension created by subduction, then a number of crustal features wouldn't be seen. These include rifting along Lake Baikal in Russia, the Rhine Valley in Germany and the broader region of stretching and volcanism in the Basin and Range Province in western North America. The Baikal and Rhine Valley rifts are manifestations of the same phenomenon on different scales. In both cases the collision of a small continent with a larger mass compresses the leading edge, but then forces one side of the larger continent aside (Italy with Europe and India with Eurasia, respectively). This creates regional tension, fracturing the lithosphere and allowing the underlying, hot asthenosphere to rise upwards into the spreading lithosphere. If you don't look at this kind of rifting from the perspective of what's happening in the overlying lithosphere—and instead try to invoke mantle plumes, you hit a wall. Although there is ample evidence of hot asthenosphere at relatively shallow depths in each case, there is no evidence for an underlying plume.

Likewise, under the Basin and Range there are no mantle plumes, yet the crust is rifting on a large scale across a region hundreds of kilometers wide. The driving force is tension created by the Pacific plate pulling on the west coast of North America. The largest force is exerted along the San Andreas Fault, but some is transmitted across the Basin and Range, clearly stretching the crust and giving the west coast that characteristic bulge that stretches from Mexico to Oregon. California will soon break off and pull away towards Alaska, and this will likely be followed by other terrains, as North America continues to move west against the Pacific plate.

Finally, back-arc spreading makes no sense at all if the spreading is driven by hot mantle rising from below. Back-arc spreading refers to a situation that is commonly observed where the overriding plate is moving slowly relative to the subducting one. As per normal, the subduction synthesizes a line of volcanic islands (an arc) at around 150 km from the subduction zone. However, in some instances lying behind the volcanic arc one finds an area of young oceanic crust and associated basaltic volcanism. Such volcanism originates in a narrow web of fractures that only loosely resemble mid-ocean ridges. New crust and lithosphere is created as the subduction zone tries to pull away from the overriding plate (roll-back). Rather than the compression one might expect at a region of plate collision, instead one sees strong tensional forces. Tension is associated with a subduction zone that is "rolling backwards" and pulling and stretching the overriding lithosphere with it.

To understand this it's important to realize what is happening with subduction. If you take a snapshot of subduction with tomography, you get a slab of colder rock (the oceanic plate) diagonally descending into the mantle. However, the impression you get of the processes is not what is really happening. Instead of a diagonal slide into the mantle, the oceanic plate descends vertically, after a relatively narrow bend at the subduction zone. The overriding plate is moving over it, so that the areas furthest from the subduction zone have descended the furthest into the mantle. Although the entire plate is moving vertically downwards into the deeper mantle, the area closest to the point of subduction has moved the least far, as it has had the least time. This gives rise to a diagonal slab, even though it is moving vertically, not diagonally!

If the overriding plate is moving forward more slowly compared to the downward velocity of the slab, then the subduction zone "rolls-backwards," or retreats from the continental (or other) margin. On Earth, this is likeliest where the ocean lithosphere is coldest, thickest and densest and where the continent is moving most sluggishly over the oceanic plate. These conditions are found along the eastern margin of Eurasia and along the coast of Sumatra, with some interesting effects. In the eastern Pacific roll-back pulled the Japanese islands from the east of China, Korea and Russia; Sumatra and the Andaman islands to its north moved west over the northern Indian Ocean and Bay of Bengal.

Of particular note is the stress-relieving effect of large earthquakes in driving this motion. Analysis of GPS data by Kelin Wang and Jiangheng He (both at the Geological Survey of Canada), together with Yan Hu (University of British Columbia) showed that Sumatra and the northern islands moved approximately 20 cm forward across the Bay of Bengal, following the 2004 Boxing Day quake. Similarly, after the massive 2011 Japanese earthquake, Japan lurched to the east, over the Pacific Basin. The southern hinge point for the movement of Sumatra and the Andaman islands is in the Sunda Strait, where the relative motion of Sumatra and Java is rifting this portion of Indonesia. Into this void come hot mantle and a copious supply of liquids and granitoid melt. The surface expression is Krakatau, one of Earth's most explosive volcanoes.

Therefore, the surface expression of subduction is more complex than would be apparent if the oceanic plate was sliding at an angle under the continent; and the consequences can be global in extent.

Mantle Plumes and Hot Spots

Although a hot, convecting mantle is not a key driving force in surface tectonism, it is critical in two ways that promote plate tectonics. Firstly, a hot mantle makes its uppermost layer soft and malleable. This allows the overlying lithosphere to slide around. Secondly, the mantle needs to shed heat, and there appear to be two different kinds of plume that act on different scales (Fig. 3.6). The smaller plumes we call hot-spots, while larger plumes are known as large low shear velocity provinces (LLSVPs). These will be examined later in this chapter.

Hot spots are the surface manifestations of hot, narrow columns of rock that rise from the core-mantle boundary towards the surface. In many cases, seismic surveying (tomography) reveals the hot conduit extending all the way to the core-mantle boundary. In others the conduit is traceable to the base of the upper mantle and transition zone at around 400–660 km down. These narrow plumes of hotter rock typically cause the overlying lithosphere to bulge and rift, allowing small clusters of volcanic centers to form. Typically, as the overriding plate moves, the plume leaves a track of volcanoes known as a "hot-spot track." These volcanos age as you move away from the plume, with volcanism usually confined to the youngest centers that are less than 5 million years old. Under Hawaii, the lithosphere is domed upwards for a few hundred kilometers on either side of the island, with the youngest volcanic center emerging on the southeastern flank of the Big Island.

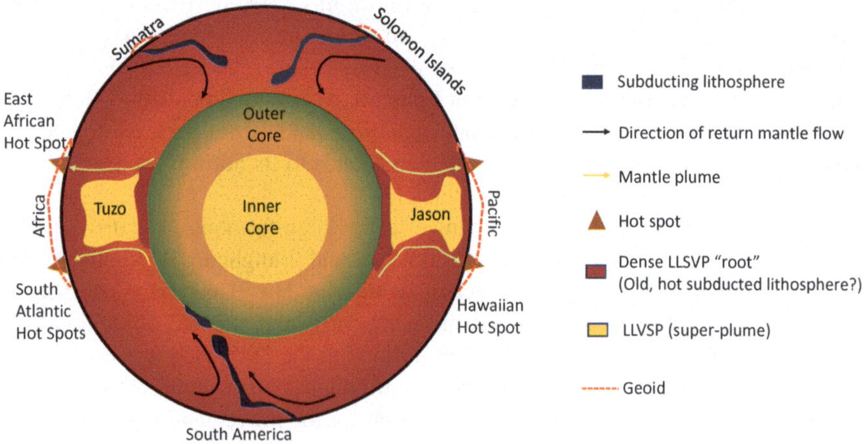

Fig. 3.6 The "Burkian" Earth. In Kevin Burke and colleagues' model of Earth, the geoid (red dashed line) follows the regions where there is greatest concentration of mass in the mantle. These are "Jason" and "Tuzo"—two broad LLSVPs that underlie the Pacific and Africa, respectively. Hot spots (brown triangles) are underlain by narrow plumes (yellow arrows) that transfer more fluid mantle and heat from the core-mantle boundary (CMB). The LLSVPs (light orange) are underlain at the CMB by denser than average material (maroon) subducted in previous epochs. Each LLSVP is a broad area of upwelling that transports heat from the hot, dense underlying area and core

Where the hot spot meets a continent, for example under Yellowstone, the surface expression is different. The Yellowstone plume is traceable along the Snake River Plain, with the oldest centers dated at around 13 million years of age, at the western end of the plain. The Yellowstone plumes arrival at around 17 million years ago are associated with the eruption of the Columbia Plateau basalts. These were erupted into the back-arc setting along the Oregon-Washington State border. However, as the plume moved away from the stretching back-arc environment, the supply of basalts partly dried up.

This didn't mean that volcanism ended; rather the heat plus partial melting under the continent drove sufficient basalts into the granitoid rock to melt it. The end result was a succession of highly explosive, caldera-forming eruptions that culminated with the formation of the current Yellowstone caldera in Idaho. Volcanism has followed a predictable pattern. Intrusion of basalt partly melts the continental granitoid, forming large underground granite plutons. The chemistry of these differs slightly from many other granitoid eruptions, with magmas containing higher than normal volumes of the element fluorine. When this magma is voluminous and buoyant enough to reach the surface a caldera is formed, with the eruption of large volumes of felsic rock known as ignimbrite.

As the supply of granite subsides and the amount of volatile substances wanes, explosive eruptions give way to rhyolite lavas. Finally, once the majority of the buoyant granitoids are erupted, basaltic magmas escape forming a veneer of darker basalts. This pattern is traceable along the entire Snake River Plain; and though Yellowstone grabs lots of headlines in the media, most likely, its time is pretty much over. The caldera has erupted explosively twice in the last 1.2 million years, and now the majority of recent volcanism is confined to rhyolite domes and an increasing number of basaltic lavas. One expects that while Yellowstone might vent its remaining magma chamber in another cataclysmic event, one should be looking to its east for the next center.

An important knock-on effect of the Yellowstone plume is the destruction of the western Rockies. Perhaps because the plume has removed the roots of the mountains along its path, the mountains are (obviously) absent in a track that runs northeastward from the Pacific. The Snake River Plain forms a natural conduit for moisture-laden air to migrate in from the Pacific Ocean. By "burning a track" across western North America, the Yellowstone hot spot has grossly modified the climate of the central northern Rockies. Far more moisture reaches inland, giving rise to a regional hot-spot in biodiversity. Such impacts will be of critical importance when we come to look at the concept of habitability in Chap. 7.

Large-Scale Impacts of Mantle Plumes

A far more personal note concerns the Icelandic plume. Although this is very similar in nature to the ones underlying Yellowstone or Hawaii, its impact on the author's country of origin is profound—and its effects clearly have implications for exoplanets.

Like other narrow plumes, the Icelandic plume is clearly definable with seismic tomography to a depth of 400 km. However, beneath this it seems to wink out temporarily as we descend towards the core. Raffaella Montelli (then at Princeton) and co-workers show it re-emerging in s-wave images, which extend below 1000 km beneath the surface and towards the core-mantle boundary. This suggests that it is very narrow—and possibly discontinuous. In turn this implies that the material rising in the plume has an unusually low viscosity as it seeps through the more viscous, surrounding deep mantle rock.

Although there is nothing particularly remarkable here, in 2017 Charlotte M. Schoonman (University of Cambridge) and colleagues implied that this fluidity has implications for a disproportionately large area of Earth's surface. In their work, they convincingly demonstrate that very fluid (and presumably anomalously hot) mantle material flows upwards under Iceland. This hot material then flows outwards in narrow fingers, which extend towards and partly under Norway and Scotland.

There is clearly no vulcanicity in Scotland more recent than the early Tertiary, but the presence of hot underlying mantle means that Scotland (and Norway) rides high in the mantle. Not only does this give both countries mountains, but it also alters the flow of air across the Atlantic Basin and into Europe. It is these kinds of effects that need to be considered when one explores habitability of planetary surfaces. Moreover, it appears that the arrival of plumes is critical in dispersing continental fragments. By lifting the continent above the plume, tensional forces can act across the dome. If there are neighboring areas where subduction is pulling on the continental margins, the continent is likely to split apart. The absence of plumes appears to make fragmentation more challenging, as the balance of forces is insufficient to fracture the 30- to 40-km thickness of continental crust.

In this regard, detailed seismic tomography has illustrated a very interesting situation under Africa, which is reminiscent of the Icelandic plume's structure. Recent work by G. C. Begg (Macquarie University) and colleagues suggests a complex picture that explains the distribution of volcanism in the central Sahara (underlying the Hoggar Massif)—and the sluggish motion of the African continent as a whole. To understand both, we need to examine the deep roots of Africa, one of Earth's youngest continents.

Underlying the Danikal Depression in the Afar Triangle of eastern Africa is a hot spot. Its arrival, some 30 million years ago, orchestrated the rifting of three basins: the Red Sea; the Gulf of Aden and the East African Rift Valley. In a classic hot-spot scenario, magma from the hot spot enters the rift network and fuels volcanism in the Afar Triangle—and perhaps in other nearby locations along the rift systems. However, sensitive tomography by Begg and co-workers reveals a far more complex and intriguing situation. The mantle underlying Africa resembles the mouth of a dentist's dream patient. Deep crustal roots extend beneath most, but not all, of the ancient Archaean shields. Such roots represent a complex mixture of frozen mantle and eclogite that were mostly synthesized in the Archean. In general, the older the lithosphere, the deeper the continental roots extend.

Much of Africa consists of stable Archaean craton, but the lithosphere underlying the eastern Sahara is not. This is known as a metacraton—a region that has characteristics intermediate between a stable continent (shield or craton) and more mobile regions that are typical of a young, active lithosphere. The ancient Archaean shields are, with few exceptions, built on very deep (150–300 km thick) roots.[1] However, the Saharan metacraton lacks such a coherent root, while the Tuareg, West African and Congo shields, which surround it, all have these roots that extend to more than 300 km beneath the surface. The Saharan metacraton has, instead, a number of thin roots that do extend to more than 200 km in depth in places. This is consistent with the metacraton being composed of a number of far smaller subsections that were stitched together during Proterozoic collisions.

Of these deeply rooted cratons, the root of the Congo is peculiar. Rather than dangling down into the mantle directly beneath central Africa, it sprawls 300 km sideways under the North Atlantic at a depth of 150–325 km. In keeping with this complex "seismic X-ray," Hoggar's volcanism emerges as something between the effects of a classic hot spot and those of a fractured geology caused by continental collision. In essence, there is no classic mantle plume rising under Hoggar. However, Begg's work shows that Hoggar is still supplied more distantly by one that underlies the East African Rift. Africa is underlain at depth by a large mantle uprising called a super-swell or super-plume, which appears to extend down to the core-mantle boundary (discussed later in this chapter). Along the edge of this super-swell—and under the Gulf of Aden and Ethiopia—lies the mantle plume (Figs. 3.6 and 3.7).

What Begg and co-workers show is how this broader area of uprising mantle is focused by the deep mantle roots of the various cratons that make up Africa. Initially, hot mantle rises in a relatively narrow zone under the East African Rift. As the hotter mantle reaches a depth of 325 km or so, it begins to encounter the deeper portions of the shield roots. These form rigid appendages within the hot plastic flow, which is then forced to flow around them towards the base of the crust. Tomography clearly reveals these narrowing fingers of hot material fanning out under northern Africa, much as the Icelandic plume appears to fan out under the North Atlantic Basin. One of these fingers percolates the upper mantle under Hoggar. This stream follows a zone of weakness that was reactivated by the Alpine orogeny (mountain building) to its north and reaches depths that are shallow enough to partly melt magma then rise under Hoggar, forming the recent volcanoes. These studies explain why there is (or was) volcanism at Hoggar, but no deep-seated hot spot to supply it!

It's worth considering another peculiarity of the African continent—the anomalously slow speed of its movement. In general the velocity of a plate can be related to its age. The oldest continental crust moves the slowest, and it seems likely this comes down to its roots. Rather than imagining ridges and underlying hot asthenosphere flowing outwards and pushing continents around, consider the continent as sailboat, with the keel analogous to the continental root. Archaean continental crust has the deepest and most extensive roots. As subduction zones pull the surface plates around, the continents move, too. However, if you imagine the upper mantle

[1] India is the best example. It appears to have no root, despite an age that extends well into the Archaean. India may have lost its root relatively recently when a hot-spot (Kergulen) partly melted, then delaminated, the micro-continent's ancient base.

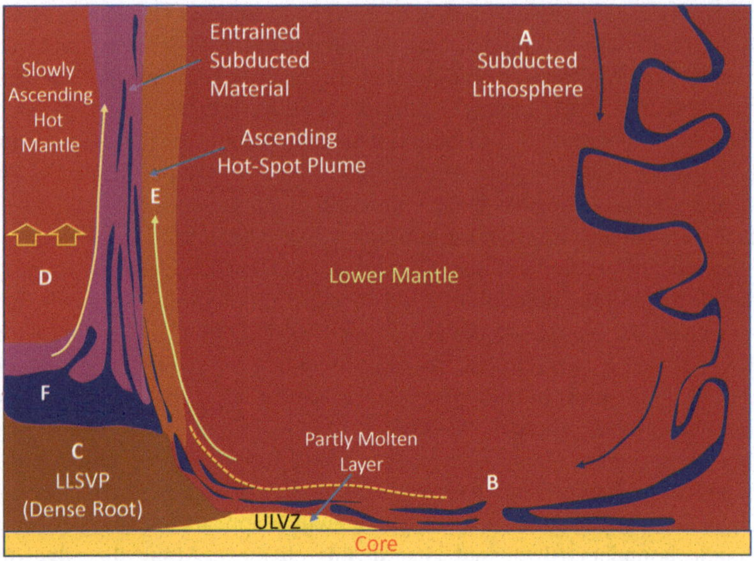

Fig. 3.7 Circulation in the lower mantle. (**a**) Subducted lithosphere cools the mantle and underlying core. (**b**) Material flows over the core surface and heats up. (**c**) The base of LLSVPs consist of much older, denser and possibly iron-rich material pooled over the core. These trap heat from the core. (**d**) Eventually, mantle overlying the LLSVP heats up and ascends, carrying subducted material which has pooled on top of the LLSVP's dense structure. (**e**) Hot spot plumes lie along the edges of the LLSVPs, where less dense mantle becomes hot and buoyant in contact with these structures. (**f**) Active plumes erode basaltic (eclogite) material from the lid of the LLSVP root. (Modified from Torsvik, et al, 2016)

(the top 400 km of viscous rock) as a treacle-like ocean, then the larger and more complex the structure of the root, the trickier it will be for the continent to "sail" through the hot, viscous sea of mantle rock.

The odd, wizened structure of Africa's roots does not appear to lend itself to the continent's speedy movement. This is lucky for us, as it allowed humankind and its primate ancestors a long time to evolve in a fortunately complex environment. Were it not for Africa's deep roots, our primate ancestors might have found themselves carted into the cooler temperate region. Had this happened 20 million years ago, humanity might never have evolved. Begg and co-workers' research clearly underscores the need to understand the deep structure of continents if we are then to make sense of the activity at the surface. In turn, this affects how we relate the action of plate tectonics to evolutionary biology. Planets with lithospheres that lack such horizontal movements may have abundant life—but they may not have the same rate of evolutionary progression, driven by the steady dispersal and coalescence of land-masses.

Plate Tectonics, Plumes and True Polar Wander

Although subduction provides the downward conduit for cold, dense mantle material, it appears that the return leg is a little more complex. Hot-spots have far too little mass, but instead a more complex system involving two-broad upwellings appears to operate on Earth. Such super-plumes are thousands of kilometers across and provide a curious link between the terrestrial mantle and those of smaller, tectonically inactive worlds, such as Mars or the Moon.

Plate tectonics shuffles continental crust on the order of 1–5 cm per year, meaning that on Earth it takes at between 22 and 110 million years to traverse 10° of longitude (a little over 1100 km at the equator). This leisurely process allows life to adapt to a slowly and fairly predictably changing climate, and remains the principle means through which continental crust migrates relative to the mantle. However, there are more abrupt phases where continents appear to shift over several degrees of longitude in a few million years. Such episodes of true polar wander occur when the lithosphere re-orientates the planet's surface and spin axis in response to the mass movement of matter in the mantle and on the surface. This is considerably faster than plate tectonic motion, but it is by no means catastrophic. However, in doing so it can redistribute millions of species in a manner that drives faster speciation, by changing the quality of the niches they occupy (Chap. 7).

The terrestrial spin axis wobbles periodically. Plate tectonics shuffles large masses of continental lithosphere in directions that are dictated by the location of subduction zones and plumes. If plate tectonic movements pile continental crust near the mid-latitudes then there will be an excess of mass in one location relative to the spin orientation of the planet. Imagine placing a lump of clay on a spinning wheel. The extra mass produces a torque on the spin. Simultaneously, the Sun and Moon pull on the rotating mass, tending to orientate the spin of all of the bodies in one plane.

However, deep within the mantle are further large inequalities in the distribution of mass. Underlying the Pacific Ocean and Africa are two super-plumes—large low shear velocity provinces (LLSVPs), which we briefly mentioned earlier. These regions are characterized by relatively low density—and presumably hotter than average—material lying on top of denser than average roots. These low-density regions extend 1000 km above the core-mantle boundary. This was the seemingly surprising observation of Harriett Lau (Harvard) and co-workers. One says "surprising" because the general structure was suggested by other researchers some years before. However, though the results from this work are a little prosaic, the method employed by Lau and co-workers was very innovative indeed. This used Earth's body tide—the displacement of the solid surface of Earth in response to the gravitational pull from the Sun and Moon. Such tides are sensitive to density differences in large volumes of mantle rock.

In earlier work Kevin Burke (University of Houston), Trond H. Torsvik (Geological Survey of Norway) and others have proposed that these LLSVPs have been stable for the last few hundred million years. Over the course of the Phanerozoic

Era, these have remained near the equator, but on opposite sides of the planet. Meanwhile, continental crust has moved progressively northwards from a location mostly near the South Pole, around three-quarters of a billion years ago, to an equatorial position 400 million years later, before finally moving to its current orientation, with most of the continental lithosphere lying to the north of the equator. In that time oceans have opened and closed in the current location of the Atlantic, but remained open on the opposing hemisphere, either as basins as wide as the Pacific or much wider, as Panthalassa.

When the motion of the continents is examined in greater detail, it's apparent that when the continents were assembled into Pangaea between 350 and 150 million years ago a significant amount of clockwise and anticlockwise rotation occurred. From 250 to 200 million years ago Pangaea wobbled anti-clockwise, then clockwise from 200 to 150 million years ago. This process repeated, but with smaller motions from 150 to 100 million years ago (anticlockwise from 150 to 140 million years ago, then backwards in the ensuing 40-million-year interval). Motion occurs around a pivot point where the center of mass of Pangaea (or its dispersing descendants) is located. Each motion attempts to redistribute the continental mass relative to the spin-axis of Earth. In turn, this is stabilized by the LLSVPs in the lower mantle and the gravitational pulls of the Moon and Sun. When the continents move one way, the lithosphere then slips backwards, to re-orientate the drifting mass. The backwards and forwards motion represents attempts by the spinning Earth to compensate for the movement of the continents at the surface.

The LLSVP cores appear to consist of dense blobs of material (Figs. 3.7 and 3.8), the origin of which may date to the solidification of Earth's magma ocean in the earliest Hadean era. A similar proposition has been made for an analogous structure underlying the Procellarum Basin on the Moon (described in detail below).

If you want to make your own model of these structures, take a hot bowl of chicken-noodle soup. The dried packet variety works the best. If you leave this to settle, the noodles and chicken/soya fall to the bottom, forming a core. The starchy suspension that forms the bulk of the mix acts as a mantle, while an upper, clear liquid layer forms our lithosphere. Soon, the mixture forms broader upwellings that are flanked on all sides by clear downwellings of fluid. These are our subduction zones. Between the zones of noodle subduction lie conglomerations of chicken and noodles at the bottom of the bowl. These are the cores of the LLSVPs and like them are usually hotter than the surrounding liquid. Now, drink the soup.

Returning to the mantle, there are two possible explanations for the persistent plume activity on Earth—and perhaps elsewhere in the Solar System and beyond. In the first scenario, when the magma ocean solidified, dense residues—mostly olivine crystals—fell from the lid of the newly formed crust and slumped onto the core. Although pressure would transform this into bridgmanite, the compositional difference and higher density compared to the surrounding mantle rock would allow it to persist. Unable to descend further because of the high density of the core, this material stayed at the core-mantle boundary as semi-solid mush. Such material would be relatively rich in radiogenic material, thus able to supply heat—but also insulate the

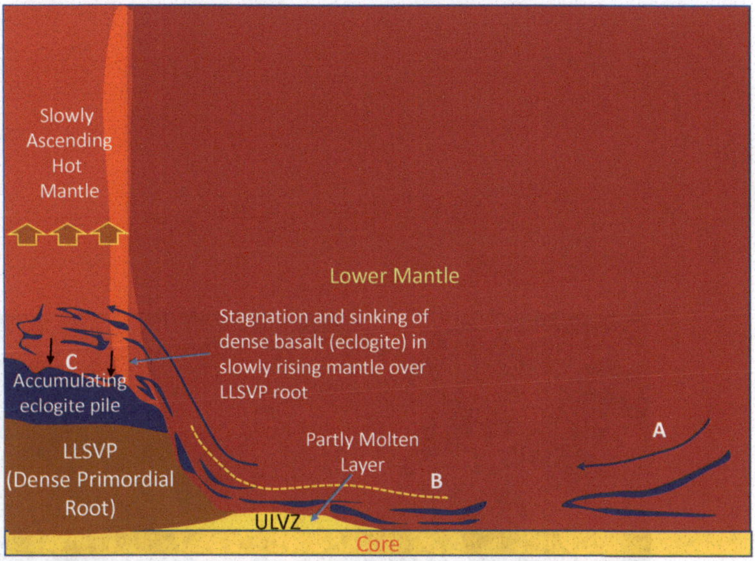

Fig. 3.8 Passive regions. Between plumes, there is slow accumulation of material above LLSVPs. There is a drizzle of subducted basaltic (ecolgite) material (**a**). This flows over the base of the mantle towards the LLSVP (**b**), before piling up on top of the dense LLSVP root (**c**). Hot, but slowly ascending mantle over the LLSVP root fails to entrain any of this material. (Modified from Torsvik, et al, 2016)

core, trapping heat beneath it. Descending slabs of material—either subducted or delaminated—would then flow over the top of these dense piles of material.

Basaltic material (effectively eclogite) could pile up on top of the olivine-sourced material, forming temporary reservoirs to supply subsequent mantle plumes. Whenever the LLSVP root became hot enough—and the supply of subducted material was great enough—a plume would launch from around the LLSVP rim. In this model's support Kevin Burke and colleagues note that the majority of hot spots lie around the edges of the LLSVPs, rather than within them, as one might presume if the LLSVP was merely a hotter region of mantle.

In the alternative scenario, LLSVPs are relatively transient phenomena that are consequences of subduction, piling material onto the core rather than onto and around primordial material. Both scenarios are viable, but the former seems a more holistic, best fit for seismic data and mantle convection models. These imply that subduction places material around the LLSVPs in very thin (50-km-thick) sheets, and perhaps as piles on top of the primordial "root." There is some evidence that the terrestrial features may be over 1.5 billion years old, implying that the LLSVPs date back to the earliest periods of subduction on Earth—and possibly predating it (Fig. 3.9).

Irrespective of the relationship between hot-spot plumes and the larger LLSVPs, their presence appears to be a common component of mantle convection in the Solar System's other telluric bodies. We examine these next, along with their impact on their host.

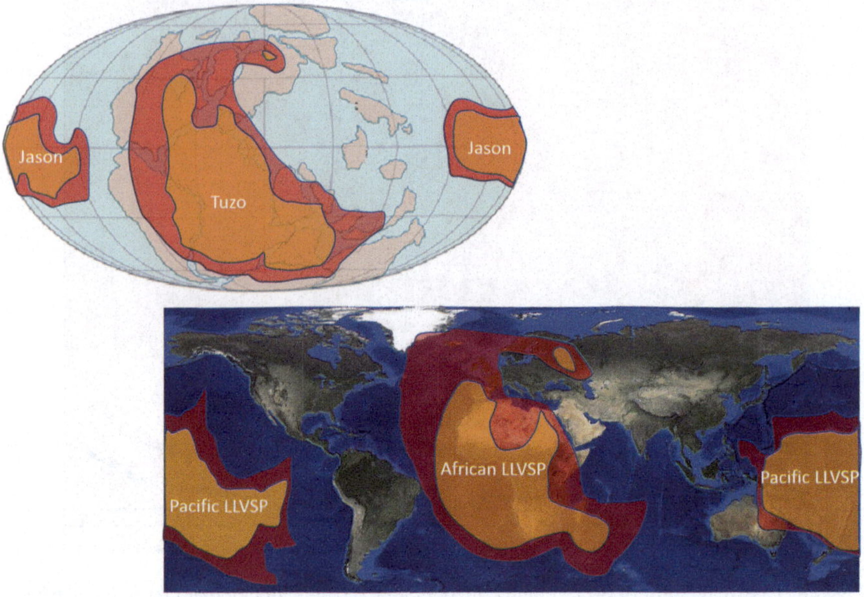

Fig. 3.9 The persistence of LLSVPs—mantle super-plumes. Evidence suggests that as with Mars and the Moon, Earth has retained broad mantle upwellings for at least 500 million years—and possibly a few billion years. These are separated by zones of subduction. At the end of the Permian one super-plume, nicknamed Tuzo, was located under Gondwanaland, with the second, Jason, found under Panthalassa. Since then, not much has changed. Tuzo still lies under Africa and the South Atlantic and Jason under the Pacific. Only the continents have danced over these. The equatorial position of the LLSVPs stabilizes the Earth's spin axis. (Satellite image: NASA)

TPW Elsewhere in the Solar System

The occurrence of TPW is a natural outcome of various geological processes wherever the surface and underlying mantle are dynamic. Once an object's lithosphere incorporates a significant amount of the underlying mantle and becomes too rigid to move TPW becomes irrelevant; no further motion is possible. As TPW has significant implications for long-term habitability and evolutionary processes, we need to understand how it occurs on other objects—particularly if it is possible in ones that lack plate tectonics. In this regard can we identify the footprint of true polar wander (TPW) elsewhere in the Solar System, and if we can, what impact has it had on these other celestial bodies? Although Venus remains shrouded in cloud (and an equivalent amount of mystery), the Moon, Mercury and Mars have surfaces that we can map and, therefore, observe potential signatures of TPW.

If we turn our attention to the Moon first, there are a number of clues that suggest that similar drivers to those of terrestrial TPW have affected this globe. These influences seem to be focused on (and underneath) the Moon's Procellarum Basin. This basin marks the site of the most recent volcanism on the Moon and forms the core of the large dark region we refer to the "Man on the Moon." Matt Siegler (Planetary

Science Institute) and colleagues used neutron spectroscopy to detect hydrogen on the lunar surface. Hydrogen is used as a proxy for water, simply because free hydrogen would not be stable and would readily escape to space. However, water, either as ice or as hydroxyl (OH), in minerals is suitably dense and adhesive enough to allow it to persist wherever the surface is cold enough and in shadow. As only the polar regions will have suitable conditions for this to be applicable, the distribution of hydrogen serves as a useful proxy for ice and hence the location of the rotation poles.

The lunar poles do indeed have hydrogen, indicative of ice. However, so do regions that lie on opposite sides of the Moon, but 5.5° south or north of the respective pole. Although there was hydrogen, indicative of near-surface and surface ice in shaded, polar craters, there was also evidence of ice at 84.9° N, 147.9° E and at 84.1° S, 309.4° E. That these deposits are on opposing sides of the Moon strongly suggests that they are the former rotation poles of the body—and therefore, the Moon has undergone TPW when it was still young and hot inside.

The most convincing explanation for lunar TPW involves a degree-1 plume—a singular plume that underlies the Procellarum Basin. In 1999 Shijie Zhong (then at MIT) showed, with others, that such a single, large plume could account for the distribution of highland and basin regions on the Moon. In this model, the origin of the dichotomy is related to how the first lunar crust formed. Here, the crust is primarily a low density rock called anorthosite. Formation of this rock requires the loss of denser minerals, such as pyroxene, olivine and ilmenite. Therefore, when the anorthosite layer forms, the other, denser minerals are left at its base. This is unstable, and these eventually fall off into the mantle, accumulating above the core, a "cumulate."

Zhong and co-workers convincingly demonstrate that this material then leads to the formation of another mantle plume. However, in this instance, it is the chemistry of the dense cumulates that is enriched in radiogenic material that preferentially heats the regions where the greatest concentrations are found. This leads to the formation of a buoyant plume, which now underlies the Procellarum Basin. This plume, while frozen today, supplied the basin with the mare basalts. High levels of radiogenic material in this basin support the conclusion that the material that formed it was unusually radioactive and capable of generating enough heat to form the basalts. In turn, prolonged volcanism bequeathed the Moon with a reasonably dense atmosphere for the best part of 1.5 billion years. Now, that would have been a sight.

In terms of TPW, Matt Siegler and co-workers convincingly show that this singular plume can explain the reorientation of the lunar lithosphere. In Siegler and colleagues' model the lithosphere gradually re-orientates to its current position as the super-plume rises under what is now the Procellarum Basin. It is the ascent of this buoyant, radiogenic and hot material that causes sufficient redistribution of mass in the mantle to re-orientate the lithosphere. As a result the old north and south poles migrate 5.5° towards the equator, taking their icy veneer with them.

The model assumes that the plume was active in the first billion years or so of the Moon's history and that subsequent to this the lithosphere and underlying mantle became too rigid to permit further reorientation. Thus a single lunar feature—the radiogenic Procellarum mantle and lithosphere—is sufficient to explain a significant number of lunar features, such as the persistent nature of magmatism on the Earth-facing hemisphere, the presence of grabens (rift-features) around the basin, and the reorientation of the lunar surface. In turn, the buoyancy of the plume comes down to the presence of radiogenic material—also known as KREEP (K, potassium; REE, Rare Earth Element; and P, Phosphorous); and this KREEP derives from the earliest history of the Moon, when the first crust was forming.

Now, consider Mars. Like the Moon, this has a crustal dichotomy with a large concentration of thick highland crust in its southern hemisphere and lowland crust to its north. This dichotomy has been variously explained by gargantuan impacts or through the action of internal forces. Although one can successfully model the Martian dichotomy in terms of such an impact, that Earth, the Moon and Mars all show dichotomies seems to require a different explanation and one that is common to each. This is where internal convection comes to the fore.

Although the Martian highlands are largely basaltic in nature—while Earth's "highland continents" are granitoid—what is now apparent is how internal convection determines the relative position of each highland region. Two sets of studies seem to show how the Martian surface has re-orientated relative to its core in response to long-lasting convection. In the first study, Shijie Zhong worked with Ondřej Šrámek (both from the University of Colorado). Like the earlier lunar simulation, computer simulations of Mars were used to demonstrate the effect of long-lasting, singular convection within the Martian mantle. These studies showed that such convection can remain stable over hundreds of millions of years, just as it did on the Moon and currently does on Earth.

Imagine one vast plume rising from the core-mantle boundary in the first 600 million years of Mars' history. Extensive partial melting produces a vast plateau near to the Martian south pole. As with the Procellarum Basin plume, this was subject to tidal forces, this time from the Sun (and to a lesser extent, Jupiter) and caused the lithosphere to rotate, carrying the mass towards the equator. As it moved magmatism continued to grow the thick crust above the plume, until eventually most of the southern hemisphere was capped in a vast region of thickened basaltic crust. Towards the tail end of this period (perhaps 3.8 billion years ago), magmatism becomes focused under part of the southern highlands. By 3.7 billion years ago, magmatism was constructing the massive Tharsis dome, over one edge of the highlands, but still well south of the equator.

In the second study, analysis of ancient drainage systems by Sylvian Bouley (Université Paris-Sud) and colleagues demonstrates that evidence from the highlands extending towards the northern basin suggest that the lithosphere underwent a final TPW event in response to this additional loading. This brought Tharsis north to its current location, overlapping the equator. The model neatly explains the south-north orientation of the volcanoes, with the youngest lying at the southern end of Tharsis, and it explains the orientation of valley networks that appear to originate in an area that would otherwise be expected to be relatively dry. By repositioning Tharsis and applying appropriate climate models, the valley systems now originate in a single, moist, southern tropical band, rather than in a waving band that moves into and out of the region of maximum precipitation. It also has Tharsis forming later than many models, with clear implications for the climate of the early planet.

When one considers TPW one might also naturally ask, why does the Man-on-the-Moon face Earth and not the other way—or indeed sideways on? Is this related to the mass distribution in (and on) the Moon, or is it simply random chance? In the Solar System, Mars has had its spin axis stabilized primarily through gravitational interactions with the Sun and Jupiter, while our Moon is orientated by Earth and Sun.

Although it is clear from Apollo measurements that the center of lunar mass is shifted towards Earth, why are the highlands primarily on the far side of our sister-world? If you produce an equipotential surface—a hypothetical smooth surface that is equidistant from the lunar center of mass, then the highlands stand proud on the far side, with the basalt-filled maria principally focused on the near side. However, this has nothing to do with the depths of the basins, as there are several deep basins on the far side that are not filled with maria. Rather, the critical feature is the depth of the basin relative to the position of the center of mass. Where the basins were excavated beneath the equipotential surface, basaltic magmas have filled them. On the far side, deep basins do not excavate to the same relative level and are, therefore, devoid of basalt (Fig. 3.10).

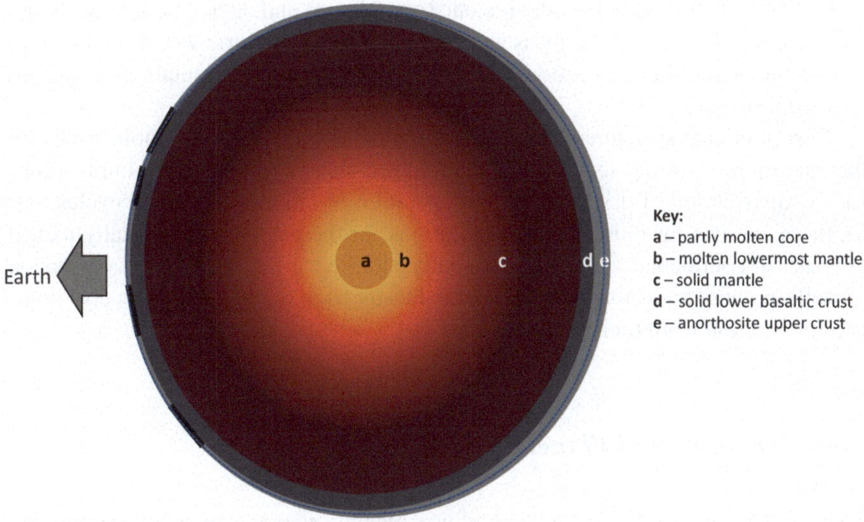

Fig. 3.10 Lunar cross-section. The small Moon is largely solid. Surveying suggests that there is likely to be a small partly molten core (1600–1700 K) less than 350 km across, of which the inner 240 km is solid (**a**); a thin, molten layer at the base of the mantle, 150 km wide (**b**); overlain by a largely solid mantle (**c**). In turn this is overlain by a thick crust, 70–150 km thick, which is largely made of the mineral plagioclase feldspar (anorthosite, **e**). The crust is thickest on the Moon's far side. This has led to the selective infilling of the basins on the Earth-facing hemisphere with basaltic lavas (black rectangular blocks). The lunar center of mass is shifted towards Earth from the lunar center by 2 km, leaving the equipotential surface (blue dashed line) on its Earth-facing side

What about the orientation relative to Earth? Modeling by Mark A. Wieczorek and Mathieu Le Feuvre (Institut de Physique du Globe de Paris) demonstrates that either orientation is equally probable after any impact that had the energy to move the Moon out of a preexisting spin orientation. Therefore, while the 1-degree plume generates the Procellarum Basin, any decent-sized impact will cause it to rotate so that the plume and overlying basaltic basin faces towards or away from Earth.

An interesting question is whether tidal forces acting on a planet, orbiting in the tidal-lock zone of a star, will also experience sufficient force to cause continental areas to orientate towards or away from their star, through TPW. The literature doesn't really address this, although some quick "spot" answers on the Internet imply that the difference in force would be insufficient to cause such a realignment. However, these assessments do not include the mass of any underlying mantle structure, *only* the continental crust. Therefore, one would like to see this modeled fully to be sure of the outcome.

If TPW was enhanced by tidal forces on habitable planets orbiting red dwarfs, it could have critical (and potentially terminal) implications for any complex life. Where complex species are advected to the dark, cold, anti-stellar hemisphere, they will be eliminated by the eons of darkness. A bit of luck might see continental crust (or plume-supported highlands on smaller planets) orientated so that one hemisphere (north or south) has the greatest preponderance of this terrain with the sub-stellar point lying along the edge of the dichotomy (or along the margin of the super-continent).

Figure 3.11 illustrates how degree one (one-plume) and degree two (two-plume) convection may relate to the mass and core volume of a telluric world. Figure 3.12 then summarizes the connection between LLSVPs and the orientation of various terrestrial bodies.

To sum up, the structure of the underlying mantle has profound implications for the operation of surface tectonics. Although this, clearly, has profound implications for the distribution of dry land and the operation of climate-regulating cycles (.ter 5), there are also far more subtle implications for the habitability of tidally locked planets. If continental crust is continually shuffled away from areas of mantle upwelling, these areas may also be shuffled away from regions which receive nourishing radiation from their star.

Plate Tectonics and Planetary Magnetism

There is the suggestion that for most telluric planets, plate tectonics is necessary but insufficient for the operation of a planetary magnetic field. The arrival of relatively cold subducted crust at the core-mantle boundary chills the liquid metal of the outer core and initiates downwellings. Such motion, in turn, affects the overall motion of liquid in the rotating core and alters the flow of charge. In turn, this will affect the orientation and strength of the magnetic field that it generates within it. If plate tectonics is absent, then further processes will be required to ensure vigorous convection in any liquid metallic core.

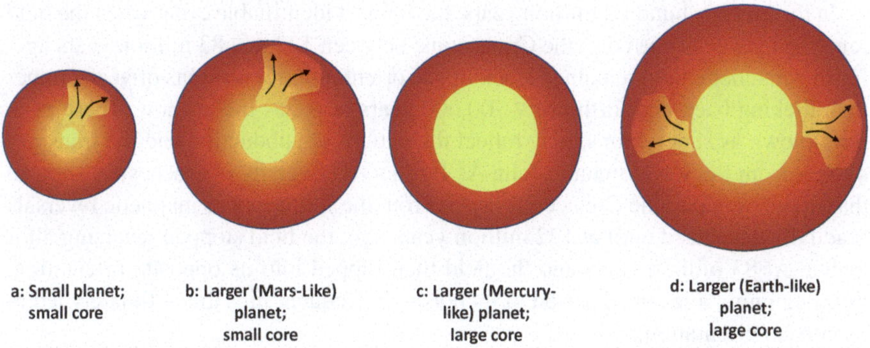

Fig. 3.11 How planetary mass and core volume might shape whole mantle convection and the fate of a planet. Small worlds (**a**) have degree-one—single plume carrying heat towards the surface. Larger planets, with larger cores, have the same pattern (**b**). In a planet with a disproportionately large core, there may be insufficient mantle depth to allow heat loss through large convective plumes (**c**). In an Earth-like planet (**d**), a massive mantle and hot core allow degree two—two dominant long-lasting super-plumes—to form. Will still bigger planets have three such plumes? Super-plumes on Earth are flanked by regions of subduction, which return mantle flow

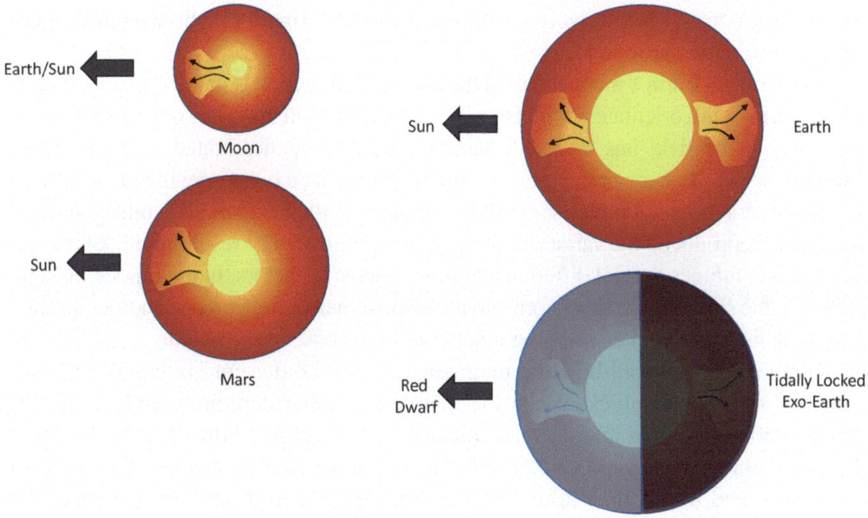

Fig. 3.12 The likely impact of LLSVPs on habitability. All Solar System bodies have their orientation linked to their moment of inertia—the alignment of mass within them. The Moon's plume underlies the Procellarum Basin, facing Earth; while Earth and Mars have plumes aligned with their equators. The location of a planet's LLSVPs will determine whether continents face their star, rendering them habitable for complex life—or not…

The terrestrial magnetic field tends to flip orientation, with an average period of 700,000 years or so, but there are periods when the field remains locked into one orientation for millions of years. What are these unusually lengthy intervals telling us about how the field operates?

In the last few hundred million years, the longest identifiable event when the field remained stable occurred in the Cretaceous, between 122 and 83 million years ago. Earth's magnetic field remained stuck in one orientation for millions of years, rather than flicking back and forth every 700,000 years or so, as it does now. These periods, known as superchrons, may reflect the pattern of subduction and the transport of heat from the lower mantle. John A. Tarduno (University of Rochester) showed that in the run up to the Cretaceous superchron, the frequency of magnetic reversals gradually decreased until at 122 million years ago, the field stopped reversing altogether. At 83 million years ago the field then flipped into its opposite orientation, thus beginning a renewed period of increasingly frequent flips that ultimately led to its current orientation.

How might mantle convection have influenced this period of magnetic stability? The Cretaceous marked a period when Tethyan and Panthalassan crust was subducting relentlessly around the margins of Eurasia. Under Panthalassa a superswell was dividing the mid-ocean ridge into three sections, generating the future Pacific plate and driving two others—the Kula and Farallon plates—to the north and east. Covering approximately half the planet, Panthalassa and the future Pacific Ocean were thus the cornerstones of a large mantle-wide circulation that would, in turn, fix convection through its contact with a significant area of the surface of the liquid metal outer core. This relatively stable setup would form an effective anchor, stabilizing Earth's magnetic field.

If you look at Earth's magnetic field the dominant direction is in the form of a dipole, with a north-south orientated field. However, superimposed upon this overall pattern are areas where the field has reversed. Most of these are concentrated in the southern Atlantic region. By considering how the liquid magnetized outer core interacts with the solid inner core—and the plastic overlying mantle—Julien Aubert and colleagues have accurately reproduced the variations in the terrestrial field. The resulting model is ingenious and combines several different and often opposing factors. In their model, irregularities in the shape of the inner core and lowermost mantle drag on one another through gravity. This interaction couples the inner core to the base of the mantle.

They then considered how the magnetic and fluid outer core interacts with this coupled core-mantle block. When the mantle and inner core are coupled, and the Coriolis effect is taken into account, these effects force the flow of liquid metal in the outer core into a giant, westward drifting, sheet-like upwelling. This flow of liquid iron and nickel then drags the magnetic field with it, and this produces the long-term movement of an area of magnetic reversal in the southern Atlantic towards the Pacific. The sustained interaction of all three planetary layers maintains the long-term pattern seen in the southern Atlantic.

Can a planetary magnetic field operate efficiently in the absence of plate tectonics? Imagine, then, a planet without this process. The core will cool by the ascent of hot plumes through the mantle, towards the surface and the conduction of heat through the lithosphere. The rate of cooling and heat transport will be far more sluggish than it is on Earth, leading to a more uniformly hot mantle. In turn, this will lead to more sluggish transport of heat from the liquid core to the mantle and a more leisurely movement of iron in the liquid core.

Within a relatively hot planet this may be sufficient to kill the generation of a magnetic field strong enough to reach the surface and shield the atmosphere from the solar wind. On a small terrestrial planet, the relatively large surface-area-to-volume ratio may allow cooling that is sufficient to drive convection. Perhaps this is the reason Mercury has a field but Venus does not.

What of the hotter, early Earth? The strength of the field in the Hadean is open to interpretation. There is precious little evidence left to interpret. However, we do know that even if there was no plate tectonic process, Earth benefitted from faster rotation more than it does now. As rotation is a critical component of field generation, more sluggish convection in the core might be offset by faster rotation. Further analysis is needed.

Different Kinds of Tectonism on Exoplanets

In terms of exoplanets, we might expect those worlds in which there are greater proportions of tidal heating, or are simply younger than Earth, to exhibit styles of tectonism similar to the early Earth. Importantly, there is no need for our current style of tectonism to move materials into and out of the mantle; something we look at more thoroughly in Chap. 4. Other options exist that will also allow planets to remain viable, long-term abodes for life.

Stagnant Lid Tectonics

Stagnant lid tectonism consists of crustal deformation without pronounced lateral motion. There are many variants of this process, some of which we discussed earlier in this chapter. These were the "hot stagnant lids" that may have functioned on the early Earth. As a planet cools, the lithosphere may react as it thickens. Almost certainly most planets that lack surface water (or other suitable solvents) evolve steadily through the thick-lid scenarios until their lithospheres become too thick to fracture and for magma to penetrate. Over time, because of lowering temperatures, the depth at which melting is possible increases. This makes partial melting even less likely, as increasing pressure increases the temperature required to cause such melting in the first place. Such worlds quietly evolve into Moon-like structures that are effectively inert.

However, in the presence of a suitable solvent—and where the temperature contrast between the mantle and surface is high enough, convection can rupture the surface. In terrestrial plate tectonics, fractures within the dense ocean crust (lithosphere) allow subduction, which then generates strong downwellings in the mantle. The critical factors are the difference in density of the cold lithosphere—in particular the upper basaltic layer, which morphs into dense eclogite at depths—and the gravitational field strength. Water both cools the top of the lithosphere but also chemically-alters the layer and makes fractures more prone to opening and sliding. Planets without water (and this is by far the most likely solvent) are unlikely to host plate tectonics. Planets with greater mass have proportionately stronger convection at a given temperature; plus they have zones of weakness that are under greater stress than planets with weaker fields.

As we've seen in terrestrial tectonics, subduction pulls the rest of the crust along, while deep, broad mantle plumes (on Earth known as Jason and Tuzo) return material to the upper mantle. Minor additional returns are made by hot-spot plumes, which also serve to cool the core. However, mantle convection is a consequence of subduction—not the other way around. What if the converse was true? What if mantle convection was able to move the continents rather than subduction? In this light, Jean Bernard (Geological Survey of Canada) has proposed that a form of horizontal tectonics was possible in the early Archaean, and this was driven by upwellings and mantle flow. These pushed on the continental keels. The deep roots of the cooling Archaean continents dipped into the underlying mantle and acted like sails that drove movement in response to broad upwellings. The continents then shuffled around, periodically bulldozing the surrounding oceanic crust, which in turn led to shallow under-thrusting of the ocean crust beneath the continents.

This variant of terrestrial tectonism, nicknamed "keel-tectonism," may operate on relatively young planets with hot but suitably viscous mantles. Although this is similar to modern tectonism, the process is driven by mantle flow rather than subduction (Fig. 3.13a). Such keel-tectonism could evolve into modern tectonism, as it has on Earth; or if a thick lid forms, a crumpling form of the granitoid stagnant lid, until the mantle becomes too viscous or the lithosphere too thick. One could, *tentatively*, describe early Martian tectonism in this regard. If one looks at the southern hemisphere, magnetic lineaments broadly parallel the edge of the Martian crustal dichotomy, where the southern highlands give way to the northern lowlands. While these resemble terrestrial oceanic crust magnetic patterns, the scale is wrong.

In a tentative model (of the author's), it is suggested that both the orientation of Valles Marinares and the lineaments are connected to the growth of the southern highlands. The thick basaltic crust of the early southern hemisphere's highlands may have been supplied by the degree-1 plume of James H. Roberts and Shijie Zhong (described earlier in this chapter). Thick basaltic lithosphere developed on top of this singular plume, which was initially focused in the southern hemisphere. This crust was sufficiently soft at its base to flow outwards, away from the plume and towards the equator. As the base flowed, the upper layers was extended by brittle fractures and faults and accommodated a steady outward growth of the crust away from the southern polar region, over the underlying lithosphere and towards the lower-lying basin to the north. The pattern of intrusion is then more like back-arc spreading, distributed over a larger area, than mid-ocean ridges, where magmatism is focused along narrow lineaments.

Every time the Martian magnetic field reversed the change was recorded in the swathe of basalts that erupted along the brittle fractures in the upper crust. Over time, the mantle cooled, and the underlying lithosphere thickened to the point that it could support even thicker crust. Crucially, this thick crust didn't spread ductily at

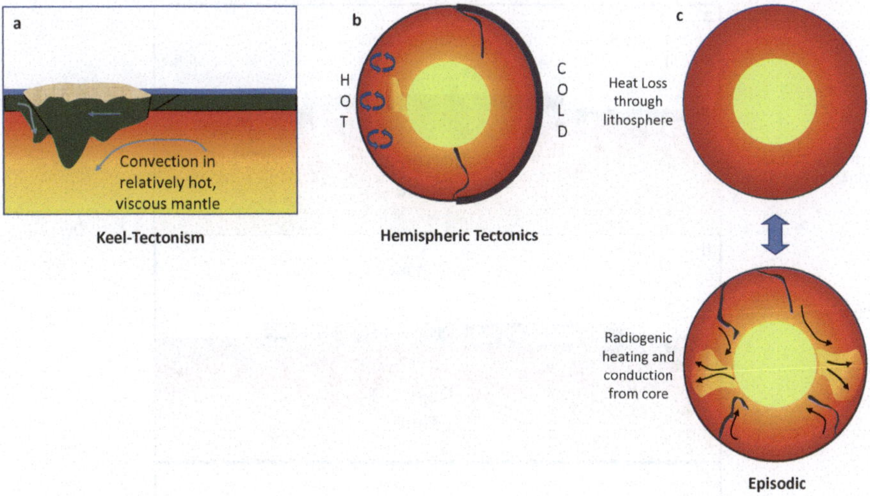

Fig. 3.13 Alternative modes of tectonism. In **a**, convection currents in the mantle push on continental roots (keels), moving blocks of lithosphere into one another. Subduction is weak and of secondary importance. In **b**, one side of the planet is hot, the lithosphere weak and vertical tectonism dominates. The other hemisphere is cold with thick lithosphere. Subduction may happen at the interface (possibly the terminator). This may be common on tidally locked, hot exo-Earths. Finally, in **c**, cooling Earth-like planets may experience episodic plate tectonics. Subduction cools the mantle below the lower limit for convection. However, radiogenic and other heating may subsequently warm it and allow it to restart

its cooling base. Valles Marinares then formed, along the edge of the highlands, in a manner driven by gravitational forces (Fig. 3.14). Now, gravitational forces are sufficient to stretch the crust and cause its failure, but insufficient to rupture the underlying lithosphere and allow magma to emerge, something that is apparent in the shallow rifting of Tibet. If this is a reasonable proposition, then Valles Marinares formed relatively late, when magmatism was increasingly confined to Tharsis.

Evidence of its validity (or relative stupidity) of this model would require a determination of the structure of the lithospheric mantle and crust and the identification of sub-surface faults. Those of Valles Marinares should combine at depth into a single thrust that terminates near the edge of the dichotomy. If such a model is reasonable, one might expect all Mars-sized bodies, in cool orbits around their host stars, to exhibit both a crustal dichotomy and potentially long (and frozen) rift structures that run broadly parallel to the edge of the highland region. The relatively small volume and (concomitant) large surface area to volume ratio means that the lower mantle can shed heat through a single convecting super-plume. Larger planets, like Earth (or Venus) require more than one plume that is manifest as two or possibly three LLSVPs, with the Antarctic plume being much smaller than Jason or Tuzo.

Stagnant lids will, therefore, be the norm on all small planets. But what about larger worlds? Are there alternatives to terrestrial plate tectonism—and is our model applicable to larger worlds? We already seem certain that plate tectonics did not

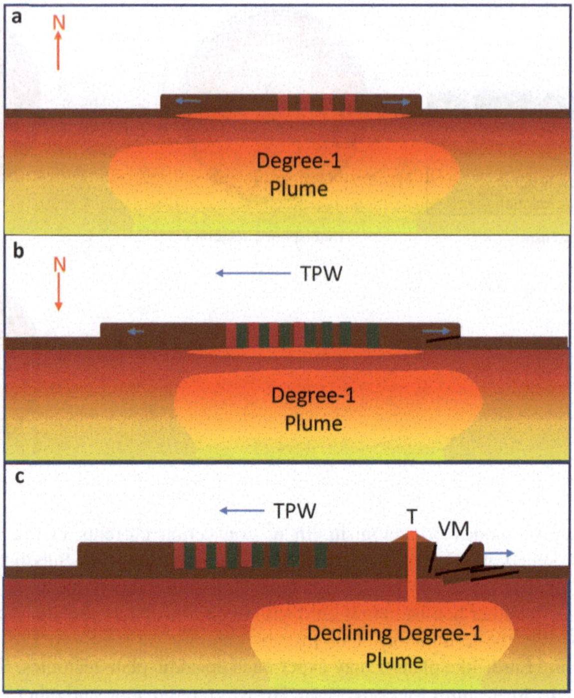

Fig. 3.14 Plume-driven tectonism on Mars? In **a**, the large degree-1 plume rises in the southern hemisphere and produces the thick lithosphere (as described by Shi Zhong). In **b**, a combination of TPW over the plume and continued injection of copious basalts produces the magnetic lineaments as the highlands grow. At this stage and in **a**, the base of the crust (brown) is hot and partly molten (stretched, red oval). By **c** (near the end of the Noachian era) the cooling plume allows the base of the crust to solidify. Continued TPW migrates the plume until it underlies the developing Tharsis (T). Migration over the plume fractures the lithosphere to make Valles Marinares (VM). Gravitational forces, acting along the edge of the highlands, promote fracturing of cool, thick lithosphere

operate in its modern incarnation throughout all of Earth's history. When the planet was younger and hotter, a thinner lithosphere favored vertical tectonics, where thick crust foundered into the mantle, or where the spreading of crust caused regional subduction around the crustal province. We should therefore expect hotter planets to exhibit similar tectonism to the early Earth rather than its modern form.

However, that's not all. What happens if plate tectonics or crustal spreading happens for prolonged periods? The end product of plate tectonics is the formation of thick, buoyant continental crust. This cannot be subducted. If a planet is hot for long enough, the surface will clog with granite. This depends on the amount of water available to stew the silica out of the mantle, but in principle once the surface has more than 70% of its area encased in granitoid crust, plate tectonics is no longer feasible. Adrian Lenandric (Rice University) has suggested that the critical area may be as low as 50% of the surface. As such, the current continental area of Earth is close to this possible value (40%).

Is plate tectonics inevitable on more massive planets? There are two counteracting forces affecting the lithosphere of planets. If you up the mass, the mantle will become denser, and that might prevent subduction. Increasing gravitational forces might also lock faults together. Countering this, increased gravitational forces may make the lithosphere increasingly prone to failure from brittleness. Arguments have flung backwards and forwards, and the impact is critical to understanding whether planets more massive than Earth will be geologically active like ours.

In 2007 Diane Valencia's group carried out modeling of super-terrans—planets with masses in excess of ours, and up to ten times that of Earth. Valencia and colleagues findings suggest that as the mass of the planet increases, the thickness of the lithosphere declines and the shear-stress acting on fracture zones increases.

Why does the shear-stress—the effect of pulling on the lithosphere—increase with increasing planetary mass? This comes down to the effect of gravity on convection. If you increase the gravitational field, convection becomes more vigorous, and this generates a greater force on the overlying lithosphere. Any effect of increasing mass on mantle viscosity is soon overwhelmed by the increase in this force. Therefore, with increasing mass, both the increase in convective force and decrease in lithospheric thickness facilitate higher rates of subduction. Indeed, while subduction is almost certainly facilitated by water on Earth, for larger planets (more than two Earth masses) even dry mantles are fluid enough and the forces great enough to drive subduction and mantle overturning.

Valencia and colleagues also demonstrate that for younger planets, plate tectonics is still feasible despite correspondingly thinner and seemingly less rigid lithospheres. The increase in gravitational force offsets the effects of youth and makes the plates rigid enough to conduct the necessary force, which in turn allows subduction.

The role of water in smaller planets (like Earth and Venus) is two-fold. It reduces the viscosity of the upper mantle and helps promote partial melting. But it also weakens faults, making them more prone to failure. Typical mantle-generated forces are on the order of 10 million Pascals (Newtons of force per meter squared). This easily exceeds that needed to crack open faults on Earth. However, while the Venusian crust is hotter, it is also much drier than terrestrial crust (and total or bulk lithosphere). This makes it more rigid than ours. However, double the mass, and the effect of a dry lithosphere disappears. Therefore, super-Venusian worlds should show plate tectonics for modest increases in mass.

Countering this argument, and published in the same year, was work by Craig O'Neill (Macquarie University) and Adrian Lenandric (Rice University), which suggested quite the opposite, that the increased gravitational field tended to increase the force on fractures, locking them shut. This made plate tectonics less likely on super-terrans. However, Jun Korenaga (Yale) reviewed these arguments, coming down on the side of Valencia's group. Korenaga's analysis, which incorporated the various mathematical models that were available, clearly shows that the effect of greater convective stress is the dominant factor.

Korenaga does point out, though, the problems with scaling planets: increasing mass increases the forces needed to fracture the lithosphere, but it increases the temperature and the rate of convection in ways that are non-linear and dependent on other factors such as the composition of the mantle and the amount of water present. Another factor that is not at all obvious is the manner and extent to which subducting plates bend. This requires force, but the amount of force is unclear. Moreover, heating of the crust during earthquakes or the presence of fluids in the fault zones dramatically affects the outcome. And how these factors interact over time as volatile substances move out of and then into the mantle affects its viscosity over time. Clearly a lot more modeling is needed.

More recent analyses by Paul Tackley (ETH Zurich) and colleagues at UCL examined the effect of planet mass and the viscosity of the lower mantle with some surprising conclusions. Although the millennial models tended to have denser and more viscous lower mantles with increasing mass, more recent analysis suggests quite the opposite: super-terrans have low viscosity lower mantles that may even host magma oceans. A mantle made of silicate perovskite would indeed be expected to become utterly rigid with increasing mass, but such simple models neglect internal radiogenic heating. Once this is plugged in, temperatures rise so high that convection (and vigorous convection) becomes inevitable. Indeed, Tackley, van Heck (also at ETH Zurich) and others show that even after many billions of years the mantle remains very hot with a low viscosity. In some instances a basal magma ocean will be present above the core-mantle boundary. Thus, viscosity changes are never close to being an impediment for mantle convection and plate motion.

Together this suggests that for terrestrial planets in tepid or cold environments, plate tectonics is an inevitability if their mass is greater than or equal to twice the mass of Earth. There remains, however, a need to determine what the lowest mass cut-off is for this process—and just how low you can go with volatile materials before plate tectonics becomes impossible.

Hemispheric Plate Tectonics

What happens if the mantle is a little too viscous; the surface too hot or the planet tidally-locked to its star? Will plate tectonics happen then? Two scenarios emerge that depend on the temperature and the presence or absence of tidal-locking (Fig. 3.13b).

In the simplest case, the planet has a dense, hot atmosphere and the surface is dry. The end result is Venus—unless the mass of the planet is sufficiently high. However, if we strip the atmosphere away and move the planet sufficiently close to its star that it tidally locks something altogether different emerges. Joost van Summeren, Clinton Conrad and Eric Gaidos (all at the University of Hawaii) modeled the effect of raising the surface temperature to hundreds or thousands of Kelvin across one hemisphere of an exoplanet.

Where an atmosphere is lacking, an asymmetric degree-1 pattern of convection develops. Lithosphere forms on the sunlit side then migrates outwards towards the dark hemisphere, where it either becomes cold and dense enough to subduct or it overrides cold lithosphere, causing it to subduct. Lithosphere on the night hemisphere undergoes plate tectonics, but that on the day hemisphere undergoes diffuse construction and outward spreading, with vigorous volcanism. Van Summeren and colleagues modeled the effect of introducing an atmosphere on their tidally locked world. When outgassing was sufficient, the entire surface heated up to the point that plate tectonics ceased. Now, the surface showed Venus-like diffuse tectonism akin to the squishy-lid that was introduced earlier. Volcanism was then distributed over the planet's surface.

A remarkable outcome was the effect of heating near to the sub-stellar point. For surface temperature contrasts of 400 K or less there was no coherent pattern. Jump that to 800 K, melting then becomes focused at the sub-stellar point with five times the amount of melt produced than is seen on Earth (100 cubic km per annum, rather than 20 cubic km, respectively). Whack the temperature contrast still higher, to the kinds of values expected for the hottest telluric planets (1200 K), and the amount of melt produced by degree-1 convection soars to 30 times the terrestrial value.

Episodic Plate Tectonics

Is plate tectonics a constant feature of planets—and what will happen to the process on Earth as our planet continues to cool? In some models plate tectonics can switch on and off as the mantle becomes less conducive to convection or the lithosphere becomes too thick and rigid to support convection. Various things are possible (Fig. 3.13c).

Consider an aging Earth. The lithosphere has become too rigid to subduct. Without subduction cooling the interior, the mantle begins to warm up through slowing but still active radiogenic heating. At some point the viscosity falls, and the upper mantle begins to convect once more. Sufficient force is available to rupture the lithosphere, and the plates begin to move once more. For tens of millions of years subduction resumes, and normality is restored. However, subduction renews cooling of the mantle until it becomes too rigid to flow. Plate tectonics ceases, and the process repeats. This cyclical onset and cessation of plate tectonics is called episodic tectonics and may operate on all telluric planets with oceans of a suitable age.

In this regard, more recent work by Joseph O'Rourke and Jun Korenaga (both at Yale) considers both the long-term evolution of surface tectonics and whether plate tectonism is an inevitability. In this analysis, yet another contrary outcome is observed. Although increasing planetary mass increases the rate of convection in the mantle, the likelihood of plate tectonics only increases marginally. Adding to Valencia's earlier work, increasing planetary mass exerts more weight on the interior at any given depth, and this reduces partial melting. The (basaltic, oceanic) crust of super-terrans is therefore thinner than Earth at any given age. However, with a greater weight at any given depth, the transition from basalt to eclogite occurs more readily—and this is a problem for plate tectonics. As we've seen, modern-day terrestrial tectonics is driven by this transition at subduction zones. In the Hadean Era this transition occurred under thick basaltic piles. However, if this happens globally on a super-terran the outcome is rather unfortunate.

As Korenaga and colleagues show, when planetary mass is high, the lithosphere will always be unstable at relatively shallow depths. Plate tectonics becomes impossible, and only delamination-dominated, squishy lid tectonics can occur. On smaller worlds with modest levels of volcanism, the crust (and lithosphere) will only be thick enough for delamination under certain circumstances. These occur globally when the planet is young and volcanism is extensive. However, in older planets, such as Earth, delamination still occurs in subduction settings when the magmatism is extensive, or in continental collision settings when the lithosphere is thickened, infiltrated with fluids and heated. The latter processes are confined to orogens (mountain-building areas), such as the Sierra Nevada and the Altiplano in the Andes. Although these scenarios are not going to be discussed here, references are provided at the end of this chapter if you wish to follow up on this. They form an interesting backdrop to the kinds of processes that appear likely to dominate the evolution of the lithosphere on super-terrans. Such scenarios are examined more closely in Chap. 7.

On massive terrestrial planets (super-terrans) as soon as a stagnant basaltic lid forms, the basal layer of the crust readily transforms to eclogite, delaminates and falls away into the mantle. This remobilizes the crust above, but it will also cause partial melting and the formation of TTGs. Therefore, soon after planet formation, the majority of the surface of a super-terran may well morph into subduction-resistant granitoid. One hopes that this illustrates the complexity of the models that are required to accurately determine the habitability of planets. It's a complex universe, but understanding it, in all its beauty and intricacy, is fundamental if we are to fully address both planetary habitability and the long-term evolution of planets.

In this regard, a reviewer of the earlier version of this book (*Under a Crimson Sun*) queried why so much time was spent in the books investigating plate tectonics, stellar evolution and the myriad of other factors that are expanded upon here. One hopes that the reason is more than apparent now.

In an attempt to quantify matters, Robert Stern (University of Texas) and colleagues examined the propensity of planets and planetesimals to undergo plate tectonics. Stern and colleagues defined various bodies using a "Tectonic Activity Index," TAI for short. Scoring was from 0 to 3, with 0 meaning no detectable surface activity for an object such as Callisto through to 3 for an object such as Io that is perpetually active. Examination of the active (or recently active) bodies shows that the majority exhibit stagnant lid tectonism. This falls into the heat-pipe mode for Io; likely squishy-lid mode for Venus; and a stagnant, thick lid, with possible delamination in the case of Mars. In Stern and colleagues estimation, the prerequisites for plate tectonics are a modestly thick and strong lithosphere, overlying a weak, plastic asthenosphere. The lithosphere has to be strong enough to transmit horizontal forces, but crucially it has to be composed of materials that become denser when colder—so conventional ice is out. This allows the lithosphere to float on the asthenosphere when it's young, but dense enough to subduct when old.

Developing an understanding of the interplay between these complex and occasionally contradictory factors will be necessary to fully model the geological evolution of planets. With biological evolution deeply intertwined and utterly dependent upon planetary geophysics, understanding the former will crystallize our understanding of the latter.

Plate Tectonics Under a Crimson Sun

Does a small orbital distance affect whether a planet will experience plate tectonics? That might sound a little peculiar, since plate tectonics is surely a process driven by temperature contrasts between the lithosphere and the underling plastic mantle. However, that's the question posed by John (J. J.) Zanazzi and Amaury H. M. J. Triaud (both at Cornell). Now, the question is not one relating to the effect of extreme insolation (discussed earlier), so these authors are not questioning whether extreme surface temperatures influence tectonism. Rather, they are interested in the effect of the large tidal forces that act on any tightly orbiting world.

Zanazzi and Triaud considered models in which their worlds were rocky, homogeneous and had a constant density. Although this might seem unrealistic, it did allow them to model the effects of tidal forces. They showed that the tidal stress from the host star was effectively comparable in magnitude to the stress on the lithosphere from mantle convection. Their work implies that planets in close orbits to red dwarfs seem more likely to experience plate tectonics than comparable worlds around Sun-like stars. This is simply because they experience more stress on the lithosphere than planets in Earth-like orbits. This has knock-on effects for habitability that we will examine in Chap. 4.

In something of an aside, Zanazzi and Triaud also show that for compact systems, such as KOI-961.03, tidal forces that act between the planets may also be sufficient to cause earthquakes. In this system, three small planets orbit the red dwarf in less than two days—well inside the inner boundary of its habitable zone. KOI-961.03 is farthest of the three from the star and is about the size of Mars, with a radius of 0.57 times that of Earth. The next planet in towards the star is KOI-961.01. This has a radius about 0.78 times that of Earth. The planet closest to the star is KOI-961.02, which has a radius 0.73 times that of Earth. All three planets should experience strong tidal interactions with their star, but also modest interactions with one another, meaning that in addition to being toasty little planets, they are also rocked by an unsettling amount of geological activity. Clearly, this kind of interaction is another variable to be discussed in any consideration of planetary habitability in such compact systems.

Conclusions

Plate tectonics is such an integral part of our existence that it's hard to consider that there are alternatives. The majority of planets will, undoubtedly, exhibit some kind of stagnant lid, either because they are too old and cold or too large and hot. Either way, the "window" for plate tectonics is likely missed on all but a select few planets that have a mass similar to ours (and likely less than two or three Earth masses) and have water.

However, this does not rule out habitability. Although plate tectonics might be needed in order to generate a strong magnetic field, it is likely not the only way. A small radius and/or a fast spin may also help drive circulation in the liquid, metallic, planetary core.

Moreover, there has come to be the idea that only plate tectonics can bring about effective mixing of the interior and exterior of the planet. However, as the next chapter illustrates, this may not be the case. Plate tectonics may be synonymous with efficient cycling of volatile materials, but it is not the only fruit in the forest.

Chapter 3 serves to illustrate the complexity of the processes that determine the manner in which mantle cooling is manifest as surface tectonics. We are not yet in a position to really evaluate the impact on habitability, but we can see that minor variation in planetary mass may have significant epistatic effects on the subsequent geological and biological evolution of the planet as a whole.

References

Tectonism

Abbott, D., & Mooney, W. (1995). The structural and geochemical evolution of the continental-crust—Support for the oceanic plateau model of continental growth. *Reviews of Geophysics, 33*, 231–242.

Ashwal, L. D., & Burke, K. (1989). African lithospheric structure, volcanism, and topography. *Earth and Planetary Science Letters, 96*, 8–14.

Basilevsky, A. T., Pronin, A. A., Ronca, L. B., Kryuchkov, V. P., Sukhanov, A. L., & Markov, M. S. (1986). Styles of tectonic deformation on Venus: Analysis of Veneras I5 and 16 data. *Journal of Geophysical Research, 91*, D399–D411.

Basilevsky, A. T., Shalygin, E. V., Titov, D. V., Markiewicz, W. J., Scholtend, F., Roatsch, T., Kreslavsky, M. A., Moroz, L. V., Ignatiev, N. I., Fiethe, B., Osterloh, B., & Michalik, H. (2012). Geologic interpretation of the near-infrared images of the surface taken at the Venus Monitoring Camera, Venus Express. *Icarus, 217*, 434–450. https://doi.org/10.1016/j.icarus.2011.11.003.

Bédard, J. H. (2018). Stagnant lids and mantle overturns: Implications for Archaean tectonics, magmagenesis, crustal growth, mantle evolution, and the start of plate tectonics. *Geoscience Frontiers, 9*, 19–49. https://doi.org/10.1016/j.gsf.2017.01.005.

Belousova, E. A., Kostitsyn, Y. A., Griffin, W. L., Begg, G. C., O'Reilly, S. Y., & Pearson, N. J. (2010). The growth of the continental crust: Constraints from zircon Hf-isotope data. *Lithos, 119*(3–4), 457–466. https://doi.org/10.1144/SP389.13.

Bindschadler, D. L., & Parmentier, E. M. (1990). Mantle flow tectonics and a ductile lower crust: Implications for the formation of large-scale features on Venus. *Journal of Geophysical Research, 95*, 21,329–21,344.

Cai, C., Weins, D. A., Shen, W., & Elmer, M. (2018). Water input into the Marianas Subduction Zone estimated from ocean-bottom seismic data. *Nature, 563*, 389–392. https://doi.org/10.1038/s41586-018-0655-4.

Condie, K. C. (1998). Episodic continental growth and supercontinents: A mantle avalanche connection? *Earth and Planetary Science Letters, 163*, 97–108.

Condie, K. C. (2002). Continental growth during a 1.9-Ga superplume event. *Journal of Geodynamics, 34*, 249–264.

Crameri, F., & Tackley, P. J. (2016). Subduction initiation from a stagnant lid and global overturn: New insights from numerical models with a free surface. *Progress in Earth and Planetary Science, 3*, 30. https://doi.org/10.1186/s40645-016-0103-8.

de Wit, M. J. (1998). On Archean granites, greenstones, cratons and tectonics: Does the evidence demand a verdict? *Precambrian Research, 91*, 181–226.

De Wit, M. J., & Hynes, A. (1995). The onset of interaction between the hydrosphere and oceanic crust, and the origin of the first continental lithosphere. In M. E. Coward & A. C. Pies (Eds.), *Early precambrian processes* (Vol. 95, pp. 1–9). Geological Society Special Publication.

Dumoulin, C., Čădek, O., & Choblet, G. (2013). Predicting surface dynamic topographies of stagnant lid planetary bodies. *Geophysical Journal International, 195*, 1494–1508. https://doi.org/10.1093/gji/ggt363.

Foley, S. F., Buhre, S., & Jacob, D. E. (2003). Evolution of the Archaean crust by delamination and shallow subduction. *Nature, 421*, 249–252.

Frost, C. D., Frost, B. R., Kirkwood, R., & Chamberlain, K. R. (2006). The tonalite-trondhjemite-granodiorite (TTG) to granodiorite-granite (GG) transition in the late Archaean plutonic rocks of the central Wyoming province. *Canadian Journal of Earth Sciences, 43*, 1419–1444. https://doi.org/10.1139/E06-082.

Gerya, T. (2014). Plume-induced crustal convection: 3D thermomechanical model and implications for the origin of novae and coronae on Venus. *Earth and Planetary Science Letters, 391*, 183–192. https://doi.org/10.1016/j.epsl.2014.02.005.

Gerya, T. V., Stern, R. J., Baes, M., Sobolev, S. V., & Whattam, S. A. (2015). Plate tectonics on the Earth triggered by plume-induced subduction initiation. *Nature, 527*, 221–225. https://doi.org/10.1038/nature15752.

Gerya, T. V., Stern, R. J., Baes, M., Sobolev, S. V., & Whattam, S. A. (2016). Plate tectonics on the Earth triggered by plume-induced subduction initiation. *Nature, 527*, 221–225.

Ghail, R. (2015). Rheological and petrological implications for a stagnant lid regime on Venus. *Planetary and Space Science, 113–114*, 2–9. https://doi.org/10.1016/j.pss.2015.02.005.

Green, M. (2001). Early Archaean crustal evolution: Evidence from ~3.5 billion year old greenstone successions in the Pilgangoora Belt, Pilbara Craton, Australia. https://ses.library.usyd.edu.au/bitstream/2123/505/2/adt-NU20030623.11023101front.pdf.

Green, M. G., Sylvester, P. J., & Buick, R. (2000). Growth and recycling of early Archaean continental crust: Geochemical evidence from the Coonterunah and Warrawoona Groups, Pilbara Craton, Australia. *Tectonophysics, 322*, 69–88.

Hashimoto, G. L., Roos-Serote, M., Sugita, S., Gilmore, M. S., Kamp, L. W., Carlson, R. W., & Baines, K. H. (2008). Felsic highland crust on Venus suggested by Galileo Near-Infrared Mapping Spectrometer data. *Journal of Geophysical Research, 113*, E00B24. https://doi.org/10.1029/2008JE003134.

Hawkesworth, C., Cawood, P., Kemp, T., Storey, C., & Dhuime, B. (2009). A matter of preservation. *Science, 323*, 49–50.

Herrick, R. R., & Phillips, R. J. (1990). Blob tectonics: A prediction of Western Aphrodite Terra, Venus. *Geophysical Research Letters, 17*, 2129–2132.

Hopkins, M., Harrison, T. M., & Manning, C. E. (2008). Low heat flow inferred from 4 Gyr zircons suggests Hadean plate boundary interactions. *Nature, 456*, 493–496. https://doi.org/10.1038/nature07465.

Jozwiak, L. M. (2014). Constraining the lithospheric thickness of Io from a modified heat-pipe model. In *45th Lunar and Planetary Science Conference*. https://www.hou.usra.edu/meetings/lpsc2014/pdf/1160.pdf.

Korenaga, J. (2010). On the likelihood of plate tectonics on super-earths: Does size matter? *The Astrophysical Journal Letters, 725*, L43–L46. https://doi.org/10.1088/2041-8205/725/1/L43.

Lopes, R. M. C., Kamp, L. W., William, D. S., Mouginis-Mark, P., Kargel, J., Radebaugh, J., Turtle, E. P., Perry, J., Williams, D. A., Carlson, R. W., & Doute, S. (2004). Lava lakes on Io: Observations of Io's volcanic activity from Galileo NIMS during the 2001 fly-bys. *Icarus, 169*, 140–174. https://doi.org/10.1016/j.icarus.2003.11.013.

Marchi, S., Bottke, W. F., Elkins-Tanton, L. T., Bierhaus, M., Wuennemann, K., Morbidelli, A., & Kring, D. A. (2014). Widespread mixing and burial of Earth's Hadean crust by asteroid impacts. *Nature, 511*, 578–582. https://doi.org/10.1038/nature13539.

Martin, P., van Hunen, J., Parman, S., & Davidson, J. (2008). Why does plate tectonics occur only on Earth? *Physics Education, 43*(2). https://www.researchgate.net/profile/Jon_Davidson/publication/30053838_Why_does_plate_tectonics_only_occur_on_Earth/links/54083bef0cf2c48563b941f3/Why-does-plate-tectonics-only-occur-on-Earth.pdf.

McEwen, A. S., Keszthelyi, L. P., Lopes, R., Schenk, P. M., & Spencer, J. R. (2004). The lithosphere and surface of Io. In F. Bagenal, T. E. Dowling, & W. B. McKinnon (Eds.), *Jupiter. The planet, satellites and magnetosphere* (Cambridge Planetary Science, Vol. 1) (pp. 307–328). Cambridge, UK: Cambridge University Press. ISBN: 0-521-81808-7. http://lasp.colorado.edu/~espoclass/homework/5830_2008_homework/Ch14.pdf.

Moore, W. B., & Webb, A. A. G. (2013). Heat-pipe Earth. *Nature, 501*, 501–505. https://doi.org/10.1038/nature12473.

Morabito, L. A., et al. (1979). Discovery of currently active extraterrestrial volcanism. *Science, 204*(4396), 972. https://doi.org/10.1126/science.204.4396.972.

Næraa, T., Schersten, A., Rosing, M. T., Kemp, A. I. S., Hoffmann, J. E., Kokfelt, T. F., & Whitehouse, M. J. (2012). Hafnium isotope evidence for a transition in the dynamics of continental growth 3.2 Gyr ago. *Nature, 485*, 627–630.

O'Neill, C., & Lenandric, A. (2007). Geological consequences of super-sized Earths. *Geophysical Research Letters, 34*, L19204. https://doi.org/10.1029/2007GL030598.

O'Neill, C., Lenardic, A., Weller, M., Moresi, L., Quenette, S., & Zhang, S. (2015). A window for plate tectonics in terrestrial planet evolution? *Physics of the Earth and Planetary Interiors, 255*(2016), 80–92. https://doi.org/10.1016/j.pepi.2016.04.002.

O'Rourke, J. G., & Korenaga, J. (2012). Terrestrial planet evolution in the stagnant-lid regime: Size effects and the formation of self-destabilizing crust. *Icarus, 221*, 1043–1060. https://doi.org/10.1016/j.icarus.2012.10.015.

Parman, S. W. (2007). Helium isotopic evidence for episodic mantle melting and crustal growth. *Nature, 499*, 900–903. https://doi.org/10.1038/nature05691.

Peale, S. J., et al. (1979). Melting of Io by tidal dissipation. *Science, 203*(4383), 892–894. https://doi.org/10.1126/science.203.4383.892.

Percival, J. A., & Williams, H. R. (1989). Late Archean Quetico accretionary complex, Superior Province, Canada. *Geology, 17*, 23–25.

Pesonen, L. J., Mertanen, S., & Veikkolainen, T. (2012). Paleo-mesoproterozoic supercontinents—A paleomagnetic view. *Geophysica, 48*(1–2), 5–47.

Polat, A., & Kerrich, R. (2001a). Geodynamic processes, continental growth, and mantle evolution recorded in late Archean greenstone belts of the southern Superior Province, Canada. *Precambrian Research, 112*, 5–25.

Puchtel, I. S., Hofmann, A. W., Amelin, Y. V., Garbe-Schonberg, C. D., Samsonov, A. V., & Schipansky, A. A. (1999). Combined mantle plume-island arc model for the formation of the 2.9 Ga Sumozero-Kenozero greenstone belt, SE Baltic Shield: Isotope and trace element constraints. *Geochimica et Cosmochimica Acta, 63*, 3579–3595.

Puchtel, I. S., Hofmann, A. W., Mezger, K., Jochum, K. P., Shchipansky, A., & Samsonov, A. V. (1998). Oceanic plateau model for continental crustal growth in the Archaean, a case study from the Kostomuksha greenstone belt, NW Baltic Shield. *Earth and Planetary Science Letters, 155*, 57–74.

Pujol, M., Marty, B., Burgess, R., Turner, G., & Philippot, P. (2013). Argon isotopic composition of Archaean atmosphere probes early Earth geodynamics. *Nature, 498*, 87–90. https://doi.org/10.1038/nature12152.

Rapp, R. P. (1999). *First origins of archean continental crust: Assessing experimentally the roles of mafic versus ultramafic sources*. New York: Dept. of Geosciences, State University of New York.

Rey, P. F., Coltice, N., & Flament, N. (2014). Spreading continents kick-started plate tectonics. *Nature, 513*, 405–408. https://doi.org/10.1038/nature13728.

Roberts, J. H., & Zhong, S. (2006). Degree-1 convection in the Martian mantle and the origin of the hemispheric dichotomy. *Journal of Geophysical Research, 111*, E06013. https://doi.org/10.1029/2005JE002668.

Roberts, N. M. W., Van Kranendonk, M. J., Parman, S., & Clift, P. D. (2014). Continent formation through time. *Geological Society, London, Special Publications, 389*, 1–16.

Rozel, A. B., Golabek, G. J., Jain, C., Tackley, P. J., & Gerya, T. (2017). Continental crust formation on early Earth controlled by intrusive magmatism. *Nature, 545*, 332–335. https://doi.org/10.1038/nature22042.

Seno, T. & Honda, S. (n.d.). Mantle convection and global sea level: Implications for the emergence of plate tectonics on the earth. http://www.eri.u-tokyo.ac.jp/people/seno/Papers/sealevel.pdf.

Stamenković, V., Noack, D., Breuer, D., & Spohn, T. (2010). On the problem of the propensity of plate tectonics on super-earths. https://www.researchgate.net/publication/224991987_On_the_Problem_of_the_Propensity_of_Plate_Tectonics_on_Super-Earths.

Stern, R. J. (2005). Evidence from ophiolites, blueschists, and ultra-high pressure metamorphic terranes that the modern episode of subduction tectonics began in Neoproterozoic time. *Geology, 33*(7), 557–560.

Stern, R. J., Gerya, T., & Tackley, P. J. (2018a). Stagnant lid tectonics: Perspectives from silicate planets, dwarf planets, large moons, and large asteroids. *Geoscience Frontiers, 9*(2018), 103e119.

Stern, R. J., Gerya, T., & Tackley, P. J. (2018b). Stagnant lid tectonics: Perspectives from silicate planets, dwarf planets, large moons, and large asteroids. *Geoscience Frontiers, 9*, 103–119. https://doi.org/10.1016/j.gsf.2017.06.004.

Tackley, P. J., Ammann, M., Brodholt, J. P., Dobson, D. P., & Valencia, D. (2012). Mantle dynamics in super-earths: Post-perovskite rheology and self-regulation of viscosity. https://arxiv.org/ftp/arxiv/papers/1204/1204.3539.pdf.

Torsvik, T. H., van der Voob, R., Doubrovinea, P. V., Burke, K., Steinberger, B., Ashwald, L. D., Trønnes, R. G., Webb, S. J., & Bull, A. L. (2014). Deep mantle structure as a reference frame for movements in and on the Earth. *PNAS, 111*(24), 8735–8740. https://doi.org/10.1073/pnas.1318135111.

van Heck, H. J., & Tackley, P. J. (2011). Plate tectonics on super-Earths: Equally or more likely than on Earth. *Earth and Planetary Science Letters, 310*(3), 252–261.

van Summeren, J., Conrad, C. P. & Gaidos, E. (2011). Mantle convection, plate tectonics, and volcanism on hot exo-earths. Retrieved from https://arxiv.org/ftp/arxiv/papers/1106.4341.pdf.

Whitmeyer, S. J., & Karlstrom, K. E. (2008). Tectonic model for the Proterozoic growth of North America. *Geosphere, 3*(4), 220–259. https://doi.org/10.1130/GES00055.1.

Zanazzia, J. J. & Triaud, A. H. M. J. (2017). Initiation of plate tectonics on exoplanets with significant tidal stress. https://arxiv.org/pdf/1711.09898.pdf.

True Polar Wander

Andrews-Hanna, J. C., Besserer, J., Head III, J. W., Howett, C. J. A., Kiefer, W. S., Lucey, P. J., McGovern, P. J., Melosh, H. J., Neumann, G. A., Phillips, R. J., Schenk, P. M., Smith, D. E., Solomon, S. C., & Zuber, M. T. (2014). Structure and evolution of the lunar Procellarum region as revealed by GRAIL gravity data. *Nature, 514*, 68–71. https://doi.org/10.1038/nature13697.

Andrews-Hanna, J. C., Zuber, M. T., & Banerdt, W. B. (2008). The Borealis basin and the origin of the Martian crustal dichotomy. *Nature, 453*, 1212–1215. https://doi.org/10.1038/nature07011.

Creveling, J. R., Mitrovica, J. X., Chan, N.-H., Latychev, K., & Matsuyama, I. (2012). Mechanisms for oscillatory true polar wander. *Nature, 491*, 244–248. https://doi.org/10.1038/nature11571.

Marinova, M. M., Aharonson, O., & Asphaug, E. (2008). Mega-impact formation of the Mars hemispheric Dichotomy. *Nature, 453*, 1216–1219. https://doi.org/10.1038/nature07070.

McCausland, J. A., & Cocks, L. R. M. (2012). Phanerozoic polar wander, palaeogeography and dynamics. *Earth Science Reviews, 114*, 325–368. https://doi.org/10.1016/j.earscirev.2012.06.007.

Mitchell, R. N. (2014). True Polar Wander and supercontinent cycles: Implications for lithospheric elasticity and the triaxial Earth. *American Journal of Science, 314*, 966–979. https://doi.org/10.2475/05.2014.04.

Mitrovica, J. X., Hay, C. C., Morrow, E., Kopp, R. E., Dumberry, M., & Stanley, S. (2015). Reconciling past changes in Earth's rotation with 20th century global sea-level rise: Resolving Munk's enigma. *Science Advances, 1*(11), e1500679. https://doi.org/10.1126/sciadv.1500679. https://www.ncbi.nlm.nih.gov/pmc/articles/PMC4730844/.

Mitrovica, J. X., & Milne, G. M. G. (1998). Glaciation-induced perturbations in the Earth's rotation: A new appraisal. *Journal of Geophysical Research-Atmospheres, 103*(B1), 985–100. https://doi.org/10.1029/97JB02121.

Nakamura, R., Yamamoto, S., Matsunaga, T., Ishihara, Y., Morota, T., Hiroi, T., Takeda, H., Ogawa, Y., Yokota, Y., Hirata, N., Ohtake, M., & Saiki, K. (2012). Compositional evidence for an impact origin of the Moon's Procellarum basin. *Nature Geoscience, 5*, 775–778. https://doi.org/10.1038/ngeo1614.

Phillips, B. R., Bunge, H.-P., & Schaber, K. (2009). True polar wander in mantle convection models with multiple, mobile continents. *Gondwana Research, 15*, 288–296. https://doi.org/10.1016/j.gr.2008.11.007.

Roberts, J. H., & Shijie Zhong, J. (2006). Degree-1 convection in the Martian mantle and the origin of the hemispheric dichotomy. *Geophysical Research, 111*, E06013. https://doi.org/10.1029/2005JE002668.

Roy, A., Wright, J. T., & Sigurdsson, S. (2014). Earthshine on a young moon: Explaining the lunar farside highlands. *The Astrophysical Journal Letters, 788*, L42. https://doi.org/10.1088/2041-8205/788/2/L42.

Siegler, M. A., Miller, R. S., Keane, J. T., Laneuville, M., Paige, D. A., Matsuyama, I., Lawrence, D. J., Crotts, A., & Poston, M. J. (2016). Lunar true polar wander inferred from polar hydrogen. *Nature, 531*, 480–484. https://doi.org/10.1038/nature17166.

Šrámek, O., & Zhong, S. (2012). Martian crustal dichotomy and Tharsis formation by partial melting coupled to early plume migration. *Journal of Geophysical Research, 117*, E01005. https://doi.org/10.1029/2011JE003867.

Tajeddine, R., Soderlund, K. M., Thomas, P. C., Helfenstein, P., Hedman, M. M., Burns, J. A., & Schenk, P. M. (2017). True polar wander of enceladus from topographic data. *Icarus, 295*, 46–60. https://doi.org/10.1016/j.icarus.2017.04.019.

Tarduno, J. A., Cottrell, R. D., & Alexei, V. (2002). Smirnov The Cretaceous superchron geodynamo: Observations near the tangent cylinder. *PNAS, 99*(22), 14,020–14,025. https://doi.org/10.1073/pnas.222373499.

Torsvik, T. H., Steinberger, B., Cocks, L. R. M., & Burke, K. (2008). Longitude: Linking Earth's ancient surface to its deep interior. *Earth and Planetary Science Letters, 276*, 273–282. https://doi.org/10.1016/j.epsl.2008.09.026.

Torsvik, T. H., Van der Voo, R., Preeden, U., Mac Niocaill, C., Steinberger, B., Doubrovine, P. V., van Hinsbergen, D. J. J., Domeier, M., Gaina, C., Tohver, E., Meert, J. G., McCausland, P. J. A., & Cocks, L. R. M. (2012). Phanerozoic polar wander, palaeogeography and dynamics. *Earth-Science Reviews, 114*, 325–368.

Wieczorek, M. A., & Le Feuvre, M. (2009). Did a large impact reorient the Moon? *Icarus, 200*(2009), 358–366. https://doi.org/10.1016/j.icarus.2008.12.017.

Wilhelms, D. E., & Squyres, S. W. (1984). The Martian hemispheric dichotomy may be due to a giant impact. *Nature, 309*, 138–140. https://doi.org/10.1038/309138a0.

Zhong, S., Zhang, N., Li, Z.-X., & Roberts, J. H. (2007). Supercontinent cycles, true polar wander, and very long-wavelength mantle convection. *Earth and Planetary Science Letters, 261*, 551–564.

Zhong, S. (2009). Migration of Tharsis volcanism on Mars caused by differential rotation of the lithosphere. *Nature Geoscience, 2*(1), 19–23. https://doi.org/10.1038/NGEO392.

Plumes

Foulger, G. R., & Pearso, D. G. (2001). Is Iceland underlain by a plume in the lower mantle? Seismology and helium isotopes. *Geophysical Journal International, 145*, F1–F5. https://academic.oup.com/gji/article-abstract/145/3/F1/2015734.

Montelli, R., Nolet, G., Dahlen, F. A., & Masters, G. (2006). A catalogue of deep mantle plumes: New results from finite-frequency tomography. *Geochemistry, Geophysics, Geosystems, 7*, Q11007. https://doi.org/10.1029/2006GC001248.

Montelli, R., Nolet, G., Dahlen, F. A., Masters, G., Engdahl, E. R., & Hung, S. H. (2004). Finite-frequency tomography reveals a variety of plumes in the mantle. *Science, 303*, 338–343.

Schoonman, C. M., White, N. J., & Pritchard, D. (2017). Radial viscous fingering of hot asthenosphere within the Icelandic plume beneath the North Atlantic Ocean. *Earth and Planetary Science Letters, 468*, 51–61. https://doi.org/10.1016/j.epsl.2017.03.036.

Lithosphere Structure and Delamination

Abdelsalam, M. G., Liegeois, J.-P., & Stern, R. J. (2002). The Saharan Metacraton. *Journal of African Earth Sciences, 34*, 119–136.

Azzouni-Sekkal, A., Bonin, B., Benhallou, A., Yahiaoui, R., & Liégeois, J.-P. (2007). Cenozoic alkaline volcanism of the Atakor massif, Hoggar, Algeria. *Geological Society of America, 418*, 321–340. https://doi.org/10.1130/2007.2418(16). GSA Special Papers.

Begg, G. C., Griffin, W. L., Natapov, L. M., O'Reilly, S. Y., Grand, S. P., O'Neill, C. J., Hronsky, J. M. A., Poudjom Djomani, Y., Swain, C. J., Deen, T., & Bowden, P. (2009). The lithospheric architecture of Africa: Seismic tomography, mantle petrology, and tectonic evolution. *Geosphere, 5*(1), 23–50. https://doi.org/10.1130/GES00179.1.

Boyd, O. S., Jones, C. H., & Sheehan, A. F. (2004). Foundering lithosphere imaged beneath the Southern Sierra Nevada, California, USA. *Science, 305*, 660–662.

Costa, S., & Rey, P. (1995). Lower crustal rejuvenation and growth during post-thickening collapse—Insights from a crustal cross-section through a Variscan metamorphic core complex. *Geology, 23*, 905–908.

Ducea, M. N. (2002). Constraints on the bulk composition and root foundering rates of continental arcs: A California arc perspective. *Journal of Geophysical Research, 107*(B11), ECV 15-1–ECV 15-13. https://doi.org/10.1029/2001JB000643.

Frimmel, H. E. (2009). Chapter 5.1 Configuration of Pan-African orogenic belts in southwestern Africa. *Developments in Precambrian Geology, 16*, 145–151. https://www.researchgate.net/publication/251455251.

Jagoutz, O., & Behn, M. D. (2013). Foundering of lower island-arc crust as an explanation for the origin of the continental Moho. *Nature, 504*, 131–134. https://doi.org/10.1038/nature12758.

Kay, R. W., & Mahlburgkay, S. (1991). Creation and destruction of lower continental-crust. *Geologische Rundschau, 80*, 259–278.

Levander, A., Schmandt, B., Miller, M. S., Liu, K., Karlstrom, K. E., Crow, R. S., Lee, C.-T. A., & Humphreys, E. D. (2011). Continuing Colorado plateau uplift by delamination style convective lithospheric downwelling. *Nature, 472*, 461–465. https://doi.org/10.1038/nature10001.

Liégeois, J.-P. (2006). The Hoggar swell and volcanism, Tuareg shield, Central Sahara: Intraplate reactivation of Precambrian structures as a result of Alpine convergence. www.mantleplumes.org.

Murray, K. E., Ducea, M. N., & Schoenbohm, L. (2015). Foundering-driven lithospheric melting: The source of central Andean mafic lavas on the Puna Plateau (22°S–27°S). *The Geological Society of America Memoir, 212*, 139–166.

Prouteau, G., Scaillet, B., Pichavant, M., & Maury, R. (2001). Evidence for mantle metasomatism by hydrous silicic melts derived from subducted oceanic crust. *Nature, 410*, 197–200.

Stein, M., & Hofmann, A. W. (1994). Mantle plumes and episodic crustal growth. *Nature, 372*, 63–68.

Tanton, L. T. E., & Hager, B. H. (2000). Melt Intrusion as a trigger for lithospheric foundering and the eruption of the Siberian flood basalts. *Geophysical Research Letters, 27*(23), 3937–3940.

Wu, F. Y., Xu, Y. G., Zhu, R. X., & Zhang, G. W. (2014). Thinning and destruction of the cratonic lithosphere: A global perspective. *Science China Earth Sciences, 57*(12), 2878–2890. https://doi.org/10.1007/s11430-014-4995-0.

Granitoids and Other Igneous Rocks

Campbell, I. H., & Taylor, S. R. (1983). No water, no granites: No oceans, no continents. *Geophysical Research Letters, 10*, 1061–1064.

Castro, A., & Gerya, T. V. (2008). Magmatic implications of mantle wedge plumes: Experimental study. *Lithos, 103*, 138–148. https://doi.org/10.1016/j.lithos.2007.09.012.

Frezzotti, M.-L., & Touret, J. L. R. (2014). CO_2, carbonate-rich melts, and brines in the mantle. *Geoscience Frontiers, 5*, 697–710. https://doi.org/10.1016/j.gsf.2014.03.014.

Kawamotoa, T., Yoshikawab, M., Kumagaia, Y., Mirabuenoc, M. H. T., Okunod, M., & Kobayashie, T. (2013). Mantle wedge infiltrated with saline fluids from dehydration and decarbonation of subducting slab. *PNAS, 110*(24), 9663–9668. https://doi.org/10.1073/pnas.1302040110.

Pitcher, A. (1983). Classification of granitoid rocks based on tectonic setting. In K. J. Hsü (Ed.), *Mountain building processes*. London: Academic Press.

Pitcher, W. S. (1993). *The nature and origin of granite*. London: Blackie Academic and Professional. xiv + 321 pp. ISBN 0-7514-0080-7.

Polat, A., & Kerrich, R. (2001b). Magnesian andesites, Nb-enriched basalt-andesites, and adakites from late-Archean 2.7 Ga Wawa greenstone belts, Superior Province, Canada: Implications for late Archean subduction zone petrogenetic processes. *Contributions to Mineralogy and Petrology, 141*, 36–52.

Puchtel, I. S., Haase, K. M., Hofmann, A. W., Chauvel, C., Kulikov, V. S., GarbeSchonberg, C. D., & Nemchin, A. A. (1997). Petrology and geochemistry of crustally contaminated komatiitic basalts from the Vetreny Belt, southeastern Baltic Shield: Evidence for an early Proterozoic mantle plume beneath rifted Archean continental lithosphere. *Geochimica et Cosmochimica Acta, 61*, 1205–1222.

Rupkea, L. H., Morgana, J. P., Hortb, M., & Connolly, J. A. D. (2004). Serpentine and the subduction zone water cycle. *Earth and Planetary Science Letters, 223*, 17–34. https://doi.org/10.1016/j.epsl.2004.04.018.

Scaillet, B., & Prouteau, G. (2001a). Oceanic slab melting and mantle metasomatism. *Science Progress, Science Reviews, 84*, 335–354.

Scaillet, B., & Prouteau, G. (2001b). Oceanic slab melting and mantle metasomatism. *Science Progress, Science Reviews 2000 Ltd., 84*, 335–354.

Straub, S. M., Gomez-Tuena, A., Stuart, F. M., Zellmer, G. F., Espinasa-Perena, R., Cai, Y., & Iizuka, Y. (2011). Formation of hybrid arc andesites beneath thick continental crust. *Earth and Planetary Science Letters, 303*, 337–347. https://doi.org/10.1016/j.epsl.2011.01.013.

Winter, J. D. (2001). *An introduction to igneous and metamorphic petrology*. Upper Saddle River: Prentice Hall. ISBN-13: 978-0132403429.

Chapter 4
Deep Cycles and Super-Terrans

Introduction

Possibly the most important consequence of plate tectonics is the cycling of materials into and out of the mantle. Such a conveyor belt delivers carbon dioxide into the interior in the form of carbonate rock and returns this gas to the atmosphere, thereby modulating the planetary greenhouse effect. One of the more subtle and poorly understood processes is the modulation of surface water—and with it the abundance of many of the ions (both metal and non-metal ions) that regulate the biological capacity of any biosphere that we hope caps our planets of interest.

Consequently, there is an increasing focus on the cycling of water and dissolved substances and a growing appreciation of how the dynamic nature of a planet's interior is linked to the vigor and sustainability of the biosphere. In this chapter, we examine how plate tectonics and the other processes that were examined in Chap. 3 help regulate the overall health of the planet's hydrosphere and atmosphere. Moreover, we examine the idea that while plate tectonics is an efficient method through which materials may cycle into and out of a planet's interior, it is not the only one in a planet's arsenal.

Subduction, Whole Mantle Convection and Element Cycling

On Earth subduction is critical to the cycling of various materials between the surface and the deep interior. Subduction clearly deposits water in the upper mantle. Were it not for this process, water would be driven out of the hot interior, leaving it dry. In turn, the manner in which the interior loses heat determines how the mantle will circulate and how quickly heat, water and other materials are driven out of it (Fig. 4.1).

This has some interesting consequences for telluric planets, some of which may not be obvious at first glance. Perhaps most obviously, the introduction of cold

D. S. Stevenson, *Red Dwarfs*, https://doi.org/10.1007/978-3-030-25550-3_4

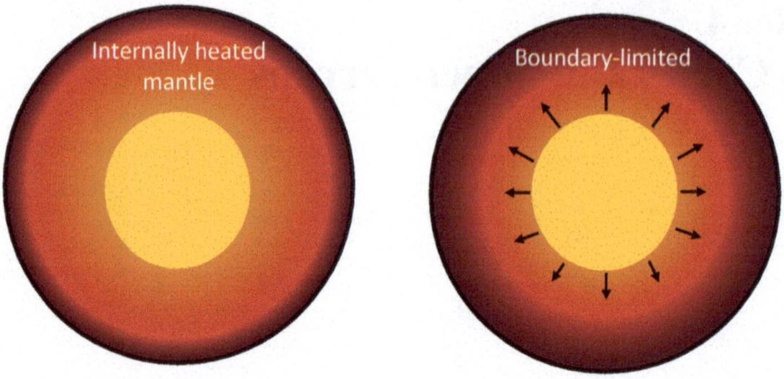

Fig. 4.1 Two models for planetary convection. In the internally heated, single-layer model, left, convection in the mantle is driven by internal heating (principally by radioactive decay). In the boundary-limited model, right, convection is driven by conduction of heat from the core and heat loss from the surface. Whichever mode is dominant determines how quickly water is driven out to the surface then sequestered by cooler igneous rock

lithosphere into the hot upper mantle results in its cooling. As we've seen this is the principle mechanism through which the Earth cools down. However, subduction also delivers water and a variety of soluble substances from the oceans above, including phosphates, halides, such as chlorine ions, as well as carbon dioxide.

If we examine the cooling effect, one might reasonably assume that as the upper mantle has cooled over time it would become more viscous and, consequently, the rate of plate motion would decline. Indeed, this was the conclusion of Martin Van Kranendonk (University of New South Wales) and colleagues, who had examined the rate at which various elements had been concentrated in rocks over the last 1.2 billion years. They found that tectonic activity appeared to increase from 3 billion years ago to a peak around 1.1 billion years ago (itself, somewhat unexpected) before declining once more. The peak corresponds to the dispersion of the Columbia/Nuna supercontinent, followed by its rearrangement and later collision as Rodinia. So far so good. In this scenario, subduction around Columbia fragments accelerates overall plate motion.

However, just a year after Van Kranendonk's findings, Ken Condie (New Mexico Institute of Mining and Technology in Socorro) and colleagues came to quite the opposite conclusion. Rather than examine mineralogy, they investigated the rate at which mountain building had occurred since the early Proterozoic. Combining this information with paleomagnetic analysis of various volcanic rocks led them to conclude that the rate of plate motion had accelerated continuously since at least the early Proterozoic. The concurrence of both sets of data is a powerful means of verifying a hypothesis, or a lone set of observations. If this is true—and many, including Van Kranendonk, will need some convincing—then it needs some explanation.

Again, if this is true, the most plausible explanation involves the subduction of water into Earth's mantle. While the mantle is cooling and should become more viscous, the lower temperatures allow more of the water delivered by subduction to persist. In a hot planet, the mass of water the mantle can hold is relatively small.

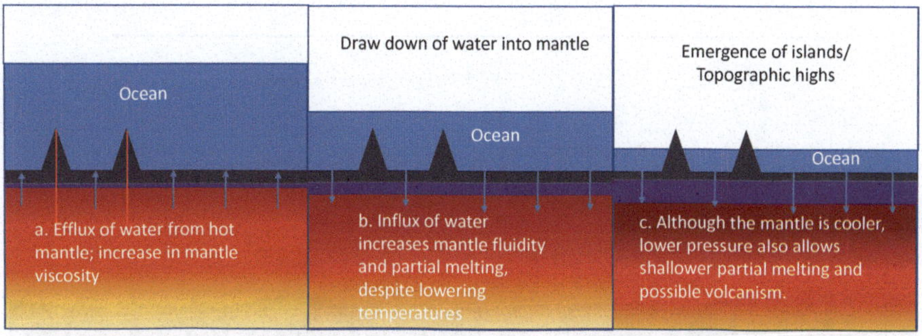

Fig. 4.2 Rescuing a drowned, capped super-terran. High pressure super-terrans should preferentially form thick granitoid crusts (Fig. 4.4), deep oceans and deep-partial melting. This favors limited volcanism (**a**). However, as the interior cools, (**b**), water can return to the mantle. This reduces the pressure on the top of the mantle, allowing greater degrees of partial melting. Influx of water may eventually expose topographic highs (**c**) and permit surface volcanism. The larger the super-terran the more limited the effect will be and the longer it would take

Therefore, while much of our planet's water may have started out in the mantle, when a planet is young and hot, it doesn't stay there for long.

However, let Earth cool, and the mantle will now be able to hold onto more water in its cooling silicate broth. Water lowers both the overall viscosity and allows increasing amounts of partial melting (Fig. 4.2). Both of these make the mantle less viscous, despite an overall cooling trend. Although such a trend cannot continue indefinitely, as long as the upper mantle remains hotter than 1100 °C, increasing the mass of water should have the desired effect. Lower the temperature much more, and the upper mantle becomes too cool to partially melt at the kinds of pressures that are encountered here (Fig. 4.2.)

To this end Jun Korenaga has attempted to formulate just how much water is delivered by subduction to the mantle per annum, while taking into account a cooling Earth. Based on various assumptions, Korenaga derives the rather formidable figure of 200–300 trillion grams per year for the last few billion years. Now, this isn't cycling; rather it is the *net gain* of water by the planet's mantle. By assuming this, Korenaga's early Earth has a drier and higher viscosity mantle than it has now. In turn, this higher viscosity allows the asthenosphere and lithosphere to gel together more harmoniously and transmit the forces needed to move plates—or blocks of proto-continental—crust around. This allows continents to stand proud of the oceans in the Archaean. In 2018, research by Cai Chen (Washington University in St Louis) and colleagues have conducted further analysis and increased the gross flow of water into the mantle, again more than Korenaga previously determined.

Korenaga assumes that water can be delivered to the deep mantle and works by convection and the steady cooling of the mantle as a whole. This model assumes that the Archaean oceans were deeper than those of today, and this suggests that while there were continental blocks, most would be largely submerged. It also appears to imply plate tectonics operated as early as the latest Archean. This concurs with oxygen-isotope evidence, uncovered by I. N. Bindeman (University of Oregon) and

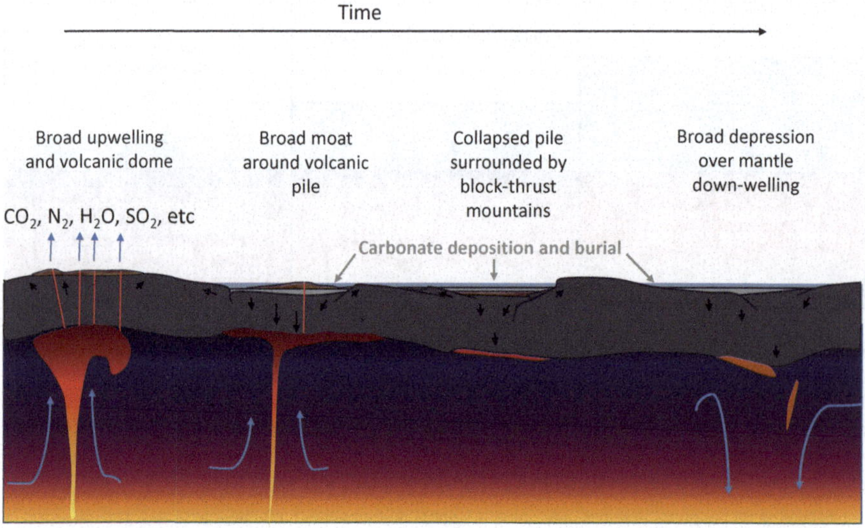

Fig. 4.3 Alternatives to plate tectonics. Mantle plumes at various stages in their evolution cause bulges and volcanism, while thick piles of volcanic material or mantle downwellings cause collapse with burial of carbonate rocks. A limited carbonate-silicate cycle is, therefore, possible. (This figure is reproduced in modified form from Stevenson, 2018a)

co-workers, which also suggests that until the end of the Archaean, most of the continental crust that existed was submerged; 10% or less was sub-aerial. Unfortunately, there is no consensus on the timing of when plate tectonics began on Earth. Prior to the initiation of plate tectonics, other forms of tectonism, and burial may have done the trick (Fig. 4.3). Again, more research is needed.

An interesting link extends to the broader consideration of habitability of the early Earth. Why wasn't it frozen over when the Sun was younger and less luminous? Minik Rosing and colleagues proposed (quite convincingly) that an early, dark Earth (faint Sun paradox) would remain warm, without implausible quantities of greenhouse gases in the atmosphere—concentrations that are not supported by the evidence available. If most of Earth's surface was under water, then it would be dark (have a low albedo). Under these circumstances it could remain warm enough to stop its surface freezing over, despite quite modest abundances of carbon dioxide or methane. Other analyses by Rita Parai and Sujoy Mukhopadhaya (both from Washington University in St. Louis) suggests that the mantle began to in-gas around 2.5 billion years ago. Although this date is earlier than that of Korenega and others it does generally follow the principle that the mantle is becoming wetter and by inference the volume of the oceans less. It does, however, concur with Bindeman's work, suggesting that the continents began to emerge from the oceans after 3 billion years ago (above). Parai's and Mukhopadhaya's work also implies that the Archaean mantle became progressively drier and likely more viscous as the era drew to a close. Subsequently, influx of water and other volatiles has reduced the viscosity in line with Condie's and Korenega's ideas.

Just to illustrate the range of views, consider the models of Tetsuzo Seno and Satoru Honda (both from the University of Tokyo). They would concur that Earth's mantle is currently gaining more water than it is losing. However, crucially, they suggest that this has only been operating for the last billion years or so. In order to explain the presence of continents in the Proterozoic—that would otherwise be under water—they suggest that plate tectonics was not in operation throughout most of this period—in agreement with other observations. Instead, they suggest that Earth had a stagnant lid prior to 2 billion years ago. With a stagnant lid, relatively little water could be transported into the mantle.

It was only when transport really got going, they suggest, that the continents emerged piecemeal from the depths. They propose that between 2.0 and 1.4 billion years ago shallow subduction brought buoyant slabs into the upper mantle, underthrusting the edges of the continents and raising them upwards. This process came to an end 1.4 billion years ago, when slabs became dense enough to sink into the mantle in the manner we see today. At this stage, continents finally emerged in a form that we would recognize. Although this variant has some advantages, it still does not address evidence for plate tectonics from 3 billion years ago, nor directly for rapid growth of continents (or submerged continental crust) in the Archaean. Nor does this conclusion agree with the isotopic evidence from Bindeman and colleague's work, for the emergence of continents at 3 billion years ago (above).

Senso and Honda suggest some solutions involving periodic avalanches of thick oceanic crust into the mantle. However, this seems a little ad hoc. The best-fit models for the continental crust seem to be those suggested in Chap. 3—heat-pipe in the Hadean followed by a squishy lid in the earliest Archaean. As the Archaean progresses, continued cooling leads to shallow subduction, then to full subduction. This gives both the correct chemistry of the continental crust—and as the next section shows, the correct sequence in diamond ages and their chemistries.

All this leads us to conclude that while the mantle is likely taking on water at a substantial rate, the timing of this process needs much closer examination. However, it is certainly appealing that the Hadean Earth may have been one vast dark ocean, dotted with volcanic islands. It was only when the volume of continental crust increased through shallow subduction and melting of eclogite that Earth began to show its hidden continents. At some point in this part of our planet's history—or perhaps as late as the Proterozoic, the oceans did begin to retreat into the cooling mantle. Earth has, thus, transitioned from an aquaplanet to the more telluric face we recognize today. Collectively, these models all suggest the same thing: that the early Earth lacked plate tectonics, and most continental crust was under water. The Hadean and early Archaean Earth was an aquaplanet.

Although this conclusion seems increasingly robust, what is less clear is when the continents emerged. Bindeman and Korenega place their emergence near the end of the Archaean. Others place it as recently as 800 million years ago. The divergence in opinion is largely a product of the models that are employed and the assumptions that go into developing them. Bindeman's and colleagues' work is certainly the more empirical of those considered here, but one might quibble over the source rocks used to determine the oxygen isotopes and how these may have been altered in the billions of years since formation. However, given a concordance between their data, the appearance of ophiolites and other chemical data (described below), it begins to crystallize the idea that continental crust did emerge from the oceans, on mass, around 3 billion years ago. If so, it is then a question of how we interpret the other observations, or refine the models to match the empirical data.

Draw-Down of Water and the Impact on Climate

The gradual (or stepwise) emergence of continents has critical implications for climate and the early evolution of life. As well as making the planet cool, the emergence of reflective (high albedo) land also provided life with an increasing number of niches. Although the ocean makes an appealing cradle for life's synthesis, a Panthalassic ocean is limiting if you want biological complexity. Almost certainly, you likely require a complex geological environment, if one is to nurture biological complexity. This is dissected in detail in Chap. 7, but the important point here is that the complex cellular life with which we are most familiar (the eukaryotes) only diversified after a billion years ago. This appears to have occurred *after* the mantle began to reabsorb water from the oceans. Is this a coincidence?

The upper mantle holds only 0.01–0.02 wt% as water, as is evident from the kinds of basalts erupted along the mid-ocean ridges. Graham Pearson (University of Alberta) and colleagues suggested that additional water may be stored in the mantle transition zone, in this case bound up in high-pressure forms of olivine known as wadsleyite and ringwoodite. In high-pressure experiments, these minerals were found to be able to hold up to 2.5% of their weight as water. Evidence to support this conclusion came from examination of small ringwoodite mineral inclusions in super-deep diamonds.

This is clearly a controversial idea; but does it tally with any other evidence or models? A seemingly abstract connection was made by two researchers who were investigating the effect of planetary mass on the persistence of oceans. In 2015, Laura Schaefer and Dimitar Sasselov (both of the Harvard-Smithsonian Center for Astrophysics) presented some analyses where they compared planetary mass and the rate at which a planet will de-gas. A critical factor in determining the outcome of their models was the manner in which the mantle overturned through convection.

If one considered the mantle as a singular block that was heated primarily from within and lost heat by conduction at the top, it was called a "single-layer" model. In this instance the stability of the mantle was determined solely by its internal properties, such as temperature and viscosity. In a second model, a planet had a mantle in which convection was driven by the input of heat at its base, through the core-mantle boundary, but also by the presence of a cold top layer (a lithosphere). This model experienced "boundary-layer convection"—convection driven by heating from the base and heat loss from above. In reality, Earth is likely to operate as a mixture of these two extremes, with a cold lithosphere, hot core-mantle boundary and internal heating through radioactive decay.

The single-layer models had a lower mantle that was much more viscous than the upper mantle. Telluric planets, which followed this model of convection and had a low mass (Earth-mass), degassed strongly early on. That is, most of their water ended up on the planet's surface within the first few tens to hundreds of millions of years. However, as the mass of the planet increased, there was an increasing delay in outgassing. This was because there was a delay in the amount of heating of the mantle through radioactive decay. These more massive single-layer models took increasing amounts of time to reach their maximum temperature and drive convection. This meant that a three Earth-mass planet, whose interior circulated in the single-layer mode, had far more water on its surface than one with five Earth masses, for most of its history (about 1.6 versus 1.4 terrestrial ocean's worth until late on). It took fully 8 billion years for the five Earth-mass world to have more surface water than its smaller cousin. Importantly, only the smallest planet considered by Schaefer and Sasselov, with one Earth-mass, released its water promptly if it operated using the single-layer mode of convection.

When Schaefer and Sasselov adopted the boundary layer model, convection was far more vigorous, and all planets lost water to their surfaces strongly and early. This implied that planets that operated this style of convection, and which had masses greater than twice that of Earth, would likely be waterworlds. A planet with five times the mass of Earth is likely to have eight terrestrial oceans' worth of water on its surface. By comparison, in the single-layer model, such planets have only 1.6 Earth-oceans' worth.

In the context of Condie and colleagues work, the cooling trend becomes important. Although these boundary-layer model worlds lost water to the surface faster, these planets also cooled faster than the single-layer models. This meant boundary-cooled planets drew their oceans back into the mantle early on—in less than 5 billion years. This left these planets both inhospitable and likely free of plate tectonics at later times. Only the Earth-mass planet has significant oceans at 10 billion years—but occupying one-fifth the current volume of terrestrial oceans. This could mean, of course, Ken Condie's and co-worker's models might be right, and there is a significant and increasing reservoir of water in our planet's mantle that might cause plate motion to accelerate.

In both the single-layer and boundary-layer models, if the length of spreading ridges is increased, then more water is returned to the mantle through the serpentinization of basalts at the ridge axis. Lower rates of plate tectonics, therefore, extends the lifetime of planetary oceans but might limit the time over which plate tectonics operates. Therefore, depending on how you take these models, whether a planet operated plate tectonics and sustains surface oceans might depend on how the mantle circulates—as well as the overall mass of the planet.

An interesting point that Schaefer and Sasselov don't consider is the effect of planetary mass on the stability of the lithosphere and its propensity to host plate tectonics. Up the mass and, as we saw in Chap. 3, then basaltic material morphs into eclogite, and the entire lithosphere becomes unstable. After a very short interval, most of the surface becomes converted to granitoid (TTGs; Fig. 4.4). This material is low density and will resist subduction, therefore inhibiting the return of water and

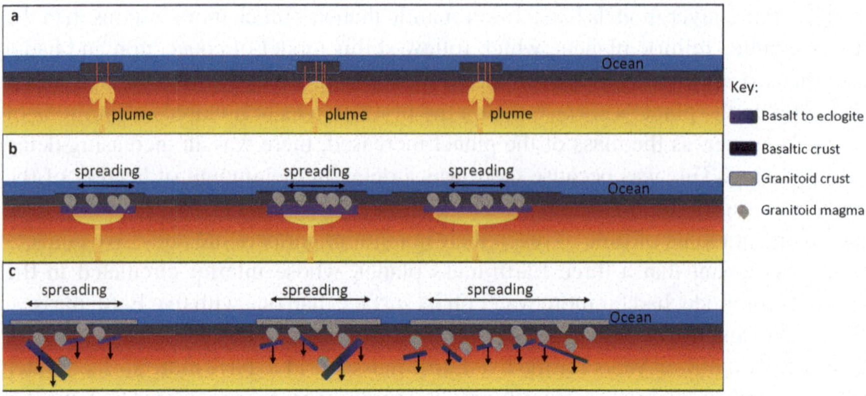

Fig. 4.4 Delamination in young super-terrans with the formation of a pan-TTG (granitoid) layer. (**a**) Plumes generate protocontinents (as happened in the Hadean Earth). (**b**) Granitoid continental crust forms from partial melting of basalt and eclogite. This is hot enough at the base to spread outwards over neighboring denser oceanic crust. In **c**, spreading continues as more granitoids are delivered to the surface. However, the thickened crust and mantle lithosphere becomes unstable, and the base falls off (delaminates). This forms more granitoids as it sinks into the mantle and partly melts. Subduction (not shown) may also occur around the edges of the spreading continents, causing further growth and spreading of continental crust, ultimately forming a thick, stagnant lid

other volatiles to the mantle. This level of modeling, it seems, remains to be done, but it will be interesting to see how the emergence and persistence of oceans is affected when the composition of the crust is taken into effect. In the absence of plate tectonics, many of the conclusions of their work become mute. Water simply isn't returned to the mantle, instead persisting on the surface and in the crust.

Oceans, Planet Mass and Kepler's Observations

How does planetary mass affect the *depth* of oceans and the potential for continents to emerge from their surfaces? Chapters 7, 8 and 9 will examine the consequences of ocean depth on life in more detail, but the principles need outlining—and connecting with observations made by Kepler. What we know presently is that there appears to be a drop in overall density for planets with more than 1.6 Earth radii. Planets more substantial than this likely have thick volatile carapaces that most likely consist of deep oceans and increasingly thick atmospheres.

What about those worlds that have radii smaller than 1.6-Earths but are still larger than Earth? Is there a simple relationship that might apply to most planets? To address this, Nicolas Cowan (Northwestern University) and Dorian Abbott (University of Chicago) modeled the effect of planetary gravity on the rates of outgassing and in-gassing that, in turn, determine ocean depth. As planetary mass increases, the rate of outgassing decreases as more water is physically forced into colder surface rocks. Simultaneously, in-gassing increases as more water is forced into reactions with olivine, forming serpentine and amphibole. Although this counters the effect of increasing mass on convection, planets with greater masses retain or return more water to the planetary lithosphere in the form of these minerals.

Schaeffer and Sasselov's models also show that the greater the mass of the super-Earth, the greater the fraction of water that is retained in the mantle. This has the curious effect of limiting the depth of oceans on any planet and suggests that at least for planets in the 1–3 Earth mass range, increasing planetary mass may not give rise to increasingly deep oceans, because more water remains trapped in the lithosphere and upper mantle. Where the mass, and hence surface gravity, goes up, the volume of water that is released onto the surface decreases. Schaeffer and Sasselov state that, at least as far as their modeling shows, any tectonically active terrestrial planet, irrespective of its mass, will maintain exposed continents if its water-mass fraction is less than approximately 0.2% of its total planetary mass.

This might seem to suggest that all planets would simply be scaled-up Earths. But, of course, this is not the case because nucleosynthesis—the process of element creation in stars—favors some elements over others. Oxygen and carbon are far more abundant than silicon and iron. Therefore, as planetary mass increases, more and more of the planet's mass will be comprised of low-mass elements, in particular hydrogen, helium, oxygen and carbon. There must come a point where the surface will be submerged first in deep oceans of hydrogen oxide (aka water), then hydrogen and helium. Kepler's observations suggested that this is around 1.6 Earth radii or around 5 Earth masses, depending on the precise composition. Earth radii of 1.6 is the absolute upper limit, with planets in the 1.4–1.6 Earth radii zones most likely being aquaplanets. Somewhere between 1.2 and 1.4 Earth radii, the surface becomes submerged for planets of Earth-like compositions. Beyond this, the mass is increasingly dominated by lighter elements and their compounds, with greater planetary masses attracting increasing abundances of hydrogen and helium from the protoplanetary material.

We can then imagine aquaplanets with thick, underlying granitoid crusts that are produced by delamination of eclogite. These are not drowned Earths but something altogether alien. Tectonic activity will not cease but may resemble the cool, stagnant lid of Mars or the hotter, squishier, thick lid of Venus. The evolutionary scenario is shown in Fig. 5.2 in the next chapter. There will be no visible continents, just a thick, hidden, pan-global granitoid veneer. However, for the planets at the lower end of this mass range (with, perhaps, 1.2–1.3 Earth-radii) hope may come from the regassing of the mantle. As Schaeffer and Sasselov suggest, over time, water returns to the mantle in most models of higher mass planets. Therefore, aquaplanets may not always remain so. They may well flood early on, when their hot mantles de-gas. However, as they cool and age beyond 5 billion years, one may see continental masses emerge from their watery cocoons. Again, the end result may look rather different than Earth. An absence of plate boundaries, and a squishy stagnant lid, may leave a planet looking like a typical science fiction world where the creators have not considered plate tectonism—a Jackson Pollock of land and sea scattered across the planetary canvas. Over time scattered islands will enlarge and merge, leaving mostly land pockmarked by small seas.

These conclusions rest on various assumptions about how much water is retained in the mantle of Earth—and this is certainly controversial. Moreover, it adds another tier to our understanding of why Earth has continents. This deep-water cycle is coupled to the presence of a buoyant continental crust that rises isostatically in the dense mantle. In turn, such thick crust has its altitude maintained by a combination of erosion, delamination and persistent growth. In concert, these processes maintain dry land on Earth and possibly on planets with masses as great as five times that of Earth.

When we combine this study with that of Nicolas Cowan and Dorian Abbott, we can begin to get a flavor of how the amount of water and the mass of the planet begin to affect how much continental crust may form and what fraction of a planet's surface may be covered with water. In Cowan's and Abbott's work, as long as the fraction of planetary mass that is water remains below 0.2 percent of the total, then any oceans will be shallow enough that continental crust may emerge above its surface. Schaefer's and Sasselov's work provides the basis of the discussion as to how effectively this water escapes, persists and is finally reabsorbed into a planet's hot mantle. Although other processes, such as bombardment, stripping by the solar wind or the action of extreme ultraviolet and X-rays (Chap. 5) take water away from the top of the atmosphere, these studies emphasize that such loss of critical volatiles must also be examined from the bottom up.

Finally, recent work identifies a more general process whereby water may be lost to the surface of a planet—and as the planet cools, not return. This goes under the generic banner "sequestration." For cool planets, i.e., those with temperate surface temperatures, water is lost through two sequestration processes. The first is obvious—freezing. Water can freeze out, forming ice caps or as permafrost. Both, obviously, require low temperatures, and the latter also requires that the geothermal gradient is not so high that the formation and persistence of subterranean ice is precluded. Terrestrial permafrost is found around the polar regions and on high mountains. The largest areas of permafrost are found around the Arctic Circle, underlying tundra. Here, the ice is insulated against high summer temperatures by overlying vegetation and thick, waterlogged, peaty-soil. Although Mars clearly lacks vegetation, there is abundant evidence for sub-surface ice. This was discussed in relation to TPW in Chap. 4, but in essence a large mass of water-ice lies buried in sediments along the equatorial band, where copious precipitation, early in Martian history, led to the growth of glaciers, as well as the development of permafrost.

At the Martian poles, a mixture of carbon dioxide and water-ice forms both visible ice-caps and a large amount of permafrost. The carbon dioxide frost readily sublimates in the hemispheric summer, but this flows in large atmospheric motion towards the opposing winter hemisphere, where it refreezes. However, in addition to the more obvious processes of freezing and re-sublimation, water has also altered the surface chemistry, and Jon Wade (University of Oxford) and colleagues have proposed that there may be substantial water sequestered into the rocks of the crust and upper lithosphere (the top 80 km or so of rock).

Depending on the figures you follow, there may be as much water in the lithosphere of Mars as there is in the form of surface and near-surface ice. The key chemistry involves ferrous oxide (FeO)-rich rocks, of which Mars has more than Earth. This allows Martian lavas to soak up more water than their terrestrial equivalents. Therefore, while we can make broad calculations regarding the absorption of water into a planet's interior, we should remain mindful that the chemical composition of the planet will influence to what extent sequestration actually occurs.

The Deep Carbon Cycle

We are all familiar with the carbon cycle as taught in high schools across the globe, but few students are as familiar with the deep carbon cycle—the fate and processing of carbon that is subducted or otherwise assimilated into the mantle. An appreciation of this cycle is fundamental to any consideration of habitability, as carbon is both the stuff of life and the principle determinant (as carbon dioxide and methane) of long-term habitability in terms of climate.

Carbon Cycling as Carbonate and Soluble Carbon Dioxide

The simplest part of this deep carbon cycle involves the delivery of carbon dioxide from the mantle to the atmosphere. Here, very unsurprisingly, a planet with a hot interior releases carbon dioxide with water vapor via volcanism. The early Earth may have had an atmosphere with 80 times its current pressure and a temperature of around 500 °C. Carbon dioxide and water vapor shielded and absorbed heat from an underlying magma ocean. Once this ocean of magma cooled and formed a crust, temperatures fell, and the water vapor became sequestered in a "Panthalassic" ocean. As soon as this process began, carbon dioxide would sequester with it as carbonic acid, then various salts of this, of which calcium, sodium and potassium carbonate would be the most abundant. Although sodium and potassium carbonate are very soluble in water, calcium carbonate (and magnesium carbonate) are not. These readily precipitate onto the ocean floor, forming limestone (and dolomite).

Unless these carbonate rocks can heat to around 700–900 °C the carbon dioxide remains bound up in the rock. Any hotter, and the carbonates thermally decompose to form the metal oxide and carbon dioxide once more. Assuming that the planet's atmosphere doesn't experience a thermal runaway, like Venus, these conditions will only be reached in the upper mantle or lower crust. Now, this is an interesting point. Thinking back to Chap. 3, the early Earth didn't have plate tectonics—indeed it probably didn't for its first 1.5 billion years. How then did the carbon dioxide return to the atmosphere in this era? A simple answer might be, "It didn't." Earth merely degassed more to broadly keep pace with the loss to the lithosphere. Studies do suggest that the amount of carbon in the mantle far exceeds that in surface reservoirs. If there was enough CO_2 bound up in the mantle, then developing surface biology and the accompanying climate remained stable. Remember, if there was a global ocean, with little land, there wouldn't be much weathering to remove carbon dioxide.

However, if land was prevalent and weathering more commonplace, we might want to consider an alternative to plate tectonics to remove and restore carbon dioxide in the atmosphere. Indeed, one agrees with the increasingly vexed sentiments of some researchers who have, for want of a better phrase, become fed up with some of their peers equating plate tectonics with carbon cycling. Although the former begets the latter, the converse is not necessarily true. What are the alternatives?

Vertical tectonics in various guises can work just as well as plate tectonics if the rate of burial is sufficient. If we imagine an early Earth with abundant volcanism, basalt and komatiite will form deep piles. Volcanism is unlikely to be continuous, allowing for the deposition and burial of carbonate rocks in these increasingly dense stacks of igneous rock. If we keep piling on the pressure—and intrude more and more basalt and komatiite into them, the increasing heat and pressure will eventually cause the rocks to decompose (Fig. 4.3). Carbon dioxide can then return to the atmosphere during the Hadean. Moving on through to the Archaean, similar burial and nascent plate tectonics will also suffice to bury and recycle carbon dioxide.

Tidal Locking and the Carbonate-Silicate Cycle

Now, here's a thought. Since the carbonate-silicate cycle operates principally through the rock cycle of weathering, erosion, transport, deposition and burial, what happens if you focus your rainfall in one area of the planet, thereby geographically restricting these processes in turn? Imagine a planet where rainfall is geographically restricted to one, relatively small region but is otherwise dry. What will happen to the efficiency of the carbonate-silicate cycle? This is precisely what is expected on tidally locked planets (Chap. 5). Can such worlds operate with an efficient cycle, irrespective of the operation of plate tectonics?

Imagine, as Adam Edson and colleagues (Pennsylvania State University) did, that we place Earth in the habitable zone of a red dwarf. Rainfall is principally focused on the SSP. Now, let's play with the location of the SSP relative to the location of the major continents. When Edson and colleagues placed the SSP over the Atlantic Ocean, the close proximity of the region of maximum heating to the continent caused precipitation and rapid weathering on the neighboring continents. This lowered the carbon dioxide concentration in the atmosphere, which in turn lowered the global temperatures to a mean value of 247 K (−26 °C). In these simulations the area of the planet's surface that remained habitable fell—unless the planet was maneuvered closer to its star.

Conversely, with the SSP over the Pacific Ocean, the long stretch of water between the SSP and the land meant that there was less erosion and consequently higher concentrations of carbon dioxide in the atmosphere. Rather than 7 parts per million (7 ppm) in the Atlantic Ocean scenario, the Pacific Ocean one had more than

60,300 ppm (a rather whopping and toxic 6.03%). Although this certainly raised the mean global temperature to 282 K (9 °C), and made more of the surface potentially habitable, one notes that the majority of the continents were rigidly frozen in ice—and if not frozen shrouded in perpetual darkness, hardly conducive for the formation of land-based food chains. Hence, the term habitable may be something of a misnomer.

Moreover, though the Atlantic Ocean scenario definitely produces a planet that is habitable in terms of temperature, 7 ppm is lower than the carbon dioxide concentration that terrestrial plants can fix through photosynthesis. The converse scenario, with more than 6% carbon dioxide, would be toxic to mammals and many other terrestrial organisms. Finally, the study neglects the operation of the return processes. How much carbon dioxide is returned by volcanism and other processes? Therefore, while interesting, Edson's article is more work in progress than an effective application.

Carbon Goes Commando in the Deep

In the deep mantle carbon adopts a different form, molecular carbon—diamond. Although this may seem like a prosaic aside, the delivery of carbon to the deep mantle (or its release from primordial reserves) is an important component of the deep carbon cycle. Aside from producing many interesting and exclusive gems, the deepest portion of the carbon cycle—that occurring in the lower mantle, transition zone and lowermost upper mantle—helps regulate melting in mantle plumes and the overall redox state of the mantle. "Redox" refers to how chemically oxidized mantle minerals are. For example iron is found either as Ferric (Fe^{3+}) or ferrous (Fe^{2+}) ions in various minerals such as pyroxene or olivine (see above, in relation to Martian lavas). Carbon may be found as carbon dioxide or carbonate in the upper mantle (where it is oxidized) or as graphite or perhaps methane where it is reduced, deeper down. On Earth there is a depth pressure limit for carbon as carbon dioxide, beneath which (with higher pressure and suitable chemistry) carbon dioxide is no longer stable.

Two kinds of chemical reactions can transform carbonates into small crystalline beauties, as carbonate is carried by convection and diffusion to greater depths. The key to these is the degree of oxidation of iron. In the uppermost mantle iron has an average of one oxygen atom per iron atom. However, as we descend, oxygen becomes increasingly stable in magnetite, with an average of three oxygen atoms per iron atom. Iron in the deep thus sequesters more and more oxygen from its surroundings. This reaction then reduces the remainder of the mantle and results in reactions that remove oxygen from carbon more likely. These are illustrated in Table 4.1.

In peridotite, the first process involves two silicates, pyroxene and olivine. A form of pyroxene called enstatite ($MgSiO_3$) reacts with magnesium carbonate (magnesite, $MgCO_3$). The carbonate mineral is reduced to carbon, while the iron in the

Table 4.1 The silicates and carbonates involved in the terrestrial manufacture of diamonds

Diamond type	Carbon source	Reductant	Product silicate	Location	Geological period
P-type	Magnesite ($MgCO_3$)	Enstatite (orthopyroxene; $MgSiO_3$)	Olivine ($MgSiO_4$)	Craton root/ Upper mantle	Throughout Earth's history
E-type	Dolomite ($CaMg(CO_3)_2$)	Coesite (SiO_2)	Diopside (Clinopyroxene; $CaMgSi_2O_6$)	Subducting ocean crust	Last 3 billion years
Super-deep	Unknown but likely carbonates, such as dolomite or primordial methane	Eclogite	Unknown because the carbon source is unknown	Subducted ocean crust	Last 3 billion years?

At suitable depth, carbonates react with silicates, either a pyroxene called enstatite or a quartz-like mineral called coesite. This forms diamond (chemically reduced carbon) and a more oxidized silicate, such as olivine or another pyroxene called diopside

enstatite ($MgSi\mathbf{O_3}$) is oxidized, leaving olivine ($(Mg^{+2}, Fe^{+2})_2Si\mathbf{O_4}$), with one additional oxygen atom. This kind of reaction will have occurred throughout Earth's history as peridotite forms the bulk of the upper mantle.

In eclogite a similar reaction occurs when dolomite ($MgCO_3$) reacts with coesite (a type of quartz, SiO_2) to form the carbon and the pyroxene diopside ($CaMgSi_2O_6$; Table 4.1). Importantly, the chemistry of eclogite necessitates formation in delaminating or subducting ocean lithosphere. This means one of two things. Either oceanic crust was thick enough early on to form eclogite, or the eclogite formed in a subduction setting. Tellingly, for terrestrial diamonds the eclogite-sourced, or E-type, diamonds are all younger than 3 billion years, while the peridotite (P)-type have ages spanning the age of Earth. This implies that plate tectonics has only operated for the last 3 billion years of Earth-history.

Other, rarer and smaller diamonds form at much greater depths. These super-deep diamonds have chemistries related to eclogite and clearly trace the subduction of old ocean lithosphere to the mantle transition zone. Importantly, this shows that carbon can be delivered to this depth. This is important, as Rajdeep Dasgupta and Marc M. Hirschmann (both from the University of Minnesota) show how at depths between 200 and 330 km down carbon dioxide contributes to the melting of peridotite. This increases very deep melting under ocean ridges and hence the global plate tectonics cycle.

An interesting observation involves one of Earth's rarest treasures, blue diamonds. These boron-bearing gems can rise with other super-deep diamonds. Interestingly, as Evan M. Smith (Gemological Institute of America) and colleagues note, these diamonds likely form when serpentinite penetrates the lower mantle, carrying boron from seawater with them. Of course, if there is boron there must also be a considerable amount of water entering the lower mantle with subducting slabs. If you follow the work of Parai and Mukhopadhaya (earlier in this chapter) then this process likely began at the end of the Archaean, when substantial volumes of water were delivered by subduction from the oceans. Subduction may well have operated at earlier times in the Archaean, but it wasn't until the late Archaean that the mantle was cool enough to oblige the influx of water to greater depths.

Vincenzo Stagno and co-workers from the University of Bayreuth carried out chemical analysis of the upper mantle under the continental cratons. They concluded that most of the frozen upper mantle that underlies the continental cratons is fairly oxidized. However, the surrounding plastic asthenosphere can hold carbon as diamond (or graphite). This chemically reduced form of carbon is then oxidized by ferric iron if it rises towards the surface, with the amount of oxidation depending on how much ferric iron is present.

The amount of oxygen present (and the overall oxidation state) has many important consequences for the planet. It tells us how well the mantle is mixed and how much water has been transported to (or is stored at) depth. The presence of abundant graphite in the upper mantle would tell us that this is poorly oxidized. However, Stagno's and colleague's work shows us that the oxidation state of the upper mantle may well be varied if the abundance of ferric iron isn't homogenous.

As regards the security of the mantle's diamonds, only the presence of abundant carbon dioxide or water can rescue these when they become entrained in convecting mantle as it rises towards the surface. At the correct depth diamonds may become trapped in the deep keels of cratons and be unable to react with the surrounding rock. These may then be "sampled" by gas-rich kimberlite eruptions that rip the material from the keel and blast it at speed towards the continental surface. In turn, the mantle can only produce these energetic, deep-sourced eruptions if conditions are right to allow carbon dioxide to come out of solution and drive diamond-containing kimberlites towards the surface. The infant Earth was almost certainly too hot and the lithosphere too thin to trap diamonds in the continental keels. It was only when the mantle cooled after the Archaean that diamonds began to become trapped under the continents; then only in the last billion years could these be "sampled" by kimberlite eruptions.

The Shallow Fate of Diamonds

To illustrate what happens to those diamonds that encounter more oxidizing conditions when they approach Earth's surface, we can look to the rarest volcano on Earth—Oldoinyo Lengai, a volcano that erupts soapsuds. Oldoinyo Lengai erupts lavas that are mostly sodium carbonate called carbonatites. Here, chemical reactions between the diamondiferous magma, ferric iron and water in the surrounding peridotite magma have thoroughly oxidized the carbon, leaving carbonate minerals. Carbonatites are peculiar. Not only are these magmas exceedingly rare they are extremely fluid, despite their low eruption temperatures (around 550 °C).

Work by Tobias Fisher (University of New Mexico) convincingly demonstrated that the huge amount of carbon dioxide emitted by this carbonatite volcano (55 billion moles of gas per year—some 2.42 billion kg) comes from a completely normal piece of mantle. Other trace gases (such as helium or nitrogen) are erupted in the same proportions as those vented elsewhere on the planet at mid-ocean ridges or other rift zones. What appears to make Lengai's magmas so carbonate rich isn't the amount of carbon dioxide in the mantle they come from. Rather it's the amount of sodium. Very small amounts of melting in the mantle selectively release sodium-rich fluids that form first the unusual nephelinite magmas, which are erupted elsewhere in the rift, for example at the more easterly and lethal volcano Nyiragongo. Under Oldoinyo Lengai, these sodium-rich magmas pick up the carbon dioxide and form sodium carbonate. As the magma approaches the surface, the vast majority of the carbon dioxide is then released as gas, but enough remains (1%) to form the soapsud lavas and ash.

These processes illustrate the complexity of carbon cycling. However, the key principle is that the mantle must be sufficiently fluid to permit some kind of turnover with the lithosphere. Although this simplest model to envisage might be the grand conveyor belt of plate tectonics, other modes of vertical tectonism will also work in its absence. The key issue on these planets is the necessity of burial, which in turn likely favors very active vertical tectonism, such as that which is indirectly evident on Venus and prevalent on Io. Substantial uplift or volcanism, allows weathering, erosion and burial, either through simple sedimentation or rapid burial by volcanism. The latter is unlikely to be prevalent on anything other than young planets. However, periodic overturning of the lithosphere might be possible with episodic tectonics. This could lead to the loss of any remaining dense crust, or the folding, thrusting and uplift of blocks of continental crust, driven by underlying plumes, superswells or general convection.

Cycling the Other Elements of Life

Aside from water and carbon, life depends on a number of other non-metallic elements. Of these, nitrogen and phosphorous are key to the synthesis of DNA (deoxyribonucleic acid) and RNA (ribonucleic acid). Phosphorous, as phosphate, forms the chemical glue that links the building blocks of these essential biomolecules, while nitrogen is an essential component of the bases—the information-storing parts of these molecules, and of amino acids, vitamins and a variety of other biomolecules (Chap. 6). Although nitrogen, like carbon, principally cycles through the biosphere, the ultimate source is in the primordial mantle rocks.

The majority of nitrogen arrives with icy planetesimals as ammonia. Ammonia is then incorporated (as ammonium ions) into the steadily heating and compressing silicate minerals of the proto-planet's mantle in place of potassium ions. Under mantle temperature and pressures, ammonia principally decomposes, with the hydrogen going into the formation of water and the nitrogen venting to the surface. For telluric planets of moderate mass, the nitrogen is stable in its diatomic form (N_2) and is sufficiently massive for gravity to retain it. If the atmosphere is dry, the nitrogen forms a subordinate component to carbon dioxide. If, however, the planet is wet (or sufficiently cold) carbon dioxide is sequestered as carbonate (or ice), and nitrogen forms the dominant component. Less abundantly, some nitrogen could be retained as ammonia in the shallower mantle, or as nitride compounds at greater depth.

An interesting consideration is how the abundance of nitrogen may have changed over time. If there was moderately abundant ammonia in the early atmosphere then this would enter the mantle in hydrous fluids. At this point we should return to the mode of tectonics the early Earth experienced. In the modern setting nitrogen is fixed biological to ammonia through nitrogen fixation and through abiotic processes, such as volcanism and lightning, to nitric acid and hence nitrate. Most of this is cycled through biological systems, but some can enter deeper fluids and be taken into the mantle.

In its modern guise, subduction will carry any ammonia (and other reduced nitrogen compounds into the mantle wedge above the subducting slab. Here, the ammonia will be rapidly oxidized to nitrogen gas and vented back to the atmosphere. However, in the Archaean the atmosphere was either mildly reducing or neutral. In this era, ammonia could have entered the upper mantle and become trapped in compounds in place of potassium, in the same manner in which it arrived on Earth. This would have steadily removed nitrogen from the atmosphere, if nitrogen fixation was widespread. In this process bacteria combine nitrogen and hydrogen (from water) to make ammonia. Interestingly, some genetic (phylogenetic) analyses suggest that this process originated around 3.2 billion years ago. If so, the atmosphere may have been losing considerable amounts of nitrogen gas to the mantle in the form of ammonia (ammonium).

For those unfamiliar with phylogenetics, DNA sequences of one or more genes are compared from different species. Differences in gene sequence are caused by mutations. Since mutation rates are fairly constant in terrestrial species, those species with the smallest numbers of differences in their gene sequences must have originated from common ancestral ones most recently. The greater the number of differences, the more time it must have taken for these to arise. Hence, those species with the greatest sequence similarity are grouped most closely together. This gives their relative relationship. Examination of the degree of divergence (difference) in the sequence then tells you when they diverged in time.

When oxygenic photosynthesis arose around 2.7 billion years ago, there was a slow but steady oxidation of the oceans. Initially, this appears to have been a local phenomenon, but around 2.45 billion years ago the bulk of the oceans rapidly oxidized in the so-called "Great Oxidation Event," or GOE. The GOE followed what appears to be the onset of subduction by a few hundred million years—depending on your interpretation of the data. Therefore, for the first 2 billion years or so of Earth's history, the mantle was sucking nitrogen back into its depths. What would this have meant for the atmospheric pressure on Earth during these periods?

In 2016 Sanjoy Som and colleagues investigated the size of fossil gas bubbles in basaltic lavas that were assumed to have solidified at sea level. These suggested that the atmospheric pressure was as low as 0.23 bar at 2.7 billion years ago. This is somewhat under one quarter of its current level and similar to the pressure at the summit of Everest. How can we reconcile these observations with the kinds of processes that operated in the early Earth—and can we understand the validity of these claims?

Although Som's conclusions might seem rather hard to swallow from our anthropocentric viewpoint, they are compatible with a number of other observations—and with the notion that there was a net loss of nitrogen to the mantle. For example, Bernard Marty (Université de Lorraine) and colleagues compared the relative proportions of nitrogen (N_2) and argon-36 in fluid inclusions that were trapped in 3.0- to 3.5-billion-year-old hydrothermal quartz. Their data suggests that the partial pressure of N_2 (the contribution of nitrogen to the total atmospheric pressure) of the Archean atmosphere could not have exceeded 1.1 bar, but crucially could have been as low as 0.5 bar.

In support of these empirical data, the surface-interior nitrogen flux estimates of Tobias Fischer (University of New Mexico) and co-workers suggest that plate tectonics provides a net outgassing of nitrogen into the atmosphere. If this is the case, there must have been less nitrogen in the Archaean and Proterozoic atmospheres than there is now. Atmospheric pressure has increased to its present value. Although these conclusions are still debatable, the analyses of them are clearly of critical importance. For, if a planet loses atmosphere to its mantle at an early stage in its history, this will affect the stability of the biosphere. This will be particularly true if this is the dominant molecule in the atmosphere and if this element is critical to the formation of biomolecules.

Sami Mikhail and Aubrey Zerkle (both from the University of St. Andrews) considered various scenarios that might explain contemporary observations of Mars and Venus in the light of these considerations. If one compares the abundance of different gases in Earth's atmosphere and that of Venus and Mars, one sees that the heavier noble gases are more abundant relative to nitrogen than they are on Earth. As these lack plate tectonics it supports the view that nitrogen has been added to our atmosphere over time from the mantle, with the Venusian and Martian levels reflecting the primordial state of all three planet's atmospheres—in the absence of plate tectonics. However (and there is always a "however"), one has to consider the possibility that Earth's atmosphere shows a net gain now—because of release of nitrogen from ammonia (and possibly nitrate) that was fixed and subducted earlier in our planet's history. In this scenario, during the Hadean and early Archaean, nitrogen was relatively abundant in the atmosphere. Over time, more and more became trapped in the upper mantle, causing air pressure to fall. This process accelerated when plate tectonics commenced and more ammonia was dropped into the mantle.

After the GOE, the upper mantle became oxidizing, and the ammonia that was present was converted at greater pace to nitrogen and released back to the atmosphere, a process that continues today. In this scenario, atmospheric pressure falls until the early Proterozoic, then rises once more. One must also remember that the mass of the atmosphere was also affected by solar activity. This was higher in the past and may have depleted the mass of atmosphere, irrespective of any chemistry that was occurring at the surface.

Although these scenarios are largely speculative, they need to be put in the context of the overall instability of ammonia in the upper mantle. Chemical reactions should convert ammonia to nitrogen (via the rocket fuel, hydrazine), as this explains the presence of nitrogen in the atmosphere of much smaller bodies such as Titan and the Tiger Stripes of Enceladus. Until the conditions at which these reactions are nailed—and the relative contribution of mantle oxidation is understood—one cannot be completely sure which scenario will turn out to be true. However, at the moment there are clear grounds to consider significant alterations in the density of Earth's atmosphere that are independent of the effects of solar radiation.

The Proterozoic Earth would have been an odd and fairly unpleasant place. Surface waters were largely oxidized, but oxygen did not penetrate far below the surface. Upwelling ocean currents would bring reduced nitrogen compounds towards the surface, where microbes that were oxidizing it lived. Here, in a very smelly environment, ammonia might have been oxidized to nitrogen gas (in a process called denitrification) or incorporated into living organisms. A considerable amount of nitrogen would have been lost from near-surface waters to the atmosphere. Under similar, sulfur-rich conditions, the element molybdenum, which is critical for the function of the enzyme that converts nitrogen to ammonia, precipitates out. This would make nitrogen fixation very difficult indeed except at the surface of the ocean—and in places where there was a continuous supply of molybdenum ions from any nearby continental crust. Would the loss of nitrogen from the oceans have resulted in increases in atmospheric pressure?

Likewise, phosphorous is a critical element for biology. As an integral glue in the sugar-phosphate backbone of nucleic acids, through the energy-storing capacity of ATP to its central role in regulating cell fate, phosphorous as phosphate is central to life's engine on Earth—and is likely also integral elsewhere. And though this element is critical to terrestrial biology's existence, it is relatively uncommon. Only massive stars produce it in any sizable quantity and then, once it is delivered to planets, becomes fairly tightly bound in rocks. The terrestrial crust contains only 0.09% by weight phosphorus, with Mars—the only other planet we can sample it—having twice this amount.

The problem for phosphate is two-fold. Most organic phosphate compounds (and salts) are initially soluble in water, if the water is acidic. In terrestrial biology, both mycorrhizal fungi and plant roots secrete organic acids (and carbonic acid), which can extract phosphate from insoluble minerals in rocks and soils. However, once liberated phosphate ions readily combine with various metals to form insoluble salts. This leads to its sequestration in rocks. Phosphate sediments that end up in the oceans remain there. The phosphate remains insoluble in the alkaline waters of our oceans, leading to its direct precipitation and burial at sea.

Interestingly, on the early Earth, with carbon dioxide-saturated waters, this would not have been the case and phosphate could have accumulated. Phosphate is then sequestered in sedimentary rocks, which are then buried and (if they are on oceanic crust) subducted. At this stage phosphate may return to the surface through tectonism, where it is then released by weathering or through the action of biological processes. Therefore, some form of surface tectonism is critical to the persistence of this critical nutrient in the biosphere.

If one looks at the Miocene (12–7 million years ago), Earth's oceans underwent a dramatic blooming, with an upsurge in all forms of aquatic life. The principle reason for this ramping up of life appears to have been increased erosion of the recently uplifted Alpine-Himalayan chain. As the mountains rose higher, and the Asian monsoon increased in vigor, the rate of erosion was increased. In turn, this delivered more material to the oceans and with it more phosphate.

There are some suggestions that a similar process operated during the Snowball Earth episodes in the Neoproterozoic. Jochen Brock (Australian National University) has proposed that alternating phases of glaciation and thawing were critical to the rise of more complex food chains. They suggest that erosion led to an increase in the level of oceanic phosphate during the Sturtian era in the Neoproterozoic, 750 million years ago. In turn, this allowed eukaryotic algae to bloom and take over the niches that were vacated by cyanobacteria. If this is true—and it's a big if—then the formation of the Rodinia supercontinent was critical to the eventual evolution of complex eukaryotes, life-forms that of course include us. Similarly, Manasvi Lingam and Abraham Loeb (Harvard) suggest that phosphate may be a key limiting factor on aquaplanets, because of limited returns to oceans by erosion. However, it will really come down to the pH of the oceans. Aquaplanets may be perfectly viable, if their oceans are relatively acidic, where phosphate can persist. After all, the early Earth may have been one such world.

The Persistence of Conditions Necessary for a Viable Biosphere

Hopefully, at this point the reader will realize that it is impossible to contemplate how viable a planet will be for life until you consider the full range of operations the environment must engage in. This involves firstly the carbonate-silicate cycle that regulates the overall temperature of the surface in concert with the energy provided by the host star. Weathering rates vary with temperature, and increased temperatures lead to more weathering, which removes carbon dioxide as calcium and magnesium carbonate. Enhanced weathering lowers the mass of carbon dioxide in the atmosphere, which then reduces the greenhouse effect, the global temperature and hence the weathering rate…

However, the carbonate-silicate cycle is dependent on the rate of burial of carbonate rocks –which may depend on plate tectonic processes. Plate tectonics and overall mantle vigor affect the rate of volcanism and the return of carbon dioxide to the atmosphere.

Mantle viscosity, and the propensity of the planet to foster plate tectonics, depends on the abundance of water in the mantle and on the surface. This depends on the composition of the planet, the effect of stellar radiation and the planet's overall mass. Whack the mass up above 2–3 Earths and the stability of the deep crust is compromised. This seems to lead to the formation of a thick granitic (TTG) lid, which will then eliminate plate tectonics, irrespective of how hot and fluid the mantle is. If this happens, the carbonate-silicate cycle—and indeed the cycling of all elements—may be compromised. Thus, the mass and composition of the planet are keys to its long-term habitability.

In the absence of plate tectonics, lower mass planets may remain drowned because they fail to produce thick, buoyant crust that can rise above the surface of their oceans. In the absence or presence of plate tectonics, older planets may lose all of their water to their mantle.

Consider some scenarios.

Scenario One: Stripping an Earth-Mass World

Take an Earth-mass planet, place it close to its red dwarf star and strip its atmosphere. All may seem lost (proverbially and literally), as some researchers suggest. But wait! The mantle still holds a reasonable reserve and can restock the atmosphere after the star settles down. However, now the aging, cooling mantle is drier than it might otherwise be and is becoming very viscous. Plate tectonics ceases. The carbonate-silicate and other mineral cycles slow down as more and more minerals are locked into the crust. The habitability of the planet then declines. However, with low-volume oceans, less carbon dioxide may sequester onto the surface. This might allow the biosphere to remain habitable for longer. Or, where carbon dioxide is already low, the surface may become too cold to support life—and cold enough to allow carbon dioxide to precipitate out, “irreversibly” freezing the biosphere—until the red dwarf leaves the main sequence. Even this seemingly simple scenario is fraught with complications that aren’t apparent on first inspection. This is illustrated in Fig. 4.5.

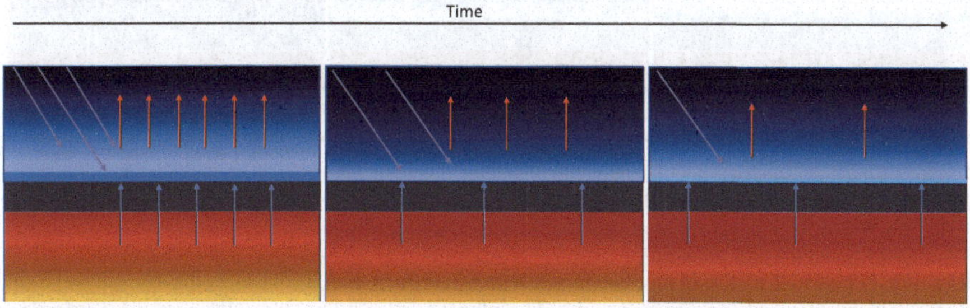

Fig. 4.5 Scenario 1. EUV strips away most of the atmosphere, but a cooler mantle retains water long enough to provide a store to replenish the losses. In some instances the planet may be stripped only to recover, while in others a steady state is achieved. If sufficient greenhouse gases are lost, then temperatures may fall, causing the remaining water to freeze, as is the case on Mars

Scenario Two: Strip a Super-Terran Before Its Mantle Cools

A more massive super-terran has its atmosphere stripped by stellar radiation from its neighboring red dwarf. Now, this planet had a hot enough interior to drive almost all of its water (and other volatiles) onto its surface at an early stage. Its mantle was dry by the time the surface was stripped. This planet cannot recover, unless a subsequent barrage of icy material replenishes its loss. The surface is still converted to a thick and largely granitic lid through partial melting of eclogite irrespective of the amount of water available. However, now the surface is locked solid, and the desiccated and cooling mantle is largely unable to mobilize it.

Venus, in this regard, is not a good analogy, despite similarities. For one, the majority of the Venusian crust is likely basaltic—with only the highlands the likely domain of granitoids, as some Venus Express data suggests. The Venusian crust is likely malleable on longer timescales, if the mantle becomes hot enough and buoyant enough, relative to the lithosphere, to rupture it. Granite is a different beast, low density and non-subductable to any reasonable extent. The surface will then resemble that of Mars (but with buoyant lithosphere), irrespective of the mobility of its underlying mantle. Figure 4.6 illustrates this scenario.

The lithosphere, like that of Io, will be largely crust, with little underlying frozen mantle. Tectonism is possible if there are inhomogeneities in the thickness of the lithosphere that can be maneuvered by convection in the mantle. The extent of this will depend on the viscosity of the mantle (the asthenosphere) and its ability to transmit force to the lithosphere; as well as the thickness and rigidity of the granitoid crust. Volcanism is also possible as long as buoyant enough magma can make its way through the potentially thick, low density crust.

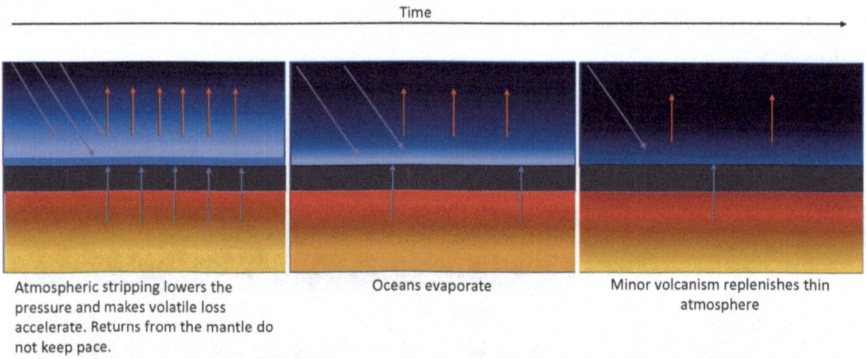

Fig. 4.6 Scenario 2. If the planet is fairly massive and the mantle remains hotter for longer, little water may be retained (or returned to it). The planet is then completely desiccated by stripping radiation from the star

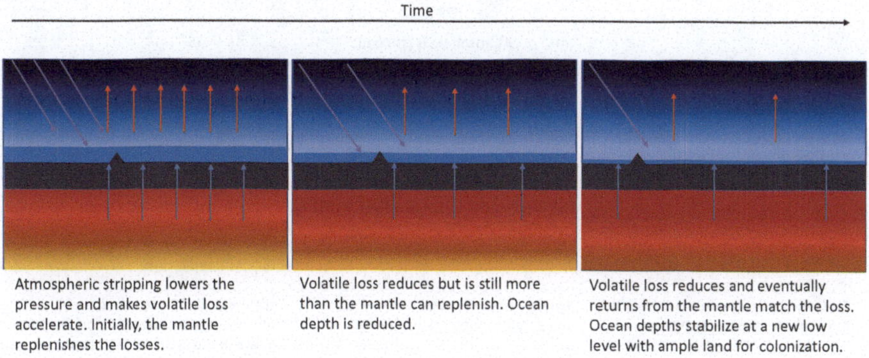

Fig. 4.7 Scenario 3. If a super-terran, like that shown in scenario 2, has more water, then stripping may improve its habitability by removing water and reducing the depths of the oceans. This exposes land and reduces pressure on the mantle, which increases volcanism

Scenario Three: Stripping a Super-Terran After the Mantle Has Cooled

In this scenario, illustrated in Fig. 4.7, a more massive super-terran planet lies equally close to its red dwarf star, and its atmosphere is being actively stripped. The planet, when young, has deep oceans and a limited carbonate-silicate cycle. However, atmospheric stripping removes more than an ocean worth of water and formerly drowned continental crust emerges from the depths. Moreover, with reduced pressure on the mantle and crust, the lower lithosphere becomes stable against the formation of eclogite. The lithosphere is able to fracture, allowing plate tectonics to begin and the carbonate-silicate cycle to operate. The habitability of the planet improves.

Scenario Four: Sequestering the Atmosphere of a Distant Super-Terran

In the final scenario we indulge ourselves, theorizing about a massive super-terran that lies in a more distant orbit (Fig. 4.8). Initially, it has sufficient reserves of greenhouse gases in its atmosphere to maintain liquid water. However, most carbon dioxide is dissolved, then precipitated, into the deep oceans. Temperatures fall and the planet freezes over. The ocean crust is unstable, and delamination of its eclogite base produces a granitic lid. The mantle becomes largely isolated from the hydrosphere above. Life is confined to hydrothermal vents, where residual volcanism is driven by hot spot activity.

For each of these scenarios (and there are obviously many more iterations) there are implications for atmospheric pressure and the persistence of magnetic fields. All

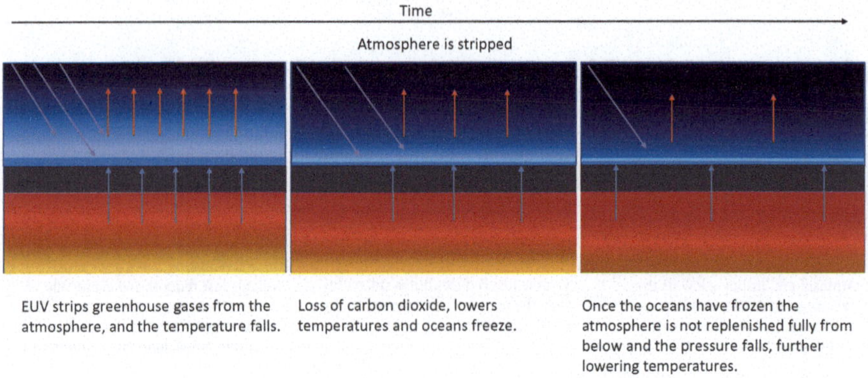

Fig. 4.8 Scenario 4. In scenario 1 the planet had a lower mass, which meant it had fewer volatiles but a cooler mantle, which degasses at a lower rate. In this scenario, the planet lies further from the star and is colder. Stripping leads to loss of greenhouse gases, and this causes the planet to cool and freeze over before all of the volatiles are lost. The atmosphere is still largely lost, but most (or much) of the water is retained as ice. A deep biosphere is possible beneath the ice, in a Europan-like scenario

of these impact habitability and with them considerations of the Fermi paradox. Not all habitable worlds are created equal. Seemingly trivial differences in planetary mass and composition can affect the stability of their biospheres, if they have them. Stick nitrogen into the pot, and things get more complex still.

Did Earth's atmospheric pressure fall from birth until the early Proterozoic because nitrogen was being drained into the mantle? If so, for scenario one a failure of plate tectonics might limit the release of nitrogen from volcanism. If nitrogen is lost to the surface, as nitrate (or fixed as ammonia), the draw-down by these processes will not be balanced by a return via volcanism. Indeed, this is true for scenarios one and four. One presumes that if there is a biosphere on this planet, the biotic cycle will be sufficient to maintain habitability from the point of view of nitrogen. The same may not be said for phosphorous, which could end up weathered and buried on the surface, making it an increasingly scarce resource for life.

As we said, if the mantle is particularly hot, perhaps because the planet has a wealth of radiogenic materials, then almost all of the mantle's reserves may end up in the atmosphere at an early stage. Formation of a thick lid will then prevent its return to the mantle, but would allow a lot of it to potentially end up in rocks and soils. Thick lids might be rather effective in maintaining a dense, nitrogen-rich atmosphere, if they come into effect at an early stage. However, if the thick lid forms when atmospheric pressure has fallen, then habitability may be compromised by low pressure, which would favor a cool climate and a general lack of available nitrogen. There are currently rather a lot of unanswered questions, and critical to our overall understanding is whether (and by how much) the atmospheric pressure of Earth has varied.

Understanding how water has cycled also remains fairly mysterious. A planet blanketed in oceans—as may have been the case in the early Earth—may lose nitrogen as nitrate. A cooling mantle, which draws water into it, will lower sea level. A lower sea-level generally means that more continental surface (or other land surface) will be exposed—and this has two effects. The first is prosaic. More dry land means more soil, and this affects the mass of nitrogen that can be held as nitrate or ammonium in this reservoir. Secondly, a planet with most of its surface under water may have less convective cloud—and hence fewer thunderstorms. If most of the tectonism is sub-surface (under water) then there will be a lower rate of conversion of nitrogen to nitrate in volcanic eruptions. Therefore, a drowned planet may have very little nitrogen in an accessible form. Both of these factors favor the preservation of a denser (and hence potentially habitable) nitrogen atmosphere—but a chemically impoverished biosphere. Clearly, there is much that is yet to be considered!

Planet Mass, Erosion and Nutrient Cycles

Consider one further factor before hanging up your hat: the relationship between planet mass and erosion. The rate of erosion is dependent on a number of factors, but one of them is the angle of slopes. Steeper slopes tend to experience landslides, which facilitates lowering of relief. However, landslides tend to block rivers and infill their valleys, which reduces their capacity to continually erode further into the bedrock (Fig. 4.9).

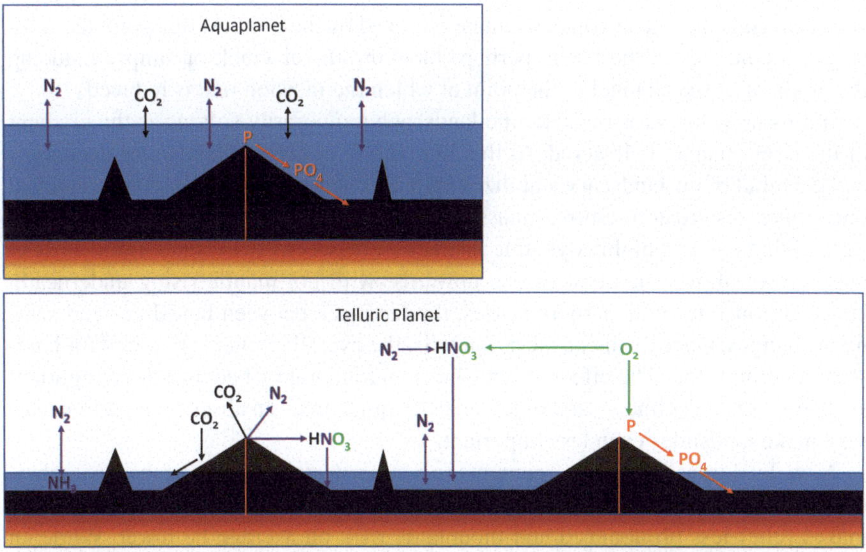

Fig. 4.9 The power of weathering and erosion. Aquaplanets have little or no exposed land (top). Nutrient-rich materials from volcanism can dissolve in seawater, but phosphate tends to precipitate out, leaving it limiting for organisms. In the telluric planet (below) volcanism and erosion contribute nutrient-rich run-off to the oceans. Nitrogen (as nitrate or ammonium) is also returned. Consequently, aquaplanets are poorer abodes for life than telluric ones. This is all the more true if the ocean is underlain by a thick lid—which seems likely

David L. Egholm (Aarhus University) examined the processes of erosion and how they determine the topography of mountain ranges. His work demonstrated the role of feedback in the control of erosion and how, in turn, this affects the appearance of the landscape. The principle agents of erosion in terrestrial landscapes (as a whole) are rivers. Although ice and (as weathering) wind act specifically in some areas, water is the principle erosive power on Earth. We must expect this to be true of any planet lying in its star's habitable zone, where liquid water will be present. Titan presents an interesting adjunct, where liquid hydrocarbons carry out the same function at much lower temperatures. Conversely on very hot, tidally locked terrestrial planets, basalts, carbonatites[1] or komatiites may function in this regard, and as detailed earlier, there is evidence that komatiites carried out erosion on the early Earth when the mantle was hotter than it is at present.

In Egholm's work, erosion is principally driven by the process of saltation-abrasion. Saltation, in geographical terms, is the action of fairly large sedimentary pieces that bounce along the surface of a riverbed. In desert environments, saltation occurs near the surface, where wind drives the movement of sand grains across it. Saltation-abrasion occurs in riverbeds and leads to the steady deepening of the channel in areas with steep gradients. Over time, increased bed erosion leads to increases in the gradient of the riverbanks and valley sides, which, in turn, leads to a greater frequency of landslides. As landslides increase in frequency, there is an initial increase in the sediment available for erosion in affected river systems, which then continue to accelerate the rate of incision in the river channel. This continues as a positive feedback until the amount of sediment delivered to the river by landslides exceeds the rate at which it can be removed by the river. At this point the sediment accumulates in the basin, perhaps blocking the channel, or simply reducing the gradient of the channel to the point at which the incision rate is reduced.

In this new, low erosion state, the landscape temporarily softens as the gradient of the river channel is lowered. At this time the frequency of landslides decreases, and the relief of the landscape stabilizes until the excess bulk of sediment is removed once more. As erosion removes mass from above, the landscape readjusts by istosasy, giving rise to a highly dynamic interaction between erosion and uplift. Loss of mass above allows the crust to rise upwards, with hot mantle rising underneath. However, once tectonic activity ceases, the feedback between landslides and sediment changes. Once the mean elevation falls, the overall gradients required for landslides declines, too. The effect of landslides on topography becomes less important as its frequency declines—and in tectonically quiet areas an absence of earthquakes may make landslides even less important.

Now, let's transport these arguments to a super-terran. Super-terrans have, obviously, more mass and hence exert a stronger gravitational pull on their surface. This makes relief less pronounced and mountains will, on average be lower. While an increased gravitational pull would favor faster incision by accelerating the downward flow of water, a lower relief would counter this. The greater mass also favors

[1] Many carbonatitites will only be stable if the temperature is less than 800–900 °C, above which sodium carbonate will thermally decompose.

faster, density-driven isostatic readjustment. The net effect of these counteracting processes matters because it will determine the draw-down rate of carbon dioxide through weathering and erosion—and this will affect the rate of transport of carbonates, phosphate, nitrogen salts and other ions from exposed rock. This is yet to be modeled, and it clearly is of critical importance in determining the viability of any potentially habitable planet.

Finally, the mass of the planet determines how hot the interior is and how quickly it will degas. If degassing produces a global ocean with little land there will be little cycling of phosphate. Phosphate will be sequestered in rocks and sediments on the ocean floor, limiting biological productivity. Only planets that exhibit dynamic surfaces will be habitable. This doesn't mean that they have to experience plate tectonics. Other forms of vertical tectonics can suffice as long as there is sufficient capacity to mobilize mineral ions from rock to biosphere and back again. Break this cycle and deposit all of one or more ingredients, and the biodiversity of the landscape must decline. This is bad enough, but think again about the crust. If it is particularly thick—as might be probable on a massive super-terran—and the crust is pressurized by the mass of water overhead, then the lithosphere will become unstable and delaminate. The resulting thick granitic (TTG) lid could render most of the surface free of volcanism—as might the high pressure of the ocean. These effects would further reduce the return of nutrients from the mantle to the surface. In this regard, the deep oceans of Ganymede and Callisto are likely biologically inert.

Conclusions

This chapter aims to put the ideas presented here in some sort of holistic framework. What is the impact of mantle forces on the habitability of a planet? This is a story that is far from completely understood. Take the simple statement that nitrogen is exolved from the mantle. The conversion of ammonia to nitrogen at temperatures of a few hundred degrees is very well understood. Yet, ammonia may well be stable under some circumstances at temperatures beyond this. So, what exactly are the conditions under which ammonia—the principle cosmological form of nitrogen—breaks down to produce the most abundant gas in the atmospheres of Earth and Titan? Titan must have a cool interior (relative to Earth), but its atmosphere is rich in diatomic nitrogen gas. Clearly, there is a bit of an issue here.

Next, did the density of Earth's atmosphere vary considerably in the first 3 billion years of its history? Was the early Proterozoic a low pressure environment in which most of the continents were drowned? Did the continents emerge early in the Proterozoic—or much later, around 1 billion years ago?

Although this leaves a lot of unanswered questions for the evolution of the planet's interior—and its mode of tectonism—there is a far bigger question linked to its habitability. For example, we know that eukaryotes emerged at the end of the "boring billion"—the era where life just plodded along, largely unchanged. Sometime after 900 million years ago the eukaryotes emerged and rapidly dominated the visible biosphere. Is this coincidental, or did the emergence of continents enhance weathering of phosphates that led to explosive diversification of life? Was the emergence of continents tied to the rise of oxygen late in the Proterozoic or is this coincidental? Finally, was the emergence of land instrumental in determining the number of niches available for the diversification of life? This is something we consider in Chaps. 7, 8 and 9.

There are a lot of unanswered questions, but we believe that this is a dynamic and rich area of research that will finally tie the concept of habitability to the underlying chemical and physical limits imposed by geology and geography.

References

Atreya, S. K. (2010). The significance of trace constituents in the solar system. *Faraday Discussions, 147*, 9–29. discussion 83–102. https://pdfs.semanticscholar.org/8b6b/6ba63e8b9b4da4d35c664267518fe08c5dfc.pdf.

Bindeman, I. N., Zakharov, D. O., Palandri, J., Greber, N. D., Dauphas, N., Retallack, G. J., Hofmann, A., Lackey, J. S., & Bekker, A. (2018). Rapid emergence of subaerial landmasses and onset of a modern hydrologic cycle 2.5 billion years ago. *Nature, 557*, 545–548. https://doi.org/10.1038/s41586-018-0131-1.

Brocks, J. J., Jarrett, A. J. M., Sirantoine, E., Hallmann, C., Hoshino, Y., & Liyanage, T. (2017). The rise of algae in Cryogenian oceans and the emergence of animals. *Nature, 548*, 578. https://doi.org/10.1038/nature23457.

Carr, M. H., & Head, J. W. (2015). Martian surface/near-surface water inventory: Sources, sinks, and changes with time. *Geophysical Research Letters, 42*, 726–732. https://doi.org/10.1002/2014GL062464. https://pdfs.semanticscholar.org/6a4b/0f49d0f096d165df36da112cdbde0997ca85.pdf.

Chen, C., Wiens, D. A., Shen, W., & Eimer, M. (2018). Water input into the Mariana subduction zone estimated from ocean-bottom seismic data. *Nature, 563*, 389–392. https://doi.org/10.1038/s41586-018-0655-4.

Condie, K., Pisarevsky, S. A., Korenaga, J., & Gardolle, S. (2015). Is the rate of supercontinent assembly changing with time? *Precambrian Research, 259*, 278–289. https://doi.org/10.1016/j.precamres.2014.07.015.

Dasgupta, R., & Hirschmann, M. M. (2006). Melting in the Earth's deep upper mantle caused by carbon dioxide. *Nature, 440*, 659–662. https://doi.org/10.1038/nature04612.

Edson, A. R., Kasting, J. F., Pollard, D., Lee, S., & Bannon, P. R. (2012). The carbonate-silicate cycle and CO_2/climate feedbacks on tidally locked terrestrial planets. *Astrobiology, 12*(6), 562–571. https://doi.org/10.1089/ast.2011.0762. Epub 2012 Jul 9.

Filippelli, G. M. (2008). The global phosphorus cycle: Past, present, and future. *Elements, 4*, 89–95. https://doi.org/10.2113/GSELEMENTS.4.2.89.

Fischer, T. P., Hilton, D. R., Zimmer, M. M., Shaw, A. M., Sharp, Z. D., & Walker, J. A. (2002). Subduction and recycling of nitrogen along the central American margin. *Science, 297*(5584), 1154–1157. https://doi.org/10.1126/science.1073995.

Korenaga, J. (2011). Thermal evolution with a hydrating mantle and the initiation of plate tectonics in the early Earth. *Journal of Geophysical Research, 116*, B12403. https://doi.org/10.1029/2011JB008410.

Lee, C.-T. A., Caves, J., Jiang, H., Cao, W., Lenardic, A., McKenzie, N. R., Shorttle, O., Yin, Q.-z., & Dyer, B. (2018). Deep mantle roots and continental emergence: Implications for whole-Earth elemental cycling, long-term climate, and the Cambrian explosion. *International Geology Review, 60*(4), 431–448. https://doi.org/10.1080/00206814.2017.1340853.

Lingam, M. & Loeb, A. (2018a). Dependence of biological activity on the surface water fraction of planets. https://arxiv.org/pdf/1809.09118.pdf.

Lingam, M. & Loeb, A. (2018b). Is extraterrestrial life suppressed on subsurface ocean worlds due to the paucity of bioessential elements? https://arxiv.org/pdf/1806.00018.pdf.

Marty, B., Zimmermann, L., Pujol, M., Burgess, R., & Philippot, P. (2013). Nitrogen isotopic composition and density of the Archean atmosphere. *Science, 342*, 101–104. https://doi.org/10.1126/science.1240971.

Mikhail, S., & Sverjensky, D. A. (2014). Nitrogen speciation in upper mantle fluids and the origin of Earth's nitrogen-rich atmosphere. *Nature Geoscience, 7*, 816–819. https://doi.org/10.1038/ngeo2271.

Mustard, J. F., Poulet, F., Ehlman, B. E., Milliken, R., & Fraeman, A. (2012). Sequestration of volatiles in the martian crust through hydrated minerals: A significant planetary reservoir of water. In *43rd Lunar and Planetary Science Conference*. https://www.lpi.usra.edu/meetings/lpsc2012/pdf/1539.pdf.

Parai, R., & Mukhopadhaya, S. (2018). Xenon isotopic constraints on the history of volatile recycling into the mantle. *Nature, 560*, 223–227. https://doi.org/10.1038/s41586-018-0388-4.

Pearson, D. G., Brenker, F. E., Nestola, F., McNeill, J., Nasdala, L., Hutchison, M. T., Matveev, S., Mather, K., Silversmit, G., Schmitz, S., Vekemans, B., & Vincze, L. (2014). Hydrous mantle transition zone indicated by ringwoodite included within diamond. *Nature, 507*, 221–224. https://doi.org/10.1038/nature13080.

Rüpkea, L. H., Morgana, J. P., Hortb, M., & Connolly, J. A. D. (2014). Serpentine and the subduction zone water cycle. *Earth and Planetary Science Letters, 223*, 17–34. https://doi.org/10.1016/j.epsl.2004.04.018.

Stagno, V., Ojwang, D. O., McCammon, C. A., & Frost, D. J. (2013). The oxidation state of the mantle and the extraction of carbon from Earth's interior. *Nature, 493*, 84–88. https://doi.org/10.1038/nature11679.

Tomkinson, T., Lee, M. R., Mark, D. F., & Smith, C. L. (2013). Sequestration of Martian CO_2 by mineral carbonation. *Nature Communications, 4*, 2662. https://doi.org/10.1038/ncomms3662.

Wade, J., Dyck, B., Palin, R. M., Moore, J. D. P., & Smye, A. J. (2017). The divergent fates of primitive hydrospheric water on Earth and Mars. *Nature, 552*, 391–394.

Walker, J. C. G., Hays, P. B., & Kasting, J. F. (1981). A negative feedback mechanism for the long-term stabilization of Earth's surface temperature. *Journal of Geophysical Research, 86*, 9776–9782.

Zerkle, L., & Mikhail, S. (2017). The geobiological nitrogen cycle: From microbes to the mantle. *Geobiology, 15*, 343–352. https://doi.org/10.1111/gbi.12228.

Diamonds

Berry, A. J., Danyushevsky, L. V., O'Neill, H. S. C., Newville, M., & Sutton, S. R. (2008). Oxidation state of iron in komatiitic melt inclusions indicates hot Archaean mantle. *Nature, 455*, 960–963. https://doi.org/10.1038/nature07377.

Bulanova, G. P., Walter, M. J., Smith, C. B., Kohn, S. C., Armstrong, L. S., Blundy, J., & Gobbo, L. (2010). Mineral inclusions in sublithospheric diamonds from Collier 4 kimberlite pipe, Juina, Brazil: Subducted protoliths, carbonated melts and primary kimberlite magmatism. *Contributions to Mineralogy and Petrology, 160*(4), 489–510. https://doi.org/10.1007/s00410-010-0490-6.

Dobrzhinetskaya, L. F. (2012). Microdiamonds—Frontier of ultrahigh-pressure metamorphism: A review. *Gondwana Research, 21*(1), 207–223. https://doi.org/10.1016/j.gr.2011.07.014.

Evans, R. L. (2008). Carbon in charge. *Science, 322*, 1338–1340.

Fischer, T. P., Burnard, P., Marty, B., Hilton, D. R., Füri, E., Palhol, F., Sharp, Z. D., & Mangasini, F. (2009). Upper-mantle volatile chemistry at Oldoinyo Lengai volcano and the origin of carbonatites. *Nature, 459*, 77–80. https://doi.org/10.1038/nature07977.

Haggerty, S. E. (1999). A diamond trilogy; superplumes, supercontinents, and supernovae. *Science, 285*(5429), 851–860. https://doi.org/10.1126/science.285.5429.851.

Pearson, D. G., & Shirey, S. B. (1999). Isotopic dating of diamonds. In D. D. Lambert & J. Ruiz (Eds.), *Reviews in economic geology: Application of radiogenic isotopes to ore deposit research and exploration* (pp. 143–171). Denver: Society of Economic Geologists.

Levander, A., Bezada, M. J., Niu, F., Humphreys, E. D., Palomeras, I., Thurner, S. M., Masy, J., Schmitz, M., Gallart, J., Carbonell, R., & Miller, M. S. (2014). Subduction-driven recycling of continental margin lithosphere. *Nature, 515*, 253–256. https://doi.org/10.1038/nature13878.

Richardson, S. H., Erlank, A. J., Harris, J. W., & Hart, S. R. (1990). Eclogitic diamonds of Proterozoic age from Cretaceous kimberlites. *Nature, 346*(6279), 54–56. https://doi.org/10.1038/346054a0.

Shirey, S. B., Cartigny, P., Frost, D. J., Keshav, S., Nestola, F., Nimis, P., Pearson, D. G., Sobolev, N. V., & Walter, M. J. (2013). Diamonds and the geology of mantle carbon. *Reviews in Mineralogy and Geochemistry, 75*(1), 355–421. https://doi.org/10.2138/rmg.2013.75.12.

Sparks, R. S. J., Baker, L., Brown, R. J., Field, M., Schumacher, J., Stripp, G., & Walters, A. (2006). Dynamical constraints on kimberlite volcanism. *Journal of Volcanology and Geothermal Research, 155*(1–2), 18–48. https://doi.org/10.1016/j.jvolgeores.2006.02.010.

Weiss, Y., McNeill, J., Pearson, D. G., Nowell, G. M., & Ottley, C. J. (2015). Highly saline fluids from a subducting slab as the source for fluid-rich diamonds. *Nature, 524*, 339–342. https://doi.org/10.1038/nature14857.

Erosion and Deposition

Egholm, D. L., Knudsen, M. F., & Sandiford, M. (2013). Lifespan of mountain ranges scaled by feedbacks between landsliding and erosion by rivers. *Nature, 498*, 475–479. https://doi.org/10.1038/nature12218.

Kirchner, J. W., Finkel, R. C., Riebe, C. S., Granger, D. E., Clayton, J. L., King, J. G., & Megahan, W. F. (2001). Mountain erosion over 10 yr, 10 ky, and 10 my time scales. *Geology, 29*, 591–594. https://doi.org/10.1130/0091-7613(2001)029<0591:MEOYKY>2.0.CO;2.

Kirchner, J. W., & Ferrier, K. L. (2013). Mainly in the plain. *Nature, 495*, 318–319.

Warrick, J. A., Milliman, J. D., Walling, D. E., Wasson, R. J., Syvitski, J. P. M., & Aalto, R. E. (2014). Earth is (mostly) flat: Apportionment of the flux of continental sediment over millennial time scales: Comment. *Geology, 42*, e316. https://doi.org/10.1130/G34846C.1.

Willenbring, J. K., Codilean, A. T., & McElroy, B. (2013). Earth is (mostly) flat. Apportionment of the flux of continental sediment over millennial time scales. *Geology, 41*, 343–346. https://doi.org/10.1130/G33918.1.

Willenbring, J. K., & von Blanckenburg, F. (2010). Long-term stability of global erosion rates and weathering during late-Cenozoic cooling. *Nature, 465*, 211–214. https://doi.org/10.1038/nature09044.

Chapter 5
Atmospheric Circulation and Climate

Introduction

As above, so below. Although the fervor of research builds over the impact of the geosphere on planetary habitability, considerably more research is going into understanding how planetary atmospheres may influence the same property. This chapter expands upon, improves the accuracy of, and produces viable models of atmospheres for planets with different orbital distances from red dwarf stars.

Although much of the work is focused on the research of a couple of groups, these results are considered alongside the work of others. Together, they paint a dynamic and rather interesting portrait of the influence of stellar mass on habitability. We blithely stated "not all habitable worlds are created equally" in Chap. 4, but this is all the more apparent in this chapter. Subtle differences in orbital period steadily alter the motion of gases in the firmaments that cover these worlds and create (as Chap. 7 will explore) dramatically different abodes for biology.

However, let's not forget the regularly neglected K-dwarfs. Although the planets that orbit these stars will almost certainly have dynamic Earthlike atmospheres, there is still the capacity for some interesting and distinctly non-Earthlike climates.

In this chapter we will examine the influence of climate on habitability and begin to relate the properties of red dwarfs (and K-class dwarfs) to the climates of the planets that orbit them.

Tidal Locking and Orbital Period

If a planet orbits close enough to its star, then tidal forces between the star and the planet will lead to tidal locking. This is a fairly general statement, but that's because tidal locking is a two-way process that depends on the mass of the planet, its internal

D. S. Stevenson, *Red Dwarfs*, https://doi.org/10.1007/978-3-030-25550-3_5

structure, the depth of any oceans and the density of its atmosphere. It also depends on the structure of the star. Red dwarfs, by their nature, are fairly homogeneous beasts—particularly those with masses less than 0.25 solar masses. The density increases fairly smoothly from the surface to the deep interior, and the whole structure convects as a single entity. Tidal forces from the planet are dissipated in the entire bulk of the star, rather than (largely) in an outer shell, as they would be for a star such as the Sun.

Tidal locking, or more accurately synchronous rotation, could in principle come in two flavors. In the first—and easily the most common—the planet orbits the star with a single orbital period matching its rotation period on its axis, as the Moon does relative to Earth. In this situation the Moon sets and rises but always presents the same face to us. Although unlikely on the current timescale of our universe, an alternative has both objects tidally locked to one another. In this case, both objects exhibit synchronous rotation. They both present the same face to one another. The analogous terrestrial situation would have the Moon locked in the same position in the sky over one geographical spot, with (as before) it presenting the same face to us—and us to it. There would be no Earthrise over the lunar surface. If we could prevent the Sun dying for another 50 billion years or so, then the (very distant) Moon and Earth would end up synchronously rotating relative to one another. However, as we know, the dying Sun will destroy both of these bodies long before then.

A fairly simple piece of math links the orbital period of a planet in a red dwarf's habitable zone to the luminosity (and hence the mass) of the star.[1] The gist of it is, the greater the mass of the star, the higher the luminosity it will have, and hence the more distant the planet will have to be from its ruddy surface in order to remain habitable. The more distant the orbit and Kepler's third law tells us how much longer the orbital period will be. Ravi Kumar Kopparapu, Eric Wolf (both of the University of Colorado) and other colleagues have produced a series of integrated models that incorporate these realistic effects. This is rather welcome because the earlier models had some peculiar inconsistencies and could be inherently contradictory.

1 $$P_{years} = \left[\left(\frac{\frac{L_*}{L_{sun}}}{\frac{F_p}{F_{Earth}}}\right)^{3/4}\right]\left(\frac{M_*}{M_{sun}}\right)^{-1/2}$$ Here, orbital period of any planet orbiting at the inner edge of the habitable zone is proportional to the mass of the star M_*, its luminosity L_* and F_* is the stellar flux on the planet. These are related to the mass and luminosity of the Sun and stellar flux received on Earth.

Circulation Basics—Horses, Trades and Super-Rotation

The Coriolis Effect

Before examining the most realistic circulation models, one needs to consider the influences that researchers have incorporated. Then you can see why there are contradictions in some of them that influence habitability. In order to do this, we need to make a quick tour of the terrestrial atmosphere so that we can make valid comparisons with those blanketing synchronously rotating planets.

Figure 5.1 shows the overall global circulation of Earth. There are two principal factors driving the terrestrial climate. The first, and most obvious, is the Sun and the energy it delivers. The second is our planet's fairly nifty rotation. Solar heating is the principal engine, which drives vigorous convection at the equator. Air rises above the region of maximum heating and spreads outwards on either side of this zone of ascent. So far, so good.

Now, remember that Earth rotates at around 1700 km/h at the equator. This means that air, which is being heated and is rising, is also moving from west to east at 1700 km/h. However, if this air moves north and south, away from the equator, it will be moving faster towards the east than the ground underneath. Therefore, relative to the surface, the air is moving from west to east, relative to the surface beneath it. Air moving aloft is increasingly deflected towards the east as it moves away from the equator. For Earth, at 30 °N or 30 °S the air is now moving due east, rather than

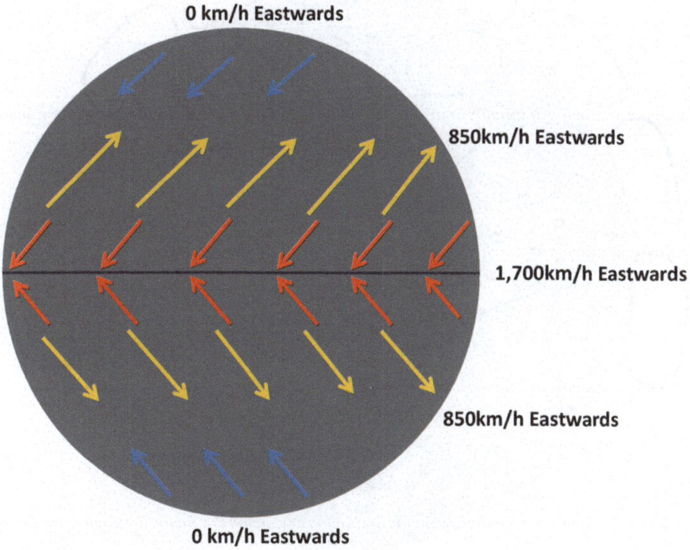

Fig. 5.1 The effect of a spinning Earth on the movement of air to and from the equator. Air moving polewards is deflected eastwards, while air moving towards the equator moves to the west. Numbers indicate the rate of rotation of the surface around the polar axis

north. At this atmospheric pileup, the air is also cold and dense enough to descend towards the surface. This forms a belt of high-pressure and dry conditions known as the Horse Latitudes.

From here, air can move back towards the equator or away from it towards the polar regions. Air moving back towards the equator is moving more slowly than the ground underneath it. This causes the air to lag behind the surface, so it appears to move westwards. This creates the northeasterly (or south of the equator, southeasterly) trade winds. If air moves towards the polar regions from the Horse Latitudes it will be moving faster than the underlying surface in a west-east direction. Therefore, as we move towards the north or south, winds blow increasingly strong from a westerly direction. This gives rise to the mid-latitude westerlies that keep western Europe mild (and often unpleasantly wet). Figure 5.2 illustrates all of these ideas. The overarching concept is that of the Coriolis effect—the deflection of air as it moves from one latitudinal band to another caused by the rotation of the body. This is not a force as it is sometimes described. Rather, it is an illustration of inertia, where bodies in motion tend to remain in the same sense of motion (retain the same velocity) unless they are acted upon by another (unbalanced) force. Air moving from one latitude to another retains its velocity—its underlying east–west sense of motion—when it moves to this new latitude.

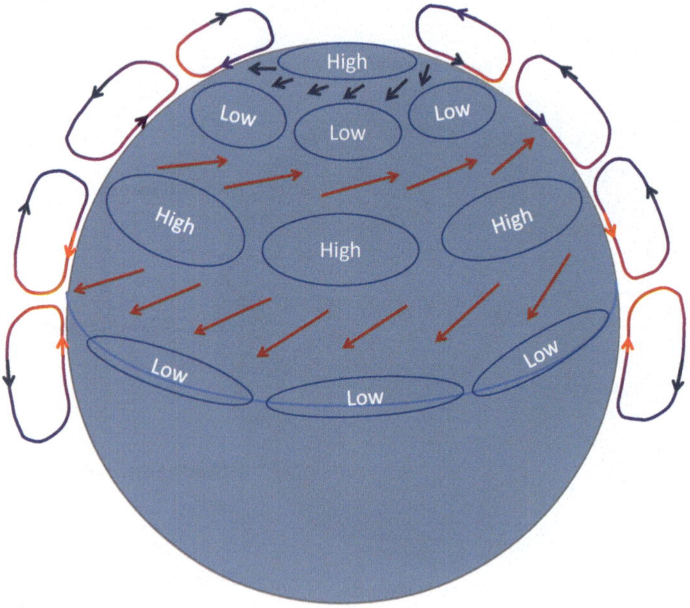

Fig. 5.2 The general pattern of wind flow and pressure on Earth in the northern hemisphere. Geographical features such as oceans and mountains strongly affect this generic pattern, as does the tilt of Earth throughout the seasons

Jet Streams

Next, there is the phenomenon of jet streams. Over certain areas, near the top of the troposphere—the lowest and densest region of the atmosphere—strongly concentrated currents of air exist. Here, air may blow at over 100 km per hour, driving smaller circulations along with them. The most prominent of these are the westerly jet streams that blow over mid-latitude regions; but there are other easterly jet streams over tropical regions that have a gross influence on the climate of the tropics. Why do these jets exist?

The answer is largely to do with the shape of the planet and the effect of heating. Changes in temperature tend to become focused into narrower regions in the atmosphere. For example, as air moves north on the surface on Earth (in the northern hemisphere), as we've seen it is soon deflected into a broad westerly airflow. This prevents direct flow and mixing with air further north. Therefore, there is a strong temperature difference between polar air and the tropical air that is moving northwards. Because the air is much colder nearer the poles, the base of the stratosphere (the tropopause) lies closer to the surface over the Arctic (and Antarctic), at around 7 km up, than it does in the mid-latitudes (at 10–13 km). The westerly jet streams congregate in this transition between the mid-latitude and polar ceiling, where the temperatures change rapidly.

Similarly, the easterly jet streams found in the tropics are associated with seasonal temperature jumps. For instance, as Asia heats up in the late spring, northern India is initially overlain by a westerly jet stream that separates the colder air to the north from the hot air to the south. However, as the area heated most strongly by the Sun tracks progressively northwards in the northern summer, soon Indian air is far hotter than the tropical air to its south. The westerly jet decays and is replaced by the summer easterly jet, with cooler air on its equatorial side and hotter air towards the polar side. The same process happens over western Africa as the Sun heats the Sahara more strongly than the Atlantic Ocean to the south of this part of the continent. In both instances, the formation of the easterly jet is critical to the success of the summer monsoon and its attendant rains. A weaker easterly jet means less air is drawn inland and upwards, giving rise to a weaker precipitation.

Other weaker, seasonal jets operate at lower levels in the troposphere and are associated with changes in temperature and the underlying topography. Critical to summer rainfall in India, the Somalian jet begins firing up in late April and brings increasingly moist air up the coast of Somalia towards the Saudi Peninsula, then across the Arabian Sea. This low-level flow delivers moisture to the western side of India and helps kick-start the monsoon, in concert with the overlying easterly jet. Between June and September, a similar low-level southerly jet brings moisture from the Gulf of Mexico and Gulf of California over the southwestern states of the United States. This brings welcome and critical summer rainfall to these otherwise parched states.

Although the presence of standard easterly and westerly jets has a limited impact on the climate of habitable red dwarf worlds, we should expect a contribution to regional rainfall from low-level (topography-driven) jets on tidally locked planets. These will be considered later in this chapter. For tidally locked planets with the shortest orbital periods, jet streams may be associated with super-rotating equatorial airflow. This is considered next.

Super-Rotation and Tropical Waves

Consider the phenomenon of super-rotation. Here, winds at the equator move faster in the direction of the spin of the planet than the planet is rotating. Elsewhere, on any planet, winds only move faster than the underlying surface because of the Coriolis effect. The air has moved away from the equator, taking momentum with it. Super-rotation is a different beast altogether, where equatorial air has gained momentum from more polar climes, or from higher in the atmosphere. This gives it a greater velocity than the ground over which it flows.

On Earth super-rotation is a transient phenomenon and most commonly associated with El Niño events. Here, the normal easterly trade winds reverse over the bulk of the Pacific Ocean and drive towards South America and the Caribbean. The phenomenon persists for several months and is accompanied by an eastward sloshing of warm water that has pooled near Indonesia. As this warm water reaches the coast of Peru, it overrides the northward-moving cold current along the coast, cutting off the supply of nutrients for local marine life. This causes the local supply of fish and other seafood to die out or migrate elsewhere down the coast. As this happens most commonly around Christmas, it gains its name from Jesus—El Niño, the boy. Its polar opposite often follows: La Niña—the girl. Here, the easterly trades strengthen and pull more cold water than normal up along the coast before driving it further west than normal, towards Indonesia. The precise mechanism that launches the El Niño is not well understood, but probably involves eastward-moving tropical waves, which are considered below.

Aside from the El Niño, the equatorial, terrestrial atmosphere displays a number of other regional waves that propagate against the prevailing easterly airflow. Understanding these is critical to understanding how the El Niño is launched and how super-rotation is driven on tidally locked planets. Some of these are transient and fairly local phenomena, while others have a remit that spans most of the tropical atmosphere that also deliver moisture to the mid-latitudes. Of particular interest are Kelvin waves, equatorial Rossby waves and the larger phenomenon known as the Madden-Julian oscillation (MJO). Of these, the Kelvin wave is the simplest to understand—and the smallest (Fig 5.3). They are also a principal component of equatorial super-rotation in hot Jupiters and (likely) many tidally locked super-terran planets.

Kelvin waves were named after their theorist, Lord Kelvin, and are found in both the ocean and the overlying atmosphere. These are alternating regions of low and high pressure that form a string along the equator and move in a west to east direction—against the general trade flow. Where pressure is low, there is enhanced

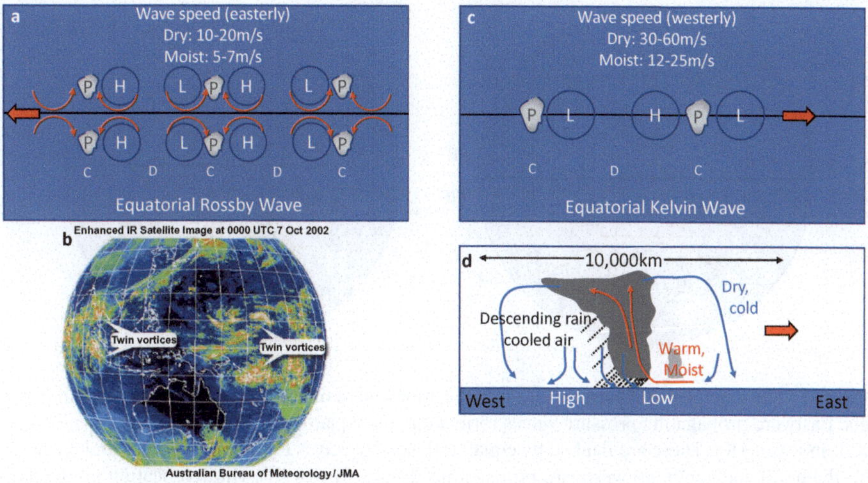

Fig. 5.3 Equatorial Rossby and Kelvin waves. On Earth, two kinds of equatorial waves are seen. Rossby waves (**a**) are paired waves that flank the equator and move from east to west, with wavelengths of 4000–10,000 km. Precipitation (P) falls where winds converge between areas of high (H) and low (L) pressure. Lower left (**b**) is an Australian meteorological infrared image showing such waves over the Pacific and Indian oceans. Kelvin waves (**c**) propagate along the equator in the opposite direction and at a faster rate. Terrestrial Kelvin waves are driven forward by precipitation (P) that falls just behind the area of lowest pressure (**d**). On tidally locked hot planets, interactions between these two kinds of waves pumps momentum from high to low latitudes, driving super-rotation

convection (indeed, convection causes the pressure to drop). However, air rising in these areas of low pressure ultimately cools, becomes denser and then falls back to the surface, forming a rear-flanking area of higher pressure. Evaporation of precipitation as it falls towards the surface helps drive cooling along the rear of the shower zone. Subsequently this cool, descending air spreads outwards, before converging with warm air in front of the band of precipitation, and this helps push the zone of uplift forward in an easterly (west to east) direction (Fig. 5.3d). You can then imagine a Kelvin wave as a rolling mass of air that pushes through the easterly trade winds. This roll is moving with an anticlockwise motion if you are looking from south to north, across the equator and down its barrel.

Experiments done by Dargan Frierson (University of Chicago) confirm that it is the pattern of evaporation and wind that are the dominant drivers of these waves in Earth's atmosphere. In this regard, Kelvin waves are important because they set up a pattern of west-to-east motion of mass and energy through the planet's tropical belt. On Earth, this effect is largely, but not completely, subordinate to the overall easterly trades so that although the waves move from west to east on monthly timescales, the predominant wind is still from northeast to southwest north of the equator, and southeast to northwest to the equator's south.

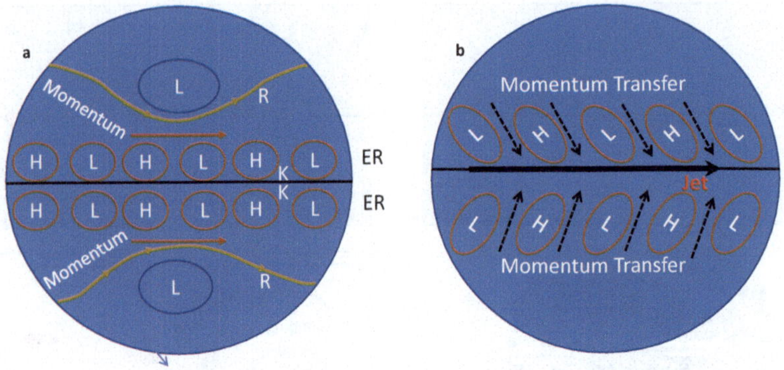

Fig. 5.4 The emergence of super-rotation in the atmosphere of hot Jupiters and super-terran planets. Eastward propagating pressure waves form along the equator (**a**). These features are known as Kelvin waves (K). These are flanked by equatorial Rossby waves (ER, orange circles and arrows). To the north and south are westward propagating Rossby Waves (R). Uneven heating on the day versus night side fixes these waves on either side of the equator. Shorter wavelength disturbances move with the mean flow from west to east. This setup causes the Rossby and Kelvin waves to develop a northwest-southeast tilt north of the equator, with a south west-northeast tilt to its south (**b**). This allows the atmosphere to transport angular momentum from the poles towards the equatorial regions by directing more and more energy into an equatorial westerly jet stream

Although these waves are subordinate to the overall flow on Earth, Kelvin waves appear to orchestrate prograde super-rotation (motion in the same sense as the planet rotates) on hot Jupiters—and perhaps other worlds. This will be discussed shortly, but the process appears to involve interactions between these equatorial waves and larger tropical features known as tropical Rossby waves (Fig. 5.4). On Earth such features come in two flavors: tropical and mid-latitude (or extratropical) Rossby waves. In these atmospheric structures, groups of waves are organized into larger structures. Terrestrial tropical Rossby waves are apparent on synoptic charts as paired areas of low or high pressure that flank the equator. The cores of the low-pressure areas contain clusters of thunderstorms and showers that are visible in satellite imagery (Fig. 5.3). Mid-latitude Rossby waves are somewhat harder to discern, but are apparent in the overall trajectories low-pressure centers take as they pinwheel through the mid-latitudes.

Rossby waves have an unusual structure that is somewhat counterintuitive. Such waves consist of a packet of smaller waves that move within the body of a larger, longer wavelength disturbance. If you consider a stone thrown into a pond, the overall pattern of ripples consists of higher frequency (short wavelength) wavelets, embedded in a larger wave structure with a long wavelength. As the larger waves (those with the greatest wavelength) expand outwards, smaller waves appear ahead of the overall expanding packet of ripples. As these larger waves move outward, the higher frequency (shorter wavelength) waves appear to move backwards through them, initially increasing in amplitude. As these higher frequency waves continue to move backwards through the advancing packet, they peak then shrink again, eventually disappearing.

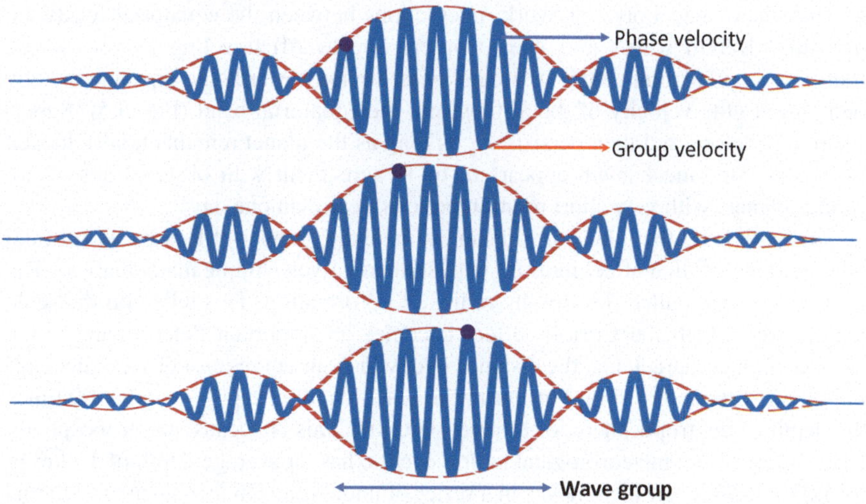

Fig. 5.5 Rossby waves are a familiar yet relatively complex idea. They consist of waves of different wavelengths that can move in different directions relative to one another. Imagine each blue wave is linked to its neighbor in a group (red dashed outline); then this group of waves will move together as a packet. This packet (or group) of waves can move towards the right (east) or towards the left (west). Larger groups of waves (those with the longest wavelengths) tend to move westward, while shorter wave groups move eastward. A low pressure area (indicated by a purple dot) moves with one wave inside the group. Rossby waves organize airflow from the tropics to the mid-latitudes. (Underlying wave image courtesy of Wikipedia Commons: Oleg Alexandrov and modified by the author)

In the terrestrial atmosphere, there are typically four to six long wavelength Rossby waves, with the number set by the radius and gravitational pull of Earth, along with the atmosphere's density and composition (Fig. 5.5). Terrestrial Rossby waves are partly anchored by geographical features, such as the Rocky Mountains. On tidally locked planets, temperature contrasts between the permanent day and permanent night hemispheres have the same anchoring effect, except that there may be only one or two waves. Saturn's polar hexagon is another very prominent packet of six waves that rotate slowly around the planet's northern pole.

All hot Jupiters are likely to be tidally locked to their star. The only exception is likely when they are young and, in a few instances, where there is a second massive planet in a close orbit. In a typical hot Jupiter, one hemisphere becomes much hotter than the other, dark side. This strong difference in heating is critical to driving super-rotation. In this instance, Adam Showman (University of Arizona) and Lorenzo Polvani (Columbia University) showed that interactions between the Kelvin and Rossby waves are fundamental. Tropical Rossby waves organize airflow somewhat further towards the poles of these planets, while Kelvin waves transport energy in a prograde direction along the equatorial band.

The large difference in heating between hemispheres results in the formation of planetary Rossby waves that become slanted in the direction of the planet's rotation.

In Showman's and Polvani's work, interactions between the equatorial (eastward traveling) Kelvin waves and these tropical (westward) traveling Rossby waves transport angular momentum in a clockwise motion towards the equator, thereby pumping up the velocity of the airflow into the equatorial band (Fig. 5.5). Super-rotation develops and then persists for as long as the planet remains tidally locked to its star. This mechanism appears to be in agreement with observations of hot Jupiter planets with very short rotation periods (less than one day).

For Earth and other planets, you can estimate the number of Rossby waves (and other parameters that affect motion) with some relatively simple mathematics.[2] The upshot is a term called the Rossby radius of deformation. This tells you the scale length over which the Coriolis effect becomes an important determinant of the motion of an air parcel, i.e., the distance over which air curves east or west, depending on the direction of spin. For Earth, the scale length is approximately 100 times the depth of the troposphere, or around 1000 km. This is because the troposphere, where most of the meteorological action occurs, has an average depth of 10 km in the mid-latitudes. Air then moves in a series of undulating (Rossby) waves with this particular scale length. On weather charts, the key weather pieces, the anticyclones and mid-latitude depressions, have approximately this diameter, which corresponds to the broad dips and ridges in the jet stream.

Importantly, for an Earth-sized planet, the Rossby radius of deformation exceeds the planet's size when the rotational period reaches approximately 5 Earth days. This means that for any planet with a terrestrial atmosphere and mass, the Coriolis effect will have a limited effect on airflow if the rotation period exceeds this value. Rather than air flowing broadly along lines of equal pressure (isobars), as it does on Earth, air flows more directly from areas of high to low pressure, across isobars.

Although this may be a little tricky to wrap your head around, the idea comes down to a few determinable properties. There will be a fixed number of long wavelength Rossby waves that are set by the radius, temperature and spin of the planet—and the depth of the troposphere. In turn, this quantity is set by the composition and mass of the atmosphere—and by the planet's gravitational pull. For terrestrial planets you can make assumptions about the composition, but the remainder of the required information comes from terrestrial observations and those by Kepler, HARPS and other instruments.

In the Solar System, many of the planets display super-rotation in their atmospheres. Venus currently experiences planet-wide easterly (retrograde) super-rotation except near its poles; Jupiter, Saturn and Titan also show it along their

[2] If you want to calculate this value, take a value "N"—the Brunt–Väisälä frequency (the atmosphere's propensity to oscillate in the planet's gravitational field); multiply it by the scale height (which can be simplified to the troposphere depth for terrestrial planets—but it's the depth over which the density decrease by the factor "e") and divide it by the Coriolis parameter, f_0; pi and another factor n, which is dimensionless: i.e., $L = \frac{NH}{f_0 \pi n}$. In turn the Coriolis parameter, f_0, is set by the planetary radius and its rotation rate (twice the rate, divided by the sine of the latitude, φ: $f_0 = 2\Omega sin\varphi$). (OK, the concept is a little tricky but the math is straightforward…)

equatorial regions.[3] Many hot Jupiters are also known to experience it, and it is likely to be largely a universal phenomenon for all such highly irradiated worlds.

Although super-rotation is common in hot Jupiter atmospheres it is unclear whether the same is true for smaller, cooler (but tidally locked) terrestrial planets. However, is it effective in the kinds of slower rotating terrestrial worlds we're interested in? This is where the disagreements start—and it appears to depend on what kind of parameters you put into your model.

Model Worlds and Their Problems

Various groups have modeled the atmospheric circulation patterns of tidally locked planets. One thing that becomes apparent is that although they usually display strong similarities, there are notable differences. These appear to be related to assumptions in the parameters used in the construction of the model, such as the strength of surface friction or the transport of heat via different kinds of Rossby wave. This section briefly summarizes them, then selects one set of models to follow for the remainder of the book.

However, before embarking on a quick tour of these models it's worth considering a few "expectations." If a planet is tidally locked to its star then one side is continually heated, while the other cools by releasing radiation into space. In the simplest scenario, hot air rises on the day side until it reaches an atmospheric lid where the air is "neutrally-buoyant." Here, the air has an insufficient difference in buoyancy to rise any further. It may have enough energy to overshoot and rise higher, using its store of momentum, but this won't carry it very far. Instead, it will spread out, cool via radiation into space and eventually become dense enough to fall back towards the surface. As it descends and the pressure increases, the temperature rises at a rate determined by the moisture content (and its overall composition). The rate of change, known as the adiabatic lapse rate, varies from 3 to 5 °C per 300 m, depending on how moist it is, more humid air heating the least.

Air that does reach the ground gains heat that is conducted and radiated into it. Remember that initially, much of the air that has descended will be in perpetual darkness as it flows over the ground. Therefore, the only sources of heat available are sub-surface heat flow and adiabatic heating that comes from its descent. Once the air crosses the terminator, stellar heating steadily increases in a manner that is proportional to the distance (or angular separation) from the sub-stellar point, SSP for short. Here, the planet's sun is directly overhead. Ultimately, the air becomes warm enough to rise once more, thus completing the loop. The region antipodal (180° longitude and latitude) to the sub-stellar point is known as the anti-stellar point, or ASP (Fig. 5.6). Air does not descend in a torrent, here; rather, in most cases the region of descent is broad and air movement leisurely.

[3] There is more on this phenomenon in Chapter 10 of the author's book entitled *The Exo-Weather Report* (Springer).

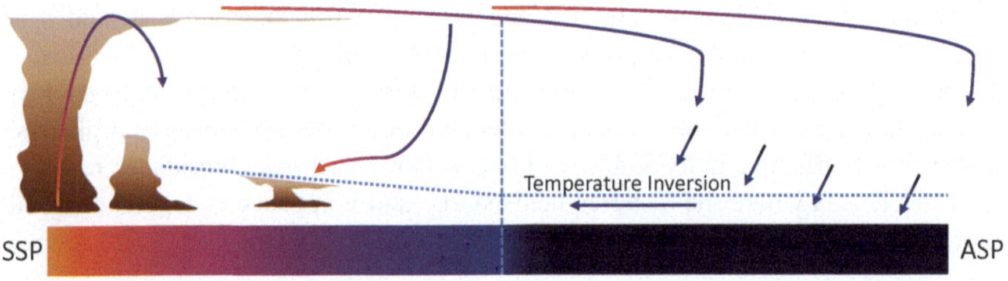

Fig. 5.6 The general atmospheric circulation pattern across a tidally locked planet. Convection is principally focused at the SSP, where temperatures are highest. Elsewhere, descending air forms a temperature inversion that suppresses convection and precipitation away from the SSP. The height of the inversion increases towards the SSP, where descending flow is weaker and the air at the surface warmer

Although this pattern of airflow is generally true, it is grossly modified by the speed of rotation around the star. For the slowest rotating planets this is likely to be (broadly) the case. We then loosely "expect" convection and precipitation at the SSP and dry conditions elsewhere. This simplistic model "makes sense" and agrees with observations of the tropics, where moist air convects, generating the doldrums and the associated thick but sporadic thunderclouds. However, no real planet sits still in space; all rotate on their axis, which is not always apparent if one considers tidally locked worlds. However, because all planets rotate, they experience a Coriolis effect. The magnitude of the effect varies with their orbital period—which in turn varies with the planet's distance from its star. Planets with very slow rotation rates (and longer orbital periods) would be expected in the habitable zone of the most massive red dwarfs (and perhaps lowest mass K-dwarfs). These will be tidally locked, hence always present the same face to their star, but will have rotation rates comparable with Venus, while presenting one face in perpetuity to their star.

However, the majority of red dwarf stars have much lower masses (0.1–0.2 solar masses). Planets in orbit within the habitable zone around these stars will have orbital periods measured in days to weeks (10–30 days is a good ballpark figure). Although these planets experience a Coriolis effect, which is substantially less than experienced on Earth, it is still sufficient to drive airflow from the SSP in different manners, which depend on the orbital period.

Most of the published models come from Ravi Kumar Kopparapu, Eric Wolf (both of the University of Colorado) and Ludmilla Carone (Max-Planck-Institute for Astronomy), as well as earlier contributions from Adam Edson and co-workers (Penn State University). Ludmilla, Ravi and others model aquaplanets—orbs with constant surface topography and frictional resistances. Winds then blow over them in a manner that is dictated by the amount of energy supplied, the orbital period (which sets the Coriolis effect for tidally locked planets), the supply of greenhouse gases (carbon dioxide and water vapor), and finally a set frictional force. Some models are oceans—true aquaplanets—and some are "swamps," which supply water vapor and a greater frictional force than an ocean. True aquaplanets are assumed to distribute energy through circulation within them, but swamp worlds do not. In most of the models, the oceans are shallow (around 50 m). Thus, while the models all reproduce the same overall effects, they differ in the effect of friction and the manner in which heat is distributed. None are truly representative of a habitable planet—at least not one that has deep oceans and a varying topography.

Ludmilla Carone's 2015 models tend to show strong super-rotation at all of the rotation periods that were considered, with planets having the shortest periods experiencing the strongest effects. Carone's models were of Earth to modest super-terran masses (1–2 Earth radii). Carone's 10- to 20-day-long orbital period planets had mid-latitude westerly jets, similar to those in Adam Edson's, Ravi Kumar Kopparapu's and Eric Wolf's models, but still retained equatorial super-rotation, which was lacking in Kopparapu's and colleagues' work.

Taken holistically, these models show a broadly continuous pattern from rapid equatorial super-rotation through broader but weaker westerly flow (assuming their planets rotate in the same manner as to Earth). Continuing to longer periods results in loss of all the zonal (west-east) flow and the formation of a singular giant (Hadley) cell that takes air from one hemisphere to another, as discussed earlier. However, in the slowest rotating models most of the returning surface airflow is over the polar regions. Here, the issue is what period is the cut-off between each state. The fastest rotation state is agreed upon, but the emergence of higher latitude westerly jets is contentious, and it depends on how the models are set up. Finally, the period at which the zonal (east-west) circulation gives way to singular cells is really not closely agreed upon.

Consider some of the relevant physics. There are a few parameters that appear to be critical in determining airflow. One is a factor we've already encountered, known as the Rossby radius of deformation. Remember that this critical value is the length scale at which the Coriolis effect becomes important for determining the motion of an air parcel. For a planet with a one-day period, this corresponds to around approximately 30° of latitude for a 90° change in direction. When planet's rotation rate exceeds this critical value, this allows the circulation to accommodate a greater number of waves per hemisphere, because the air is deflected into an east-west orientation at a lower latitude.

If you want to get your head around this think about Mars. It has roughly the same rotation period as Earth (24 h) but has half its radius. This means that it rotates at half the speed that Earth does in order to complete a day in the same time as Earth. Consequently, the tropical Hadley cells—those large convective cells that begin at the equator and sweep outwards—reach twice as far to the poles on Mars as they do on Earth. If you drop the rotation rate of an Earth-sized world to two days, the effect would be the same. You are halving the deflection caused by the Coriolis effect. Doubling the diameter has the opposite effect for the same period because you are doubling the rotation rate, hence the extent of the deflection. A large deformation radius means that air travels further before it experiences significant deflection.

In Edson's and colleagues' work for both dry and aqua planets, the transition between the fast-zonal flow to hemisphere-to-hemisphere (meridional) flow occurred when the Rossby deformation radius exceeded half the planetary radius.

Planets with rotation periods less than the cutoff point (i.e., the faster rotators) had more than one Rossby wave per hemisphere. Slower ones had a single standing wave, fixed around the sub-stellar point, where heating was greatest. The presence of this single standing wave is evident as parallel areas of high pressure flanking the equator around the SSP (Fig. 5.7). Air super-rotates between these high-pressure areas, across the SSP in a manner directly equivalent to the El Niño, where air super-rotates along the Pacific equatorial band between opposing blocks of high pressure.

Edson's work also showed that the transition between the fast and slow-rotation states depended on whether the planet was dry or wet (an aquaplanet). For dry planets, the airflow pattern adopted multiple rapidly transitioning forms when the rotation period lay between 5 and 10 days. The period at which this transition occurred was reduced by roughly 1 day in their aquaplanet model.

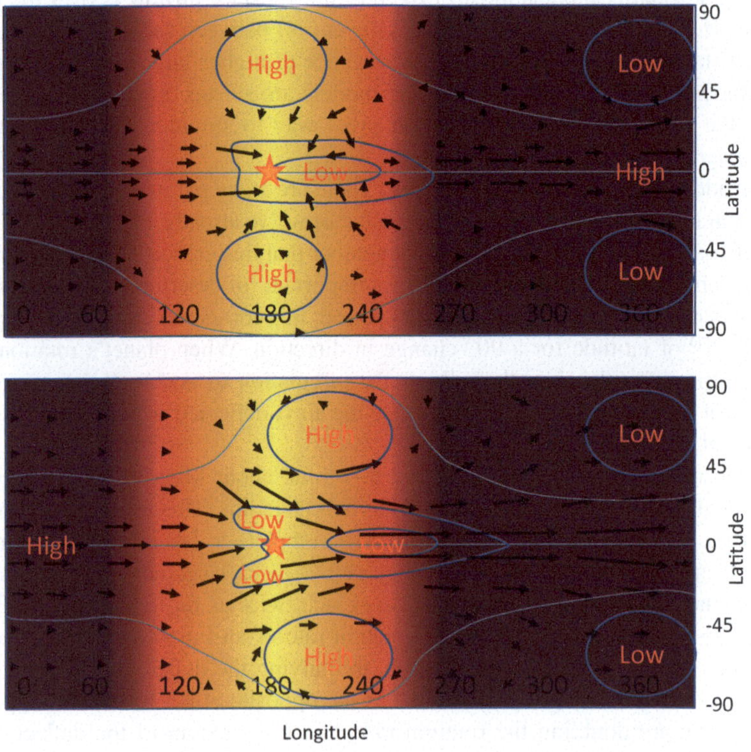

Fig. 5.7 Super-rotation and carbon dioxide concentration. Super-rotation is a dominant feature in these 2014 models of habitable, tidally locked planets. In the slowest rotating planets, the westerlies are 0.5–1 m per second (or 1–4 km per hour)—a very modest breeze. However, for a fixed orbital distance, increasing the concentration of carbon dioxide increases the degree of super-rotation in Wang's models. Top: 355 ppm CO_2; bottom, 200,000 ppm CO_2

In the models of Merlis and Schneider—and later ones by Ravi Kopparapu, Eric Wolf and others—the air returns predominantly across the poles, with weaker, converging flows around the equator. However, as the rotation rate is increased, super-rotation around the equator comes to dominate this return flow. Concordantly, with the strengthening of the equatorial westerlies, less air is delivered by convection above the SSP. The large (Hadley) cells that dominate the tropical regions are relatively weak and flanked by temperature inversions (areas where the temperature is either static or rises with altitude) in Carone's short-period planets. Thus, convection is confined to a broad but displaced region downstream of the SSP.

If one summarizes these models in the context of the overall circulation then we have slow-rotators where temperature differences drive convection in the slowest rotating planets (those with the longest periods). Here, there is a single cell spanning both hemispheres of the planet. As one cranks up the rotation rate (shortens the period) standing tropical Rossby waves develop on tidally locked planets that pump energy into the equatorial band through interactions with west-to-east migrating Kelvin waves. This is broadly consistent between all the models. The mid-latitudes remain more of an issue, as we mentioned previously.

One knock-on effect is the area of the planet's surface that experiences the highest temperatures. On the slowest rotating worlds, the highest surface temperatures are experienced where heating is strongest—at the SSP—irrespective of the model that is discussed. However, on planets with the shortest periods, equatorial super-rotation brings cooler air into the SSP from the west, thereby shifting the region of highest temperatures to the *east* of the SSP (i.e., downwind). This is also agreed upon by all research groups and is naturally consistent with common sense. However, in Carone's work, those planets with intermediate rotation rates (5–12 days depending on the planet's diameter) cause the formation of extra-tropical standing waves. These Rossby waves – which are broadly equivalent to those seen on mid-latitude Earth—shift the area of highest surface temperatures to the *west* of the SSP (i.e., upwind). This happens because the extra-tropical standing wave that develops brings cooler air from the polar regions directly towards the equator at the SSP. Therefore, the choice of model matters when we consider those planets orbiting within the habitable zone of stars of 0.10–0.15 solar masses.

Thinking now about precipitation, if we consider only the convective rainfall on a slowly rotating planet, this will be restricted to an area spanning 15° around the SSP although this is wider (up to 30°) in some models. Precipitation will be slight elsewhere, meaning any biosphere would be limited in extent—confined in large part to other available sources of moisture. For planets with super-rotation, precipitation might be smeared out along the equatorial band—particularly to the east (downstream) of the SSP, because super-rotating winds will deliver moisture and higher temperatures in this region. This assumes, of course, that the SSP does not lie over the heart of a supercontinent, where moisture may be limited. More on these in Chap. 7, where there will be a number of gedanken experiments looking at some possible iterations. Some of the outcomes may be surprising.

Other Sources of Precipitation

Such considerations of precipitation neglect obvious terrestrial counterparts: frontal and orographic precipitation. If one was to only consider convection, then we would only expect substantial precipitation around the terrestrial equator, where heating is strongest. However, substantial precipitation occurs in the mid-latitudes, where converging air masses produce uplift, cooling and condensation of any resident moisture. On slowly rotating planets with airflow that simply converges on the SSP, frontal rainfall is likely to be limited, as the air will converge with similar temperature and moisture content. However, one might envisage scenarios where some air is moving over warmer ocean and crosses onto a continental surface that is warmer. Differences in friction and temperature might cause the colder air to undercut the warmer continental air, resulting in uplift and precipitation. Although this coastal "front" is relatively stationary, there are precedents for this kind of situation on Earth, and it can give rise to coastal showers at a variety of latitudes where the land is warm.

Elsewhere, changes in air flow can produce frontal boundaries if air of one origin is able to undercut or be overrun by air from another source. In some locations there are quasi-stationary features, where there is a strong southerly flow into mid-latitude westerly flows (or corresponding northerly flows into westerly flows in the southern hemisphere). One example is the Mei-Yu front. This forms a roughly stationary feature that migrates north and south with the seasons over eastern China, Korea and southern Japan. Northward moving tropical air converges with the mid-latitude westerlies, forming a wobbling band of precipitation. Similar features form over the Atlantic side of South America and again on the Atlantic side of the United States. Here, the feature forms a seasonal band in the spring and summer months for the same reason. Usually evident as a discontinuous warm front that lies between northern Florida and Massachusetts during the warm seasons. This is illustrated in Fig. 5.8.

Topography and Precipitation

Similarly, if the planet is habitable it likely has mountains, as these will be associated with the carbonate-silicate cycle. Such mountains will deflect and cool converging air, and if the air is sufficiently moist, it will form clouds and precipitation on windward slopes. Although slowly rotating planets are expected to have descending air over much of the region distant to the SSP, this is unlikely to cap uplift on windward slopes, particularly if the mountains are high and the airflow strong. Such an inversion rises as one approaches the SSP and weakens, as the rotation period decreases (as airflow from the SSP is weaker). Therefore, in regions close to the SSP, or where mountains are particularly tall, we should expect precipitation. Figures 5.9 and 5.10 illustrate these ideas.

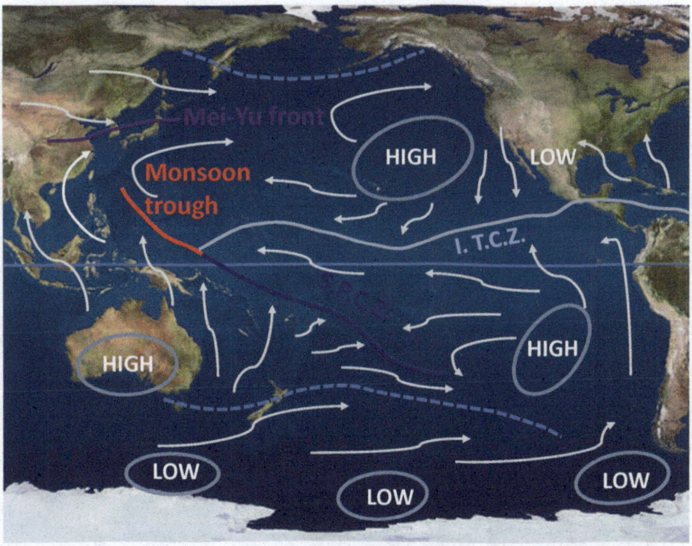

Fig. 5.8 Slow-moving regions of convergence produce copious rainfall. In the northern summer the Mei-Yu front demarcates the tropical airflow from the westerlies of the mid-latitudes. To its south lies the monsoon trough, where the easterly trade winds meet the summer monsoon flow towards India. South of the equator the South Pacific Convergence Zone (S.P.C.Z.). Like the Mei-Yu front this feature connects the tropical circulation with the westerlies that lie further towards the poles. The major equatorial band of convergence is the Inte-tropical Convergence Zone (I.T.C.Z.). Arrows indicate average direction of air flow

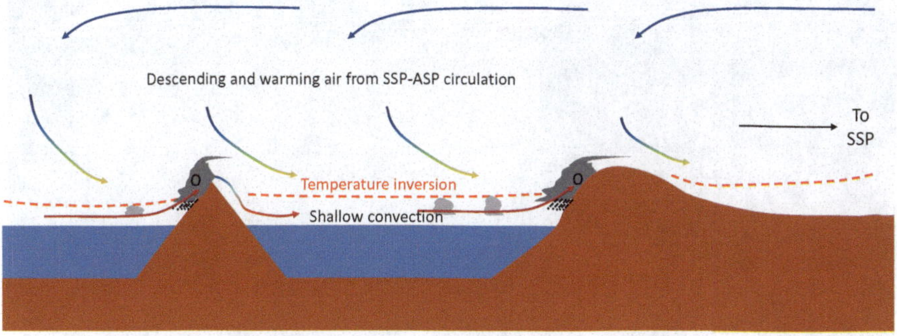

Fig. 5.9 Orographic rainfall on a tidally locked planet. Air flow towards the SSP will undoubtedly encounter obstacles. These include islands and mountain chains. Here, moist air will be forced over the obstacle, causing precipitation on the windward slopes and rain shadows on the SSP-facing slopes. This is "inversion breaking," and it deposits increasing amounts of moisture above the temperature inversion as one approaches the SSP and weakens its effect

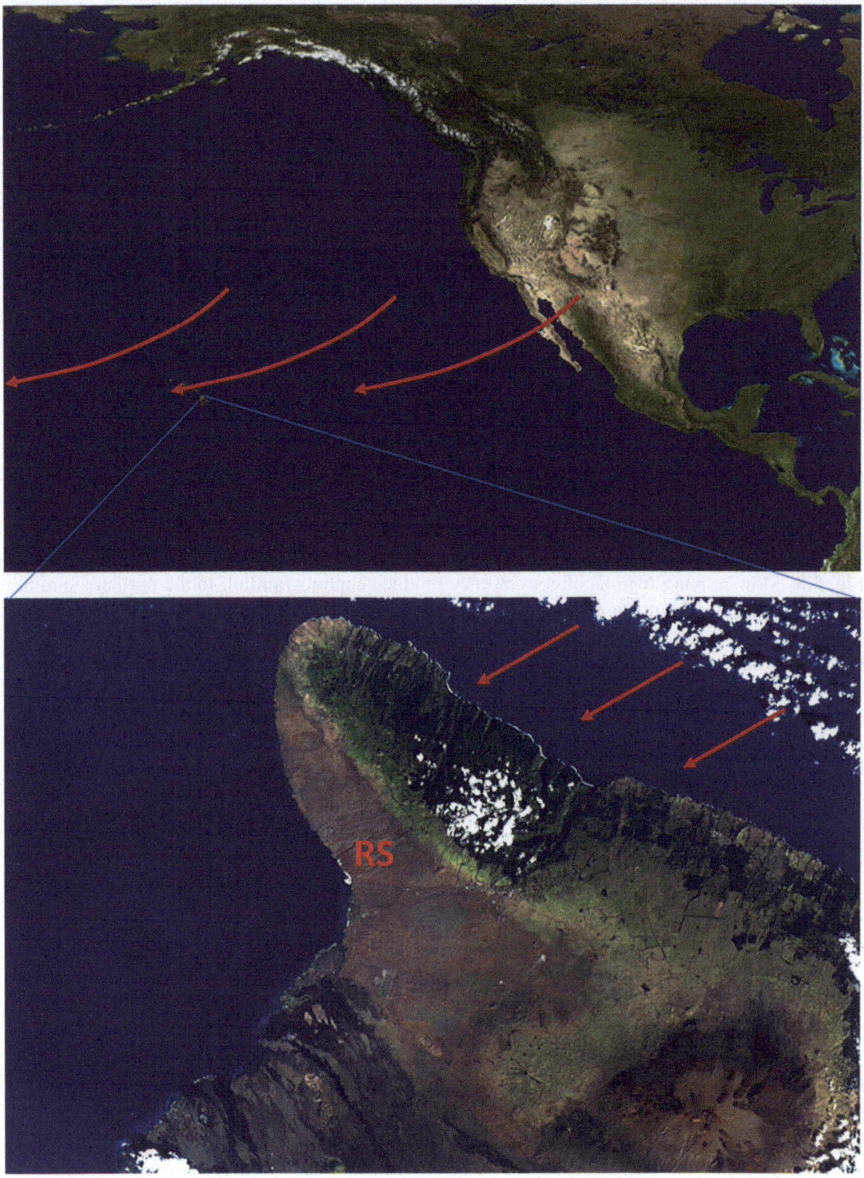

Fig. 5.10 Trade winds and orographic rainfall in Hawaii. Winds blow across the Pacific ocean in the northeasterly trades (red, curved arrows, top image). As these sweep over the mountainous slopes of Hawaii (bottom image), winds (red arrows) are uplifted, the air cools and moisture condenses, producing copious rainfall (evident as clouds on the green, vegetated slopes). To the southwest of these mountains, the land is brown and parched in a rain shadow (RS). (Both images courtesy of NASA)

Finally, before taking the plunge, we need to consider the relevance of these models. Giving a planet terrestrial insolation means that it has to be at an appropriate distance from its star. However, this sets the orbital period—and for tidally locked planets, it also sets the rotation period.

There are no planets in the habitable zone of their star with periods less than 5 days. The lowest mass a star can have—and be a star—is 0.075 solar masses. Any less massive and the nuclear engine fails to ignite. The circumstellar habitable zone for a star of this mass would have planets with orbital periods in the region of 5–6 days. If you are unsure, look back at the equation in the footnotes that relate stellar mass to orbital period. The smallest star with a well-characterized system of planets is TRAPPIST-1. With a mass of 0.08 solar masses, this is almost as small as you can get. Its innermost (and *potentially* habitable) planet has a period of 4.05 days (planet d). This has an equilibrium temperature—the temperature set by the balance of incoming and re-radiated radiation—of 282 K (9 °C). Earth, by contrast, has an equilibrium temperature of 255 K (−18 °C). Those interior to it are, clearly, too hot, and planet d is, at best, problematic (Chap. 9). Therefore, suggesting that a tidally locked planet with a period of less than 5 days is habitable is clearly unrealistic. Indeed, if we have an equilibrium temperature a good 27 °C hotter than Earth's, we should expect it to undergo a rapid loss of water. Earth is expected to begin this process when it has a surface temperature only 15 °C hotter than it is now. Therefore, considerations of habitability need to include the effect of greenhouse gases—particularly water vapor, which is a necessity for life (Chap. 6), and a greenhouse gas, whose atmospheric concentration varies with temperature.

Finally, differences in the depth, density and composition of the atmosphere are often not modeled, and a simplified terrestrial atmosphere is assumed. The composition of this is usually nitrogen with a splash of water vapor, the concentration of the latter varying with temperature. Clearly, if the planet has a lower or higher mass than Earth, this may not be true. The proximity to the central star, which may strip or otherwise modify the atmosphere, will also clearly have a bearing on this factor. The concentration of carbon dioxide not only affects the temperature and the ability of the planet to support photosynthesis (Chaps. 6 and 7), but it also influences the propensity of the atmosphere to super-rotate as well. Higher concentrations of carbon dioxide promote super-rotation. Therefore, excluding it from models is questionable.

Choice of Models

Taking everything into account, this book now examines and uses the models of Jacob Haqq-Misra (NASA's Astrobiology Institute's Virtual Planetary Laboratory), Ravi Kopaparanu (NASA's Goddard Space Flight Center and previously the University of Colorado), Eric Wolf (University of Colorado) and colleagues. These models were chosen because of their consistency and resolution. As well as incorporating the effects of Rossby waves, they also use another term, known as the

"Rhines length." This, like the Rossby deformation radius, is the scale at which planetary rotation causes east-west elongation, or stretching, of circulations to form atmospheric jets. On Jupiter, for example, the rotation rate smears the broadest circulations around the planet, forming counter-flowing easterly and westerly jets streams separated by bands of cloud.

The production of the jets is related to a property of turbulence, known as the so-called "inverse cascade." Here, small turbulent structures (vortices) merge to form larger ones. As planets have a fixed size, the cascade cannot produce structures larger than some characteristic scale, which is the Rhine's scale. When the largest turbulent structures reach a certain size—the Rhines length—the energy they carry begins to flow into Rossby waves instead of producing ever larger structures. Now, the elongation ceases, and structures, such as jets or smaller vortices (circulations), merge into the larger Rossby waves.

In the Rhines models, there is still space to elongate the circulation pattern adjacent to the SSP. Here, you see broad flows into the SSP from different directions that are associated with circulation around broad structures in the circulation of air in the day hemisphere. These gradually wink out as the rotation period increases and the flow becomes hemispheric—i.e. from the dark hemisphere to the lit hemisphere at the surface and vice-versa aloft.

Using these two quantities, Jacob Haqq-Misra and co-workers define three broad types of planetary circulation. Those with rotation periods less than 5 days show equatorial super-rotation, as in the models of Edson and Carone (Fig. 5.11). Here, the Coriolis effect is strong enough to deflect air moving away from the equator and tropical Rossby waves pump up the circulating air until it super-rotates. These have a Rossby radius of deformation that is smaller than the radius of the planet. Such habitable planets will orbit the lowest mass M-dwarfs (less than 0.1 solar masses).

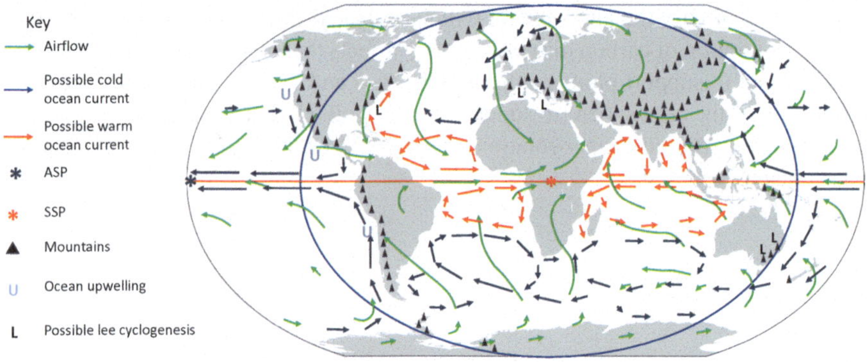

Fig. 5.11 In this gedanken experiment, Earth is a fast (east–west) synchronously rotating planet, with a period of less than 5 days. This falls into the rapid rotator regime. Airflow is strongest to the west of the SSP. There is a broad region of convergence around the SSP and along the entire equatorial band. This suggests that convection and associated rainfall is plausible over a broad area. Planets in this regime orbit at the inner edge of the habitable zone around the lowest mass stars. (Base map: Creative Commons, Wikipedia)

By contrast, those planets with periods longer than 20 days show the hemisphere-to-hemisphere flow from the SSP to the ASP, which is linked to the size of the Rossby radius of deformation and the Rhines length. Such slow-rotating planets have a radius and length that is larger than the radius of the planet. This means that a single wave fits on the surface, and there is insufficient distance to stretch eddies into jets. For planets at the inner edge of the habitable zone, the star must have a temperature of 3300–4500 K. These stars constitute the most massive M-dwarfs and least massive K-dwarfs. The lower edge of this boundary between slow rotators and other mechanisms is considerably shorter than seen in Carone's models (35 days), but planets with longer rotation periods are otherwise the same.

In between these two regimes, the authors of this study find a third regime where the planetary radius is smaller than the Rossby radius, but there is still capacity for turbulence-driven jets at mid-latitudes and substantial flow around the equator. This is qualitatively similar to Edson's work, but the jet-like behavior at mid-latitudes is weak, and the strongest equatorial flow is from the east into the SSP (Fig. 5.12). These planets have rotation rates between 5 and 20 days. The stars with such planets will lie broadly in the middle of the M-dwarf mass range and have surface temperatures between 3000 and 3300 K. This group of stars is important, as the planets that orbit them include Proxima b, the nearest (potentially) habitable planet.

These models are still considered swampy aquaplanets and have no ocean circulation; though this is not as exciting as a true-Earth-like world, they do form a suitable starting point from which to consider the other variables mentioned previously. Their models were applied to planets at the inner edge of the habitable zone and have consistent rotation periods and insolations (the amount of energy they receive from their star). If one wants to extrapolate to longer periods for a given star, then the amount of radiation they receive goes down. This may, in turn, reduce the

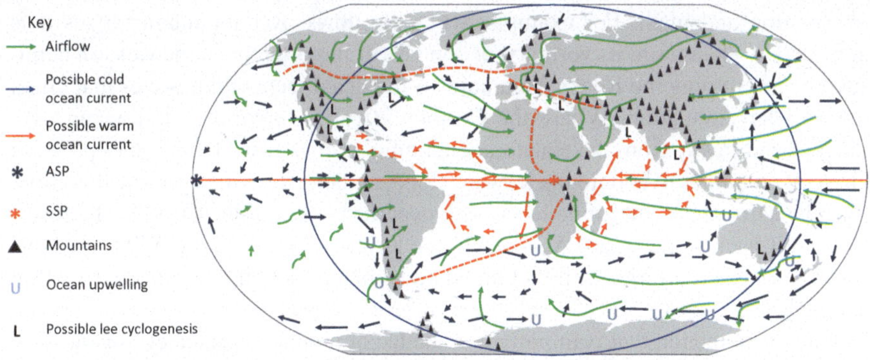

Fig. 5.12 In this gedanken experiment, Earth is a moderately fast (east–west) synchronously rotating planet, with an orbital period of 5–20 days. This falls into the Rhines regime. The major differences relate to enhanced east-west zonal flow at high and equatorial latitudes, particularly in the eastern hemisphere in both the atmosphere and oceans, but far weaker flow in the western hemisphere. Regions where stationary (Mei-Yu-like) fronts may form are indicated with dashed red lines. (Base map: Creative Commons, Wikipedia)

strength of the circulation, regardless of its structure. Therefore, planets lying further towards the outer edge of their star's habitable zone may have weaker circulations and ones that trend towards direct circulation from SSP to ASP.

Making Realistic Models: Other Factors to Consider

Land

The presence of land gives us geographical niches for life and in particular intelligent life. Although life almost always will evolve first in water—i.e., oceans—it can't develop art, science or technology while under (or on) water. The reasoning will be expanded upon in Chap. 7, but for now consider this as a "done deal.".

Next, land gives us a carbonate-silicate cycle. We will need this if we are to regulate surface temperatures efficiently.

Finally, and most prosaically, land is a likely outcome of geological processes that are needed to produce the above. Internal tectonism drives mountain building and the formation of continental crust. Therefore, on all but aquaplanets, land is a likely outcome of normal geological evolution.

Mountains

Although mountains are likely on any planet with land, mountains have a number of important impacts on atmospheric circulation. Although none of these has been modeled for tidally locked planets, we can make assumptions based on terrestrial observation and math. For example, when air flows over mountain ranges it is deflected by changes in its vorticity, or spin. Most importantly, in the western hemisphere, air crossing the Rocky Mountains is deflected, generating waves that affect the broader circulation. Where the Coriolis effect is important air flowing over ranges bends away from the equator, generating a ridge of warmer, higher pressure air. As it continues over onto the leeward slope, air deflects in the opposite direction, forming a trough. As the air deflects north over the ridge tops the Coriolis effect acts to bend it back south again. As a result, the terrestrial jet stream (and broader airflow) is sent into a series of broad undulations that affect climate across the entire hemisphere.

One phenomenon that is important in the lee of mountains is lee cyclogenesis—the formation of low-pressure areas in the lee of mountains (Figs. 5.11, 5.12, and 5.13). Although the Alps illustrate a specific kind of this effect, more generally, the formation of a trough in the jet stream favors the kind of spin that generates low-pressure areas. Although this phenomenon is fairly muted over North America, the same is not true over Argentina. Here, frontal troughs can cross the Andes and pick

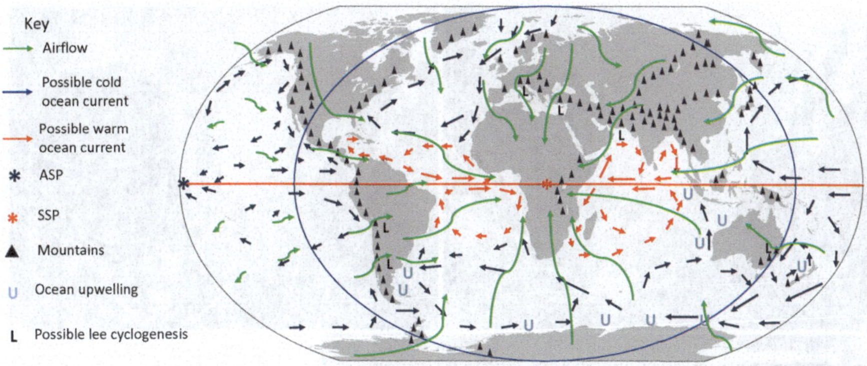

Fig. 5.13 Earth as a slow, synchronously rotating planet, showing ocean circulation and airflow. The orbital period exceeds 20 days for an Earth-sized planet, and this limit increases with planetary radius. Airflow is strongest to the east of the SSP. (Base Map: Creative Commons, Wikipedia)

up suitable spin in the lee of the mountains. The developing low-pressure area has an ample nearby source of water vapor from the southern Atlantic. Many such reinvigorated low-pressure areas bring heavy seasonal rainfall to Argentina, Uruguay and occasionally further towards the equator as they flow along the coastline.

The Rockies also illustrate another couple of effects. In the first, the mountains cause uplift, which cools the air and condenses most of the moisture. Windward Pacific slopes are thus drenched by orographic rainfall, while leeward slopes lie in a rain shadow. Moreover, the air that initially rises over the windward slopes cools at the moist adiabatic rate, while the air that descends the lee slope is now dry, having shed its moisture over the windward slopes. Therefore, air flowing down the lee slope is typically warmer than air that initially ascended the windward slopes.

If air rises over a mountain range that is laden with moisture it may cool from 20 to 14 °C with a rise of 1000 m (the moist rate). Dry now, it descends and warms up by the dry lapse rate. This takes its temperature to 24 °C. Obviously, the greater the difference in relative humidity and the greater the ascent and descent, the greater the effect. This föhn effect has critical local impacts on temperature in mountainous regions. In the case of the Alpine föhn, warm, moist air may move northwards from the Mediterranean basin (Fig. 5.14), depositing rain and snow on the windward, Italian side. However, on crossing the Alps the now dry air gives clear and much warmer weather on the northern, leeward slopes. Similarly, in North America the Chinook (meaning "snow-eater") drops onto the high plains from the Rockies, destroying winter snow as it goes.

More generally, the Rockies form a barrier to more temperate maritime air. In the winter, this barrier effect combines with the induced dip in the jet stream to bring cold Arctic air south over most of the continent. In the summer months, the continent heats up faster than the ocean, and temperatures soar. The barrier effect is critical in determining the climate of central North America. This is illustrated in the action of the Yellowstone hotspot. As the North American plate has slipped south-

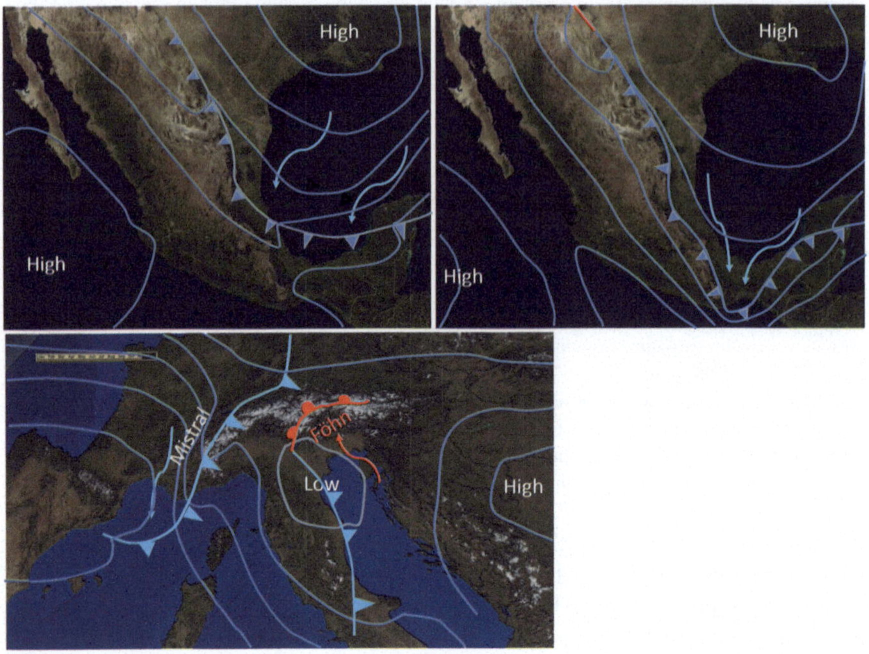

Fig. 5.14 Mountain winds. Top two panels show Mexico's gap wind, the Tehuantepecer, while the bottom panel shows how the Alps coordinate both the cold Mistral valley wind and Chinook-like Föhn. The formation of low pressures over northern Italy is another consequence of airflow over the Alps. (Image courtesy of NASA)

westward, the hotspot has carved a path through the Rockies from Washington State to Idaho—the Snake River plain and Yellowstone caldera. This has made a natural and utterly critical conduit for moisture to migrate 1000 km inland and bring moisture to areas that would otherwise be parched). Contrast this with the precipitation to the lee of the Rockies north and south of this conduit.

Although one should not concern oneself with seasons on tidally locked planets, we should expect some of these effects to be manifest on the planets of red dwarfs and be common on those of K-dwarfs.

On Earth, mountains only really provide constant shadow on pole-facing slopes—and the extent varies with the seasons. However, this is a global and immutable phenomenon on tidally locked planets for all slopes facing away from the SSP. The effect will obviously grow with angular separation (distance) from the SSP, as the angle that light will hit the surface will decrease with increasing distance. This will lead to some areas that lie in the lee of mountains being in permanent shade—cooler and poor abodes for photosynthetic organisms.

Consider a north-facing wall in a garden as a good analogy. Far fewer plants will tolerate the darker and cooler conditions than those inhabiting south-facing slopes. Near the terminator, mountains will produce a glum, permanent and very cold shade. Here, with winds blowing from the dark hemisphere, the landscape will likely be frozen until erosion, landslides or plate tectonics takes the land into light once more.

Next, consider the Alps. Here, the specific shape of the mountains facilitates a unique form of lee-cyclogenesis. Here, the Alps form a curved barrier, as an arc that extends from the French coastline at Nice through Switzerland and Austria. The Rhone Valley marks a channel through which cold continental air can flow in the winter and spring. Imagine, then, a cold front—the boundary between colder, southward-moving air and warm, humid air over the Mediterranean. As the cold air advances, the Alps block its passage into the Mediterranean. However, the Rhone Valley is open along the western edge of the range. Now, the cold front can push south faster along this conduit than it can over the rest of the Alps. The front then backs up, curving around the range. Warm air is drawn northwards over the eastern Alps as cold air moves down the western side. A combination of flow over the Alps, generating a trough in the atmosphere, and the curving effect of the colliding air masses, regularly generate low pressure areas over northern Italy and the Balkans. Cold air funneling down the western flank of these intensifying lows can generate vicious mistral winds, which howl into the western Mediterranean.

Lastly, there are gap winds. The mistral (above) is one such wind, driven by the flow of cold, continental air down and through the Rhone valley. Another and more spectacular example occurs in Mexico at the Chivella Pass. These gap winds blow through the Sierra Madre Mountains, funneling cooler, drier air from the North American continent through the Chivella Pass and into the Pacific. Such northeasterly gap winds, specifically the Tehuantepecer wind, have important climatic effects. These include influencing the formation of hurricanes and typhoons. They may also contribute to worldwide climatological events, such as the El Niño. Therefore, we need to consider such gap winds when we model the climate of potentially habitable planets.

Fronts

Fronts are an endemic feature of terrestrial air flow along the mid-latitudes of Earth. These are where "weather occurs" in its most dramatic sense. With the Rhines rotators there are good opportunities for convergence at mid-high latitudes along front-like boundaries. Here, there are strong changes in wind flow from easterly to northerly or southerly as air converges on the SSP from regions of zonal flow. In other circulation regimes, "fronts"—boundaries in air masses—will only be present transiently and likely associated with turbulent flow from land to ocean or vice versa. Otherwise, the motion of air is so predictable and unchanging that there will only be limited opportunity for air of different density and humidity to converge. Mountains may be a means by which air might periodically become dammed behind ranges. However, this is not likely to have a dominant (or even significant) effect on the climate on most planets that orbit red dwarf stars.

In the next section we've adapted the models of Jacob Haqq-Misra and co-workers to take into account topography and the distribution of landmasses. Precipitation was not formally modeled in the work of Haqq-Misra and colleagues.

Therefore, very conservative predictions are used in these extrapolations that are based on the airflow and temperature profiles. Therefore, these remain qualitative and should be taken lightly at this stage. It will be some time before computers can resolve the kinds of features that we should see if we are to truly contemplate habitability in a realistic planetary scenario.

Models with Terrestrial and Pangaeal Continents

The following models were presented in the International Journal of Astrobiology in 2018. The circulation models of Jacob Haqq-Misra and co-workers were adapted, incorporating some of the effects described above. Also (tentatively) included was ocean circulation, as was likely with these atmospheric models. The latter will remain problematic until higher resolution computer models with realistic continents are produced. Until then, general models by Jodie Cullum (University of East Anglia) are applicable.

The Climate of the Slow-Rotating Anthropocene Earth Model

In the slow-rotating model of Haqq-Misra and co-workers air flow moves across the poles towards the SSP (Fig. 5.15). There is a relatively gentle converging westerly and easterly flow towards the SSP. Broadly the flow is radial. Positioning continents with the current terrestrial configuration makes some changes, principally where mountains—in particular the Alpine-Himalayan chain—deflect this converging flow.

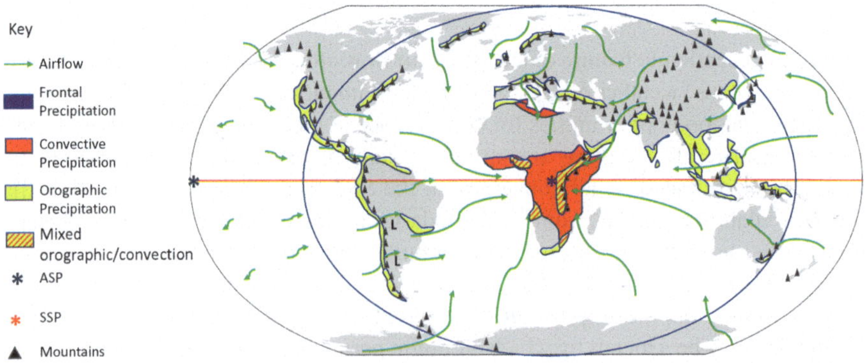

Fig. 5.15 Earth as a slow, synchronously rotating planet, showing ocean circulation and airflow. Possible precipitation is shown, which includes the effects of orography. Airflow is strongest to the east of the SSP. Precipitation is limited in these conservative models, because there is a strong, overlying temperature inversion that prevents uplift to areas of strong heating or orographic forcing. (Base map: Creative Commons, Wikipedia)

The climate to the north of the Alpine-Himalayan chain should be dry, as the air flowing over this region is crossing continental crust and has a cold, dry source. Air is also capped above the surface by a temperature inversion—a boundary where air temperatures rise or remain constant with height. Such inversions limit convection and thus the propensity for precipitation. Temperatures are likely to rise progressively towards the SSP, but as continental crust has a relatively low specific heat capacity and is constantly illuminated, temperatures may be rather high throughout, depending on the insolation. This leaves precipitation rather up in the air. In a simple model, the air flowing retains little moisture and has a progressively lower relative humidity as it heats up, likely making our tidally locked Asia (and North America) dry. The passage of cold air over warm, wet ground will, however, generate the so-called "lake-effect." Over the Great Lakes and Lake Baikal in early winter, cold air flowing over the lakes picks up moisture and generates copious convective rainfall or snowfall on the windward side of the lake. In turn, air that is heating from below and picks up moisture from internal bodies of water could generate shower activity. As the wind direction is largely fixed, limited areas that lie downwind of the lakes may have their own temperate and humid climate.

Our tidally locked planet may have a Mediterranean climate over the equivalent body of water. Again, depending on the temperature of the air flowing south towards the SSP over Africa, areas on the southern edge of the Mediterranean may have sufficient rainfall to support vegetation. However, one must make a comparison with the winter and summer rainfall patterns. In the winter, southward moving cold air picks up heat and moisture, which generates rainfall. However, in the summer, it is not the same. Now, there is a combination of an upper level temperature inversion, caused by descending air from the Azores High, and a hotter central European landmass, which means that the air does not convect, and rainfall is virtually absent. Therefore, the effect of our tidally locked Med will depend on the temperature of the air that flows over it and the temperature of the water in the basin.

Likewise, the development of lee cyclones over our tidally locked Italy and Balkans will depend on whether the air is suitably moist and able to convect. That said, the movement around the mountains may be sufficient to cause rotation and uplift. However, as air flows over Africa it should be strongly convecting. Rainfall may be muted along the Saharan and sub-Saharan areas as the airflow will be dry, having sourced over Asia. The western coastal areas of north Africa may get some convective rainfall, as the air has crossed and picked up moisture from the Atlantic. The Atlas Mountains may also drive orographic rainfall, as might other upland areas in the central Saharan region.

The remainder of the continent should get copious rainfall, which will be concentrated in the Congo region. One final area to consider that might be drier would be along the southwestern coast of Africa. Here, the general airflow towards the SSP might bring cooler surface waters northwards. As is currently the case in this region, the colder waters might generate a temperature inversion aloft—and these will reduce the availability of moisture, making this region relatively dry.

Elsewhere, the possibility of precipitation is likely to be low. Any overlying temperature inversion will prevent or limit convection over Australia and South America.

Orographic rainfall might be possible, particularly along the Andean margin as airflow from the Pacific is deflected strongly over the Andes. This may also be true of the Snowy Mountains of eastern Australia, although the effect will be more muted. The easternmost promontory of Brazil may get modest convective rainfall, depending on the depth of any inversion and the amount of convection that's possible as air is forced over the warm plateau.

The concentration of carbon dioxide will alter the pattern to an extent. Planets with greater concentrations of carbon dioxide will obviously be warmer and convection may be more vigorous and widespread on the day hemisphere. However, in general models tend to show greater extents of super-rotation in the atmospheres of planets with greater concentrations of carbon dioxide (and these include models of Earth). Therefore, one might see more focused precipitation on windward slopes to the west of the SSP and less to the east.

The Climate of the Slow-Rotating, Pangaeal Earth Model

In this model the continents are grouped in the Pangaeal formation they had near the end of the Permian era (Fig. 5.16).

Most of the continental area is grouped around the SSP, with the Appalachian Mountains forming the focal point for convergence. Here, precipitation is heavy and focused by both high temperatures and topography.

Although the general pattern of precipitation mirrors that of the Anthropocene model, the arrangement of continents on either side of the Paleo-Tethys Ocean, and the presence of mountains along the northern margin, favors both convective rainfall

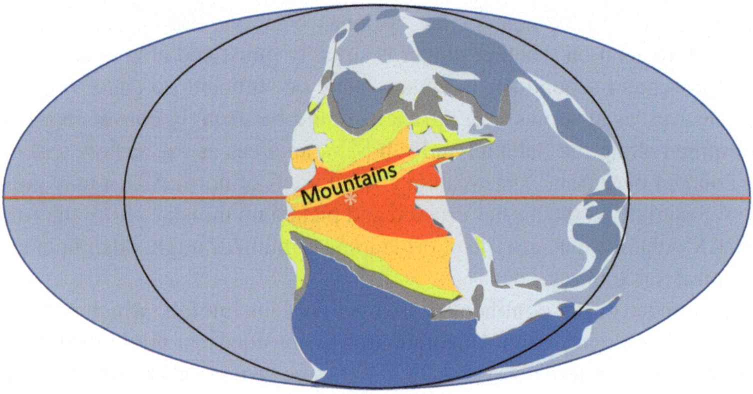

Fig. 5.16 A late Permian Pangaea as an alternative arrangement for a planet with an orbital period of greater than 20 days. This model world is simplified to color-code the areas with heaviest precipitation (red), grading through to blue with little or no precipitation. Mountainous areas flanking the northern and eastern Tethys Ocean or along the eastern margin of Panthalassa can receive orographic precipitation if winds converge onto them from the seas

and orographic rainfall along the southern edge of Asia. General convergence as air approaches the western Tethyan margin may also more generally increase rainfall across the basin, particularly as one approaches Gondwanaland.

Marine transgressions (where the oceans have flooded the land) over Asia and Europe also increase the supply of moisture to air flowing towards the SSP at the surface. Therefore, one would expect a broader area of convective rainfall over central Laurasia.

Shifting the SSP to the far east of the Paleo-Tethys leaves most of the supercontinent dry. Precipitation in any meaningful form is limited to the Chinese platforms. Orographic rainfall is possible along parts of the Paleo-Tethyan margin, but with most of Pangaea at or beyond the terminator, most of the habitable land is frozen and dry.

Figure 5.17 illustrates the same world, but with Pangaea shifted to the west so that most of it is near or beyond the terminator.

The Climate of Rhines-Rotating Anthropocene Earth

The climate of these planets is rather dynamic. Although the general airflow is still towards the SSP, the radial pattern is skewed, with stronger convergence along the eastern and western margin of the SSP. Moreover, there are westerly winds blowing at high latitudes around large anticyclonic circulations. Placing continents on this general circulation is likely to favor the formation of warm-cored anticyclones over the continental areas if the surface is relatively cold at these latitudes. Here, the colder surface would facilitate the descent and outflow of air coming from aloft. Such planets might be nearer the outer edge of their circumstellar habitable zone (Fig. 5.18).

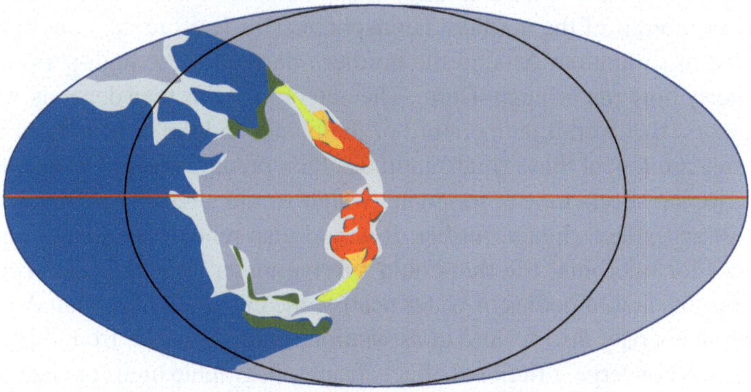

Fig. 5.17 A Pangaeal slow-rotating model, where Pangaea is shifted westwards. As before, but on this occasion we see most of the continental surface uninhabitable except for extremophiles. Precipitation is mostly over the ocean or on the Chinese micro-continental fragments. Again, red areas have the highest precipitation and blue the least

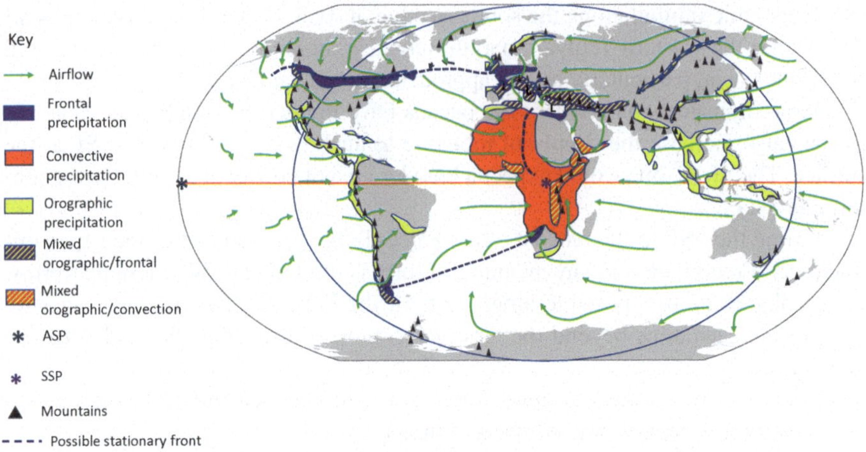

Fig. 5.18 Precipitation on a Rhines regime synchronously rotating planet. Regions where stationary (Mei-Yu-like) fronts may form are indicated with dashed dark blue lines. Such stationary fronts provide narrow bands of precipitation, heaviest on the equatorial side of the boundary. More sporadic convective rainfall may occur in areas near the SSP than is shown. But with low atmospheric moisture, this will be unreliable. Precipitation is limited, globally, because there is a strong overlying temperature inversion that limits uplift to areas of strong heating or orographic forcing. (Base map: Creative Commons, Wikipedia)

However, if the planet is warmer and the continental surface hotter surface pressure will be lower. Areas of higher surface pressure might be displaced onto neighboring, cooler oceans, where descent and outflow of near-surface air is promoted. Most likely, areas of descending air will be driven by the overall circulation, irrespective of the layout of the underlying landmasses, should they exist.

For the sake of argument, we'll stick with higher pressure as is shown in Fig. 5.18. Here, fortuitously for the modeling but not for habitability, these areas correspond to the cores of North America and Asia, in a manner similar to what would be seen in the winter months of the northern hemisphere. This leads to dry conditions, with very limited precipitation. Around the northern flank of these high-pressure zones some westerly flow can bring moisture. The outcome of this then depends, critically, on two factors: the overall temperature of the planet and the proximity to water. If temperatures are low at these (high) latitudes then precipitation will be as snow. In this scenario, ice sheets may accumulate, which would alter the surface albedo (its reflectivity) and affect climate further. It would also render these areas uninhabitable, except for microbial life that could survive under or in the ice. Secondly, if there are no substantial bodies of water nearby, the climate will remain dry.

These planets may also feature quasi-stationary Mei-Yu-like frontal boundaries where airflows converge. Frictional effects from topographic highs or changes in the strength of friction along coastlines should facilitate the formation of these fronts, because airflows will experience the strongest deflections. Therefore, these worlds might have relatively narrow bands of frontal rain (or snow) that may be significant.

Consequently, in general, these planets may offer the best chance of finding life over the largest area of their surface, where land is present. This is lucky, as these worlds include Proxima b.

The Climate of Rhines-Rotating Pangaeal Earth

If we rearrange the continents, then the effects are hopefully by now somewhat predictable. The orientation of the Paleo-Tethys enhances convergence towards the SSP, which we've located over the Appalachian chain. Stronger equatorial winds facilitate greater opportunities for orographic rainfall across windward slopes along the northern Tethyan margin.

However, the continental interiors will be drier on either side of the SSP than in the slow-rotating model. Mid-high latitudes westerlies bring moisture inland, and with these and frontal boundaries the potential for orographic precipitation (above) increases. The presence of inland seas (in what is known as a marine transgression) across central Asia allows further sources of moisture to the mid-latitudes and the potential for further orographic (or even weak convective rainfall) – if the presence of the overlying temperature inversion is not prohibitive. However, we should be cautious here; these worlds are still drier than Earth.

Shifting Pangaea westwards relative to the SSP renders most of the supercontinent uninhabitable, as before. Now, most of the land will be dry, cold or in permanent darkness. Only the Chinese platforms remain in areas with significant moisture. However, the stronger westerly flow along the equator, and across the length of the Paleo-Tethys, does allow the possibility of orographic rainfall on some windward coasts, particularly along the northern Paleo-Tethys coastlines, where there is active mountain building and topographic relief.

The Climate of a Fast-Rotating Anthropocene Earth

These planets lie close to their star with short orbital (and hence rotation) periods. Faster rotation strengthens equatorial super-rotation, which has two effects, illustrated in Fig. 5.19. Firstly, it weakens convection at the SSP, while smearing it out along most of the equator. Secondly, in Carone's models, air flowing out above the SSP generates secondary, upper-level cells and areas of descending air near to the SSP. This generates a temperature inversion around the SSP but weakens the global inversion elsewhere. This reduces rainfall at the SSP but allows broader convection away from it, where conditions are favorable.

These areas are concentrated where convergence is strongest, principally in the aquaplanet models to the northeast of the SSP. Simpler models have a lobster-claw pattern around the SSP. However, Haqq-Misra's and colleagues' models have a

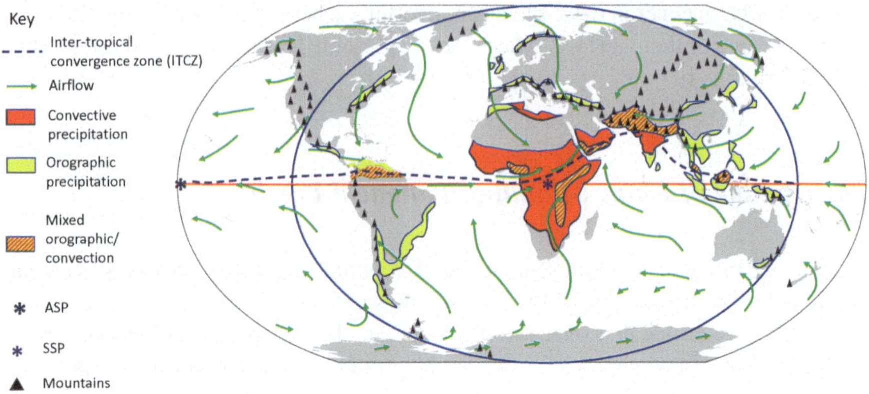

Fig. 5.19 Wind and precipitation on a fast rotating planet (orbital period less than 5 Earth days). Winds are the same as in Fig. 5.12, but showing both convective rainfall (red) and orographic rainfall (yellow). More sporadic convective rainfall may occur in areas near the SSP than is shown. But with low atmospheric moisture, this will be unreliable. Precipitation is limited, globally, because there is a strong overlying temperature inversion that limits uplift to areas of strong heating or orographic forcing. However, forcing of converging air over our exo-Himalayan chain enhances convection and precipitation. General convergence along the equatorial band *may* lead to the formation of an ITCZ-like region, with showery precipitation. (Base map: Creative Commons, Wikipedia)

more distorted pattern. Superimposing this on a terrestrial distribution of continents has some fortunate consequences—both from the habitability aspect and from the modeling point of view. The circulation pattern drives more air flow towards southern Asia, in a manner somewhat akin to the summer monsoon pattern. With the Alpine-Himalayan chain in place, convection and orographic effects combine to produce strong uplift along a broad area of southern Asia—as well as at the SSP over Africa.

The northwest of Africa may be somewhat drier than in the Rhines model, but that will depend on the precise track of air towards the SSP. A more westerly aspect will maintain higher levels of precipitation.

Secondly, airflow along the equatorial band—and even into the dark hemisphere on the eastern flank of the SSP—carries significant moisture, if not direct precipitation from convection. This can in principle deliver significant orographic rainfall on any windward slopes. Across the terminator, light snowfall might be possible in a similar manner, which would then accumulate as ice sheets. The original models of Haqq-Misra lack this effect, as these are for aquaplanets, planets that clearly lack topography. They do, however, permit fog in the dark hemisphere along the equatorial band, which presumably would be freezing fog that would deliver precipitation as ice granules. Although we are obviously not going to get plants here, we could get subsurface life across broad areas because of the supply of moisture. More on this in Chap. 7.

The Climate of a Fast-Rotating Pangaeal Earth

Finally, we turn our attention to the Pangaeal model. In this, the distribution of land is again poor with regard the delivery of precipitation to the continents. However, one major plus is the zonal (west–east) flow along much of the equator. Although most precipitation falls on the Chinese platform, the easterly flow to the west of the SSP could deliver orographic rainfall (or snowfall) over Pangaea to the east of the Appalachian Mountains. Thus, this area may be generally habitable in terms of its precipitation as well as the temperature.

Generic Conclusions from These Models

One thing that is immediately apparent is how little of the surface of tidally locked planets may be habitable for most species, compared to our planet with its terrestrial climate. The reason has nothing to do with temperature, as this may be broadly similar to that of Earth. The problem is precipitation. Although Earth has a strongly zonal (east-west) orientated wind structure, those of tidally locked planets generally do not, and any zonal flow that does occur is largely restricted to low latitudes. This leaves the bulk of the surface (that which is not at low latitude) bereft of rainfall. Moreover, the rain-bearing areas on Earth migrate north and south with the seasons, meaning that almost all of the planet sees precipitation at some point in the year. This is not true of tidally locked planets, where areas of precipitation are fixed.

The aquaplanet models have virtually no precipitation away from the SSP. Those of Merlis and Schneider have all but a whiff of precipitation confined to 15° on either side of the SSP in slow-rotating models. Elsewhere, the balance is strongly in favor of evaporation. This leaves tidally locked planets largely desert, scrub or, perhaps where there are orographic sources of rainfall and surface water, savannah, steppe or tundra, depending on the temperature. The climate models for Proxima b, which were produced by the Met Office, have broader areas with precipitation than Merlis and Schneider. However, this is still confined to regions within 60° of the SSP—and much of this precipitation is snow.

Remember that even where it doesn't rain, rivers may carry water to otherwise parched areas. Think of the Nile weaving its long, green cloak through the Sahara, on its way to the Mediterranean. Tidally locked planets may also have significant areas of continental crust near or below sea level, which would allow a greater probability of moisture becoming available as fog/dew. However, this seems like slim pickings compared to Earth, with its copious supply of mid-latitude moisture through migrating frontal systems. Therefore, even with equitable temperatures, tidally locked planets are far poorer abodes for life than equivalent asynchronous-rotating planets with similar overall temperatures.

If we want something more benign, the Rhines model worlds may be the best bet. These host turbulence-driven and largely zonal systems at mid-latitudes. They could deliver moisture across continental areas if the landscape has the appropriate configurations. If so, the precipitation models for the Rhines-rotating planets will be a tad too conservative, allowing more of the mid-latitudes of these worlds to blossom.

General Climate and the Long-Term Habitability of Planets

Before we embark on the final part of this chapter, it's worth reminding ourselves about the kinds of stars we are interested in, as these affect how the climate, as a whole, responds. As we move from F through G and K to M stars, they become cooler, and, therefore, redder. The effect is not merely prosaic. Surface water-ice, atmospheric clouds, carbon dioxide and water vapor strongly absorb in the near-infrared. Therefore, when one thinks about the overall climate of the planets of red dwarfs, we need to turn the thermostat down relative to other, hotter stars. To reach a particular equilibrium temperature, a planet will lie further from its star than one orbiting an early K- or G-class star. A planet orbiting a red dwarf that receives the same amount of radiation as we do will be warmer than Earth.

Taking this into account, we can now look at the temporal domain and consider about the long-term stability of climate on tidally locked planets. In the aquaplanet models four broad and stable climate states are identified in the work of Eric Wolf and colleagues. The coolest is the snowball state. This has an equilibrium temperature that is less than 235 K. The water belt (that most closely analogous to that of Earth when heavily glaciated) has an equilibrium range of 235–250 K. A largely temperate (warm Earth) state has an equilibrium temperature of 275–315 K—analogous to Earth in the next few hundred million to two billion years. And finally, there is the moist greenhouse condition, which is loosely analogous to Earth in 2–3 billion years' time, depending on how much water is lost. This has an equilibrium temperature in excess of 330 K. In between each are unstable regimes that rapidly transition from cool to warm as either ice melts and decreases albedo (reflectivity) or water evaporates and adds to the burden of greenhouse gases.

An interesting case in point in Wolf's models are the nature of the rapid transitions and what then stabilizes the climate in its new state once the transition has been crossed. Take, for example, the transition between the water belt and temperate states. If the temperature falls, then ice cover increases. However, ice does not then go on to completely mask the planet, because at some point it encounters land that already has a reflectivity (albedo) similar to ice. This occurs when ice sheets encounter desert. At this stage there is no further change in albedo, and positive feedback ceases. Similarly, in the transition from water belt to temperate, once ice is confined to the polar regions the increase in absorption of energy ends and the planet stops warming.

The interesting thing about all of these models (and various ones have been produced by various groups) is the consistency in the speed of each transition. Such rapid transitions were also spotted in the models of the snowball Earth, where the transition from frigid to torrid was expected to be fast. Here, several times in the last 300 million years of the Neoproterozoic, Earth repeatedly froze, largely but not completely, over. Glacial deposits were produced in each episode, but these are overlain by further deposits indicative of extreme heat and high carbon dioxide. The indication was that each glacial surge was terminated by a spike in carbon dioxide and a rapid rise in temperature from subzero to around 30–40 °C. The length of the transition was less than a millennium, with melting of highly reflective ice exposing dark rock, which led to a positive feedback. Unfortunately, this means that for planets, whether tidally locked or not, if they lie near the outer edge of the habitable zone and are frozen, the transition to habitability won't be pleasant—or long-lasting.

Another factor in the rapid melt—which is only applicable to biologically viable planets, is the effect of anaerobic respiration. Such processes operate in water-logged soils, such as the Tundra and in parts of the Taiga in particular. Organic matter, when buried, becomes sodden because of low evaporation rates in the summer months. Now, bacteria convert organic matter to carbon dioxide and methane, which can become trapped in the water-logged soil. If temperatures rise sufficiently to melt any permafrost the gas escapes to the atmosphere. As both are strong greenhouse gases, their release amplifies any warming effect. In turn, this melts more ice and permafrost, further amplifying warming. Martin Kennedy (University of California, Riverside) and co-workers convincingly report that such an effect may have been critical in driving Earth out of its frigid state. Yannick Donnadieu (Laboratoire des Sciences du Climat et de l'Environnement, CNRS-CEA) and colleagues report that erosion and runoff may have been the triggers for its instigation.

In a similar regard, Kristen Menou (University of Toronto) modeled the effects of reducing the amount of energy the planet received from its star. They showed that Earth-like planets, which received less sunlight than the current Earth, repeatedly cycled between unstable glaciated and deglaciated climatic states, in a manner analogous to the Neoproterozoic planet. Menou's models were then adapted by Jacob Haqq-Misra and colleagues and fortunately shown to be a little too pessimistic. When modified to include the effects of weathering and a biosphere, the climate, though still prone to instability, was more stable and less prone to sporadic bouts of glaciation.

These factors emphasize the impact of feedbacks in the long-term stability of planetary climate. In this regard the circulation regimes need to be taken in context. Climate will evolve as the star grows more luminous with age. Although this is a relatively trivial factor for red dwarfs, which evolve very slowly, it becomes significant over the longer history of the universe. Secondly, carbon dioxide levels will progressively fall over time, as increasing amounts of gas are taken up as carbonates and not returned by geological processes. Next, over geological timescales the oceans will decrease in volume. This will expose more land and likely increase the rate at which carbon dioxide is absorbed into rock. Of course, there will also be less water available to support an active hydrosphere. Conversely, aquaplanets may transition to terrestrial planets, as formerly drowned land becomes exposed. A weak carbonate-silicate cycle may then kick into gear and the circulation pattern of the atmosphere alter as friction becomes more important and ocean circulation more fragmented. There is a lot to consider!

The Role of Ocean Circulation in the Control of Climate

Thinking just about the oceans, the circulation pattern will clearly impact the manner in which the atmosphere circulates, transports heat and regulates the abundance of greenhouse gases.

Firstly, the area of open ocean will determine the surface temperature. A frozen ocean is clearly far more reflective (has a higher albedo) than open, dark ocean. Open ocean is also in contact with the overlying circulating atmosphere; a frozen one is not. Therefore, determining how much area remains ice free is critical in determining how heat energy is transported across the globe. An ocean with a greater proportion of its surface exposed will clearly transport far more heat from the SSP than one where the surface is largely frozen. Secondly, the manner in which the ocean circulates is critical in determining both the transport of heat energy and the efficiency of any carbonate-silicate cycle.

In the fast-rotating, tidally locked aquaplanet models of Yongyun Hu and Jun Yang (Peking University) there exist two overlapping circulations within the ocean that should serve to transport nutrients effectively around the planet. This is vital if the oceans are to remain vibrant and potentially viable for aquatic life. Detailed modeling work suggests that oceans with depths greater than 2 km will have two circulations. In the top-most layer, above 2 km, warm waters move from the equator towards each pole. This circulation is driven by the zone of equatorial westerly winds that drive water downwards at the equator then outwards to each pole, where the waters rise up once more. Beneath these hemispheric cells, if the ocean is deep enough, is a thermohaline-driven circulation. In this thermohaline cell, water moves downwards from the poles towards the equator. The driving mechanism lies with the formation of ice. As seawater freezes, denser, salty water is displaced from the growing ice crystals because salt does not fit readily into the structure of water-ice. This cold, salty water descends under its own weight towards the ocean floor and generates a flow from the poles towards the equator, i.e., in the opposite direction to water driven by the westerlies in the atmosphere nearer the equator. If, however, the oceans are relatively shallow, this arm of the global ocean circuit will be broken.

Further modeling by Jodie Cullum, David Stevens and Manoj Joshi (University of East Anglia, UEA) show that planets with slow rotation ironically had more vigorous mixing between the polar and tropical regions than fast-rotating models. Mixing was driven by the temperature contrast between these latitudes and facilitated by weaker east-west (zonal) flow than in the fast-rotating planets. Mixing was also enhanced where various barriers to east-west flow were in place. Although their work did not formally show circulation models, the gross transport rates were highest for planets with continental barriers and long orbital periods.

More recent work by Jodie Cullum and co-workers' UEA group examined the role of salinity in maintaining habitability of planets. In this work the authors identify a range of possible salinities (the concentration of dissolved—ionic—substances) over which the ocean would continue to circulate effectively and thus help regulate global temperatures and oceanic oxygenation. The terrestrial ocean circulation is driven by the sinking of cold, dense water at the poles. This flows outwards then rises back to the surface at the equator to replace warm water moving away towards the polar regions.

Cullum and colleagues investigated other scenarios where evaporation drives an increase in salinity of the surface waters of the tropical ocean, in a manner analogous to the circulation of Hu's and Yang's model planets that were discussed earlier. This highly saline water then sinks to the ocean floor, before flowing along the abyssal plain towards the polar regions. In this ocean model, less dense but colder polar water is pulled towards the equator along the surface and warms, and evaporation drives the increase in density once more. In general, planets with saltier oceans tend to show the latter circulation pattern, deemed a negative circulation pattern, which is driven by salinity rather than temperature (the positive pattern of circulation that is seen on Earth).

What are the consequences for the climate of a strong ocean circulation, and how does changing the concentration of carbon dioxide, or the amount of stellar radiation, affect the climate? Surprisingly, some of the effects might well confound your expectations. In general Hu and Yang's models show that increasing the concentration of carbon dioxide increases the area of open water. Now, this isn't because the planet as a whole warms up but rather because the atmosphere can hold more heat energy and thus transport it more efficiently from lit to dark hemispheres. Therefore, although the day side barely warms at all, increasing the concentration of greenhouse gases, particularly carbon dioxide, to around 2% from an early industrial terrestrial 0.035% causes the dark hemisphere to warm substantially. Indeed, it causes the ice to melt across the entire globe.

Climate Impacts on the Overall Habitability of Planets

In terms of the atmosphere, the strength of the super-rotating westerly jet increases substantially and extends downwards towards the ocean beneath. In the atmosphere with early industrial levels of carbon dioxide the atmospheric motion at the surface is only weakly from west to east, except closest to the western edge of the SSP. However, once the carbon dioxide concentration is increased the increasingly deep overlying westerly jet stream drives a very noticeable westerly airflow around the equator, and ultimately it is this that causes the dark hemisphere to warm up. This is analogous to the short-orbital period/fast rotation models of Jacob Haqq-Misra and others, which were described earlier.

Moreover, if we consider the end-Permian mass extinction, ocean circulation appears to have played a pivotal role in exterminating most of the planet's complex life. Here, an already warm ocean became super-heated and stagnant, with microbial life filling its depths with methane and hydrogen sulfide. Ultimately, it was the hydrogen sulfide that killed off most of the ocean and possibly land-based life. For a tidally locked planet, the active circulation described here should minimize the risk of ocean anoxia (lack of oxygen) and euxinia (the build-up of hydrogen sulfide), which would likely cause the death of most aerobic (oxygen-loving) life forms.

Increasing carbon dioxide concentrations has other effects that might be surprising. You might expect that more carbon dioxide and a warmer world overall might mean more convection and hence more clouds. However, Yongyun Hu and Jun Yang's work suggests that as carbon dioxide increases, it is the effect of the increasingly strong westerly jet stream and ocean currents that matter most. Instead of enhancing convection, the strengthening westerly jet and ocean currents mix cooler air more thoroughly around the planet, along with cooler water in the oceans. Therefore, as was mentioned above, global temperatures are more even throughout. With a lower contrast in temperature, convection is actually suppressed by increasing carbon dioxide. This is despite a higher tropopause, which might favor taller convective clouds. Thus, the planet is more cloud-free than if it has less carbon dioxide.

What about habitability? Here, there are some contradictory outcomes. Now, you might presume that a larger area of open water would allow more habitability. And that might be true for a planet right in the middle of the habitable zone, but overall, more open water decreases the planetary albedo—the planet becomes darker. This was suggested for the early Earth as a mechanism to warm it with a dimmer Sun. The authors conclude that the end result is that a darker planet will absorb more energy from its star, and this, in turn, will cause the inner edge of the stellar habitable zone to move outwards away from the star. After all, a darker planet will be hotter at any given amount of stellar heating, and this will lead to the planet overheating and experiencing a thermal runaway sooner. This narrows the habitable zone.

Now, it must be stressed that this conclusion is still open to interpretation. Look, for example, at another piece of work by Jun Yang in the preceding year. Yang and colleagues concluded that convection under the SSP will produce enough cloud to cause negative feedback. What this means is that the extra cloud reflects incoming radiation and thus reduces the likelihood that the planet will overheat. This prevents the dreaded thermal runaway and keeps the planet habitable. Therefore, in this work convection broadens the habitable zone because the clouds it creates are reflective. It also moves the warmest area of the planet back towards the SSP despite the pervasive westerly jet stream around the equator.

In this set of published models, oceans were included, but they were not truly global in extent, with continents partly blocking the movement of water around the globe. Thus, it is only on true waterworlds that the habitable zone will be affected. However, as these planets would be rotating faster than Yang considered, the effect is not pronounced. Indeed, faster rotation in later models (Yang being then a member of Haqq-Misra and co-worker's team) conclude that with faster rotation, the cloud cover is reduced. Here the planets heat up—the opposite conclusion!

Turn again to another study and something slightly different comes out. In the work of Jérémy Leconte and colleagues, thermal runaway is averted because of negative feedback as the planet warms. In their model, as temperatures rise, the Hadley cells strengthen. Although this leads to increased cloud cover near the equator, it causes a reduction in cloud cover elsewhere, where air descends, warms and can hold more water vapor.

In Yang's work, reducing cloud cover is thought to have a destabilizing effect—reducing planetary albedo because oceans are darker. However, Yang's work suggests this is dependent on the type of planet. If you increase the amount of radiation a tidally locked planet receives, then the amount of convection increases and the planet gets cloudier. This makes it more reflective and has two effects. More cloud cover makes the dayside more reflective and the night-side less; this evens out temperature contrasts, increasing habitability.

By contrast a non-tidally locked planet, such as Earth, experiences the opposite effect. Stronger heating, in Yang's model, means less of a temperature gradient and weaker Hadley cells. This causes greater warming as fewer clouds mean reduced reflection of incoming radiation and more absorption of radiation by the planet's surface. Conversely, Leconte's work suggests that on an Earth-like world greater warming leads to *stronger* Hadley circulation and (overall) *fewer* clouds, but those changes to albedo are offset by increases in the ability of air to hold moisture. With less water escaping to the stratosphere, a runaway greenhouse is held off until the amount of radiation the planet receives increases to 375 Watts per meter squared—or approximately 100 Watts per meter squared more than Earth currently receives from the Sun. This broadly corresponds to an average surface temperature of 65–70 °C for an Earth-like planet.

Leconte's work also shows that over the Horse Latitudes, where the air is far from saturated because it is warming and drying out, the atmosphere acts like a giant radiator and cools the planet down. Thus, in the cloudless areas, energy is lost, and the planet can absorb more energy before it experiences a thermal runaway. More recent work by James Kasting also reaches the same conclusion. Although Leconte's work only looked at the troposphere, Kastings included more water vapor and examined what happened in the stratosphere. At similar temperatures to those seen by Leconte, the stratosphere fills with water vapor and the planet dries out. These two pieces of work suggest that planets can remain habitable for longer, with greater amounts of heating, before they expire. The loss of water, via the stratosphere, limits the greenhouse effect produced by water vapor—while reducing the volume of water at the surface that is necessary for life.

Moreover, Leconte's work also contradicts Yang's conclusions in other ways. Their models have increasing cloud cover, accelerating the greenhouse effect. Here, the water droplets and ice crystals in clouds absorb more energy than they reflect and, therefore, increase warming. How come this is so different? Well, it's down to the type of clouds in the models (Fig. 5.20). Yang has mostly cumulus and cumulonimbus convective clouds, while Leconte favors an increase in cirrus and cirrostra-

Fig. 5.20 Clouds matter. If planets produce large, convective, cumulus and cumulonimbus clouds, like those above (left), they will reflect much of the incoming radiation and remain cooler and habitable for longer. However, should they produce mostly high ice clouds, such as this unusual example of cirrus above (right) or the cirrostratus ice cloud above it, then they will preferentially trap more heat at night time than they release during the day. This will cause the planet to overheat more quickly than if it had no cloud at all. (Photographs by the author)

tus because they suggest under hotter conditions more high-level clouds form, while lower clouds are evaporated. Cirrus and cirrostratus usually absorb rather than reflect, meaning that as they increase in abundance, the greenhouse effect increases in sync. Consider the 9/11 attacks in the United States. When planes were grounded, there was a 1.8° increase in the diurnal (day-night) temperature range—primarily because more radiation was escaping at night, with fewer plane contrails to absorb it. Despite less reflection in the daytime, the dominant effect was absorption of outgoing radiation at night, not more reflection during the day.

All of this is rather confusing. It comes down the role of clouds. If their dominant effect is to reflect energy, as Yang concludes, then habitable zones are extended where tidally locked planets have limited ocean circulation. If Leconte is correct, then increasing cloud will have the opposite effect—but this depends on the types of clouds a tidally locked planet will form. In Leconte's models, decreasing cloud will stabilize the planetary temperature and limit the chances of a thermal runaway because, overall, there will be fewer cirrus and cirrostratus-like clouds. At the moment, the best we can do is speculate.

Wildcards: Atmospheric Collapse, Erosion and Sequestration

The long-term stability of a planet's atmosphere depends crucially on a number of factors—two of which were described in Chap. 4 under the banner "sequestration." Water is continually removed from the atmosphere of cold planets by condensation and then sequestration as ice. On Mars, most of the planet's surface water and atmospheric water is contained in this reservoir. Moreover, on planets with warm interiors and active surface geology, water is returned to the deep crust and mantle. This short section considers these influences on the long-term stability of a planet's atmosphere.

Atmospheric Stripping

Water is also lost through the process of photolysis, where water is split by ultraviolet radiation into hydrogen and hydroxyl (then oxygen). This process is dominant in the middle atmosphere—the stratosphere—where there is greatest penetration of UV. The lower atmosphere is usually too dense to permit high intensities of ultraviolet radiation, although this is not always the case. In Earth's atmosphere, ozone had both direct and indirect protective effects on our water reserves. Directly, ozone absorbs ultraviolet radiation at the kinds of energies in which the water molecule is most readily split. In an atmosphere that was free of molecular oxygen, ozone might be expected to be very limited in extent. However, the release of oxygen by the photolysis of water vapor or carbon dioxide can produce limited amounts. Formation of ozone is offset if there is a lot of hydrogen present—just as would be expected from the release of oxygen by photolysis of water. Therefore, if the stratosphere has abundant water, ozone levels will be low.

Secondly, the formation of ozone (or several other possible molecules), after the absorption of ultraviolet radiation, raises the energy and hence the temperature of the stratosphere. This makes it warmer than the underlying layer. Such air cannot convect—giving the stratosphere its name. On most planets, there will be a stratosphere generated by the absorption of ultraviolet radiation. This helps preserve the amount of water in the lower (tropospheric) layer by trapping it below the level that ultraviolet radiation can penetrate.

Mass loss can also be driven by interactions between the stellar wind and the atmosphere if there is no magnetic field, or through charging and acceleration of particles along magnetic field lines. In 2010 J. Zendejas and colleagues (Universidad Nacional Autónoma de México) generated models to illustrate the loss of water from planets orbiting red dwarfs. These were compared with studies of mass loss from the Venusian atmosphere to determine their authenticity. Where the planet has no magnetic field, this study found that the planets lost most of their atmosphere to the stellar wind in 1 billion years. Planets orbiting more massive red dwarfs took the longest to lose their atmosphere.

However, for planets orbiting close to very active (usually the lowest mass) red dwarfs, mass loss was more extreme because it lasted for longer periods of time—and because the planet is physically closer to its star. Remember, the lowest mass red dwarfs are still dropping onto the main sequence in the first billion years of life. Therefore, they are very active for this period. Furthermore, in some instances the star may still be physically contracting long after the planet has formed, so that the planet lies inside the inner edge of the circumstellar habitable zone. Such planets, which include those of TRAPPIST-1, would be expected to be at serious risk of atmospheric stripping as a result of these factors.

Likewise, Peter Wheatley (University of Warwick) and co-workers found that even with a magnetic field the TRAPPIST-1 planets and their ilk were not necessarily immune to stripping. Here, TRAPPIST-1 is magnetically active and emits relatively intense ultraviolet and X-rays (hereby, abbreviated to XEUV for X-ray and extreme ultraviolet). The primary atmosphere would likely be removed. Although the loss of the primary atmosphere may not be a bad thing, the prolonged nature of the assault means that any secondary atmosphere—the kind that might host biological life—would also be severely impacted. If the planet has a substantial hydrosphere, loss of some of this may have benefits over a longer period of time (Chaps. 4, 7 and 9).

Working with Wheatley and others in 2017 Vincent Bourrier (Observatoire de l'Université de Genève) extended the analysis to show that TRAPPIST-1b and c might be in a state of thermal runaway, with water loss accentuated by rapid greenhouse warming. This raises the level of the tropopause (the boundary between the convecting lower troposphere and the upper stratosphere) and allows faster photolysis and water loss. Indeed, all of the planets interior to TRAPPIST-1g might have lost up to 20 Earth-ocean's worth of water in the 8 billion years of their existence. TRAPPIST-1e and h may have lost less, depending on how their orbits altered during the system's formative years (Chaps. 1 and 10). Such figures clearly depend on a number of other variables, which include late addition of volatile materials and further outgassing from the mantle. This was considered earlier, but we will consider the biological impacts of this in Chaps. 7, 8 and 9.

Tidal Heating

A penultimate and perhaps most surprising consideration is tidal heating. In Chap. 1 we saw how the warm Neptune GJ436b was surrounded by a large cloud of hydrogen. Vincent Bourrier (Observatoire de l'Université de Genève) and colleagues suggested that the peculiar orbit had tidally heated the planet's thick envelope, sufficiently warming it to cause its expansion beyond the planet's gravitational fence—the edge of its Roche lobe. Likewise, Rory Barnes (University of Washington) modeled the impact of orbital shape and evolution for terrestrial (telluric) planets orbiting red dwarfs. Where the orbit is initially eccentric, tidal forces with the host star circularize the orbit in part by transferring orbital energy to the envelope of the planet. This can heat the atmosphere strongly, depending on how quickly the process occurs. In the more extreme scenarios, all of the water in the atmosphere is driven off into space, in a manner currently being experienced by GJ 436b. If so, one might find that planets that lie within their star's habitable zone are at best analogous of Venus, rather than Earth. Those planets orbiting the lowest mass stars are at particular risk, because they have the tightest orbits and hence experience the strongest effects.

Sequestration and Collapse

In Chapter 4 we looked at processes that return water vapor, carbon dioxide and nitrogen to the interior of a planet (principally its crust and upper mantle) as it cools. These processes can be considered, generically, as sequestration. Under appropriate conditions this can result in the catastrophic loss of habitability as one or more key components is gone from the atmosphere. For terrestrial planets, orbiting G-class stars the time the planet resides in the circumstellar habitable zone is broadly similar to the time it will take these components to return to the mantle in sufficient quantity to dam the planet's habitability. However, this is not true for planets in orbit around K- and M-class stars.

Working with a variety of researchers, Manfred Cuntz (then at the University of Texas) modeled the evolution of habitability for planets orbiting stars of various masses. Taking into account the effects of chemical weathering planets orbiting low mass stars (those with masses less than 0.9 Suns) became uninhabitable for complex life once carbon dioxide concentrations became limiting, and this occurs before their star evolved far enough along the main sequence to cause overheating. Therefore, habitability of planets in orbit of all red and orange dwarfs have their habitability determined by how long they can retain suitable concentrations of carbon dioxide to support photosynthesis. However, the precise time depends on the extent of continental surface that was exposed to weathering. The lower the area, the longer carbon dioxide concentrations remained high enough to support photosynthesis. Taken to its conclusion, aquaplanets remain habitable for longest where the central star does not drive a runaway greenhouse.

Looking at this another way, for planets lying near the outer edge of the habitable zone, progressive loss of carbon dioxide through weathering could lead to a sustained lowering of temperatures. For planets orbiting red dwarfs, this was thought to be a risk for such worlds, because one hemisphere was continually cold and in darkness. However, most models show the surfaces never become so cold that both water vapor and carbon dioxide snow out. Therefore, there is no risk of atmospheric collapse where the temperatures fall low enough for this to occur. Moreover, near the outer edge of the habitable zone, while temperatures could fall low enough to snow out most moisture, the process cannot amplify beyond the point where the surface freezes. For at some stage, any continental surface will be buried in ice, and the opportunity for further loss of carbon dioxide terminated. Temperatures will then stabilize at the new low point, until increased stellar output reverses it.

One caveat is whether this general scheme is applicable to super-terrans. In Chaps. 3 and 4 we saw how the propensity for such worlds to host plate tectonics is likely dependent on the depth at which eclogite forms. The more massive the world the shallower this will occur, until catastrophic conversion of any oceanic crust to eclogite commences. In this environment plate tectonics seems unlikely, and there will be rapid generation of continental crust across the planet. These worlds may have a burst of silicate weathering, then a long period of protracted decline once erosion removes most elevated structures. Here, carbon dioxide may persist at high enough levels for photosynthesis – and to maintain a temperature climate even in the absence of a global ocean.

Although many of these deliberations seem logical and reasonable, they should caution against any optimism regarding the habitability of planets that orbit in their star's so-called habitable zones… The universe is neither sympathetic to our beliefs nor our hopes.

Conclusions

The climate of tidally locked planets is very different to the kind of systems we are familiar with on Earth. The simplest analogy is the monsoon, which affects Asia. During the summer monsoon, cooler maritime air flows northwards and northwestwards into southern and eastern Asia from the surrounding ocean basins, bringing prolonged rainfall to India, southeast Asia and China. In the analogous situation, air flows from the cold, dark hemisphere towards the SSP, where heating is generally strongest. Here, air convects upwards, flows outwards and across most of the rest of the surface, then gently descends back towards the surface to complete the loop.

Those planets with the shortest orbital periods experience increasingly strong super-rotation in the direction of orbital motion. Such worlds exhibit limited areas where rainfall is guaranteed and generally lack the kinds of frontal structures and mobile circulations that we experience in the mid- and high-latitudes. Therefore, these areas are likely to be disproportionately dry compared to Earth. This strongly inhibits their ability to host complex biological life, as we shall see in Chap. 7. However, if we are generous and include the effects of topography, rainfall becomes likely on windward slopes and coasts. The overall effect on rainfall (away from the SSP) then depends on the extent of topography and the strength of any overlying temperature inversion that will block convection in locations distant to the SSP.

Finally, we put planetary climate in the big mixing pot of astrophysics. All planets that orbit red dwarfs experience strong tidal heating early on and are exposed to fluxes of radiation far in excess to those we experience on Earth. In such situations, the ability of a planet to preserve habitability will depend on both its capacity to resist the effects of stellar radiation and on the strength of its internal constitution – how much water and other volatiles it can hide for the first few billion years of its life. In the end, if you are a planet orbiting a red dwarf in its habitable zone, it may be better to begin life as an aquaplanet, with a thick ocean and a large reserve of volatile materials. Then nature can take its course, quietly revealing a hidden landscape of continent and basin as the waters recede. More on this in Chaps. 7 and 8.

References

Abbot, D. S., Cowan, N. B., & Ciesla, F. J. (2012). Indication of insensitivity of planetary weathering behavior and habitable zone to surface land fraction. *The Astrophysical Journal, 756*, 178. https://doi.org/10.1088/0004-637X/756/2/178.

Armstrong, J. C., Barnes, R., Domagal-Goldman, S., Breiner, J., Quinn, T. R., & Meadows, V. S. (2014). Effects of extreme obliquity variations on the habitability of exoplanets. *Astrobiology, 14*(4), 46. https://ntrs.nasa.gov/archive/nasa/casi.ntrs.nasa.gov/20150008371.pdf.

Arnold, N. P. (2014). Atmospheric superrotation in warm earth climates. Doctoral dissertation, Harvard University. http://nrs.harvard.edu/urn-3:HUL.InstRepos:11744437.

Barnes, R. (2017). Tidal locking of habitable exoplanets. https://arxiv.org/pdf/1708.02981.pdf.

Barnes, R., Mullins, K., Goldblatt, C., Meadows, V. S., Kasting, J. F., & Heller, R. (2013). Tidal Venuses: Triggering a climate catastrophe via tidal heating. *Astrobiology, 13*, 225–250. https://arxiv.org/pdf/1203.5104.pdf.

Basa, J. (2007). Lee cyclogenesis. Seminar, University of Ljubljana. http://mafija.fmf.uni-lj.si/seminar/files/2006_2007/Lee_cyclogenesis_-_Joze_Basa.pdf.

Boutle, I. A., Mayne, N. J., Drummond, B., Manners, J., Goyal, J., Lambert, F. H., Acreman, D. M., & Earnshaw, P. D. (2017). Exploring the climate of Proxima B with the Met Office Unified Model. https://arxiv.org/pdf/1702.08463.pdf.

Carone, L., Keppens, R., & Decin, L. (2015). Connecting the dots II: Phase changes in the climate dynamics of tidally locked terrestrial exoplanets. Preprint. http://arxiv.org/pdf/1508.00419v1.pdf.

Clift, P. D., Hodges, K. V., Heslop, D., Hannigan, R., Long, H. V., & Calves, G. (2008). Correlation of Himalayan exhumation rates and Asian monsoon intensity. *Nature Geoscience, 1*, 875–880. https://doi.org/10.1038/ngeo351. http://tectonics.asu.edu/page5/files/page5_1.pdf.

Cullum, J., Stevens, D., & Joshi, M. (2014). The importance of planetary rotation period for ocean heat transport. *Astrobiology, 14*(8), 645–650. https://doi.org/10.1089/ast.2014.1171. https://www.liebertpub.com/doi/10.1089/ast.2014.1171.

Cullum, J., Stevens, D. P., & Joshi, M. M. (2016). Importance of ocean salinity for climate and habitability. *PNAS, 113*(16), 4278–4283. https://doi.org/10.1073/pnas.1522034113.

Cuntz, M., von Bloh, W., Schröder, K.-P., Bounama, C., & Franck, S. (2011) Habitability of super-earth planets around main-sequence stars including red giant branch evolution: Models based on the integrated system approach. https://arxiv.org/pdf/1107.5714.pdf.

Donnadieu, Y., Goddéris, Y., Ramstein, G., Nédélec, A., & Meert, J. (2004). A 'snowball Earth' climate triggered by continental break-up through changes in runoff. *Nature, 428*, 303–306.

Edson, A., Lee, S., Bannon, P., Kasting, J. F., & Pollard, D. (2011). Atmospheric circulations of terrestrial planets orbiting low-mass stars. *Icarus, 212*(1), 1–13. http://www.meteo.psu.edu/~sxl31/papers/ICARUS.pdf.

Faherty, J. K., Tinney, C. G., Skemer, A., & Monson, A. J. (2014) Indications of water clouds in the coldest known brown dwarf. Preprint. http://arXiv:1408.4671v2.pdf.

Foley, B. J. (2015). The role of plate tectonic-climate coupling and exposed land area in the development of habitable climates on rocky planets. https://arxiv.org/pdf/1509.00427.pdf.

Forget, F. & Leconte, J. (2013). Possible climates on terrestrial exoplanets. https://arxiv.org/ftp/arxiv/papers/1311/1311.3101v2.

Frierson, D. M. W. (2007). Convectively coupled Kelvin waves in an idealized moist general circulation model. *Journal of the Atmospheric Sciences, 64*, 2076–2090. https://doi.org/10.1175/JAS3945.1.

Fujii, Y., Del Genio, A. D., & Amundsen, D. S. (2017). NIR-driven moist upper atmospheres of synchronously rotating temperate terrestrial exoplanets. https://arxiv.org/pdf/1704.05878.pdf.

Geerts, B. & Linacre, E. (1999). Potential vorticity and isentropic charts. http://www-das.uwyo.edu/~geerts/cwx/notes/chap12/pot_vort.html

Gerbier, N., Koschmieder, H., & Zierep, J. (1960). Technical note No. 34 the airflow over mountains. Report of a working group of the Commission for Aerology prepared by P. Queney, Chairman – G.A. Corby. Edited and coordinated by M.A ALARA of the WMO Secretariat. https://library.wmo.int/opac/doc_num.php?explnum_id=1734.

Gohm, A., & Mayr, G. J. (2004). Hydraulic aspects of föhn winds in an Alpine valley. *Quarterly Journal of the Royal Meteorological Society, 130*, 449–480. https://doi.org/10.1256/qj.03.28.

Haqq-Misra, J. & Kopparapu, R. K. (2014). Geothermal heating enhances atmospheric asymmetries on synchronously rotating planets. https://arxiv.org/ftp/arxiv/papers/1410/1410.0186v1.

Haqq-Misra, J., Kopparapu, R. K., & Wolf, E. T. (2017). Why do we find ourselves around a yellow star instead of a red star? *International Journal of Astrobiology, 17*(01), 77–86. https://doi.org/10.1017/s1473550417000118. https://arxiv.org/ftp/arxiv/papers/1705/1705.07813.pdf.

Haqq-Misra, J., Wolf, E. T., Joshi, M., Zhang, X., & Kopparapu, R. K. (2017). Demarcating circulation regimes of synchronously rotating terrestrial planets near the inner edge of the habitable zone. https://arxiv.org/ftp/arxiv/papers/1710/1710.00435.pdf.

Haqq-Misra, J., Wolf, E. T., Joshi, M., Zhang, X., & Kopparapu, R. K. (2018). Demarcating circulation regimes of synchronously rotating terrestrial planets within the habitable zone. *The Astrophysical Journal, 852*(2), 67. https://doi.org/10.3847/1538-4357/aa9f1f.

Haqq-Misra, J., Kopparapu, R. K., Batalha, N. E., Harman, C. E., & Kasting, J. F. (2016). Limit cycles can reduce the width of the habitable zone. *The Astrophysical Journal, 827*, 120. https://doi.org/10.3847/0004-637X/827/2/120. https://arxiv.org/pdf/1605.07130v1.pdf.

Heath, M. J., Doyle, L. R., Joshi, M. M., & Haberle, R. M. (1999). Habitability of planets around red dwarf stars. *Origins of Life and Evolution of the Biosphere, 29*, 405–424. Preprint. http://www.as.utexas.edu/astronomy/education/spring02/scalo/heath.pdf.

Heng, K. & Kopparla, P. (2012). On the stability of super earth atmospheres. https://archive.org/details/arxiv-1203.1922.

Hu, Y., & Yang, J. (2014). Role of ocean heat transport in climates of tidally-locked exoplanets around M dwarf stars. *PNAS, 111*(2), 629–634.

Hu, Y., Yang, D., & Yang, J. (2008). Blocking systems over an aqua planet. *Geophysical Research Letters, 35*, L19818. https://doi.org/10.1029/2008GL035351.

Joshi, M. M., Haberle, R. M., & Reynolds, R. T. (1997). Simulations of the atmospheres of synchronously rotating terrestrial planets orbiting M dwarfs: Conditions for atmospheric collapse and the implications for habitability. *Icarus, 129*, 450–465.

Joshi, M. M., & Haberle, R. M. (2011). Suppression of the water ice and snow albedo feedback on planets orbiting red dwarf stars and the subsequent widening of the habitable zone. *Astrobiology, 12*(1), 3–8. https://doi.org/10.1089/ast.2011.0668. https://arxiv.org/ftp/arxiv/papers/1110/1110.4525.pdf.

Kadoya, S., & Tajika, E. (2016). Evolutionary tracks of the climate of earth-like planets around different mass stars. *The Astrophysical Journal Letters, 825*, L21. (5pp). https://doi.org/10.3847/2041-8205/825/2/L21.

Kennedy, M., Mrofka, D., & von der Borch, C. (2008). Snowball Earth termination by destabilization of equatorial permafrost methane clathrate. *Nature, 453*, 642–645. https://doi.org/10.1038/nature06961.

Kopparapu, R. (2013). A revised estimate of the occurrence rate of terrestrial planets in the habitable zones around Kepler M-dwarfs. *The Astrophysical Journal, 767*(1), 5.

Kopparapu, R. (2016). A catalog of Kepler habitable zone exoplanet candidates. *The Astrophysical Journal, 830*(1), 1. https://doi.org/10.3847/0004-637x/830/1/1.

Kopparapu, R. K., Hébrard, E., Belikov, R., et al. (2018). Exoplanet classification and yield estimates for direct imaging missions. *The Astrophysical Journal, 856*(2), 122. https://doi.org/10.3847/1538-4357/aab205.

Kopparapu, R. K., Ramirez, R., Kasting, J. F., et al. (2013). Erratum: habitable zones around main-sequence stars: New estimates (2013, ApJ, 765, 131). *The Astrophysical Journal, 770*, 82. https://doi.org/10.1088/0004-637X/770/1/82.

Kopparapu, R. K., Wolf, E. T. Haqq-Misra, J., Kasting, J. Y. J. F., Meadows, V., Terrien, R., & Mahadevan, S. (2016). The inner edge of the habitable zone for synchronously rotating planets around low-mass stars using general circulation models. https://arxiv.org/ftp/arxiv/papers/1602/1602.05176v1.pdf.

Kopparapu, R. K., Wolf, E. T., Arney, G., Batalha, N. E., Haqq-Misra, J., Grimm, S. L., & Heng, K. (2017). Habitable moist atmospheres on terrestrial planets near the inner edge of the habitable zone around M-dwarfs. *The Astrophysical Journal, 845*(1), 5. https://doi.org/10.3847/1538-4357/aa7cf9. https://arxiv.org/ftp/arxiv/papers/1705/1705.10362v2.

Krishnamurti, T. N., Sahu, D. K., Kumar, V., Krishnamurti, R., Ghosh, T., et al. (2017). Extreme orographic rains from streams of moist marine layer wind systems with a possible geoengineering application. *International Journal of Hydrology, 1*(7), 00039. https://doi.org/10.15406/ijh.2017.01.00039.

Leconte, J., Forget, F., Charnay, B., Wordsworth, R., & Pottier, A. (2013). Increased insolation threshold for runaway greenhouse processes on Earth-like planets. *Nature, 504*, 268–271.

Leconte, J., Wu, H., Menou, K., & Murray, N. (2015). Asynchronous rotation of Earth-mass planets in the habitable zone of lower-mass stars. *Science, 347*(6222), 632–635. https://doi.org/10.1126/science.1258686. https://arxiv.org/pdf/1502.01952.pdf.

Licht, A., Cappelle, M., Abels, H. A., Ladant, J. B., Trabucho-Alexandre, J., France-Lanord, C., Donnadieu, Y., Vandenberghe, J., Rigaudier, T., Lécuyer, C., Terry, D., Jr., Adriaens, R., Boura, A., Guo, Z., Soe, A. N., Quade, J., Dupont-Nivet, G., & Jaeger, J.-J. (2014). Asian monsoons in a late Eocene greenhouse world. *Nature, 513*, 501–506. https://doi.org/10.1038/nature13704.

Liu, B. & Showman, A. P. (2012). Atmospheric circulation of hot Jupiters: insensitivity to initial conditions. Preprint. http://arxiv.org/pdf/1208.0126v2.pdf.

Luetscher, M., Boch, R., Sodemann, H., Spötl, C., Cheng, H., Edwards, R. L., Frisia, S., Hof, F., & Müller, W. (2015). North Atlantic storm track changes during the Last Glacial Maximum recorded by Alpine speleothems. *Nature Communications, 6*(6344), 1–6. https://doi.org/10.1038/ncomms7344.

Menoua, K. (2015). Climate stability of habitable Earth-like planets. *Earth and Planetary Science Letters, 429*, 20–24. https://doi.org/10.1016/j.epsl.2015.07.046.

Merlis, T. M. & Schneider, T. (2010). Atmospheric dynamics of earth-like tidally locked aquaplanets. https://arxiv.org/pdf/1001.5117.pdf.

Messager, C., Parker, D. J., Agusti-Panareda, O. R. A., Taylore, C. M., & Cuesta, J. (2010). Structure and dynamics of the Saharan atmospheric boundary layer during the West African monsoon onset: Observations and analyses from the research flights of 14 and 17 July 2006. *Quarterly Journal of the Royal Meteorological Society, 136*(s1), 107–124. https://doi.org/10.1002/qj.469.

Nasreen Akter, N., & Tsuboki, K. (2017). Climatology of the pre-monsoon Indian dryline. *International Journal of Climatology, 37*(11), 3991–3998. https://doi.org/10.1002/joc.4968.

Noda, S., Ishiwatari, M., Nakajima, K., Takahashi, Y. O., Takehiro, S., Onishi, M., Hashimoto, G. L., Kuramoto, K., & Hayashi, Y.-Y. (2017). The circulation pattern and day-night heat transport in the atmosphere of a synchronously rotating aquaplanet: Dependence on planetary rotation rate. *Icarus, 282*, 1–18. https://doi.org/10.1016/j.icarus.2016.09.004.

Penn, J. & Vallis, G. K. (2017). The thermal phase curve offset on tidally- and non-tidally-locked exoplanets: A shallow water model. https://arxiv.org/pdf/1704.06813.pdf.

Ribas, I., Bolmont, E., Selsis, F., Reiners, A., Leconte, J., Raymond, S. N., Engle, S. G., Guinan, E. F., Morin, J., Turbet, M., Forget, F., & Anglada-Escudé, G. (2016). The habitability of Proxima Centauri b I. Irradiation, rotation and volatile inventory from formation to the present. https://arxiv.org/ftp/arxiv/papers/1608/1608.06813v2

Schinder, P. J., Flasar, F. M., Marouf, E. A., French, R. G., McGhee, C. A., Kliore, A. J., Rappaport, N. J., Barbinis, E., Fleischman, D., & Anabtawi, A. (2011). Saturn's equatorial oscillation: Evidence of descending thermal structure from Cassini radio occultations. *Geophysical Research Letters, 38*, L08205. https://doi.org/10.1029/2011GL047191/full.

Schubert, W. H., Ciesielski, P. E., Lu, C., & Johnson, R. H. (1995). Dynamical adjustment of the trade wind inversion layer. *Journal of the Atmospheric Sciences, 52*, 2941–2952. https://doi.org/10.1175/1520-0469(1995)052<2941:DAOTTW>2.0.CO;2. https://journals.ametsoc.org/doi/pdf/10.1175/1520-0469%281995%29052%3C2941%3ADAOTTW%3E2.0.CO%3B2.

Showman, A. P., Menou, K., & Cho, J. Y.-K. (2007). Atmospheric circulation of hot Jupiters: A review of current understanding. http://www.lpl.arizona.edu/~showman/publications/showman-etal-santorini-paper-submitted.pdf.

Showman, A. P., & Polvani, L. M. (2011). Equatorial superrotation on tidally locked exoplanets. *The Astrophysical Journal, 738*(1), 24. https://doi.org/10.1088/0004-637X/738/1/71.

Telferner, A. (1990). Lee cyclogenesis resulting from the combined outbreak of cold air and potential vorticity against the Alps. *Meteorology and Atmospheric Physics, 43*(1–4), 31–47.

Tsuda, T. (2014). Characteristics of atmospheric gravity waves observed using the MU (Middle and Upper atmosphere) radar and GPS (Global Positioning System) radio occultation. *Proceedings of the Japanese Academy, Series B, 90*, 12–27. Preprint available at: http://www.ncbi.nlm.nih.gov/pubmed/24492645.

Turbet, M., Leconte, J., Selsis, F., Bolmont, E., Forget, F., Ribas, I., Raymond, S. N., & Anglada-Escudé, G. (2016).The habitability of Proxima Centauri b II. Possible climates and observability. Article published by EDP Sciences. https://doi.org/10.1051/0004-6361/201629577.

Wheatley, P. J., Louden, T., Bourrier, V., Ehrenreich, D., & Gillon, M. (2016). Strong XUV irradiation of the Earth-sized exoplanets orbiting the ultracool dwarf TRAPPIST-1. https://arxiv.org/pdf/1605.01564.pdf.

Wolf, E. T., Shields, A. L., Kopparapu, R. K., Haqq-Misra, J., & Toon, O. B. (2017). Constraints on climate and habitability for earth-like exoplanets determined from a general circulation model. https://arxiv.org/ftp/arxiv/papers/1702/1702.03315.pdf.

Wolf, E. T., Shields, A. L., Kopparapu, R. K., Haqq-Misra, J., & Toon, O. B. (2017). Constraints on climate and habitability for earth-like exoplanets determined from a general circulation model. *The Astrophysical Journal, 837*(2), 107. https://doi.org/10.3847/1538-4357/aa5ffc.

Wordsworth, R. (2014). Atmospheric heat redistribution and collapse on tidally locked rocky planets. https://arxiv.org/pdf/1412.5575.pdf.

Yang, J., Boué, G., Fabrycky, D. C., & Abbot, D. S. (2014). Strong dependence of the inner edge of the habitable zone on planetary rotation rate. Preprint. http://arxiv.org/pdf/1404.4992v1.pdf.

Yang, J., Cowan, N. B., & Abbot, D. S. (2013). Stabilizing cloud feedback dramatically expands the habitable zone of tidally locked planets. https://arxiv.org/pdf/1307.0515.pdf.

Yang, J., Cowan, N. B., & Abbot, D. S. (2013). Stabilizing cloud feedback dramatically expands the habitable zone of tidally locked planets. Preprint. http://arxiv.org/pdf/1307.0515.pdf.

Yang, J., Richard Peltier, W. M., & Hu, Y. (2016). Monotonic decrease of the zonal SST gradient of the equatorial Pacific as a function of CO2 concentration in CCSM3 and CCSM4. *Journal of Geophysical Research – Atmospheres, 121*, 1–17. https://doi.org/10.1002/2016JD025231.

Zendejas, J., Segura, A., & Raga, A. C. (2010). Atmospheric mass loss by stellar wind from planets around main sequence M stars. Preprint. http://arXiv.org/pdf/1006.0021v1.pdf.

Chapter 6
The Origin and Early Evolution of Life

Introduction

Terrestrial life emerged in a murky and violent period of history that has left little trace of its existence. Astrobiologists have been left to infer its *likely* origins from what meager and indirect evidence nature has left for us to decipher. We know, for example, that one of the building blocks of cells—a group of chemicals called amino acids—are found in the nebulae from which planets condense. Amino acids are also ubiquitous in a class of meteorites called carbonaceous chondrites. These observations imply that they *could* have been delivered to Earth very early in its history, but it does not say that they were.

Amino acids can also be made in volcanic vents. Therefore, even at this simple level, there is no clear consensus on the origin of some of life's ingredients, never mind how life itself originated. This chapter then attempts to paint a portrait of a young planet on which chemistry came alive. In doing so we summarize much of the content and ideas of the author's earlier book *The Nature of Life and its Potential to Survive* (Springer, 2017). However, the aim is to integrate these into the context of life on planets that orbit our little cousins, the red dwarfs. Chapters 7, 8 and 9 will carry these principles forward and discuss the full implications of life on a world with a red sun. In the meantime, we will focus here on the three key principles of life: how it originated, its means of generating energy, and its evolution.

Astrochemical Possibilities

We begin with the building blocks of living things on Earth. All living material on our world contains a common family of chemicals—biomolecules. Given the manner in which our planet assembled, this implies that all life has a common origin, not only on Earth but, at least in terms of chemistry, elsewhere in the universe. These common chemicals are the following: proteins; carbohydrates; co-enzymes (many

D. S. Stevenson, *Red Dwarfs*, https://doi.org/10.1007/978-3-030-25550-3_6

of which we call vitamins); nucleic acids; fats (lipids) as well as a host of small organic molecules; ions; and, of course, water.

In the last fifty years or so, chemists have used an increasingly diverse set of tools to probe the near vacuum of space. Although relatively few experiments have taken place in outer space per se, astronomers have used observations at infrared, microwave and radio wavelengths to probe the cosmos for information on what it is composed of.

On Earth we've been privileged to obtain meteorite fragments that convey messages from the darkest epochs of the Solar System's infancy, where most of the ingredients for life were distilled from the inky blackness of space. In the last few decades, increasingly intricate experiments have sought to create the likely conditions found in the universe's dark factories, the giant molecular clouds where stars are born. More recently, astronomers have tasted the stuff of comets through the briefly successful Philae lander mission on Comet 67P. Simultaneously, until October 2016, the parent Rosetta mission sampled the comet's outflows while it orbited the Comet 67P.

Of the interstellar chemicals that we know about, various inorganic and organic ions are detectable in outer space along with amino acids and water, and were probably present when Earth formed. Those that were not were likely delivered soon afterwards by colliding comets and asteroids. However, of the 20 or so amino acids that terrestrial biology uses only a handful have been detected in space thus far. This does not equate to evidence of absence, merely an absence of evidence. However, in 2003 Yi-Jehng Kuan of the National Taiwan Normal University announced that he discovered glycine—the simplest of the 20 most common amino acids in terrestrial organisms. Using the National Radio Astronomy Observatory (NRAO) in Arizona, Kuan identified the spectral signature of glycine in three giant molecular clouds—massive clouds of dust and gas from which stars form. These finds are listed in Table 6.1

Later, in 2014, Dr. Arnaud Belloche (the Max Planck Institute for Radio Astronomy) used the ALMA (Atacama Large Millimeter Array) to identify a likely building block of all amino acids—isopropyl cyanide. Like Yi-Jehng Kuan, this was also detected in the Sagittarius B2 molecular cloud, at the galactic center, suggest-

Table 6.1 The molecules of life

Biomolecule	Found where?
Amino acids	GMCs; meteorites and comets
Nucleobases	Meteorites; GMCs
Alcohols	GMCs, comets, meteorites
Simple sugars (polyols)	Meteorites
Glycerol[a]	GMCs (in lab)
Fatty acids	Meteorites

Where have the molecules associated with life (biomolecules) been found? Abbreviation: *GMC* Giant molecular cloud

[a]Glycerol has been synthesized under GMC conditions in the laboratory

ing that the steps needed to manufacture amino acids are scattered widely through space. Superseding, or perhaps supplementing, these discoveries—and predating them by nearly 30 years—was the discovery of nearly 75 amino acids in the Murchison meteorite, which fell to Earth in September 1969. Although only half of the naturally occurring amino acids have well-defined construction pathways outside living cells, the sheer abundance of amino acids in general makes it highly likely that they were largely delivered in this manner to the infant Earth.

Although life may have used a whole battery of amino acids that were delivered during the construction of our planet, these would soon run out once the bombardment of Earth fell away. Life would then have to make its own, in the absence of any other mechanism to replace those that were consumed. In this situation, only those that were the most straightforward to make would be retained. Anything primitive life couldn't manufacture would soon be discarded or replaced once the abiotic supply dwindled.

The Murchison meteorite belongs to the class of meteorites known as carbonaceous chondrites. These are chemical goldmines. Since the discovery of the Murchison meteorite analysis has revealed nearly every biomolecule necessary for life. Although many of these are present in small quantities, the sheer number of such objects that must have impacted the infant Earth in its first 700 million years makes it certain that Earth would have been heavily stocked with the raw materials for life. The Murchison meteorite contains the building blocks of nucleic acids, including the all-important nucleobases, amino acids and simple sugars. Indeed, you would be hard-pressed to think of a molecule the meteorite does not have. The only group of biomolecules that is missing is the lipids (fats). However, the building blocks of these (carboxylic acids and components from which glycerol can be made) are present. Figs. 6.1 and 6.2 illustrate this chemical richness.

How is this bewildering array of compounds manufactured in the absence of life? The answer lies with three simple compounds: carbon monoxide, ammonia and an unstable form of hydrogen known as the trihydrogen cation. Of these, the trihydrogen cation (or hydronium ion) is central to the synthesis of larger molecules. Although the discoverer of the electron, J.J. Thomson, identified this molecule as early as 1911, it wasn't until the 1960s that its significance was realized. Then, it became obvious that it must be a major component of nebulae.

In the freezing near vacuum of outer space, large hydrogen-rich clouds coalesce under the influence of gravity. These giant molecular clouds (GMCs, for short) are bathed in ultraviolet radiation from hot stars and supernovae explosions. This splits hydrogen molecules (H_2) into molecular hydrogen ions (H_2^+) and electrons. The H_2^+ ions then combine with further molecules of hydrogen (H_2) to form a three-hydrogen molecule with a positive charge: H_3^+ (plus a spare hydrogen atom). This molecule is unstable under dense terrestrial conditions, where it will readily interact with other materials. However, in the cold vacuum of space it is relatively free to wander around unhindered. Place it near another molecule, however, and its positive charge allows it to either combine with these materials or cause them to become reactive in their own right, through the transfer of electrons.

Another key molecule that is ubiquitous in outer space is carbon monoxide. This molecule is assembled in the outflows of red giant and supergiant stars. As these

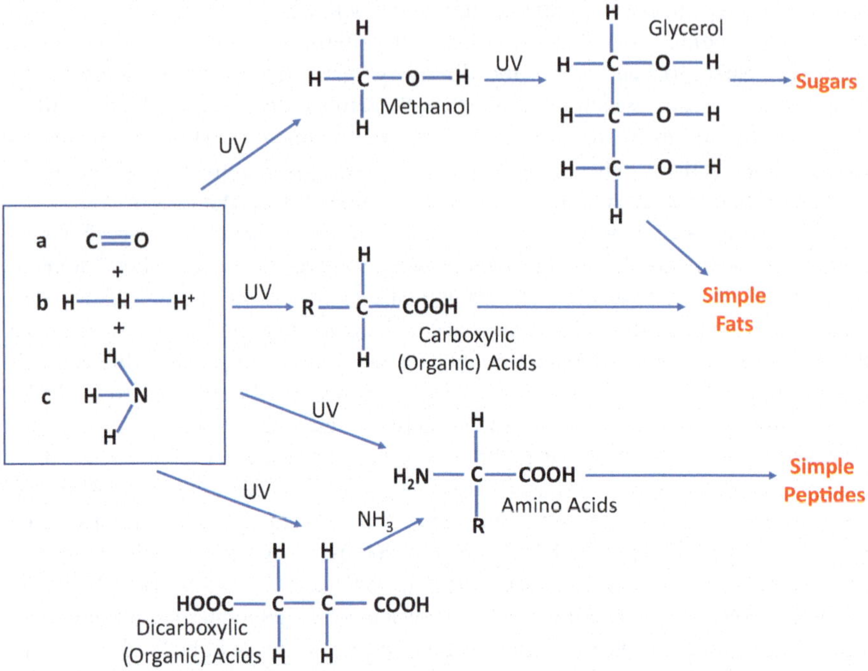

Fig. 6.1 Manufacturing life. Carbon monoxide (CO, **a**), trihydrogen cation (H_3^+, **b**) and ammonia (NH_3, **c**) are all ubiquitous in interstellar space. Add some UV, and, at very low temperatures and pressures, these can react with dust grains to form a variety of organic molecules that are integral to life on Earth—and likely elsewhere

stars evolve towards their oblivion, this material disperses and is readily incorporated into the sorts of gas clouds where the triatomic hydrogen cation is found.

What about ammonia? First identified in 1968, ammonia is found fairly abundantly in interstellar space, where it lies within dense, cold clouds of molecular hydrogen. Here, radio emission indicates that there is roughly one ammonia molecule for every ten million hydrogen molecules. Ammonia has also been found lurking within the cooler outflows of some giant stars. Ammonia appears to form directly, via the combination of hydrogen and nitrogen ions. This probably forms ammonium ions first, which then combine with free electrons to produce ammonia and atomic hydrogen.

Mix these molecules together in a giant molecular cloud—a million solar mass aggregate of dust and gas—and you can readily manufacture most of the materials found within the Murchison meteorite. Although many require a nice solid surface to come together, nature provides dust in abundance in the outflows of aging stars. About 4.6 billion years ago, when the Solar System was coalescing, nature had already endowed it with a vast treasure trove of chemicals from which life could form.

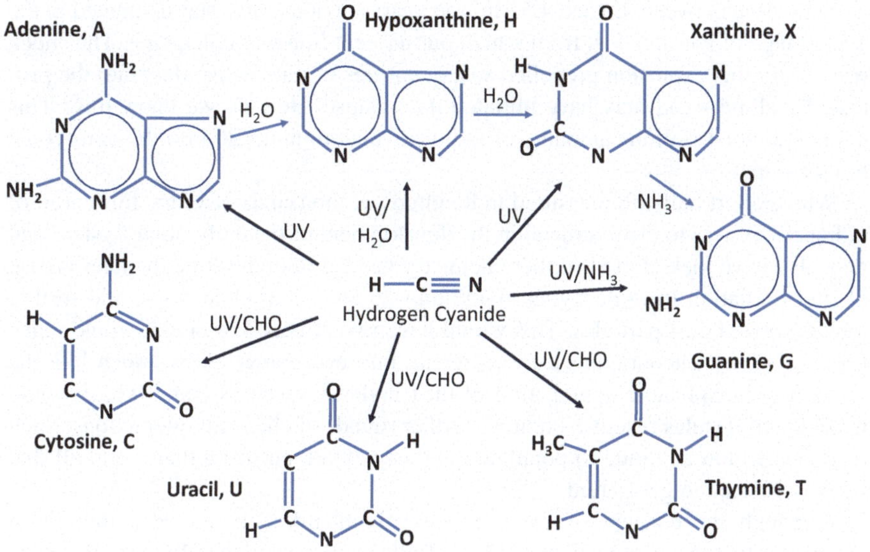

Fig. 6.2 The wonderful world of hydrogen cyanide. The infamous poison is readily made in outer space when ultraviolet light (UV) or cosmic rays bombard nitrogen (or ammonia) and carbon monoxide. The resulting three atom compound can polymerize (combine) in a number of different ways to form the core information molecules (nucleobases) in DNA, along with their derivatives xanthine and hypoxanthine. All of these molecules have been found in the Murchison meteorite, and some have also been detected in comets and interstellar clouds

Ribose is the central molecule of RNA and (as deoxyribose) DNA. This simple molecule should be formed with relative simplicity. However, until very recently it hadn't been detected in space. This led to the idea that simpler sugars may have preceded ribose in the structure of whatever molecule first held the information of life. However, in 2016, researchers at the University of Sophia Antipolis in France announced that they had created the molecule under the conditions of outer space. Cornelia Meinert showed that a mixture of methanol and water, frozen onto dust grains, would combine into ribose when exposed to ultraviolet light. Both methanol and water are abundant in interstellar space, and one would hope that ribose will be detected in the near future, confirming Meinert and co-worker's discovery. Quite rightly, you should ask, if ribose can be created under the conditions found in interplanetary space, why had it not been detected in the Murchison, or similar Allende, meteorites? The likeliest answer is the most prosaic: we didn't have the technology to detect it.

Meinert and colleagues were only able to detect it after upgrading their machinery. Specifically, astrochemists use gas chromatography to separate and identify molecules. However, if ribose behaves like other molecules in the brew that the experiment produces, it won't be distinguishable in the final mixture. Meinert's group used a new means of identifying their compound and hit the jackpot. Sadly, for other groups—including those examining the meteorites, ribose was likely always present; it just couldn't be detected.

Sometime between 4.6 and 4.55 billion years ago a massive star detonated in the assemblage of gas and dust from which our nascent Sun was collapsing. The shock wave from this explosion propelled vast quantities of radioactive dust into the protostellar cloud—and may have initiated its collapse. How do we know this? This dying star shed copious amounts of radioactive aluminium-26 into the embryonic Solar System.

With a short half-life measured in hundreds of thousands of years, this cargo of radioactivity was to prove crucial in the development of the early Solar System and may also have kick-started further chemistry needed to orchestrate the later arrival of life. Radiation from decaying aluminium-26 and other elements would further ionize gas and dust particles. This would have two effects. For one, it would make the dust sticky, encouraging it to aggregate into ever-larger blobs, much like the seemingly inexplicable aggregation of dust under a student's bed. Two, it would make the molecules reactive, eliciting further rounds of chemical interactions. Such reactions would continue to populate the dusty cloud, ensuring that it had all that life would need to get started.

Although the concept of the late heavy bombardment—the idea that Earth received a late pummeling some 4.2–3.9 billion years ago—is falling by the wayside, one needs to recall that even a couple of billion years after Earth formed, it still was being hit by asteroids several tens of kilometers across. Life emerged somewhere during this extended period of declining activity. Life was likely snuffed out regularly, only to rebound—or driven into the deepest domains on the planet, only to re-emerge, phoenix-like from its ashes. On each occasion it then had to innovate to proliferate once more. Certainly, by 3.8 billion years ago living organisms that had not secured a means to produce amino acids, co-enzymes and nucleic acids would have become extinct as their supply of raw materials was at best impoverished. Such co-enzymes, which are illustrated in Fig. 6.3, often are derived from the same building blocks (nucleobases) used to make RNA and DNA. Life uses what it has available. As nucleobases seem easy to manufacture in outer space, these are what life has crystallized around.

However, relying on a declining delivery from interstellar space also implies that life had invented either photosynthesis or volcanism-driven chemosynthesis by 3.8 billion years ago. If it hadn't there would be no cells to fossilize a hundred million years later. Work published in late 2018 shows that amino acids can also be synthesized, as carbonate is chemically reduced by hydrogen in alkaline undersea vents. Bénédicte Ménez (Université Paris Diderot, CNRS) and colleagues show that serpentinization of olivine in mafic crust is associated with the production of the amino acid tryptophan. Thus, it seems that the early, volcanically active Earth was an abundant factory of amino acids.

As Fig. 6.4 illustrates, the geological engine of the planet has an additional critical back-up role. Volcanism can also generate the kinds of molecules organisms need to respire, but it also produces a natural chemical battery that life seems to have taken on board. Geothermal vents—particularly those in the oceans—have strong chemical gradients across the rock between the relatively alkaline volcanic fluids and the surrounding acidic oceans. Such gradients are critical to the transfer

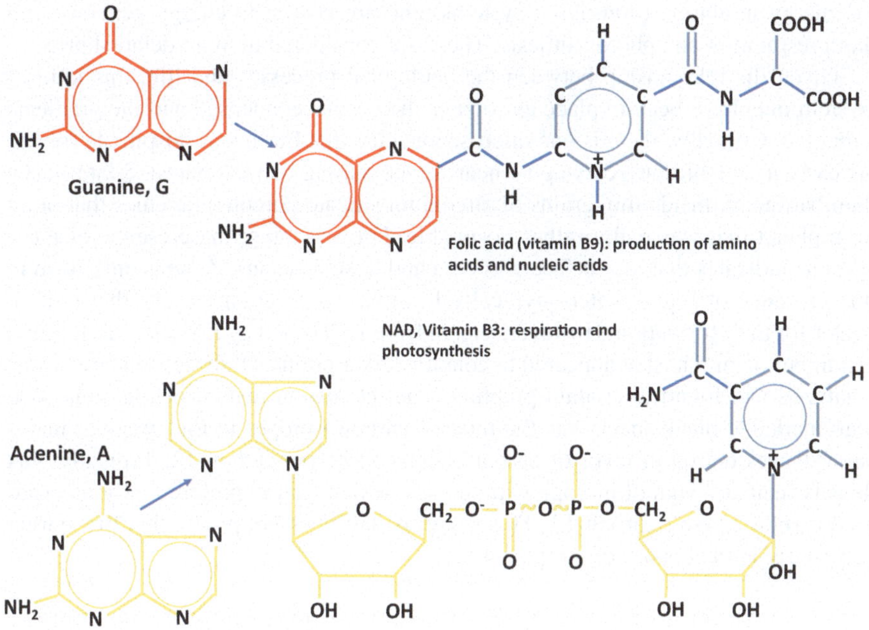

Fig. 6.3 The role of nucleobases in central metabolism. The working end of B9 is derived from the base guanine (red), while the majority of the B3 molecule is a component of RNA (purple). The remainder of the molecule is derived from vitamin B3 (blue)

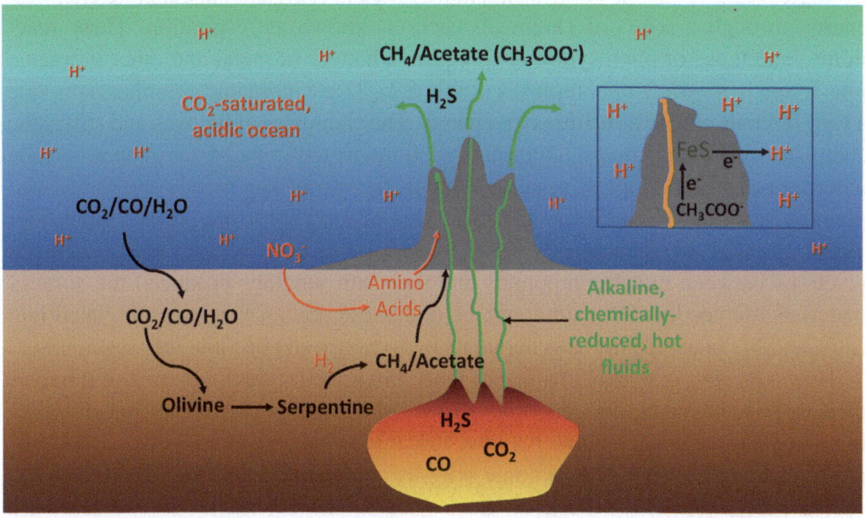

Fig. 6.4 Geology's great incubator. Seafloor volcanism naturally produces the chemicals organisms need to respire—small organic acids such as acetate, as well as amino acids. Although today's terrestrial ocean is saline and fairly alkaline, the early ocean would have been a lot more acidic. A pH gradient between the ocean and the crust of the planet would produce a natural battery. This system is still employed by organisms on Earth to power respiration and deliver energy in photosynthesis. Life, therefore, naturally evolved from non-living geological systems. Life is geology come alive

of energy in abiotic (non-living systems) and are central to energy generation in both respiration and photosynthesis. These are considered in more detail shortly.

Given the relationship between the geological processes and life, this kind of system may have been in place as soon as the oceans condensed and the gradients emerged. Critically, there is one microscopic clue that Earth was hospitable for life as early as 4.1 billion years ago—near the beginning of the contested late heavy bombardment. Inside tiny grains of zirconium silicate (zircons) lie clues that point to a planet that was really rather hospitable. For one thing, the presence of these zircons indicates that the surface was cool and hosted oceans. Zircons only form in the presence of liquid water—typically forming inside granite rocks that require water for their formation. However, a team from UCLA led by Elizabeth Bell found a number of zircons that appeared to contain darker grains. Of these, one zircon was analyzed and found to contain graphite. The interesting feature of this otherwise unremarkable black speck was the ratio of carbon isotopes in this graphite inclusion. It was biased in favor of carbon's lighter isotope, carbon-12. Typically, this bias is seen as a sign of biological processes, which tend to preferentially incorporate carbon-12 over carbon-13. This suggests, but does not prove, that life started very early in the history of our planet.

Respiration

We now move on to consider how life sustains itself against the ravages of nature. Biological self-sufficiency and sustainability means being able to generate chemical energy through a series of chemical reactions known as respiration. These reactions—and those of photosynthesis—require a battery of metal and other inorganic ions found in a variety of complexes (Table 6.2). These inorganic complexes provide a very obvious clue to how life got its wings. The chemistry of life was and still is the chemistry of Earth. Energy-liberating reactions in the crust of the planet became incorporated into biological systems through the use of key metals and their compounds. You can imagine early life clustering around hot vents on the ocean floor where a rich-broth of chemicals was spewed into the early oceans. Here, chemical reactions between metal compounds and their surroundings produced the sorts of materials life needed. Over time, one by one, these processes were assimilated into cells, and life became able to generate its own energy and become self-sustaining.

However, you can also imagine a more integrated model, where life emerges piece by piece from abiotic reactions. Processes that we regard as biological initiate in cavities in rocks and sediments near volcanic vents; these gradually assume the appearance of life when they become self-sustaining and when the systems begin to evolve independently of the geological system. The emergence of life is then a gradual process, not a singular event.

To get an idea of how this would happen, one has to take a peek at respiration. The process consists of a series of chemical reactions that go through three stages. In the first, known as glycolysis, glucose or other small sugars (monosaccharides)

Table 6.2 Metal ions and their minerals in the biology of cells

Mineral	Metal complex	Enzyme metal complex	Reaction catalyzed
Mackinawite (Fe/Ni S)	$(FeS_2Fe)4H_2O$ $FeS_2Ni)4H_2O$	Rieske protein (Hydrogenase e) parts of cytochromes bc_1, b_6f	Electron transport in respiration (plants, animals and bacteria)
Greigite (Cuboidal)	$2(FeS_2Fe)$	Ferredoxin	Electron transport in photosynthesis
Greigite $NiS_2(Fe_4S_4)S_2Fe$	$NiFe_5S_8$	Acetyl-CoA synthase/ carbon monoxide dehydrogenase	Conversion of carbon monoxide and carbon dioxide to acetate to serve as a carbon source in some archaea and bacteria
Greigite Twin $(Fe_4S_3)S_4(S_3Fe_4)$	$(Fe_4S_4)^{2+}$&MoS_4^{2-}	Nitrogenase	Fixation and conversion of nitrogen gas to ammonia
Hollandite (or tunnel manganite)	Ba $(Mn^{4+,}$ $Mn^{2+})_8O_{16}$ or $CaMn_4O_8$	Oxygen-evolving complex in Photosystem II	Splitting of water to release oxygen gas, electrons to reduce chlorophyll and protons to make chemical gradients

Many inorganic materials are used by biological systems and are integral to the survival of life on Earth. This suggests that life evolved from systems that were closely entwined with geological formations and activities

are progressively split to produce a three-carbon organic acid (pyruvate). In the process, the intervening molecules are oxidized, and some of the energy they contain is recovered as ATP (adenosine triphosphate). The process is inefficient, and very little energy is extracted—around 2% of the available chemical energy.

In the second phase the pyruvate is shunted into another cellular compartment, the mitochondrion, which is itself the biological remains of an oxygen-utilizing bacterium with which our ancestral eukaryote cell fused. This is discussed in the next section of this chapter. Once in the mitochondrion, pyruvate is stripped down to two components: carbon dioxide and hydrogen. The former is released into the atmosphere, while the latter is stripped into its components. The nucleus of the atom (a hydrogen ion, or proton) is shunted across one of the mitochondrial membranes as the accompanying electron is fed through a series of metal-containing carriers and other small molecules towards oxygen. This component is called the electron transport chain (or ETC, for short). This is illustrated in Fig. 6.5.

In the mitochondria (and across the inner cell membrane of some bacteria) the hydrogen ions build up, forming a chemical gradient. A special motor protein, called ATP synthase, forms a channel through which these ions can flow back into the cytoplasm of the bacterial cell, or inner core (the matrix) of the mitochondria. As they do, the heart of ATP synthase spins like a dynamo, and ATP is synthesized from its components. This process returns 17-times the amount of energy from the original molecule that glycolysis could muster. In essence, this is efficient enough to allow most of the sophisticated biological processes we see around us, such as the growth and reproduction of large organisms.

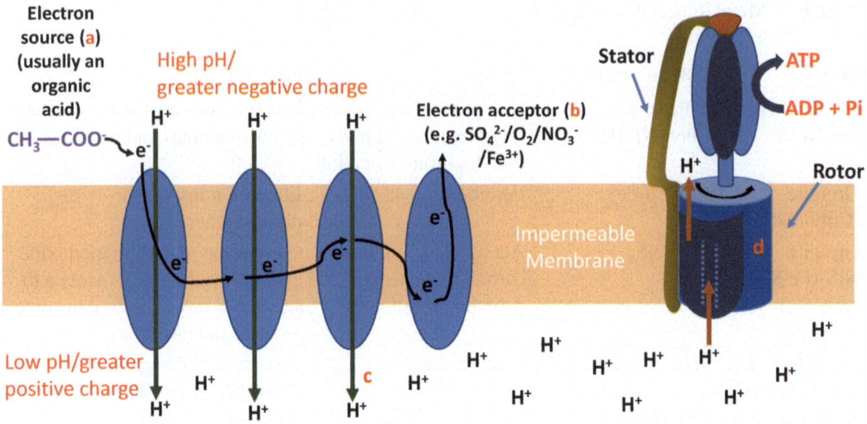

Fig. 6.5 The underlying mechanism of energy generation in a cell. Cells take compounds that are quite reduced (electron sources), such as **acetate**, shown at left, (**a**) and effectively "wire them up" to another substance that accepts electrons (an electron acceptor), such as oxygen or ferric iron (**b**). When electrons (e^-) move from one substance to another they carry protons (hydrogen ions, H^+) with them. By arranging electron-conducting proteins and substances in their otherwise insulating, fatty membranes, they can transport protons from one side of the membrane to the other. This generates an electrical and chemical gradient, just like water being stored behind a dam (**c**). Simultaneously, a beautifully constructed protein, called ATP synthase (**d**), acts as a channel to allow the protons to flow back across the membrane, down their chemical and electrical gradient. As the protons pass through the channel the protein spins like a dynamo. Instead of generating electricity, this creates the molecule **ATP** from its constituent parts, **ADP** and **phosphate** (**Pi**)

Although oxygen serves as the molecule that receives the electrons from "food molecules," there are many anaerobic (oxygen-free) variants. Each is less efficient than the oxygen-utilizing process, but uses effectively the same mechanism. In each case there is an oxidized molecule that will receive electrons from the chemically reduced "food molecule." As long as this "terminal electron acceptor" is available, the process keeps going. These electron acceptors include sulfate, manganese ions, ferric ions and nitrate. Reduction (the addition of electrons) to nitrate is the bane of farmers' lives, as in waterlogged or poorly aerated soils any bacteria will ditch the requirement for oxygen and reduce nitrate with which the farmer has lovingly fertilized his or her fields. Figure 6.6 describes how the quantity of energy available by these different reactions is determined by their chemistry.

All of this seems a far cry from the chemistry of the seafloor or other hydrothermal vent where we believe life began, so how would this conjecture have become a reality? Imagine an organism living in or very near a hot ocean floor vent. The vent spews out materials needed by the cell. Between the vent rocks and fluids, and the surrounding cold, dark ocean, there is a large chemical gradient of various ions, but notably hydrogen ions. Ocean floor vent water is usually relatively alkaline, while the surrounding oceans at that time would have been acidic, thanks to all of that carbon dioxide that was dissolved in them.

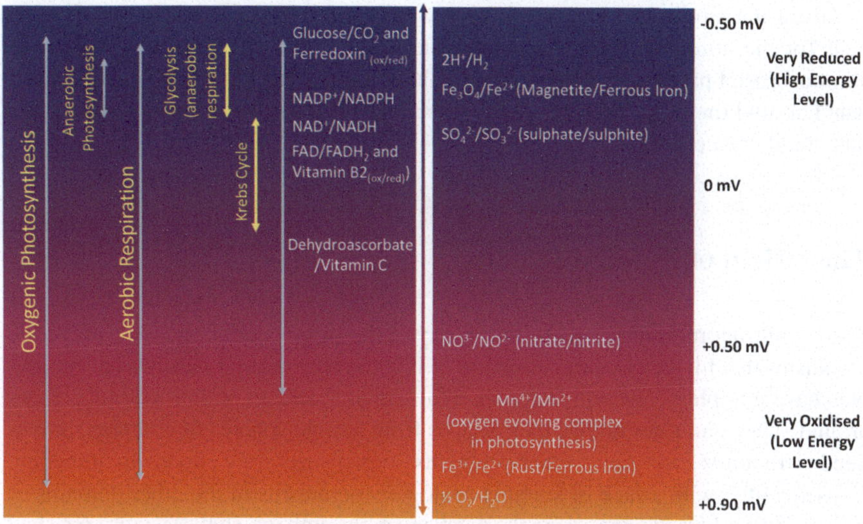

Fig. 6.6 The ultimate source of energy in chemical reactions. Chemists ascribe redox values for chemical reactions that involve the gain or loss of electrons. "Redox" is an abbreviation for reduction and oxidation. Reactions are split into two halves (half reactions) with values given as voltages—potential differences in electron energy. For example, the oxidation of hydrogen (H_2 to $2H^+$) has a very negative value, while the reduction of oxygen (½O_2 to O^{2-}) is very positive. The difference between the two components tells you how much energy is available to the electrons in this reaction. Thus, the reaction between hydrogen and oxygen (H_2 and O_2) releases a lot of energy, while the reaction between hydrogen and sulphate (H_2 and SO_4^{2-}) releases a lot less

In what is clearly a matter of biological opportunism, hot ocean floor vents also provide a wealth of nutrients in the form of small organic acids. Indeed, these are precisely the kinds of molecules used by mitochondria and bacterial cells to power their electron transport systems. They include lactic acid, ethanoic acid and pyruvic acid, which we came across earlier. In all of these systems, the molecules that supply the electrons are those produced by vents. As such one then envisages microbes initially scavenging these small organic materials as fuel. However, incorporation of the abiotic (geological, non-living) systems that were manufacturing them into their own machinery also brought them close to a simple system for generating energy.

Over time, increasing competition for these chemical resources forced primitive life to move ever closer to the remaining sources of these materials—the ion complexes themselves. Perhaps initially holding these metal compounds close to the cell membrane, these would eventually become consumed inside invaginations within the cell's outer layer. Finally, continued natural selection would force these metal compounds to become isolated within compartments inside the cell. At this stage not only does this non-biological machinery become part of the cell, but now the cell can regulate how the machinery works, by limiting the chemical reactions. This gives the cell an enormous selective advantage over "less evolved" competitors because it now controls how it transfers energy from materials in its surroundings.

Given the similarity between these processes and those occurring in bacterial cells (or the mitochondria of our eukaryote cells), it seems improbable that these abiotic, crustal processes did not beget the later biological one. Life, it seems, originated around the hot volcanic vents on the seafloor, only later to evolve into forms that could exist elsewhere.

The Origin of Eukaryote Cells

A key anthropomorphic concern in biology is the origin of eukaryotes. Unlike the organisms that fill the two other domains of life, the bacteria and the archaea, eukaryotes have complex cells with internal compartments and an ability to shuttle these around. They can also modify their shapes from simple spheres to complex, articulated structures we (and other complex organisms) rely upon to function. Consequently, with a lack of suitable analogous organisms in the other domains of life, the difficulties in uncovering the mechanism through which these cells arose had become somewhat insurmountable. However, various recent discoveries have finally shattered the intellectual ceiling that held back our understanding of our origins.

To cut a long story short, the origin of eukaryote cells comes down to one process—endosymbiosis. In this mechanism, an ancestral cell consumed another bacterial cell, which was later to become the mitochondrion. This cell is the primary eukaryote from which all others were derived. Perhaps half a billion years later, another cell, this time a photosynthesizing cyanobacterial cell, was consumed by some of these "animal" cells and became the chloroplast we see in plant cells today.

In support of this general pathway, both mitochondria and chloroplasts have their own circular (prokaryote-style) chromosomes and use prokaryote-style ribosomes—the machinery used to synthesize proteins. This is despite the fact they live in eukaryote cells. Moreover, the genetic code of their genes is closely aligned with that of prokaryotes. Cyanobacteria make the very close match for chloroplasts, while mitochondria are tied by their DNA sequences to alpha-proteobacteria.

As for the underlying process of endosymbiosis, it's a no-brainer. We can see it happening today in a variety of eukaryotic organisms. Take, for example, dinoflagellates—the principal cause of toxic "red tides." These are known as tertiary endosymbionts, because they contain the remains of three episodes of this process. The primary episode involves the consumption of the cyanobacterium by the ancestral eukaryote (the cell that already had a mitochondrion). This leaves a red algal cell. The secondary episode is where this algal cell is consumed by the ancestor of the dinoflagellates, producing a peridinin dinoflagellate. And finally, later in some instances, this peridinin dinoflagellate fuses with another cell called a haptophyte to produce the final lineage of dinoflagellates, known as a fucoxnthan dinoflagellate. That's quite a history for one organism, but one that graphically illustrates the process that gave rise to the ancestral cell of all current eukaryotes.

Although the underlying process of endosymbiosis is fine, there remains an issue with the scheme—at least when it comes to the origin of the *first* eukaryote—the mitochondria-containing entity. Was it simply a case that the ancestral eukaryote consumed the proto-mitochondrial cell; was it more a marriage of convenience; or was the precursor of the mitochondrion a pathogen of the other cell? Although the simple endosymbiosis route was initially favored, it is one of the latter two processes that now seem likely.

There was always the question of what was the (mitochondria-free) proto-eukaryote cell like? Did something like it still exist, lurking in some deep muddy pool or buried under thousands of meters of water? It now appears as though the proto-eukaryote never existed. Instead, close analysis of a clade (grouping) of archaea has revealed the likely ancestor of part of the eukaryote genome. In 2015 the identity of *Lochiarchaeota* was published by Anja Spang and co-workers from the University of Uppsala. Although not cultivated in a pure form, microorganisms were recovered from the deep-sea volcanic vent, Loki's Castle, at the bottom of the Arctic Ocean. Here, growing in complete darkness and extracting nutrients from the volcanic springs, Lokiarchaeota was sampled and some of its DNA analyzed. Using a technique known as metagenomics, DNA was extracted from a mixed source, in this case sediment. These DNA fragments, making up 92% of its genome, were then sequenced, and the order of its genetic code determined using complex computer algorithms. Although this method has its risks and flaws, the extracted DNA in this instance was clearly free of contaminating eukaryote DNA—yet it held a small percentage (approximately 3%) of eukaryote-like genes.

In all, the organism had a total of 5381-protein-coding sequences—a significant number. Of these nearly one third (29%) were related to genes in standard bacteria and 26% to other archaeal genes. 32% were novel, with no homologue in other types of cell recovered at that stage. The odd combination of genes suggested extensive gene-sharing between the bacteria and archaea in this bleak environment—something that is likely a necessity for their survival under these extreme conditions. The key finding in this discovery was the discovery that 3.3% of the genes were clearly relate to those involved in shaping the eukaryote cell (so-called *cytoskeletal* proteins) and those involved in the shuffling of membrane-bound compartments. Lokiarchaeota is thus the *Holy-Grail* of eukaryote evolutionary biology. This organism links the formerly disparate eukaryote lineage to the archaea and illustrated how simple cellular organisms developed the machinery our cells use to orchestrate their activities.

In 2017 a further analysis of the same sediments by Katarzyna Zaremba-Niedzwiedzk (Uppsala University) and colleagues identified additional, often closely related, organisms. Again, metagenomics was used to probe the community of organisms found in this volcanic mud. Analysis led researchers to group their discoveries in a super-phylum—a large aggregation of related archaeal organisms. Given the location of the find (the Norwegian Sea) and the origin of the investigating team in Sweden, this cohort of microbial cousins was grouped under the name "Asgard." In Norse mythology Asgard was one of nine worlds that was home to the Æsir tribe of gods, Loki included.

Members of the super-phylum are closely related to eukaryotes in terms of the genes involved in membrane trafficking—the movement of vesicles that we mentioned above. This new discovery effectively nailed eukaryotes (including us) as rather complex archaea. A humbling reclassification, which remains somewhat unpopular. It now seems that the third domain of life, eukaryote, is a misnomer.

Taken together, the eukaryote ancestor was an archaeal cell that was able to manipulate its cell membrane and traffic material into and out of its cell. One needs to bear in mind typical school and university teaching of microbial cells structure remains firmly focused on the kinds of microbes that infect us or our eukaryote cousins. This excludes a large variety of prokaryote species that have complex internal membrane structures and exhibit the kind of cellular complexity we normally consider reserved for the eukaryotes.

For example, cyanobacteria have photosynthetic thylakoids, magnetotactic bacteria have magnetosomes, while various nitrifying and methylotrophic bacteria have power-generating membranes analogous to the internal membrane of our mitochondria. Finally, the planctomycetes divide their genetic material from the cytoplasm in a manner analogous to the nuclear membrane of all eukaryotic cells. By limiting our world view, we restrict our perception of what is possible, then consequently fail to consider the full breadth of evolutionary options. A re-emphasis on teaching microbiology more broadly may change this.

Taken in perspective, once the mitochondrion had been acquired and stabilized within the ancestral eukaryote cells, evolution had a major helping hand in driving the diversification of eukaryote lineages. It isn't clear when this occurred, but it seems to have been prior to 1.6 billion years ago, when the first eukaryote microfossils appear in the geological record. However, the timing and subsequent evolution of the eukaryotes becomes easier to understand if we consider the all-critical geological context.

Between 1.9 and 1.3 billion years ago Earth was still lumbering through a particularly dull episode in its biological and geochemical history known as the "boring billion." During this phase there was a downturn in atmospheric oxygen levels as the planet belched forward significant quantities of oceanic and continental crust. The continents appear to have been largely coagulated in a persistent supercontinent, known as Columbia. The southeastern margin of the United States (Laurasia) was growing rapidly seaward, through the addition of large masses of granite.[1] Oxygen levels may have been suppressed by the rapid addition of copious amounts of iron to the oceans through undersea volcanism. Meanwhile, cyanobacteria were still adding oxygen to the waters at a local level.

Under these circumstances, there was enormous selective pressure for organisms to use any oxygen that was available. Imagine a community of microbes living near undersea vents, perhaps in shallow water near the margin of a volcanic arc. Cyanobacteria produce oxygen locally, but this is readily soaked up by various non-biological processes. Any alpha-proteobacterium living nearby is able to grab this oxygen and use it to liberate more than a dozen times the amount of energy from biomolecules that standard anaerobic respiration manages. Either through the pathogenic invasion of the archaeal cell by the alpha-proteobacterium, or through the consumption of the latter by the former as food, the two cells ended up fused together.

[1] This is discussed in more detail in the author's Springer book, *Granite Skyscrapers*.

Now, the alpha-proteobacterium had shelter from other potential eukaryote predators—and the archaeal cell had a vast new supply of energy available to it. As happens in our bodies today, when the immune system fails, the inadvertent preservation of the invading bacterium allows one to coexist in the cytoplasm of the archaeal cell, then both cells benefit.

Evolution now has a cell that can use oxygen, if it is available, and likely exists in a community of other organisms that provide an abundance of food—alongside that spewing from the nearby volcanic vents. All it needs now is more oxygen in the atmosphere to allow it to proliferate across Earth's surface…

The Emergence of Photosynthesis

Photosynthesis is perhaps *the* critical biological process on Earth. Without photosynthesis, life would likely still be a dismal affair, confined to hot vents on the ocean floor or hunkering down for millennia in the vain hope some stray cosmic rock might deposit a meal nearby. Therefore, we are compelled to investigate this critical biological processes if we are to understand how life really took off on Earth—and could do elsewhere.

The first stage in the rise of the process was likely the loss of nutrients from the formation of Earth. By 3.9 billion years ago, the rate of impacting fell precipitously. This would seriously curtail the delivery of the raw materials for cells—or whatever structures life had adopted at this time. However, though impacts constituted a declining source of useful chemicals, deep-sea vents continued to deliver organics such as methane, small alcohols, organic acids and sulfur-rich materials. In these crucibles, life would likely have taken a new course.

Although the early chemistry of terrestrial life was the chemistry of outer space, it would soon switch to using whatever Earth could supply. Perhaps very soon after life emerged, the oceans became so depleted in useful chemicals that organisms had to construct them using alternative sources. This process—chemosynthesis—requires an input of energy. The most obvious location for this process to begin is at the ocean vents that are still imbibing the oceans with usable organic chemicals. As we've already loosely discussed, the likely starting point was the development of electron transport chains that funneled chemical energy between molecules. These chains likely started with an enzyme that removed electrons from the ubiquitous organic acids that were spewing from the ocean vents.

Initially, these systems would have functioned in a respiration-like capacity. These would liberate energy, directly through the production of a chemical gradient, which was then used to generate the energy-storing molecule ATP (adenosine triphosphate). These systems would almost certainly have been made of proteins. However, as we already know, inorganic compounds, such as iron-sulfur crystals are embedded in living systems. Crucially, this means that there is a very real possibility that early life was directly wired-up to the materials in the hydrothermal vents. While many of these linkages were largely broken as these

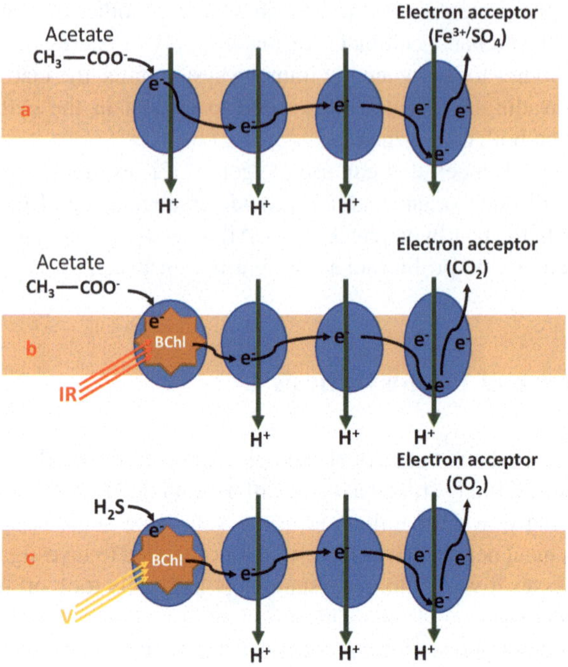

Fig. 6.7 The evolution of oxygenic photosynthesis from respiration. At first, **a**, respiration takes electrons from organic acids and feeds these to an electron acceptor. This generates a chemical gradient and ATP). In **b**, infrared radiation is captured by chlorophyll molecules and electrons are fed to carbon dioxide, reducing it to glucose. Electrons in chlorophyll are replaced from these same small, organic acids used in respiration. In the final step, **c**, increasingly energetic visible radiation replaces infrared radiation. Hydrogen sulfide is split and modern photosynthesis is born. Hydrogen sulfide is eventually replaced by bicarbonate then water

compounds became incorporated into cellular life, some bacteria remain connected to their environment, providing an insight into the likely early mechanics of the process (Fig. 6.7).

The trick with photosynthesis is making the link between the pre-existing electron-conducting proteins and the external energy source. Bacteria and plants make this connection using a family of molecules we collectively refer to as chlorophyll. Although the chlorophyll molecules of plants absorb energy in the visible portion of the electromagnetic spectrum, there exist far greater varieties of these molecules that absorb infrared radiation. On the dark ocean floor, visible light is confined to transient bursts from volcanism—or is generated by living organisms. As it is not abundant, it was almost certainly not available for living organisms in any meaningful way. However, around hydrothermal vents, where temperatures can exceed 300 °C, infrared radiation is abundant, at least over short ranges. At these temperatures infrared radiation is available at 800–900 nm, precisely in the range where certain chlorophyll molecules absorb radiation most strongly. Figure 6.8 illustrates some of these forms of chlorophyll.

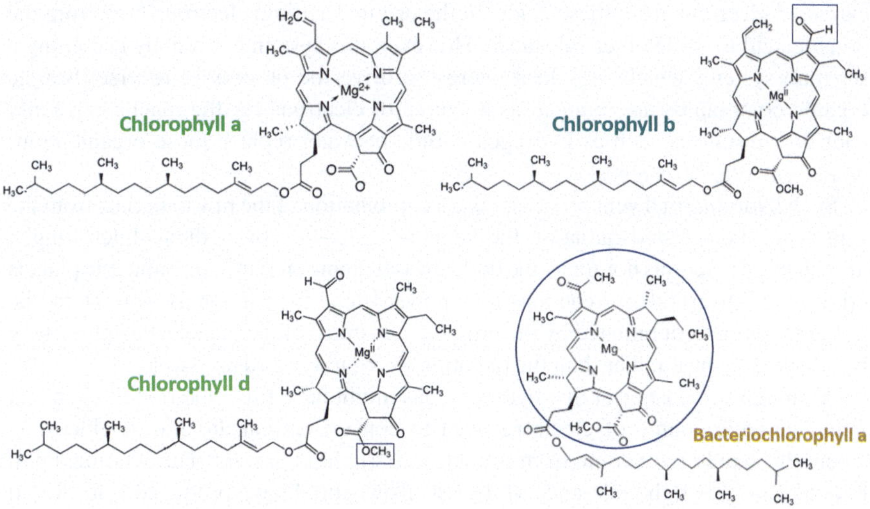

Fig. 6.8 The structure of chlorophyll 1, b and d; and bacteriochlorophyll a. The core part of the molecule is a magnesium ion that can lose electrons when struck by photons of light. The parts of the molecule around this affect what wavelengths of light each type of chlorophyll can accept. The differences are subtle and highlighted by the boxes. Bacteriochlorophyll has a couple of changes in the circled central portion of the molecule

One of the rather surprising discoveries is how easy it is to synthesize the rather complex-looking chlorophyll molecules. Under very simple conditions that are likely found in undersea hot vents, the central molecule of chlorophyll—a porphyrin ring—can form from very simple precursors. Stefan Fox (University of Hohenheim) and co-workers succeeded in making the building blocks of these porphyrin molecules from a mixture of salts that would be found in the oceans, along with a single amino acid that is present in the Murchison meteorite. By simply heating this mixture in a manner that was likely to occur in a deep-sea volcanic vent, out popped the core of the porphyrin—the heart of chlorophyll. This is a critical experiment, as it succeeds in producing the molecules that are integral to photosynthesis and respiration in precisely the kind of environment where these processes are likely to have originated.

Chlorophyll may not have functioned initially to capture radiation for biological use, but its ability to lose and gain electrons makes it immediately useful if it can be linked to a pre-existing electron transport chain. In essence, this is what happens in most photosynthesizing organisms. Infrared radiation (or visible radiation in some) ejects electrons from chlorophyll. This process is called photooxidation. Indeed, one of the most telling observations is that green non-sulfur bacteria use infrared radiation to drive the oxidation of their bacteriochlorophyll molecule, BChla. In these bacteria this oxidized molecule then recovers electrons from precisely the kinds of small organic acids that are produced at hydrothermal vents. This implies that photosynthesis evolved from respiration, where electron transport chains extracted electrons from these acids. In the original setting, electrons flow from the organic acid to some other substance. However, this system evolves by becoming a separate system, which uses heat energy to drive the process in reverse. Simple organic compounds are retained as a source of electrons for the machinery. Only later do substances such as hydrogen sulfide or water replace these organic compounds, as Fig. 6.7 suggests.

Such hydrothermal vent processes use a combination of the raw materials from the vent—and the infrared radiation the vents provide—to power them. Interestingly, though these reactions dominate the dark abyssal plains on Earth, on habitable planets orbiting red dwarfs these reactions may power life at the surface as well. Here, the stellar radiation is dominated by the infrared, and much of the planet's biosphere may be powered by heat rather than the radiation *we* would regard as visible.

Although some suggest that hydrogen gas might have filled the role of chemical reductant—the source of electrons used to convert carbon dioxide to glucose—where this would have come from is unclear. Earth has too weak a gravitational pull to hold onto this light gas; and, while organisms could have produced it locally, it seems more likely the materials supplied by hydrothermal vents would have filled this niche.

Modern photosynthesis uses the electrons that are liberated from chlorophyll in one of two ways. In the most primitive incarnation of the process, these electrons flow briefly out of chlorophyll into other compounds before returning to the chlorophyll. This short, electrical circuit is used to produce a chemical gradient that is then used to produce ATP. This process is known as cyclic photosynthesis because of this behavior. In other instances, the energized chlorophyll can steal back electrons from hydrogen sulfide, while directing these along the electron transport chain to a compound called NADP—a molecule made in part from one of RNA's building blocks. This NADP molecule is then used to convert (reduce) carbon dioxide to glucose. As electrons flow in a linear pattern, rather than along a closed circuit, this is known as non-cyclic photosynthesis. Electrons flowing from chlorophyll to NADP can also generate ATP by the same process as in the cyclic system.

The second non-cyclic system uses organic molecules, hydrogen sulfide or water to source electrons. This is known as photosystem II (PSII). Although this system is better able to steal electrons from various substances, it cannot generate enough energy to chemically reduce the all-critical compound NADP. However, put the two systems together, and now the organism can reduce NADP and make ATP in one fell swoop. This biological innovation occurs in only one organism—cyanobacteria—and its descendants, the chloroplasts. In most cyanobacteria and all chloroplasts, water is split to produce electrons, which are funneled into photosystem I (PSI). Here, they receive a further boost of energy, which is sufficient to produce a chemically reduced form of NAPDP. Cyanobacteria's success has ensured that it was able to access water as a source of chemical energy and come to dominate the oceans and (indirectly) the surface of the land, as well.

Photosystem II uses somewhat more energetic radiation in the visible part of the spectrum than the original anaerobic systems. However, on its own chlorophyll a, which makes up the light-harvesting engine of the system, is not powerful enough to split water and produce oxygen. The organisms that can split water use both chlorophyll a and a chemical catalyst to power the full reaction. As we discussed previously, this reaction has only evolved once on the planet. Although many enzyme systems have evolved over the last 3 billion years, splitting water has only emerged once. Almost certainly, this reflects the difficulty of dragging oxygen away from hydrogen. The water-splitting reaction requires such a large input of energy that evolution "devised" such a suitable system only once. To produce a measly 32 g of oxygen gas from water, an organism has to use over 300,000 Joules of energy. That's a fairly horrific energy cost for an organism.

Immediately, the question arises, if it is so difficult to come up with a system that can split water, how did it happen at all? Although we understand the root of the process in organic molecules, evolving it to the point where it splits water is a bit of a puzzle. Understanding how this game-changing reaction came about has involved solving some perplexing observations that go back to the 1970s. In some largely forgotten research, Helmut Metzner found that the presence of bicarbonate appeared to stimulate the production of oxygen from cells held in suspension. Moreover, it appeared that at least briefly that source of oxygen could have been the bicarbonate. This led to the suggestion that bicarbonate might have played a more fundamental role in photosynthesis early in the evolution of the system that makes oxygen (Fig. 6.9).

Although the first bacteria used infrared radiation to power chemosynthesis, their fate was tied to the availability of suitable organic acids or hydrogen sulfide. Thus, they were restricted to hydrothermal vents on the ocean floor. The problem with this is that such vents are only active for a few years at a time, and there is likely to be

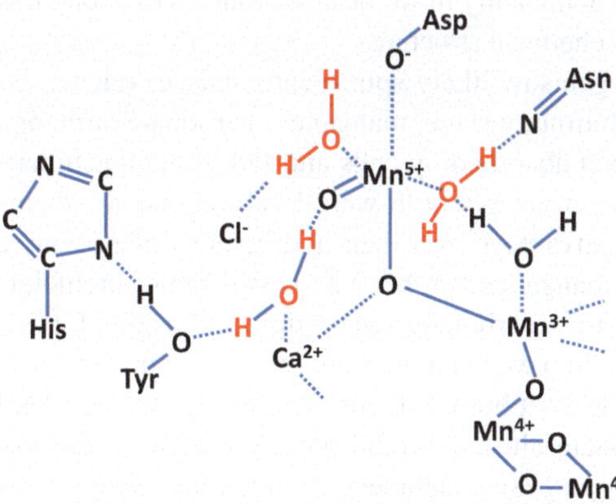

Fig. 6.9 The core crystal within the Earth's most important enzyme: water oxidase. This enzyme consists of the mineral hollandite ($CaMn_4O_8$), which has been captured inside a protein. Light drives electrons from nearby chlorophyll forming the "oxygen-evolving-complex". These losses are replaced by electrons that are stolen from the water molecules (red) attached to highly oxidised metal (Mn^{5+}) ions in the hollandite crystal. Oxygen gas is released and the Earth's Great Oxidation Event gets underway

extensive competition for this resource. On Earth, you can see this in effect every spring. The oceans experience a brief bloom of anaerobic photosynthesizing bacteria before the oxygen-evolving species kick in. One reason is the availability of the materials they need for photosynthesis. Water is clearly more abundant than hydrogen sulfide and, therefore, much more useful. Studies suggest that anaerobic photosynthesis produces only around one-tenth of the useful biological materials produced by oxygenic forms. Any organism that can use water will, therefore, have a far greater advantage over its rivals.

The likeliest scenario for the evolution of oxygenic photosynthesis involves a gradual shift first to ferrous iron that was dissolved in seawater and then to bicarbonate ions. In the first move, mutations led to the production of chlorophyll a from other, infrared-absorbing forms. This chlorophyll molecule is able to generate a far more powerful oxidizing force than its predecessors are. Once chlorophyll a emerged, cells could generate a strong oxidizing power that could be used to smash small molecules apart. The incorporation of the manganese-bicarbonate compounds was the next step. With these manganese compounds available, chlorophyll a could work its magic, smashing strong chemical bonds to oxygen and releasing the gas to the atmosphere for the first time. However, the transition was likely gradual, involving a number of steps. Each transition took the cell one step closer to splitting water.

The portion of photosystem II that releases oxygen uses a cluster of four manganese ions and one calcium ion bound to oxygen. This chemical structure is found in nature as rancieite or hollandite ($Mn_4CaO_9.3H_2O$) and—in principle—could simply have been taken up by early cells, but there are some problems with this straightforward assimilation. Although similar to manganese bicarbonate it is not the same. Instead, it appears that this complex of four manganese ions, a calcium ion and water *evolved* from the manganese bicarbonate form. With an atmosphere rich in carbon dioxide, this molecule was found abundantly in Archean seawater. Therefore, this was a likely precursor to the current chemical structure.

Initially, such organisms likely split bicarbonate to release carbon dioxide and oxygen. In this evolutionary stage, manganese carbonate or manganese bicarbonate ions would have been absorbed by cells and fed to the machinery that replenished chlorophyll with electrons. As both were likely abundant, such organisms would have had a massive advantage over their hydrogen sulfide-loving cousins.

However, using manganese in this way is still rather inefficient, and the system needs to reload with bicarbonate every time chlorophyll is oxidized by light. Therefore, there would have been pressure on the organism to come up with a more permanent fix inside its photosynthetic machinery. Consequently, later, the two manganese-bicarbonate clusters would have been "taken on board" and incorporated into the photosynthesis machinery, in much the same way as iron sulfide was taken into the machinery of respiration. Gradually changes to the structure of the manganese compound would then have occurred. One by one, the bicarbonate ions would have been replaced by oxygen and the structure become locked into the protein. Instead of oxidizing free-floating manganese bicarbonate, it would have become able to split bicarbonate then water.

It seems likely that life transitioned from iron and sulfur-based photosynthesis to bicarbonate photosynthesis between 3.5 and 3 billion years ago. Around 3 billion years ago the changing environment likely forced the hand of these organisms. The driving force for this final transformation was probably the amount of carbon dioxide in the atmosphere. This was in steady decline as more and more of it became locked up in carbonate rocks at the bottom of the oceans.

As atmospheric carbon dioxide levels continued to fall, the supply of bicarbonate, dissolved in the oceans, began to run low. Simultaneously, in the dark depths of the oceans the availability of calcium ions likely increased. Although bicarbonate and calcium react together to form calcium carbonate rock, this reaction would slow as carbon dioxide levels declined. In today's world, humans are busily reversing this trend, as we pump more and more carbon dioxide into the atmosphere. Calcium ions were pouring into the oceans from one of our planet's new features—continents. As the Archean progressed, more and more land emerged above the surface of the ocean. Plate tectonics was sculpting the first continents, and weathering and erosion of these released calcium ions into solution. This was very fortuitous for life. Calcium ions dramatically boost the ability of the manganese clusters to split water. Therefore, the fall in carbon dioxide levels allowed photosynthesis to move from its inefficient cracking of bicarbonate to a much more powerful splitting of water. As this happened, bicarbonate first moved to an accessory role, holding the cluster together before it, too, was replaced by the manganese oxide seen today.

Somewhere between 3.2 and 2.8 billion years ago, this final transition occurred, and modern photosynthesis was born. Although this did not impact atmospheric oxygen for another few hundred million years, the stage was now set for the explosive development of life and the irreversible rusting of the young Earth. In that murky gap, oxygen steadily removed local supplies of organic carbon and sulfides. Once these were gone the oxygen-producing bacteria burst onto the scene.

Another, often neglected reason for the delay in the rise of oxygen was the propensity of many oxygen-evolving cyanobacteria to use hydrogen sulfide if it was available. After all, if you could get your hands on hydrogen sulfide it was far easier to split than water. Therefore, it seems likely that while bacteria could produce oxygen relatively quickly after the emergence of photosynthesis, they probably didn't do it all the time.

Only when the supply of this noxious gas declined—both because of dwindling volcanism and because the oxygen produced by bacteria was consuming it—did oxygen production really switch on. The time it took the environment to run out of hydrogen sulfide and organic carbon effectively fixed the time it took oxygen to pour first into the oceans then the atmosphere. Figure 6.10 sums up the process of photosynthesis, illustrating where all of these bits fit together.

It's worth bearing in mind the impact of the environment. Since oxygen production was likely dependent on the amount of hydrogen sulfide that was present, every now and again there would be surges and rapid retreats in the amount of oxygen delivered to the oceans. When volcanic activity was at its most intense, the amount of hydrogen sulfide in the environment would likely increase in step. Volcanic activity might then slow oxygen production. This sort of effect may explain why there appear to have been brief burps in atmospheric oxygen levels in the few hundred

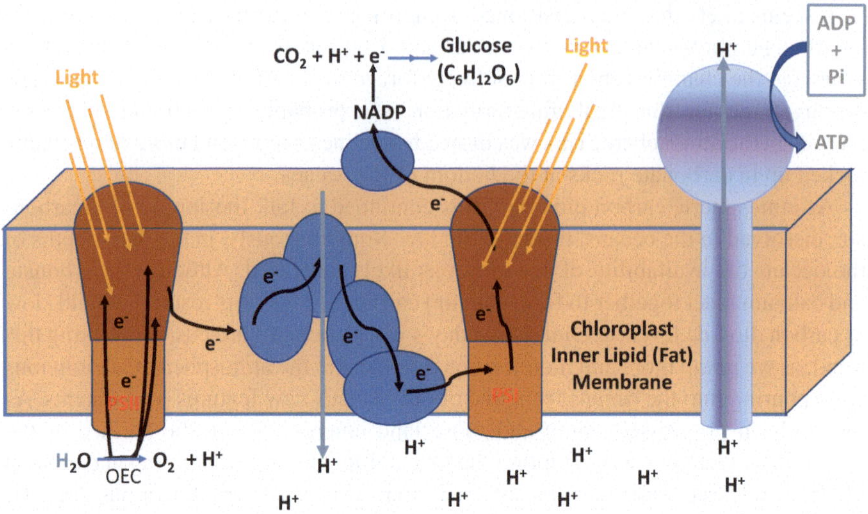

Fig. 6.10 A simplified view of photosynthesis. Light drives electrons from a series of receptor molecules (**PSI** and **PSII**) and oxidises them. Liberated electrons funnel through carrier proteins and other molecules (blue) and are added to carbon dioxide via a set of enzymes (not shown). This chemically reduces it to produce glucose. In the process hydrogen ions (**H^+**) are pumped across the chloroplast membranes, generating a chemical gradient that is used to make **ATP** (as before). Electrons (**e^-**) that are lost from the receptor molecules, which include chlorophyll, are replaced from water, or in some organisms from hydrogen sulphide. When water is used, the water molecule is split to produce oxygen gas. Splitting hydrogen sulfide produces sulfur. OEC is the "oxygen-evolving-complex" from the previous figure

million years that preceded the great oxidation 2.45 billion years ago. Only when the supply of sulfide declined and the amount of calcium in the oceans stabilized could photosynthesis begin to pollute the atmosphere with oxygen.

These processes have important ramifications for life on other planets. A large number of newly found worlds are significantly larger and more massive than Earth. Where these planets are rocky, there will likely be a greater extent of volcanism than is seen on Earth, at least for the first few billion years. Moreover, such extensive volcanism may well last longer than it did on Earth. Although this sort of geological activity may favor life in general by supplying it with nutrients, it could hold back the development of oxygenic photosynthesis and the rise of atmospheric oxygen. This is both because the greater supply of nutrients removes the necessity for innovation, but also because hydrogen sulfide is easier to split than water and thus will preferentially be used by photosynthetic organisms. This is discussed more fully in Chap. 7.

Finally, all of these stages must be held against the backdrop provided by the atmosphere. Before the rise of atmospheric oxygen, there was no ozone layer, and thus the atmosphere and the planet's surface were bathed in ultraviolet radiation. Although it is possible that other molecules provided shielding from this harmful radiation, it is likely the surface was still a very hostile environment for life. This implies that photosynthesis only gradually worked its way upwards from the depths to the surface. Only the top few tens of meters of ocean have sufficient light to drive the reactions we take for granted.

Thus, until life could survive at these depths, oxygen-liberating photosynthesis was likely held largely in check—this despite the enormous advantage organisms had that split water over other compounds. One might imagine some sort of slow but steady migration of photosynthesis upwards along the slopes of volcanic islands and the growing bulk of continents. These might have hopped towards the surface via hydrothermal springs until there was sufficient shielding overhead to allow them to emerge into the full glare of the Sun. Perhaps methane was abundant enough in the atmosphere that it formed protective oily materials on the ocean surface. Such chemical slicks are produced under the glare of ultraviolet radiation from the Sun and are abundant on Titan. Whether these existed or life simply used the vast depths of the oceans as protection until atmospheric oxygen levels rose is unclear and may likely remain so.

Regardless, the rise of photosynthesis was a game-changer on Earth. It gradually replaced the supply of organic materials from space with a sustainable source of chemically reduced and "edible" carbon. Without this step-wise innovation life would not have survived. Without its creation of oxygen gas, life would likely never have elaborated into the wonderful, multicellular forms seen today. Although bacteria can form intricate and rather sizable colonies, they do not produce anywhere the sophistication of structures that requires a high density of free oxygen gas.

The only question that remains for biologists is how the two photosystems were married together. To reiterate, though one is optimized for converting carbon dioxide into glucose, the other is optimized to split the abundant solvent, water. At present it looks as though there was one system that used small organic acids to produce electrons. This evolved into two separate systems, each with its own specialization. At some point, perhaps 3.4–3.0 billion years ago, two bacteria came together in a colony, on some dimly illuminated rock, in a shallow sea. Bacteria regularly swap genes, by a process called lateral gene transfer. Sometimes this process gets carried away and entire chunks of the chromosome get transported from one cell to another. It seems that one such bacterium, with the carbon-dioxide-reducing photosystem I, transferred its genes to another with Photosystem II. The resulting "hybrid" bacterium became the ancestor of the modern-day cyanobacterium and the chloroplast. If that seems unlikely, just watch the news about the spread of antibiotic resistance. The same underlying mechanism is used. In the cramped confines of bacterial colonies, it is very easy to see how genes move around. In this instance the brief dalliance between these cells was the revolutionary step that brought oxygen to our atmosphere and food to our plate.

Putting this in context, consider how photosynthetic organisms collect energy from the Sun. Terrestrial systems have a "reaction center" that harvests energy using a collection of plant pigments, which vary from the red and orange β-carotenes to the green of chlorophyll itself. These are arranged in a large complex of molecules called the light-harvesting complex (or LHC, for short). Figure 6.11 illustrates how these pigments collect energy from the Sun and deliver this to a central molecule of chlorophyll. This central chlorophyll then delivers the energy to the systems that produce ATP and reduce NADP.

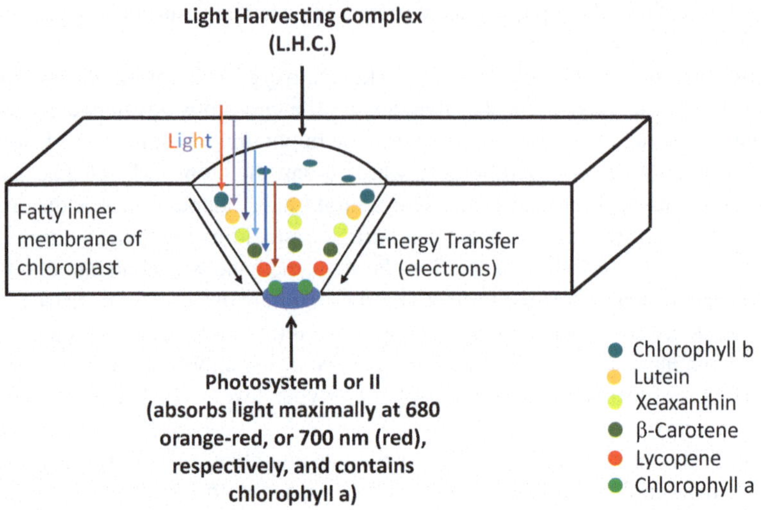

Fig. 6.11 Stylized visualization of the light-harvesting complex in the chloroplast of higher plants or cyanobacteria. A collection of light-absorbing pigment molecules (colored dots) is arranged in a pair of ring structures alongside a pair of proteins called a photosystem. These photosystems consist of a pair of specialised proteins containing hundreds of chlorophyll molecules. Electrons liberated from the L.H.C. funnel through the different pigment molecules ultimately ending up in the reaction centre. Each pigment can absorb a different wavelength of light and become oxidized. Electrons flow from these towards the central photosystem which is also oxidized by light. Electrons lost in this way are replaced by others from water (or other compounds in different bacteria)

It's important in this context to remember that the Sun releases most of its energy in the middle of the visible spectrum; red dwarfs do not. Most of their energy is liberated in the near-infrared. This certainly limits the amount of energy that would be available to drive oxygenic photosynthesis, but it is not so restricting as to make the process untenable. Evolution, therefore, has at least the opportunity to synthesize organisms that can make oxygen gas from water. Chapter 7 will examine the implications of this in more detail. The propensity of planets to host oxygenic or other photosynthesis will depend on the energy the planets receive from their star. Figure 6.12 illustrates the energy available from a red dwarf. Will this support the evolution of oxygenic systems like the Sun has on Earth?

The Entropy of Evolution

Okay, so now we have respiration and photosynthesis, but how do we turn little cells into complex organisms that are capable of understanding and latterly changing their world? This is where entropy comes in. Given the propensity, even in academic texts, to equate entropy with disorder this may seem rather surprising. Everywhere

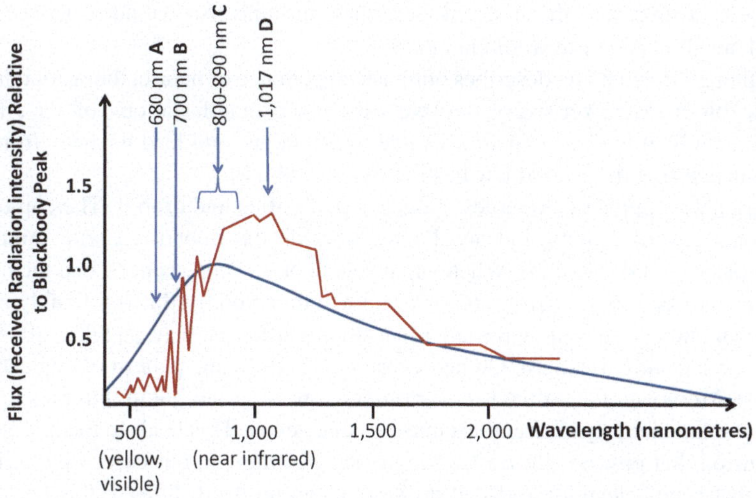

Fig. 6.12 The effective radiation released by an M8V 0.1 solar mass red dwarf compared to its idealised, or blackbody, curve. (**a**) The peak absorption of light by Photosystem I; (**b**) Photosystem 2 of plant chloroplasts, and the bacterial photosynthetic pigments, (**c**) bacteriochlorophyll a and, (**d**) bacteriochlorophyll b. The latter two absorb maximally in the infrared portion of the spectrum, rather than the visible wavelengths of light. As might be expected on evolutionary grounds, neither absorption peak is affected by atmospheric absorption

you look or listen someone, somewhere is saying entropy is a growth in disorder. What they mean is that entropy and disorder are one and the same. However, this is untrue. The word entropy means something rather different, and understanding its true meaning is fundamental to deciphering how life evolves.

In order that we understand entropy we have to look more widely at the process of thermodynamics. Thermodynamics is one of the grand theories that describe a very general property of energy. In everyday life we observe some basic processes such as apples falling to the ground or hot things cooling down. Although gravity is described by a number of theories, notably those of Einstein and Newton, these remain fundamentally incomplete and cannot describe all of the features of gravity in every situation. However, they are precise enough for us to understand most of what the universe can throw at us. The flow of energy from hot to cold or bright to dark is effectively complete. Born in the heat of the Industrial Revolution, thermodynamics was fundamental to driving our economic and industrial growth. By defining, mathematically, how energy flowed between regions of hot and cold our appreciation of entropy defined what we could do with the heat energy we released.

The first law of thermodynamics is one we all understand. It can be summarized as energy can neither be created nor destroyed. Energy can only be altered in its form. We see light bulbs release electrical energy as heat and light to liberate our homes from darkness; we feel heat spreading from the chemical reactions in a fire; or, more distantly, the Sun's nuclear furnace breathes life onto our world. Without the basic process that the first law describes, the universe wouldn't work. Stars would not shine, and life would not exist.

Although the first law describes our basic experience of nature, the second law is a little bit trickier. Yet every day we experience manifestations of its effects. Unfortunately, it is these experiences that misguide us, and lead us away from the fundamental role the second law has in driving evolution.

Here are a couple of examples. Take a cup of coffee and drop it. The cup obeys Newton's law of gravity and accelerates towards the floor (as Earth, minutely, accelerates towards the cup). When cup meets floor, the cup shatters, and the newly liberated coffee spreads across its surface. Each time you try this (preferably not in your own kitchen) the cup will shatter in a slightly different way, and the coffee will end up in a different pattern. On no occasion does the same pattern of cup carnage emerge. Moreover, the coffee is never observed to re-group, jump into the cup and the cup spontaneously reassembles into a usable vessel. This leads to the reasonable conclusion that process *always* lead to greater *disorder*. In our case, we observe a mess that needs cleaning; sadly, it does not clean up itself. Indeed, this particular example is often used as an example of entropy. However, as we shall see, the increasing entropy of the cup and its contents does not equate with increasing disorder. This is a misconception.

Now, look at another example where the concept of entropy is misunderstood. You're boiling water in a kettle. The emerging steam wafts over a cold window and begins to condense into droplets. The formation of increasingly large droplets from swirling microscopic beads of water is *increasing order*, not causing disorder. Similarly, inside your cells, proteins are assembled from amino acids. Once more, this is increasing order. Randomly coiled protein chains assemble into loops and coils after the amino acids are bonded together. This is increasing order, not disorder.

Since we equate entropy with disorder this casual observation appears to violate its principle: entropy either stays the same or increases. Does this mean the second law of thermodynamics only applies in some circumstances? Absolutely not! The second law is universal. The problem with our casual observation is that we only see one piece of the puzzle. In essence, we see a closed snapshot of the processes that are occurring. Although a smashing cup and dispersing cup of coffee might represent a visceral demonstration of thermodynamics, the manufacture of a triple helix collagen protein from individual amino acids is an altogether subtler affair. Nonetheless, thermodynamics is as fundamental to the manufacture of collagen as it is to spilt coffee.

What happens when your cell manufactures collagen? Every time an amino acid is needed, it has to be loaded onto the tRNA and delivered to the ribosome. Loading of the amino acid onto the tRNA requires the breakup of one molecule of ATP to produce AMP (adenosine monophosphate) and two free phosphate ions. This reaction liberates chemical energy from the molecular bonds in the ATP. Once the amino acid has been attached to the tRNA, much of the energy held in the ATP molecule is discarded. This is, effectively, heat energy, which warms the cell, slightly. The amino acid is then delivered, with the tRNA, by another protein that also requires the energy molecule GTP (guanosine triphosphate, another building block of RNA). Therefore, a lot of energy is used in the manufacture of proteins.

All of the chemical reactions that release heat increase the motion of other particles within the cell—notably water molecules. Therefore, the release of heat energy increases their *disorder* by randomizing their motion. Indeed, the manufacture of proteins, fats, carbohydrates and nucleic acids all results in an increase in the overall *disorder* of the contents of the cell. Each of these processes by necessity releases additional heat energy. This comes from the construction of chemical bonds in the various molecules involved in each process. Therefore, the construction of materials—representing a reduction in the disorder of these molecules—results in a net *increase* in disorder of all of the substances around it. However, it is here where the use of the word *disorder* first runs into trouble—hence the reason it is italicized throughout this paragraph.

Although the heat released by these processes alters the distribution of water molecules, they are not becoming "disordered." Rather, they are being given energy that allows them to move around more. Rather than becoming *disordered*, they are having their positions randomized. Here "random" and "disorder" are different. The energy allows them to adopt positions and motions that they would not otherwise have had. This means that they can adopt different "states." Because we are talking about small particles and small differences in position, these are called particle "microstates." *Increasing entropy represents a growth in the number of possible microstates*—effectively the heat gives the particles more opportunities to be in different places at any one time. This is entropy.

Similarly, in our earlier example, when water vapor condensed onto a cold window pane, the process of condensation releases heat energy—and this heat energy increases the motion of neighboring particles in the air. Overall, the extra heat allows the number of positions of other molecules in the air to increase, even while the position and motion of water molecules in the drop decreases. Indeed, the water and other molecules in the air increases the number of their microstates more than the water molecules in the drop decreases theirs. There is always an increase as a whole in the entropy of the particles in the area.

In this example physicists regard the condensing water droplets as a closed system. However, in reality water vapor condenses on a piece of glass, and air molecules surround the window. This system exists on the surface of Earth and Earth within the void of surrounding space. Ultimately, the condensing drop of water releases its energy to the cold vacuum of space. The system in which the water condenses, or that in which the protein folds into its final form, are always open, in the sense that the energy is never completely trapped but can escape and influence its surroundings. The universe is only closed if it is static, i.e., not expanding—and this is clearly untrue. An expanding universe experiences a growth in overall entropy as its volume increases and there is more capacity for particles to adopt various and increasing numbers of microstates.

How does this affect biology? For one thing, biological systems are adept at increasing the level of entropy of the particles around them. Therefore, though we observe DNA molecules copying, cells dividing and organisms growing as an increase in order (a decrease in their entropy), the reality of life is that we are vast entropy-creating machines. All that growth, repair and reproduction releases copious amounts of heat that vastly increases the amount of entropy in the world we inhabit.

In essence this is entropy. It's not so much an increase in *disorder of particles*, rather it's an increase in the *number of possible ways in which particles can be arranged*. By interpreting entropy as a distribution of possibilities, we can fundamentally reassess our understanding of evolution. Some quibble over the precise context of entropy—for example, if it should only ever be used in connection with energy—but the reality is that many systems can display entropy, even if it is not in a strict thermodynamic sense. Mathematically, entropy can be defined as the "sum of the number of available states the particles can be in multiplied by a couple of constants."[2] This can apply to particles in a chemical system or the number of possible microstates of information, or indeed any system where there will be a distribution of probability.

The conventional relationship with energy and temperature comes down to its often-overlooked definition. Temperature is a measure of particle energy—or, more accurately, their average energy. If particles have a lot of energy, they move about more than if they have little energy, much like children hyped up on caffeinated soft drinks. Entropy then becomes a measure of particle energy. Particles that have a lot of energy can do a lot more moving around and hence they exist in many more possible microstates. Particles collide with one another in different and unpredictable ways, which shares energy between them. This means that in a hotter object there is a greater spread of particle energies than in a very cold object. Taken to the extreme, the third law of thermodynamics states that entropy is zero at absolute zero. Here, there is no particle (atomic) motion, and hence particles exist in only one state. With only one possibility, there is zero entropy.

When it comes to the second law of thermodynamics, the overall entropy must either increase or stay the same. Therefore, in life, expect any organism to expend more energy than it needs to carry out a particular process. Some energy will always be "lost" to the surroundings so that the net amount of entropy goes up (or in some cases stays the same). You can see a direct impact on the manner in which organisms utilize energy. If one burns glucose, we can get 2,880,000 Joules (2880 kJ) of energy per mole of glucose. A mole represents a rather large 600 billion, trillion (6.03×10^{23}) molecules of glucose. Respiration, for all its sophisticated intricacy, retrieves around 1037 kJ per mole of glucose—or around 36% of the total available energy. The remaining energy is shed as heat. Every time a particular reaction in respiration occurs, some energy is shed, moving particles around that are not directly involved

[2] Although it looks complex, calculating the entropy for a system is quite easy. $S = -k\Sigma p_i lnp$ is a simple equation that defines a version of entropy known as the Gibbs entropy. "Σ" is the sum of all the possible states, or microstates, the particles in the system can be in; "*p*" represents their probability and "in" is the natural log; "*k*" is the Boltzmann constant. Shannon entropy is a modified version of this equation, which relates to information. We'll make something of a big deal of Shannon entropy in Chaps. 7, 8 and 9.

in the transfer of energy to ATP. The author once added further chemical steps to the Krebs cycle to try and improve it's yield of energy. This gave another couple of ATP molecules per molecule of glucose. Adding in a couple of extra oxidation steps in place of the Krebs cycle, you can get another ATP or two out of the process every time NAD dropped off a pair of electrons. However, despite this, the efficiency was only increased by around 5%. Life, it seems, has pretty much sorted out the best available patches for the process of energy capture.

Moreover, in terms of particles, the process of respiration increases the overall entropy as one molecule of glucose (and six molecules of oxygen) becomes six molecules of carbon dioxide and six molecules of water. Each of these liberated molecules can move around independently and thus has a larger number of possible states than the original solo glucose molecule.

Aside from the fairly obvious notion that any organism that wants to survive is going to have to put more in than it gets back, what is the consequence of the second law of thermodynamics for the flow of information from generation to generation? Here, the concept of entropy takes on two faces. In its simplest incarnation, every time a molecule of DNA is copied, energy must be released, and the overall entropy of the cell and its surroundings increases. However, we can also look at the information DNA contains as a kind of physical state. Here, mutations can be thought of as an increase in the entropy of the system. Each mistake is an increase in the possible information content as the genetic language evolves. It is in terms of information entropy that evolution really takes off.

The Lady's Not For Turning: Why Evolution Never Goes Backwards

Chromosomes consist of long strings of code, usually in the form of the molecule DNA. The code consists of triplets of letters: A, C, G or T (adenine, cytosine, guanine and thymine). Changes to the underlying code are called mutations and can affect any run of information in the genome with fairly even predictability.

Now, while most people consider mutations to be mostly harmful, the reality is that the majority have no effect whatsoever. This is because the triplet code is "redundant," or "degenerate." Changing one code to another does not alter the amino acid that the code signifies. Several different but related codes are assigned to the same amino acid in all but two cases. Therefore, when the code is changed, the amino acid stays the same.

Now, let's think about how this code could change over time. Assume that the rate of mutations an organism will experience is roughly similar everywhere in its DNA. If that is true and a mutation strikes, then it is unlikely to strike twice in the same place. Why is this so? Figure 6.13 illustrates this point. A stretch of fifteen nucleotides may mutate at one position. Since mutations are random events, there is very little chance that another will strike in the same location, rather than strike at another location along this stretch of sequence.

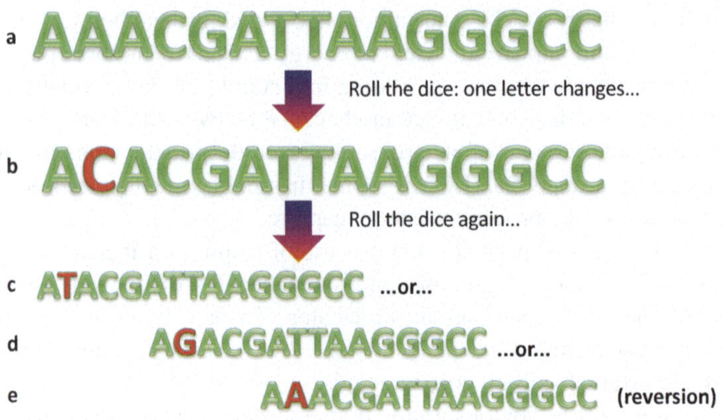

Fig. 6.13 Why the lady's not for turning—at least not often. A mutation is a change in a DNA sequence and it is the foundation of evolution through natural selection. Mutations happen randomly in nature. Let's say a mutation happens once every thousand years. In this piece of DNA a letter in the code (**a**) changes to one of the other three possibilities (**b**). Should a mutation happen again (**c**), is it more likely that this will change the original altered letter (**C**) back to its first form (**A**, reversion) or change it to one of the other three possibilities (**c**, **d** or **e**)? Clearly it is more likely to become one of the other three possible letters rather than its original form. However, even this is an ideal case. In reality any of the 15 letters shown could change rather than the one that has already changed. Therefore, the odds of the mutation reverting are *virtually* nil. If you're interested the odds of it doing so are 4^{15} for this molecule. "4" is the number of possible base options and 15 is the length of this molecule. A typical bacterial chromosome is 2 million nucleotides long so the odd of a reversion mutation is $4^{2,000,000}$ against

For the prospective gambler who fancies his or her chances, the odds are even more stacked against you. For, even if you do hit the same location twice you only have a one in three chance that the mutation will restore the original sequence.

For this reason, and this reason alone, DNA sequences are compelled to drift over time. While it is *physically possible* that mutations will restore the original DNA sequence, it is physically very, very improbable. To paraphrase the late British prime minister, Margaret Thatcher, "The Lady (of evolution) is not for turning." Evolution is a one-way street, but one where the sequence is sampling the vast number of possible microstates (in this case the possible code). Thereby, the progressive drift in sequence over time becomes a graphical illustration of the power of entropy. Rather than produce a decay in the structure of the DNA molecule, entropy is then embodied in the manner through which DNA sequences change and evolve over time. Although we are not looking at a strict thermodynamic interpretation of entropy, rather this is a statistical interpretation, which follows the scheme of J. Willard Gibbs or, better still, its reinterpretation by Claude E. Shannon as information entropy, or Shannon entropy.

In this light we can think about the evolution of one species from a common ancestor and the elaboration into extended family trees. Each base in a molecule of DNA has four possibilities. As far as the physics is concerned, there is no preference for any particular base at any one position. Therefore, because there is more than

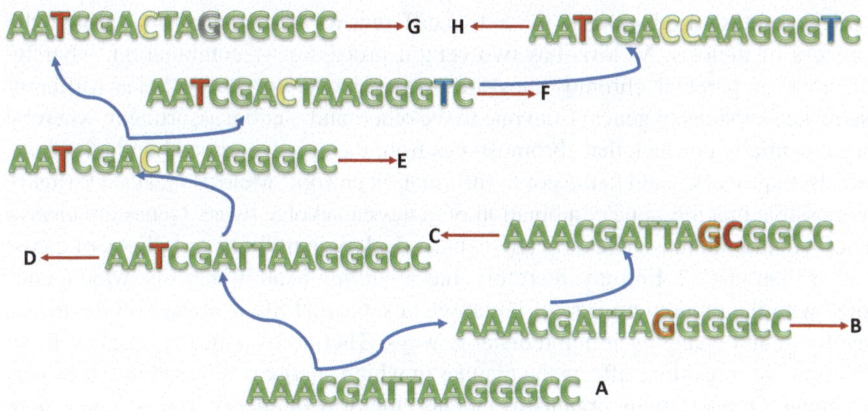

Fig. 6.14 Bifurcating a tree of life. What does evolution mean at the level of DNA? Imagine that the string of letters (A) is part of the genetic material, or genome, of an organism. The original sequence (A) mutates in different ways, giving rise to sequences B and D. There is little chance that the sequence will mutate in the same way in different organisms of a species as the process is utterly random. This process continues, giving rise to sequences C through to H. Because there is little probability that any one sequence will be the same in different organisms these changes will ultimately give rise to separate species. Evolution is thus simply a process of random chance that is coupled to the outcome: will the species produced by progressive mutation survive, reproduce and pass on its new sequence of DNA? If not, that branch of the tree withers and dies. However, as most mutations have no effect most branches continue to grow. In this way the tree of life expands until every available niche is filled

one possible base, nature is obliged to try them all out (Fig. 6.14). The sequence change illustrates the increase in the number of "available states" for each position along the DNA molecule. This is the manifested "entropy" change.

To put this in the broader biological perspective we need to remember that DNA isn't simply copied once, with this singular copy passed generation to generation. Reproduction tends to happen more than once in complex organisms, and bacteria reproduce in a continuous fashion through binary fission. Moreover, in the transition from fertilized egg to adult, there are countless rounds of cell division, each of which may give rise to new and interesting mutations. This means that successive generations receive multiple varieties of sequence, each produced by the steady process of genetic drift—the slow accumulation of mutations. Each time DNA is copied, different sequences are passed on. These sequence varieties are the branches in the evolutionary tree of life.

This is most clearly apparent in our children. No matter how many children you have, you will never see two exactly the same. Even "identical twins" have some differences that render them distinguishable. The environment or selection pressures within the species then decide who wins and who loses. The winners go on to procreate and create further branches in the tree as the steady process of mutation and selection works its magic. The losers? Well, they don't…

Eukaryotes introduce variation in two different ways: mutations and through the process of meiosis. Meiosis has two central processes—recombination, whereby sections of parental chromosomes are swapped over to move alleles (different sequence versions of genes) from one to the other; and random assortment, whereby it's essentially pot-luck that chromosomes from the grandparent end up in the gamete. Both processes add to the pot of information entropy. Meiosis makes it virtually impossible that the same combination of genes can evolve twice. Genes are always shuffled around, like a deck of cards—but a deck with millions or billions of cards, rather than just 52. Entropy, therefore, has a guiding hand in meiosis, which, coupled with the random process of mutation, ensures that life is always on the move, evolving and changing in unpredictable ways. Thermodynamically, each of these "moves" or transitions affects the manner in which energy is released into the environment. Overall, living organisms increase the flow of energy from substances in our food to the surrounding universe, in the form of heat. All that evolution has served to diversify the types of energy-liberating machines that are present.

However, before we conclude we need to re-visit a particular form of entropy—Shannon, or information, entropy. For it is this that will have the biggest bearing on the future of life on potentially habitable worlds. We can consider the Shannon entropy of living things in terms of the DNA (their genetic Shannon entropy) or in terms of species (the Shannon Index of Diversity, Chaps. 7, 8 and 9). Each is a measure of the kind of possibility (or more precisely, the probability) that an organism will embody particular genetics. In Chap. 7, we will see how this measure allows us to probe biodiversity and relate it to the kinds of worlds we are interested in; Chaps. 8 and 9 apply the principles to some of the worlds we now know exist.

Conclusions

This chapter has been a snapshot of the kinds of biological processes that operate on Earth and are believed to operate generally. Here, we've examined some of the ideas for the origin and the subsequent evolution of life. Although this has been a very abridged tour, the key ideas are as follows: that life can generate, or extract, usable energy from its surroundings—this is respiration; that it can sustain the creation of usable energy and substances once the primordial supplies are exhausted—this is photosynthesis; and finally, that life evolves in a manner that is consistent with the underlying laws of nature—this is principally the second law of thermodynamics and the principle of information (or Shannon) entropy that derives from this.

These key principles are then carried forward into Chap. 7 and beyond, when we consider the possible abodes for life on tidally locked and other worlds. The point is that life becomes an inevitability when one considers the chemistry and its universality. Moreover, the fact that information is not passed coherently and consistently from generation to generation necessitates the process of evolution. It's then a question of time. Will a planet be habitable for long enough not merely to play host to the transition from the geological to the biological, but will it be able to sustain the diversification of information we call evolution through natural selection? That is where the content of the last three chapters really matters—and it's where the next chapter aims to provide some answers.

References

Ackerman, S. H., & Tzagoloff, A. (2005). Function, structure, and biogenesis of mitochondrial ATP synthase. *Progress in Nucleic Acid Research and Molecular Biology, 80*, 95–133.

Allen, J. F. & Vermaas, W. F. J. (2010). Evolution of photosynthesis. In *Encyclopedia of life sciences (ELS)*. Chichester: John Wiley & Sons, Ltd. https://doi.org/10.1002/9780470015902.a0002034.pub2

Attwater, J., Wochner, A., & Holliger, P. (2013). In-ice evolution of RNA polymerase ribozyme activity. *Nature Chemistry, 5*, 1011–1018.

Beagle, S. D., & Lockless, S. W. (2015). Electrical signalling goes bacterial. *Nature, 527*, 44–45. https://doi.org/10.1038/nature15641.

Bell, E. A., Boehnke, P., Harrison, T. M., & Mao, W. L. (2015). Potentially biogenic carbon preserved in a 4.1 billion-year-old zircon. *PNAS, 112*(47), 14,518–14,521.

Bird, J. G., Zhang, Y., Tian, Y., Panova, N., Barvík, I., Greene, L., Liu, M., Buckley, B., Krásný, L., Lee, J. K., Kaplan, C. D., Ebright, R. H., & Nickels, B. E. (2016). The mechanism of RNA 5′ capping with NAD+, NADH and desphospho-CoA. *Nature, 535*, 444–447. https://doi.org/10.1038/nature18622.

Blount, Z. D., Borland, C. Z., & Lenski, R. E. (2008). Historical contingency and the evolution of a key innovation in an experimental population of *Escherichia coli*. *PNAS, 105*(23), 7899–7906.

Brewin, N. (1972). Catalytic role for RNA in DNA replication. *Nature: New Biology, 236*, 101–101.

Cech, T. R. (2000). The ribosome is a ribozyme. *Science, 289*(5481), 878–885. http://web.biosci.utexas.edu/psaxena/BIO226R/articles/ribosome.pdf.

Cernak, P., & Sen, D. (2013). A thiamin-utilizing ribozyme decarboxylates a pyruvate-like substrate. *Nature Chemistry, 5*, 971–977.

DasSarma, S., & Schwieterman, E. W. (2018). Early evolution of purple retinal pigments on Earth and implications for exoplanet biosignatures. *International Journal of Astrobiology*, 1–10. https://doi.org/10.1017/S1473550418000423.

Dismukes, G. C., Klimov, V. V., Baranov, S. V., Kozlov, Y. N., Das Gupta, J., & Tyryshkin, A. (2001). The origin of atmospheric oxygen on Earth: the innovation of oxygenic photosynthesis. *PNAS, 98*(5), 2170–2175.

Forterre, P. (2011). A new fusion hypothesis for the origin of Eukarya: better than previous ones, but probably also wrong. *Research in Microbiology, 162*, 77–91. https://doi.org/10.1016/j.resmic.2010.10.005.

Forterre, P. (2013). The Common Ancestor of Archaea and Eukarya Was Not an Archaeon. *Archaea, 2013*, 372396. https://doi.org/10.1155/2013/372396. https://www.hindawi.com/journals/archaea/2013/372396/.

Fox, D. (2016). What sparked the Cambrian Explosion? *Nature, 530*, 268–270.

Fox, S., & Strasdeit, H. (2013). Abiotic synthesis of porphyrins and other oligopyrroles on the early Earth and Earth-like planets. *EPSC Abstracts, 8*, EPSC2013-104.

Gounaris, Y., Litinas, C., Evgenidou, E., & Petrotos, C. (2015). A hypothesis on the possible contribution of free hypoxanthine and adenine bases in prebiotic amino acid synthesis. *Hypothesis, 13*(1), 1–8.

Grosberg, R. K., & Strathmann, R. R. (2007). The evolution of multicellularity: A minor major transition? *Annual Review of Ecology, Evolution, and Systematics, 38*, 621–654. https://doi.org/10.1146/annurev.ecolsys.36.102403.114735.

Hedges, S. B., Blair, J. E., Venturi, M. L., & Shoe, J. L. (2004). A molecular timescale of eukaryote evolution and the rise of complex multicellular life. *BMC Evolutionary Biology, 4*, 1–9. http://www.biomedcentral.com/1471-2148/4/2.

Hill, H. G. M., & Nuth, J. A. (2003). The catalytic potential of cosmic dust: Implications for prebiotic chemistry in the solar nebula and other protoplanetary systems. *Astrobiology, 3*(2). http://www.uni-leipzig.de/~biophy09/Biophysik-Vorlesung_2009-2010_DATA/QUELLEN/LIT/A/B/3/Hill_Nuth_2003_catalytic_potential_cosmic_dust_prebiotic_chemsitry_protoplanetary_systems_astrobiology.pdf.

Huber, C., & Wächtershäuser, G. (2006). α-Hydroxy and α-amino acids under possible hadean, volcanic origin-of-life conditions. *Science, 314*, 630–632.

Jarosewich, E. (1971). Chemical analysis of the murchison meteorite. *Meteoritics, 6*(1), 49.

Kaiser, R. I., Maity, S., & Jones, B. M. (2015). Synthesis of prebiotic glycerol in interstellar ices. *Angewandte Chemie (International Ed. in English), 54*(1), 195–200. https://doi.org/10.1002/anie.201408729.

Knoll, A. H. (2011). The multiple origins of complex multicellularity. *Annual Review of Earth and Planetary Sciences, 39*, 217–239. Downloaded from www.annualreviews.org by Harvard University on 04/28/11. For personal use only.

Koonin, E. V. (2010). The origin and early evolution of eukaryotes in the light of phylogenomics. *Genome Biology, 11*, 209–220. http://genomebiology.com/2010/11/5/209.

Kuan, Y.-J., Charnley, S. B., Huang, H.-C., Tseng, W.-L., & Kisiel, Z. (2003). Interstellar glycine. *The Astrophysical Journal, 593*, 848–867.

Kuhn, H. (1972). Self-organization of molecular systems and evolution of the genetic apparatus. *Angewandte Chemie (International Ed. in English), 11*, 798–820.

Lathe, R. (2005). Tidal chain reaction and the origin of replicating biopolymers. *International Journal of Astrobiology, 4*(1), 19–31. https://doi.org/10.1017/S1473550405002314. Seminar available at: http://star-www.st-and.ac.uk/~kdh1/abs/lathe_r.talk.pdf.

Lenton, T. M., Boyle, R. A., Poulton, S. W., Shields-Zhou, G. A., & Butterfield, N. J. (2014). Co-evolution of eukaryotes and ocean oxygenation in the Neoproterozoic era. *Nature Geoscience, 7*(4), 257–265. ISSN: 1752-0894.

Liu, Y., Wang, Z., Liu, J., Levar, C., Edwards, M. J., Babauta, J. T., Kennedy, D. W., Shi, Z., Beyenal, H., Bond, D. R., Clarke, T. A., Butt, J. N., Richardson, D. J., Rosso, K. M., Zachara, J. M., Fredrickson, J. K., & Shi, L. (2014). A trans-outer membrane porin-cytochrome protein complex for extracellular electron transfer by *Geobacter sulfurreducens* PCA. *Environmental Microbiology Reports, **6***(6), 776–785.

Logue, J. S., & Morrison, D. K. (2012). Complexity in the signaling network: insights from the use of targeted inhibitors in cancer therapy. *Genes & Development, 26*, 641–650. https://doi.org/10.1101/gad.186965.112.

Lundin, D., Berggren, G., Logan, D. T., & Sjöberg, B. M. (2015). The origin and evolution of ribonucleotide reduction. *Life (Basel)., 5*(1), 604–636. https://doi.org/10.3390/life5010604.

Martin, L. L., Unrau, P. J., & Müller, U. F. (2015). RNA synthesis by in vitro selected ribozymes for recreating an RNA world. *Lifestyles, 5*, 247–268.

Meinert, C., Myrgorodska, I., de Marcellus, P., Buhse, T., Nahon, L., Hoffmann, S. V., d'Hendecourt, L. L. S., & Meierhenrich, U. J. (2016). Ribose and related sugars from ultraviolet irradiation of interstellar ice analogs. *Science, 352*(6282), 208–212. https://doi.org/10.1126/science.aad8137.

Ménez, B., Pisapia, C., Andreani, M., Jamme, F., Vanbellingen, Q. P., Brunelle, A., Richard, L., Dumas, P., & Réfrégiers, M. (2018). Abiotic synthesis of amino acids in the recesses of the oceanic lithosphere. *Nature, 564*, 59–63. https://doi.org/10.1038/s41586-018-0684-z.

Menneken, M., Nemchin, A. A., Geisler, T., Pidgeon, R. T., & Wilde, S. A. (2007). Hadean diamonds in zircon from Jack Hills, Western Australia. *Nature, 448*, 917–920. https://doi.org/10.1038/nature06083.

Paul, N., & Joyce, G. F. (2002). A self-replicating ligase ribozyme. *PNAS, 99*(20), 12,733–12,740.

Perez-Bercoff, R. (Ed.). (2013). *Protein biosynthesis in eukaryotes* (Volume 41 of NATO Science Series A). Boston, MA: Springer Science & Business Media.

Schirrmeistera, B. E., de Vosb, J. M., Antonellic, A., & Bagheria, H. C. (2013). Evolution of multicellularity coincided with increased diversification of cyanobacteria and the Great Oxidation Event. *PNAS, 110*(5), 1791–1796.

Shannon, C. E. (1948). A mathematical theory of communication. *Bell System Technical Journal, 27*(3), 379–423. https://doi.org/10.1002/j.1538-7305.1948.tb01338.x. http://worrydream.com/refs/Shannon%20-%20A%20Mathematical%20Theory%20of%20Communication.pdf.

Shannon, C. E. (1948). A mathematical theory of communication. *Bell System Technical Journal, 27*(4), 623–666. https://doi.org/10.1002/j.1538-7305.1948.tb00917.x.

Spang, A., Saw, J. H., Jørgensen, S. L., Zaremba-Niedzwiedzka, K., Martijn, J., Lind, A. E., van Eijk, R., Schleper, C., Guy, L., & Ettema, T. J. G. (2015). Complex archaea that bridge the gap between prokaryotes and eukaryotes. *Nature, 521*(7551), 173–179. https://doi.org/10.1038/nature14447.

Spiegel, D. S., & Edwin, L. T. (2011). Life might be rare despite its early emergence on Earth: a Bayesian analysis of the probability of abiogenesis. *PNAS, 109*, 395–400. https://arxiv.org/pdf/1107.3835v1.pdf.

Stevenson, D. S. (2002). Co-evolution of the genetic code and ribozyme replication. *Journal of Theoretical Biology, 217*, 235–253. https://doi.org/10.1006/yjtbi.3013.

Szathmáry, E., & Smith, J. M. (1994). The major evolutionary transitions. *Nature, 374*, 227–232. https://doi.org/10.1038/374227a0.

Szilárd, L. (1929). On the decrease in entropy in a thermodynamic system by the intervention of intelligent beings. http://www.sns.ias.edu/~tlusty/courses/InfoInBio/Papers/Szilard1929.pdf.

Szathmáry, E., & Smith, J. M. (1995). *The major transitions in evolution*. Oxford: Oxford University Press. New edition in 1998. ISBN-13: 978-0198502944.

Turka, R. M., Chumachenkob, N. V., & Yarus, M. (2010). Multiple translational products from a five-nucleotide ribozyme. *PNAS, 107*(10), 4585–4589. http://www.pnas.org/content/107/10/4585.full.pdf.

Watson, R. A., & Szathmáry, E. (2016). How can evolution learn? *Trends in Genetics, 31*(2), 147–157. https://doi.org/10.1016/j.tree.2015.11.009. Available through Researchgate.

Weiner, A. M., & Maizels, N. (1987). tRNA-like structures tag the 3′ ends of genomic RNA molecules for replication: Implications for the origin of protein synthesis. *PNAS, 84*, 7383–7387. http://www.pnas.org/content/84/21/7383.full.pdf.

West, J., Bianconi, G., Severini, S., & Teschendorff, A. E. (2012). On dynamical network entropy in cancer. *Scientific Reports, 2*, 802. http://arxiv.org/pdf/1202.3015v1.pdf.

White, H. B., III. (1976). Coenzymes as fossils of an earlier metabolic state. *Journal of Molecular Evolution, 7*, 101–104.

Yamagata, Y. (1999). Prebiotic formation of ADP and ATP from AMP, calcium phosphates and cyanate in aqueous solution. *Origins of Life and Evolution of the Biosphere, 29*, 511. https://doi.org/10.1023/A:1006672232730.

Zaremba-Niedzwiedzka, K., Caceres, E. F., Saw, J. H., Bäckström, D., Juzokaite, L., Vancaester, E., Seitz, K. W., Anantharaman, K., Starnawski, P., Kjeldsen, K. U., Stott, M. B., Nunoura, T., Banfield, J. F., Schramm, A., Baker, B. J., Spang, A., & Ettema, T. J. G. (2017). Asgard archaea illuminate the origin of eukaryotic cellular complexity. *Nature, 541*, 353–358. https://doi.org/10.1038/nature21031.

Zuo, Y., Xing, D., Regan, J. M., & Logan, B. E. (2008). Isolation of the exoelectrogenic bacterium *Ochrobactrum anthropi* YZ-1 by using a U-tube microbial fuel cell. *Applied and Environmental Microbiology, 74*(10), 3130–3137. https://doi.org/10.1128/AEM.02732-07.

Zuo, Z., Peng, D., Yin, X., Zhou, X., Cheng, H., & Zhou, R. (2013). Genome-wide analysis reveals origin of transfer RNA genes from tRNA halves. *Molecular Biology and Evolution, 30*(9), 2087–2098. https://doi.org/10.1093/molbev/mst107. http://mbe.oxfordjournals.org/content/30/9/2087.full. Wuhan University.

Chapter 7
The Niche, Its Hypervolume and the Entropy of Existence

Introduction

Although life may be ubiquitous, the ultimate question is whether intelligent (or even complex multicellular) life is equally abundant. Moreover, given a few certainties, such as an energy and nutrient source, will life inevitably follow a path to complexity? Although Chaps. 8 and 9 deal with some specifics, this chapter will focus on developing some underlying ideas and principles that are likely to be applicable to all biological systems, no matter their origin or overall design. By the close of this chapter, you should be able to consider any number of planets in terms of their habitability for life in general—and in terms of whether complex and intelligent biology becomes likely. As a consequence, we can then address the so-called Fermi Paradox—why has ET not (officially) phoned us up?

As such, this chapter is the critical chapter in this book. For here, we synthesize the grand plan, which derives (or evolves) from the principles outlined in the preceding chapters. This chapter is then the model in which the habitability and evolutionary potential of *all* planets should be framed.

The Niche

The key to understanding this model is the term "the niche." A biological niche is a generic term that describes the location in which an organism exists. Unlike the term "habitat," each organism is assigned its own niche; and the "niche is defined in terms of its physical, chemical and biological characteristics. The principle one niche per organism then applies. Humans, for example, occupy different niches, depending on the kind of ecosystem in which they find themselves. For example, if you live in the swamps of lowland Papua New Guinea then you may live in a tree house, above water level. Here, you have a variety of foods, water and access to the outside world. Humans in Berlin tend not to live in tree houses, but they require the same underlying

D. S. Stevenson, *Red Dwarfs*, https://doi.org/10.1007/978-3-030-25550-3_7

nutrition, environmental warmth and of course, water. As Berlin tends to be a tad colder than Papua New Guinea in the winter, humans also have insulated houses with heating. Similarly, an English oak requires basic nutrients from its soil, reasonable warmth and rainfall and a variety of interactions with microbial organisms—notably mycorrhiza—via their roots. Remove one or more of these components, and the well-being of the tree is compromised.

Although the principle of one niche, one organism, is sound, different organisms can compete for niches, with the end result being one organism per niche. This is part of the principle of natural selection.

If we look at organisms in this general way, we can define their niche in terms of a few parameters, such as temperature, humidity, nutrient supply or light intensity. The niche is then bounded by conditions in which the organism cannot survive at all. However, within the habitable domain, there are a range of conditions that may allow existence but not reproduction, for example. This is illustrated in Fig. 7.1.

Therefore, though conditions may be habitable, they may not be survivable in terms of the species. If it cannot reproduce there is no future, and extinction becomes inevitable. Thinking a little bit more along these lines we come to a simple but powerful way of defining the niche: its hypervolume. Hypervolume defines the total set of conditions in which survival is possible. In its simplest incarnation, the one we will focus on has three parameters: light intensity, temperature and humidity. Consider these as axes extending in three planes at 90° to one another. Manifest as a cube, the niche hypervolume is a box in which an organism can prosper. The range of conditions in which reproduction can occur is a narrower zone within the habitable hypervolume. This is illustrated in Fig. 7.2.

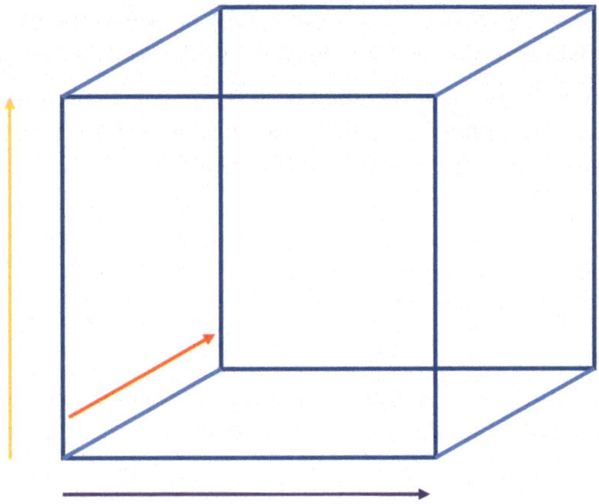

Fig. 7.1 The niche hypervolume concept. Imagine the conditions in which an organism of your choice can survive. Take, for example, a petunia. This plant needs a certain amount of light, a relatively narrow range of temperatures, and a relatively small amount of water. Anything outside this range results in first the petunia's ill-health, and then, if things get worse, its death. The niche hypervolume concept illustrates these three parameters, or conditions, as three mathematically definable quantities, shown by the three colored arrows. Petunias exist only inside this volume, defined by the length of the arrow

Fig. 7.2 Niche volume's finer detail. Survival is not enough, in itself. As conditions deteriorate from optimal, first reproduction, then growth, then simple maintenance of the organism is lost. The extreme ends of the range of any condition are areas where organisms simply exist but cannot prosper. These are the inferior and superior lethal range (ILR and SLR, respectively)

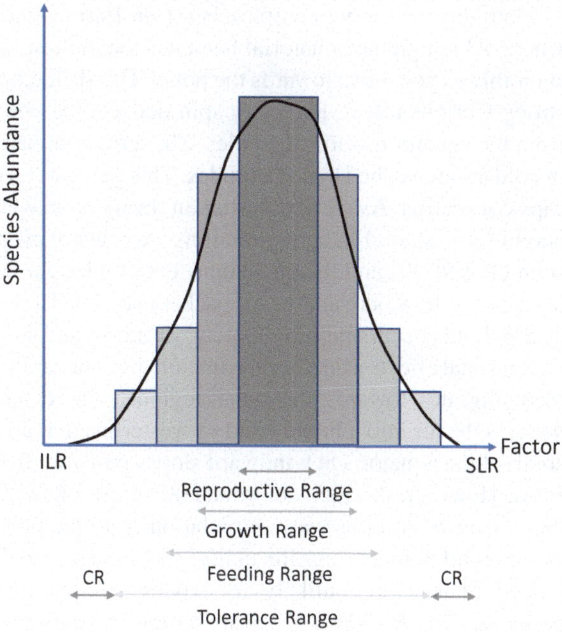

The axis for temperature, for example, extends from the lowest to the highest temperature in which this organism is found—or over which it can reproduce. Likewise, for plants, the hypervolume axis extends from the lowest to the highest light intensity a plant can tolerate and continue to photosynthesize. When this method is applied to Earth some simple and reproducible patterns emerge that will be familiar to all of us, the intelligent reader. Life in the terrestrial hypervolume (niche volume) is best defined by three variables. Specifically, this is all complex, multicellular life on land, so does not include oceanic life and microbial life, in general. Yet, here, these organisms have their own sets of habitable conditions, that, in turn, can be fashioned into their own niche volumes. It's just that they will be defined by a different set of criteria, such as temperature, light and nutrient availability.

This caveat aside, the surprising truth about terrestrial, land-based complex life is that its overall distribution can be defined using only three variables: temperature, humidity and another called net primary productivity, or NPP for short[1]. Although the first two will be familiar to everyone, the latter is likely novel to anyone who does not study biology. The NPP is what biomass plants (or other producer organisms) create after you take into account the mass of material they respire in order to exist. The net primary productivity for land plants is broadly proportional to the light intensity and declines from a peak at the equator to a minimum at each pole. Temperature, unsurprisingly, follows the same trend.

[1] For plant and animal species two other terms (Potential Evapotranspiration – PET – and precipitation) are often more useful terms.

Humidity is a more complex beast on Earth—and this is likely also true elsewhere. Although the equatorial band has the highest humidity, this does not decline smoothly as you move towards the poles. This difference comes down to our planet's strong Coriolis effect. Planetary spin deflects the winds that move (broadly) away from the equator towards the poles. The terrestrial pattern is dictated by the sinking of cold air above the Horse Latitudes. This generates a temperature inversion, which caps convection. As we saw earlier an inversion is where air temperature rises with height (or is stable) rather than falling. Any warm air packet rising into this layer is soon choked off, and the air flattens out at a height determined by how warm that layer is and how buoyant the air packet was. The Coriolis effect dictates the latitude at which this occurs (approximately 30° north and south of the equator).

Seasonal effects aside, approximately half the air that sinks at the Horse Latitudes then migrates towards the polar regions, where once again the Coriolis effect deflects the air into a broad band of westerlies that are focused around 60° north or south of the equator. On windward slopes moisture from these winds falls as rain or snow. However, the focus of winds 60° north of south of the equator also dictates that air flows into this warm circulation from the polar regions. This generates the frontal rainfall that brings the majority of moisture to the mid-latitudes. Most tidally locked planets are unlikely to experience this form of precipitation—except, perhaps, along Mei-Yu-like (near) stationary fronts.

However, to sum up, the terrestrial pattern of rainfall means that there is a secondary zone in the mid-latitudes where humidity is relatively high. When all of these factors are taken together, you get the most productive land, in terms of biomass—the total mass of living organisms—at the equator and secondary peaks across the mid-latitudes.

It is important to be clear, here. The presence of extensive continental crust and mountains on Earth grossly distorts this pattern. Overall, the mid-latitudes are rich in species, but there are areas in the lee of mountains or that are located distantly to oceans that are as species poor as the sub-tropical deserts. Therefore, though the general pattern holds, there is substantial regional variation in these broad domains, in a manner that mirrors gross precipitation. This is illustrated in Figs. 7.3, 7.4, and 7.5.

Likewise, Figs. 7.4 and 7.5 applies the same principle to Rhines-rotator and slow-rotator planets respectively. The principle is straightforward to apply, as long as you trust the circulation models for the planets. To recap a little from Chap. 5, remember that these circulation models are only directly applicable to aquaplanets, where surface friction is relatively constant at all points. This is obviously not true of a planet like Earth, with a roughly one-third to two-thirds split of land and sea. On Earth, surface friction is broadly responsible for air flow across isobars—lines of equal pressure. Unsurprisingly this is strongest over land. Over the oceans on Earth, air flows more smoothly along lines of equal pressure, with only slight convergence or divergence from areas of low or high pressure, respectively.

If we slow the rotation down, air flows broadly across isobars with minimal deflection. Obviously, this scales with both the rotation period and the diameter of the planet. A rotation period of 24 h for a planet with twice Earth's diameter (almost certainly an aquaplanet if young) means the surface rotates at a given latitude with

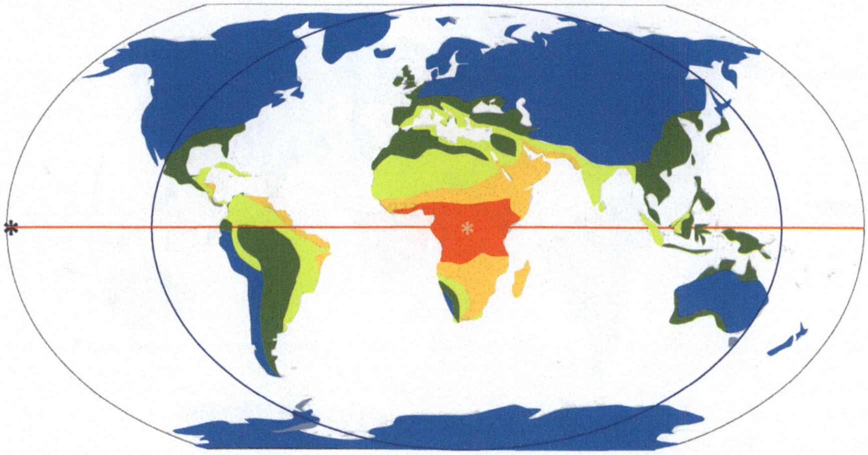

Fig. 7.3 A fast rotating, tidally locked Earth with a rotational period of less than 5 days. The SSP, ASP, terminator and equator are shown as before. Color-coding is qualitative and probably optimistic away from the SSP. This assumes that there is some precipitation along the equatorial band, where winds converge. Conversely, the extent of the blue area over Eurasia may be pessimistic, depending on the ability of mountains to block moist airflow from the Indian Ocean. (Base map: Creative Commons, Wikipedia)

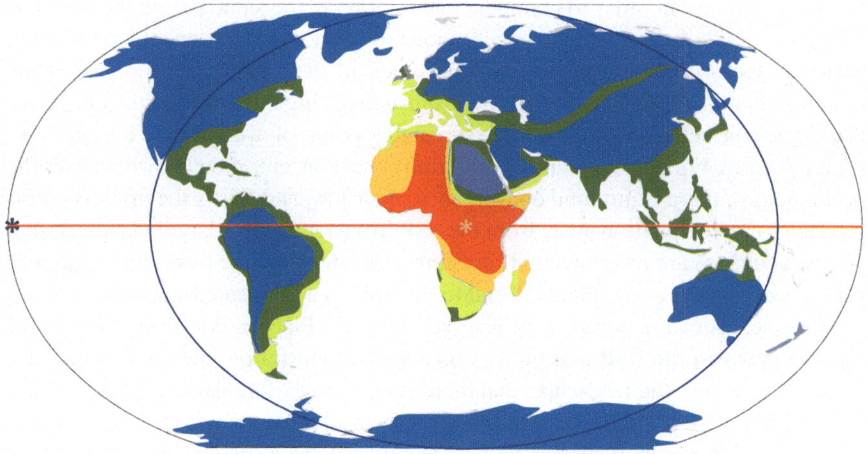

Fig. 7.4 As for the previous figure but for a tidally locked Earth with a rotational period of 5–10 days. The SSP, ASP, terminator and equator are shown as before. Possible frontal precipitation, where there are significant changes in wind direction, increases the capacity of some mid-latitude regions to support plants, as does broad north-south convergence over northern Africa. (Base map: Creative Commons, Wikipedia)

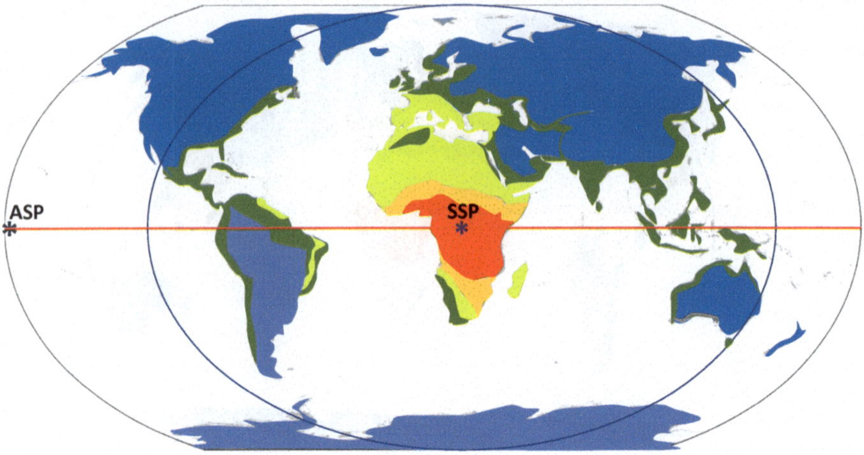

Fig. 7.5 An adaptive map as an attempt to model overall climate and habitability of a slow-rotating world. Adapted from Stevenson (2018b) this map qualitatively illustrates the niche volume for such a synchronously rotating planet. In this context red areas have the highest rainfall, through orange, yellow and green to blue, where blue is likely to be desert. Away from the SSP, almost all precipitation will be orographic and as such the potential distribution of niches is likely optimistic, with most of the planet dry. As with the other models, this remains qualitative, at present, as no such worlds have been mathematically modeled in this way. (Base map: Creative Commons, Wikipedia)

twice the velocity in order to complete one rotation in the same period. However, if the planet is tidally locked then synchronous rotation with the star sets the rotation period to the orbital period—and that becomes all that matters for a planet in the circumstellar habitable zone. Airflow is from areas of high to low pressure, and there is a single convecting core at the SSP (sub-stellar point) of every tidally locked slow-rotating world. For an aquaplanet, aside from some concentration of airflow over the polar regions, there is minimal convergence of airflow, and this is the primary limiting factor on precipitation away from the SSP. Transfer this to a largely land-covered planet, and there are no sources of moisture. Persistent airflow from the dark hemisphere leaves most of the planet, distal to the SSP, in a desiccated state.

Imagine instead a planet with a mixed surface. Place a continent the size of Eurasia between the SSP and the terminator and then follow the logic. Cold, dry winds blow across the landscape, and there is no source of moisture for tens or hundreds of kilometers, bar the occasional river. Such a landscape will be barren desert. If we add, ad hoc, small seas or large lakes, then if the air is suitably cold, it may pick up moisture on the downwind side of the lake. This could lead to localized showers. However, such lake precipitation on Earth is seasonal. If you take the Great Lakes, for example, they are warm in the summer, then cool in the autumn. If there are cold inbreaks of air in the late autumn or early winter, before the lakes freeze, these pick up moisture and dump it downwind in sheets of rain. However, this process only occurs because the lakes are warmer than the overriding air mass. This will not be true of the landscape on a tidally locked planet.

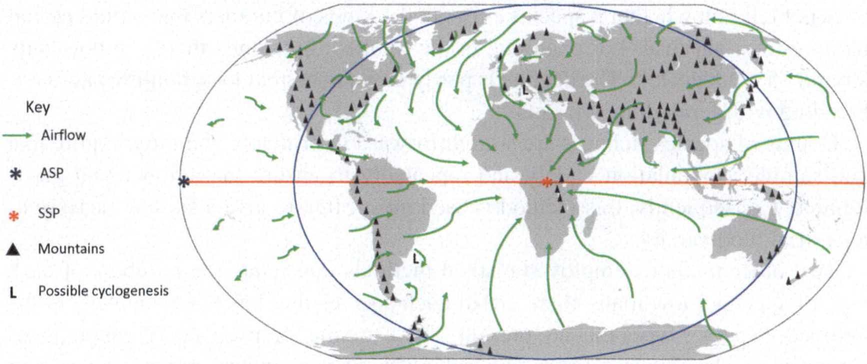

Fig. 7.6 The effect of friction on airflow. The slow-rotating (greater than 20 Earth day planet) has strongest flow south across the north Atlantic. Friction slows airflow over Eurasia and North America, generating a counter-circulation over southern Europe and the Great Plains. In the right circumstances this might generate a precipitation-bearing low over southern or central Europe. Topography is an important player in determining climate, which is otherwise fixed on these planets

After a few months in this state the lake and the air will have reached an equilibrium temperature because of the unflinching flow of air from one direction. There will be no convection, only the steady evaporation of moisture into the overlying air as it crosses the lake. Now, the only possibility for precipitation that involves moisture from this source is if it experiences rapid uplift—most probably driven by topography. Consequently, it's a matter of blind luck if the landscape has a fortuitous "design" that places moisture sources upwind of suitable topography. Life is then confined to such locations as the topography allows.

Another effect of friction is to slow airflow over land, compared with the oceans. In this regard, if we use Figs. 7.3, 7.4, and 7.5 as a guide, we should expect strong airflow down and across the Atlantic towards the equator in the north. With airflow slower over the land, air will tend to swirl in a cyclonic vortex with northerly flow over the North Atlantic, and much more sluggish, friction-driven southerly or westerly flow over Eurasia. Opposing and equally sluggish anticyclonic southerlies would dominate over North America. This kind of effect is not included in current climate models matters, as it will shift temperatures and precipitation. Figure 7.6 illustrates the possible effect on a slow-rotating world. Take this as a thought experiment for now.

Species Diversity on Earth: A Primer for Extraterrestrial Biodiversity

We can investigate the diversity of life on Earth in a number of ways. The first is simply to try and count how many different species exist in a particular location. This gives a number known as the species richness. Although this is interesting, it does not tell you whether a species exists in the kinds of numbers that would permit its long-term survival. For example, in the Scottish highlands there are now only around 35 wildcats left. The decline is partly down to habitat loss, hunting and interbreeding with feral, domestic cats.

Counts of species richness are straightforward in principle and only require that you sample a population widely and repeatedly to ensure the number you get is reliable. Consequently, this method is used most often to give a simple measure of terrestrial biodiversity.

Two other methods employ statistical methods that relate the numbers of each type of species (how many there are of each species that has been counted) to the number of species types that are present. These are the Simpson and Shannon diversity indices. The Simpson index is a little difficult to calculate, while the Shannon index simply sums the probability of finding each type of species.[2] Both of these methods produce reliable data, if you can carry out extensive sampling and make accurate counts of each of the type of species that is present.

Although this is feasible on Earth, it is not always practical, given the diversity and often challenging nature of sampling species in difficult terrain. Moreover, where animal species are highly mobile, there is the question of tagging and identifying each member of the species so that it is not counted more than once. Therefore, though measures of species richness clearly have their flaws, they remain the most widely used measure.

Can We Determine Species Diversity on Other Worlds?

A short, immediate answer might be, "No." A more interesting answer and increasingly likely answer is, "Maybe – at least to an extent…" We definitely cannot describe the species that will be present on any exoplanet, as evolution may take a variety of paths from an initial starting point. Each path is determined by the kind of genetic material an organism has, the rate at which it mutates and finally, the opportunities life has to fill (the available niches). However, the final point provides an interesting handle on the situation. Can we determine the number of niches available for life? To this question the answer is a partial "Yes."

This may seem surprising, but if we take the measurement of the Shannon diversity index, something interesting and perhaps unexpected emerges. If we determine the Shannon diversity index for terrestrial land plants, for example, we find that it is proportional to another measurement: the Shannon landscape diversity index. What

---

[2] Shannon index, $H'_a = \sum_{i=1}^{a} \frac{n}{N} \ln \frac{n}{N}$ and Simpson's index, $D = [N(N - 1)/\Sigma n(n - 1)]$, where N is the community size and n is the population size for each species in the community. A population is defined as the number of individuals of a species in an area at a given time. A community is the sum of all of the members of each population in that area at that time.

is this? Essentially, it's the same measure of land structure that we use to consider species diversity.[3] Here's how it works.

Take an area of land, say 1-km square. If you then analyze the layout and structure of the landscape you can divide the area into "patches"—smaller areas that have particular topography or soils or precipitation, etc. The Shannon landscape diversity index simply looks at the probability of encountering any one of these patches of land in the given area. For example, if one patch type occupied 90% of the area, the probability of encountering it randomly would be 0.9. If there were ten other patches, each occupying the same proportion of the remaining area, they would have a probability score of 0.01, or 1% (0.1 divided by ten). The Shannon landscape diversity is then the sum of these probabilities, multiplied by its natural log (see footnote 1).

Therefore, without knowing anything about the species on a planet we can begin to make assumptions about the kinds of life forms or biodiversity that are possible given its landscape. This is pretty amazing in its own right—if, of course, we assume that the terrestrial biodiversity pattern is applicable to life elsewhere. This may seem problematic, but if we consider what life is, and what it needs to survive and proliferate, then these assumptions are reasonable. Take the three basic requirements for the function of any biological entity: a source of organic carbon, an energy source and a suitable solvent to move biomolecules and ions around in.

Of all the possible solvents, water is by far the most abundant and the most suitable. It has a modest pH, reasonable boiling and freezing points, a high specific heat of capacity (which allows it to buffer temperature changes), and high latent heats of fusion and evaporation (energy requirements to change state). It also produces a high surface tension. And finally, the frozen state is less dense than the molten (liquid) state. The high latent heat of fusion—the energy needed to melt it—allows water to remain liquid for longer, while the high latent heat of evaporation means that water acts as an effective cooling agent when it evaporates. Moreover, ice (or rather ice I) is less dense than water's liquid state, which means that ice (I) floats on top of water, forming an insulator in cold conditions and allowing organisms to survive in the underlying liquid pool.

Each of these key properties is related to the ability of water to hydrogen bond with itself (cohesion) and other substances (adhesion). Small differences in the charge of hydrogen and oxygen atoms in the molecule mean that each molecule is attracted to its neighbors, causing them to become sticky.

Energy can be supplied to living organisms in a variety of ways, as can the necessary biomolecules. These were discussed in Chap. 6. Life needs biomolecules to build itself and to propagate and to have the energy to allow chemical reactions to occur. The final property of living things that is effectively unique to them is the ability to evolve through natural selection. This is a property of the biomolecules themselves, of the manner in which they replicate, and of thermodynamics—entropy—which determines the rate at which information varies when it is copied.

[3] Simply swap "n" for "a" for the area of a patch of land with a characteristic; and "N" for "A" for the total area of land that is being considered. Otherwise, it is the same equation.

Taken together, we have a fairly predictable system we call life. This "system" does not depend on a particular chemical, nor does it depend on the type of organism. Therefore, we can be bold and use the Shannon diversity index and landscape diversity index to make reasonable assumptions about life.

Rodrick Wallace (New York State Institute of Psychiatry) and this author (Carlton le Willows Academy) are currently taking these indices and applying them to address the issue of habitability of planets orbiting red dwarfs (and stars, in general). The key determinant is water. It's not that tidally locked planets lack water; it's more that the distribution of usable water (in the form of precipitation) is markedly different from that found on Earth. Take the terrestrial pattern. The equatorial region broadly experiences daily convective rainfall. This falls in the region of maximum convergence (the meeting) of the trade winds. Air rises strongly along a line a few degrees wide, the intertropical convergence zone (ITCZ). In the mid-latitudes, the Coriolis effect and temperature gradient organizes the westerly winds into mobile zones of convergence. Here, a combination of frontal and orographic rainfall keeps these zones habitable for a broad variety of life.

If you go to a tidally locked planet, most of the precipitation is confined to a region 15–30° wide around the SSP. However, this (potentially) leaves the rest of the planet bone-dry, situated in regions with less pronounced heating and, crucially, under a temperature inversion that prevents strong uplift. Think of the terrestrial trade wind belt, where there is abundant shallow convection. However, this broad region of atmosphere is capped by an inversion that prevents clouds from growing to the point where they can produce widespread precipitation.

Fortunately, there are useful exceptions. One of the wettest places on Earth is on the island of Oahu (Hawaii). The layout of the volcanic peaks intercepts the persistent northeast trade wind and brings a drenching that continues all year round on eastern windward slopes. Similar conditions are found on the eastern slopes of the Cape Verde Islands, the Galapagos, the windward islands of the Caribbean and elsewhere where the wind is forced to rise over the mountains. To the west of these lines of peaks is a rain shadow. However, on the windward slopes there is considerable precipitation and with it high biodiversity. This suggests that on tidally locked planets, away from the SSP, localized regions of high biodiversity are possible on windward slopes. This contrasts with the majority of the planets, where limited precipitation gives these worlds a far lower evolutionary and biological potential than Earth. The variation in color of the landscape in Fig. 5.10 illustrates how important precipitation is.

If we combine the effect of rain shadow and the effect of the shading of wet areas on windward slopes, we get a landscape that remains fairly impoverished in terms of habitable niches compared to Earth. The shading effect of mountains is to an extent unique to red dwarf planets. Once mountains are in place, shading is a permanent feature of their aeon-long days—until erosion whittles the peaks away. On Earth, barring north- and south-facing slopes, the effect will vary throughout the day, as the Sun moves across the sky.

Therefore, if the slopes are steeply angled so that light is blocked from the star, it will in turn have a low biodiversity. Fortunately, mountains are not continuous

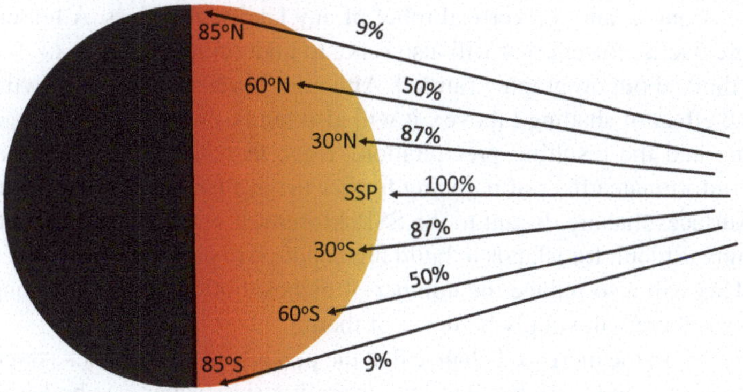

Fig. 7.7 From the sub-stellar point (SSP) the effect of distance on the amount of available radiation at a tidally locked planet's surface. Zones of decreasing sunlight extend in a concentric pattern around the sub-stellar point, which lies directly under the overhead Sun. Any plants growing nearest the terminator might lie sideways to maximize exposure of the unflinching light to their leaves, or they may grow along the ground, with leaves angled towards the light

barriers—there are gaps—so there will remain areas where there is both light and precipitation. The effect will also be limited close to the SSP, as the angle at which the Sun illuminates the landscape will also be greatest with the least opportunity for slopes that are sufficiently steep to obscure sunlight (Fig. 7.7). It will only be nearest the terminator where mountainous slopes may permanently obstruct illumination by the planet's sun, with the effect reducing in severity as one approaches the SSP, where illumination is more direct.

Super-Terrans, Gravity and Habitability

A final point worth considering is the effect of planetary mass on the number of niches on telluric planets. The effect is subtler but no less important for planets with limited possibilities for precipitation.

If we think back to the idea of niche volume then the climatic patterns of tidally locked planets present further problems if the mass of the planet is considerably higher than that of Earth. We've seen how precipitation is affected by the predominant flow of air into the sub-stellar region. The get-out clause that we discussed is the likelihood of orographic rainfall. Although this is likely on windward slopes, consider the effect of gravity. If, for now, we assume that high-mass telluric planets have oceans but not ones so deep that the continental surfaces are drowned, then we need to think about the relative relief of the mountains. On Earth, with a modest mass and a thick continental lithosphere (the solid outer portion of the planet), the continents stand 3 km above the abyssal plain, with topography that extends up to 8 km above sea-level. However, if you increase the mass of the planet, gravitational forces are stronger, and the vertical relief of any land will be less. A hotter mantle and a more ductile lower crust will also serve to undermine mountains.

Now, think about orographic rainfall. Although lower vertical relief will tend to reduce the effect of shading (above), it will also tend to weaken the effect of uplift on cooling and the resulting precipitation. Thus, increasing planetary mass may have the unfortunate effect of reducing further the availability of water on the continental surfaces that are distant to the SSP. Moreover, with greater mass, it will be increasingly difficult for islands to build high enough to penetrate the surface of any oceans. This will also reduce the number of niches, both on land and along island margins—as there will simply be fewer of them.

Couple that to the increased chance that the planet will form a thick, stagnant lid, and super-terrans suddenly become less appealing (less super…) abodes for complex life.

Another, rather subtle effect of increasing planetary mass is the effect that this has on tree height (Fig. 7.8). Terrestrial vegetation is limited in height by the counteracting forces of transpiration pull and gravity. Transpiration pulls water up the tree by a process called cohesion tension. Here, water molecules stick to one another by hydrogen bonding. As one molecule is pulled out of the leaf (through stomata), other molecules to which it is attached are pulled up the xylem from the roots. On Earth, the tallest trees are (or were) around 122–130 m. As shown in a recent paper

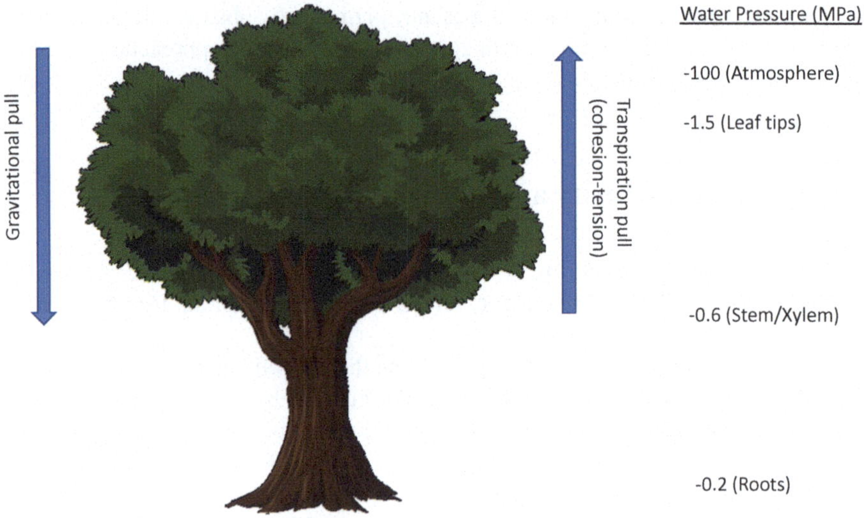

Fig. 7.8 Trees act as moisture pumps. Transpiration pulls water up the xylem, but gravity tries to restrain this flow. Tree height will therefore vary with planetary mass, with larger planets having lower vegetation. In turn, lower vegetation heights lead to lower surface friction and faster wind speeds compared to similarly vegetated parts of Earth. Most plants have a leaf tip pressure of −2.0 MPa. The lower limit is −1.0 MPa. Beneath this value, transpiration fails. (Tree image from http://detiru.net/wp-content/uploads/2016/07/how-to-draw-a-tree-4.png)

by the author, this height will always exceed that on a super-terran if cohesion tension is used to draw water up from the roots.

That's not all; if tree height is lower on super-terrans, then wind-speeds will be greater. Roughly, increasing the height of vegetation by 10 m reduces wind velocities by approximately 0.3 m per second. With lower vegetation heights on larger planets, the impact will be higher wind speeds compared to the corresponding terrestrial situation.

Some Assumptions on How Habitable a Planet Can Be

The key determining factor in truly understanding whether a planet will be habitable or not for complex life—that is, not merely microbes—is whether there are sufficient numbers and varieties of niches to fill. Organisms fill niches in a manner that is as dependent on the structure of the environment as it is on the underlying process of evolution through natural selection. Evolution is not going to blindly produce a human being on a world that is a searing desert, a block of ice or where land is submerged under several kilometers of water. There has to be a habitable surface with a variety of structures that are amenable for life to fill. After all, you will not need eyes without light, or the energetically demanding use of fingers and opposable thumbs if there is nothing to grip onto. Remember that if your habitat is 2 km down in an ocean there will be little, if any, light, bar the occasional and widely scattered glimmer from undersea volcanism.

If your habitat demands swimming, your body will need to be streamlined if you are to locate food and avoid predation. This is not that life adapts to its surroundings in a Lamarckian manner, merely that there if there is no requirement for a biological structure, such as a hand, an organism will be unlikely to survive if it is wasting energy making one during its development. Therefore, even if by chance the genetic material of a fish was to undergo a mutation that caused it to grow a hand, the resulting fish would be a poorer swimmer stuck with a structure that had no biological benefit. The outcome would likely be this handy-fish would make a tasty, if unusual snack, for another organism. Limbs on aquatic organisms are only useful if they help them swim.

However, stick some dry land into the mix, and fish with hands are more useful. Now this aberrant limb becomes an asset that allows it to explore this new vista. Lungfish and their ilk can go on land, breathe air—with a higher concentration of oxygen than water—and perhaps most importantly, evade aquatic predators. Evolution is blind, but if the geology of the planet provides niches, evolution has more to grip onto and synthesize new life forms that are complex. Moreover, given the right circumstances, which we will look at shortly, niches may begat other niches. Consider parasites as a case in point. The niche of the tapeworm is the gut of a suitable mammal; no mammal, no tapeworm.

Fish Farming and Aquaplanets

One might be interested in determining the potential biodiversity of aquaplanets. Here's where something odd happens—at least in science. In researching this area (for this book and for two papers) the author came across some work on fish farming and aquaplanets. Not that we believe that there are fish farms on waterworlds, but here there is an odd convergence of research areas. Rolando Cardenas (Universidad Central—Marta Abreu|| de Las Villas, Cuba) has developed a mathematical tool referred to as the aquatic primary habitability index. This model is, in turn, based on a pre-existing theory, known as quantitative habitability theory (QHT, for short). The premise is straightforward—that you can determine habitability based upon a number of environmental variables. This is precisely what we have done using the niche volume concept. However, when it comes to aquaplanets, the number of variables is small, and the impact of these variables (and the model itself) can be tested in terrestrial settings.

Rolando Cardenas applied the aquatic primary habitability index to Cuban coastal waters, where fishing and tourism were important. Here, they used a mathematical relationship that incorporated light, nutrient availability and temperature. This was instead of light, humidity and temperature that we considered earlier. The end result was a meaningful description of the productivity of the Cayo Coco region of the Cuban coast. One might add that oxygen concentration should be included as an additional factor, because this varies with a variety of environmental conditions.

More generally the SETI group has taken these habitability models to produce an index known as standard primary habitability (SPH, for short). This gives a gross measure of habitability that is again based on some (potentially) determinable variables, such as temperature, nutrients or relative humidity. The end result is a gross picture of planetary habitability that is similar to the output of the niche volume system we used. At a global scale, SETI describes a planet in terms of the kind of life it might host. This is summarized in Table 7.1.

Perhaps, surprisingly, when one applies these criteria to Earth, it comes in at the lower end of the habitability classification. The SPH index is 0.39, making it a

Table 7.1 SETI classifications of planets according to the dominant vegetation type

Planet type	SPH	Dominant vegetation type	Name of planet type
A	>0.8–1.0	Dense vegetation	Amazonian
B	>0.6–0.8	Mixed vegetation	Serengetian
C	>0.4–0.6	Scrubland	Mediterranean
D	>0.2–0.4	Grassland	Pampian
E	>0.0–0.2	Sparse vegetation	Saharan
F	0.0	No vegetation	Not viable

Planets covered in dense vegetation are Amazonian. Progressively drier worlds bear the names associated with lower levels of vegetation cover. The SPH index is similar to the niche volume concept (Fig. 7.1) and classifies planets based on a number of climatic characteristics that are needed for life. Where do you think Earth falls using this measure?

Pampian planet. This means that our planet is relatively dry and incapable of supporting dense vegetation over most of its surface. The word "most" is important. Remember that this is a gross simplification that is applicable to most of the planet's land surface. In reality, Earth has many areas with low levels of precipitation that give rise to desert, or with slightly more precipitation, tundra, steppe and savannah. The distribution of biomes comes down to the size and distribution of landmasses and the mode through which air circulates. The only way one can have an Amazonian planet is if the sub-aerial land—the land that is exposed to the air—is largely made up of small islands that are all adequately bathed in humid air.

Of course, even in this simplistic scenario, for rainfall to support such lush vegetation, air must rise. What goes up must come down, and where the air descends back to the surface the areas underlying this column of descending air will be intrinsically dry. A whole terrestrial planet cannot be evenly drenched in rainfall; some areas will always be dry. Therefore, this suggests that the SPH system needs to be applied regionally, not globally, in light of atmospheric circulation models.

Oxygen and Niche Amplification

A final consideration is energy. In all living organisms cells are able to carry out "biologically meaningful operations," such as building nucleic acids or transporting materials, using the energy from respiration. A subset of these organisms uses electron-transport-chains (ETCs) to deliver electrons from "food molecules," such as small organic acids, to another material. In many organisms this electron acceptor is a soluble ion, such as nitrate, sulfate or iron oxide (iron III oxide). The process of electron transport oxidizes the food molecule and allows the organisms to access far more energy from its food than is available from the simplest form of anaerobic respiration, glycolysis. However, a relatively small number of species on Earth can also use oxygen as their electron acceptor. This aerobic respiration process allows far more energy to be liberated from molecules such as glucose.

Although the process of aerobic respiration is restricted to most eukaryotes and a subset of prokaryotes, it is fundamental to the diversification of life on Earth. Superficially, oxygen allows a planet to build a UV-shielding ozone layer. This opens up a vast array of geological niches on subaerial (continental and island) surfaces, as well as the surface of oceans, where light may be present in abundance. However, more subtly, the advent of an oxygen-rich atmosphere also has profound implications for the process of weathering. Weathering and erosion in such an atmosphere liberates phosphorous as phosphate, and this is a critical and usually limiting nutrient for life on Earth. Thus, the rise of oxygen in the Proterozoic had fundamental impacts on the biological potential of the biosphere of our world. That much of Earth may have been underwater until the Proterozoic has significant implications for how nutrients cycled globally and how this affected the evolution of life on Earth. Remember that Earth, as we see it now, is not always how it was.

However, this is not all. When one comes down to developmental biology, aerobic respiration releases sufficient energy from glucose and fats to allow organisms that use it to develop more complex internal structures and develop muscles that allow vigorous movement. Movement, in this context, is not restricted to gross bodily motion, which may also allow predation or its evasion, but it also allows for peristalsis and the development of larger body plans with longer intestinal tracts. This increases the efficiency of digestion. Oxygen also facilitates the growth of more complex central nervous systems, which are energetically expensive organs to operate.

At a fundamental level, as Rodrick Wallace has pointed out, niche amplification is likely a consequence of aerobic respiration in an oxygen-rich atmosphere. What is niche amplification? This is the process through which the presence of a niche for one organism facilitates the development of other niches that allow further diversification of life. You can dumb this down to the earlier idea whereby the absence of mammals would result in the absence of tapeworms.

Consider the flow of energy through a food chain. Each organism in the chain requires energy to survive, but each is an inefficient user of the energy available to them. Energy losses occur through the generation of heat, egestion and through movement, in the case of animals, motile bacteria and many protists. If there is little energy available from food, and the efficiency of energy transfer is low, then food chains will be shorter than if the converse is true. Oxygen allows for a seventeen-fold enhancement in energy production, through aerobic respiration, compared with the simplest respiration process: glycolysis. Therefore, when oxygen is present, food chains can be longer and niche numbers may amplify, as the presence of one organism facilitates the evolutionary selection and development of others.

Oxygen, thus, changes everything for a planet.

Is there anything that makes oxygenic photosynthesis more likely? Perhaps. We know relatively little about the origin of the process. However, the two-part structure of the protein complex that liberates oxygen suggests two separate processes merged to produce the oxygen-liberating process we examined earlier. Although there is only one operating system on Earth, this does not tell us if the process is easy or difficult to initiate, as it may have emerged more than once but the other versions were out-competed, with the organisms that had them becoming extinct.

However, the protracted rise in atmospheric oxygen meant that its impact on Earth was the best part of 2 billion years in coming. After an initial rise in the early Proterozoic, oxygen levels slumped back and stayed low until the end of the geological era. However, once oxygen began its final ascent to its current abundance, biological evolution went on the rampage, with waves of eukaryote innovation in the later Proterozoic (the Ediacaran) and then in the Cambrian eras. It seems that rising oxygen directly fueled the evolutionary rush and the rise of the current biological empire that dominates Earth. This, however, may be more geological than biological in timing. We should then expect that planets with abundant oxygen will be the ones that have the largest number of energetic life forms. Oxygen free (anoxic) planets that are otherwise habitable are most likely filled with sludge-living microbes that hide from their star's light.

What about the nature of the environment? In Chap. 6 we looked at how the environment and the nature of tidally locked planet will limit this process. As Fig. 7.9 illustrates the proportion of radiation these stars emit that is energetic might

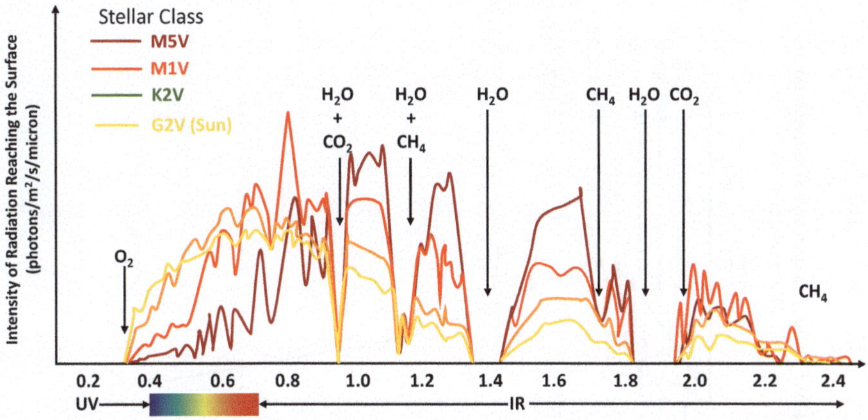

Fig. 7.9 The relative intensity of radiation reaching the surface of an Earth-like planet in the stellar habitable zone of an orange (K2V) dwarf and two different red dwarfs (M1V—hotter; M5V-cooler). Although the red dwarfs are dimmer in terms of visible radiation, they emit a lot of radiation at near-infrared wavelengths (700–800 nm). This has some peculiar consequences, as much of this radiation is directly absorbed by water vapor and other gases in the atmosphere. At these wavelengths, little or no radiation if available to power photosynthesis. This figure is modified and simplified from Fig. 2, available at http://www.giss.nasa.gov/research/briefs/kiang_01/ (Nancy Kiang, 2007)

be a limiting factor. However, as we've seen, even with 10% of the intensity of these shorter wavelength radiation than sunlight red dwarfs will undoubtedly produce enough to allow oxygenic photosynthesis (Fig. 7.10). Scattering of radiation by the atmosphere and the lack of variation in the angle of illumination are other issues life will simply have to work around.

However, like a perpetual sunset, as one approaches the terminator not only will there be less radiation available, proportionately less of this will be from the energetic blue end of the spectrum. Depending on the composition of the atmosphere and the evolutionary track of life on the planet, oxygenic photosynthesis may become limited as one tracks away from the SSP towards the terminator. This is speculative, but again, not unreasonable. If true, this could affect how long it takes the planet to synthesize an oxygen-rich atmosphere, once such processes develop through natural selection.

PAR and the Red Edge

The principal source of oxygen in Earth's atmosphere is clearly oxygenic photosynthesis. However, the rate of photosynthesis in any plant varies significantly with wavelength. Simple experimentation can be used to compare the rate of photosynthesis with the irradiating wavelengths of light a plant or a bacterium

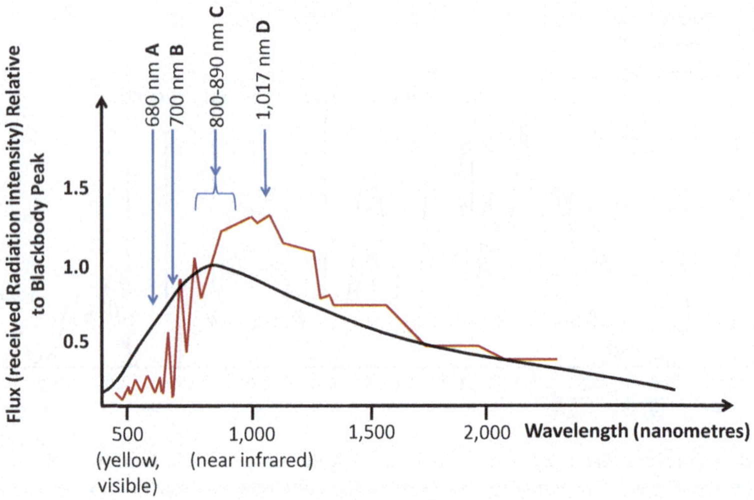

Fig. 7.10 The effective radiation released by an M8V 0.1 solar mass red dwarf compared to its idealized, or blackbody, curve. (**a**) The peak absorption of light by Photosystem I (680 nm); (**b**) Photosystem 2 (700 nm) of plant chloroplasts, and the bacterial photosynthetic pigments, (**c**) bacteriochlorophyll a and, (**d**) bacteriochlorophyll b. The latter two absorb maximally in the infrared portion of the spectrum, rather than the visible wavelengths of light. As might be expected on evolutionary grounds, neither absorption peak is affected by atmospheric absorption. The jagged red line illustrates the wavelengths as would reach the surface—simplified from Fig. 7.9—for an "average" red dwarf compared with the blackbody curve (black line)

receives. Monochromatic sources (usually light emitting diodes—LEDs) may be used to deliver light in particular ranges of wavelength. The rate of photosynthesis can readily be measured by quantifying the amount of oxygen the organism releases in a given time when illuminated. By comparing wavelength with oxygen release, an action spectrum is produced. In essence this is a simple means of quantifying the effect of light on a photosynthetic organism.

Most higher plants show a pattern where the rate of photosynthesis is highest at the blue and red ends of the photosynthetic spectrum and lowest in the green portion. Given that plants and photosynthesizing cyanobacteria are green, this is to be expected. Most of the green light is reflected. Without absorption, the plant cannot capture the Sun's energy. There is a rapid fall off in absorption in the infrared portion of the spectrum, giving rise to the so-called "red edge." Sounding like some form of troublesome political persuasion, the red edge may be used to distinguish life-bearing habitable worlds in future. By analyzing variation in light reflected from planetary surfaces across the red and infrared portions of the electromagnetic spectrum, astronomers may be able to observe the signatures of Earth-like biospheres. However, bear in mind many other pigments are used in photosynthesis, across the full swathe of life on Earth.

Shiladitya DasSarma (University of Maryland School of Medicine) and Edward W. Schwieterman (University of California, Riverside) propose one example. The photosynthetic pigment chlorophyll, though derived from simple molecules that

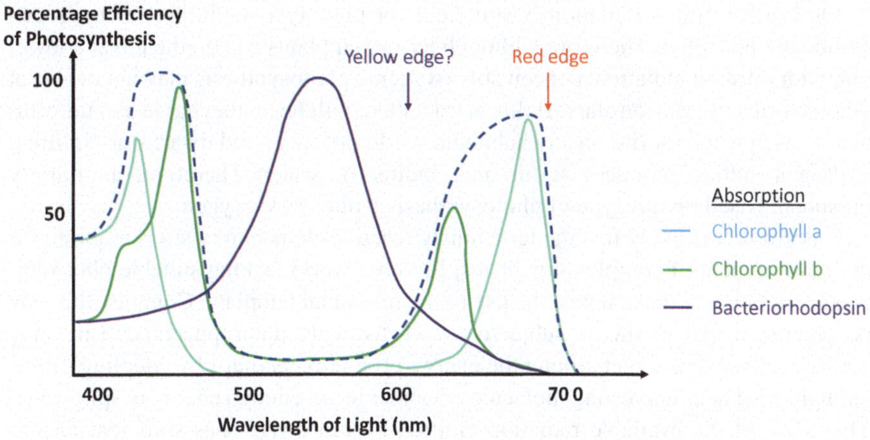

Fig. 7.11 The variation in the rate of photosynthesis with wavelength of green plants on Earth. This "Action Spectrum" (dashed line) relates wavelength to the rate of photosynthesis. The maximum absorption of useful energy occurs at the red and blue ends of the spectrum. Plants are green because little light is absorbed at green wavelengths. This produces the "red edge," beyond which no reactions occur. Bacteriorhodopsin has a very distinct absorption and would give a diametrically opposed pattern of photosynthesis

were likely abundant on Earth, is a lot more complex than molecules such as vitamin A—the core of a pigment called rhodopsin. This pigment is purple and is found in a variety of photosynthetic organisms, as well as the mammalian eye. If rhodopsin was abundant on the early Earth, the planet might have been distinctly purple rather than green. The red edge becomes the yellow one. Figure 7.11 illustrates both the red and yellow edges.

Although the Sun emits most of its radiation in the core of the visible spectrum (principally blue-yellow), red dwarfs, as is obvious, emit the bulk of their radiation in the red-infrared portions of the spectrum. Plants (or any photosynthetic life) will need to adjust accordingly to absorb the maximum amount of the available energy. However, the good news is that if blue light is essential for oxygenic photosynthesis, even very cool red dwarfs, such as TRAPPIST-1, emit around 10% of their energy at the blue end of the visible spectrum.

On Earth many species of bacteria absorb light at wavelengths extending well into the infrared. For example, the exquisitely named purple non-sulfur bacterium, *Rhodospirillum rubrum,* and the equally challenging tongue-twister *Rhodopseudomonas capsulate,* have pigments that absorb infrared radiation at 870 nm. Other bacteria contain the molecule bacteriochlorophyll b, which absorbs light maximally at 960 nm—well into the infrared. However, these organisms do not split water to release the vital hydrogen gas needed to drive the production of glucose. Instead they use the lower energy of infrared radiation to split molecules such as hydrogen sulfide (H_2S). Importantly, the radiation emitted by red dwarf stars is primarily in the infrared portion of the spectrum—exactly that which is needed for non-oxygenic photosynthesis.

On Earth infrared radiation is sufficient for photolysis (splitting) of hydrogen sulfide but not water. Therefore, although terrestrial plants can do the job at a lower rate with infrared radiation, conceivably oxygenic photosynthesis may not evolve at all on worlds of crimson stars. If this is true, there will be an inevitable and unfortunate consequence for life on any habitable world orbiting a red dwarf star. Splitting hydrogen sulfide produces sulfur and, indirectly, water. Therefore, an Earthly biosphere based on this type of photosynthesis would lack oxygen.

Since it seems likely that the ferociously reactive element oxygen is required for the development of complex organisms, life on a world lacking suitable photosynthesis might never make it past the use of the microbial template. Complex life may be precluded. All of this is subjective, and distinctly anthropocentric. Life may evolve mechanisms to get around the energy problem—perhaps by coupling different light- and heat-harvesting molecules together to get enough energy to split water. That 10% of the available radiation emitted by red dwarfs has sufficient energy suggests that "coping mechanisms" are evolvable. However, be aware that given the abundance of lower energy radiation on planets such as Proxima b might mean that there is never the selective pressure to drive the evolution of oxygenic photosynthesis. Observations of planets such as Proxima b will be critical in evaluating this kind of argument—and this will be possible in the near future. If abundant oxygen is present and a red edge can be resolved, oxygenic photosynthesis will be firmly on the table.

Tidal Locking and Angle of Incidence and Their Impact on Photosynthesis

A fixed source of light in the sky imposes unique geographical constraints on vegetation cover. Plants located under the point of greatest illumination (the sub-stellar point) will have to compete for the radiation coming from directly overhead. Shading of leaf tissue by any overlapping leaves will limit, or prevent, photosynthesis occurring in those tissues, *if* it is driven by visible light.

Light and shade are permanent features, never moving from hour to hour. Thus if plants are to compete for light, they will have to physically move around shading obstacles. This form of *shade-avoidance* could lead to some intriguing evolutionary arms races as plants battle for supremacy in their command of light. As the distance from the sub-stellar point (SSP) increases, illumination will come from progressively shallower angles, leaving land plants the job of angling their stems and leaves so that the leaf surface is positioned at ninety degrees to the incident light. Once again, shading by any geographical feature, as well as any intervening plant or animal, will prevent photosynthesis in these tissues. The implications of this are considered in Chap. 9.

If photosynthesis is driven by infrared radiation (technically, this is not *photo*-synthesis) then every surface, irrespective of illumination by the star, can be a source of energy—if it is in thermal equilibrium with its surroundings (effectively, at the

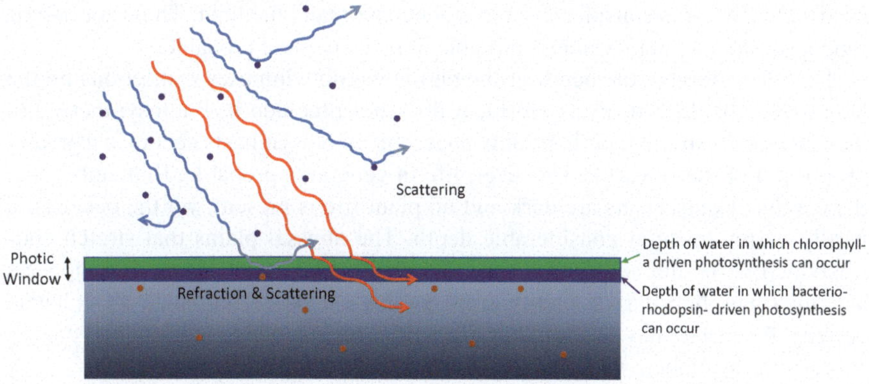

Fig. 7.12 The scattering of light by particles in air and water. Blue light is scattered far more effectively than red light as the wavelength of blue light most resembles the size of the particles. Conversely, waves of red light essentially weave around these. The effect explains why our sky is blue and sunsets red. In a column of water light is also effectively refracted so it quickly becomes dark as you descend. This limits the depth at which photosynthesis can occur (the photic window)

same temperature). Consequently, any plant on any world orbiting a red dwarf is thoroughly bathed in infrared radiation. Shading from the star becomes an irrelevance. Therefore, in principle, if plants are adapted to absorb sufficient infrared radiation they could survive even when shaded from direct sunlight.

In water, the effect of scattering and refraction of visible light is even more pronounced than it is in air (Fig. 7.12). To survive aquatic plants and photosynthesizing single-celled organisms will need to grow nearer the top of the water column than those organisms on Earth. Once the rate of photosynthetic production falls to 1%, that found at the surface of our planet, production of food molecules by photosynthesis matches that consumed by respiration. This is known as the compensation point.

At this point any plant or microbe attempting to make a living from its sunlight could not make enough food to sustain itself. The depth above which this transition is reached is called photic window. On Earth, this is typically up to 200 m in depth in open ocean. However, nearer coastal areas, where runoff from adjacent land brings sediment or nutrients for microbial growth, this can be less than 15 m in depth. Unfortunately, light in the green and yellow parts of the spectrum are the most penetrating, while useful red and blue light are effectively blocked by this murk. Infrared radiation is even less penetrating in water. Therefore, microorganisms adapted to the sole use of this type of radiation are likely to be restricted to the very top of the water column, or to adjacent land.

This gives bacteriorhodopsin a distinct advantage in driving photosynthesis on planets where the water may be turbid or where the proportion of blue light is limiting. Although less abundant from red dwarf stars compared to K- and G-class stars, yellow and green remain relatively common wavelengths in the light emitted by red dwarfs. Indeed, it is considerably more abundant than blue light. Thus, not only is rhodopsin-driven photosynthesis possible, it is likely more probable.

Therefore, though the depth of the photic window imposes restrictions on the habitability of the red dwarf world, it does not preclude it. Photosynthetic life may be more restricted; perhaps only anaerobic photosynthesis occurs in a greater proportion of these worlds. However, life in general is probable. Remember that the depths of our oceans are dark and no plant life is present, yet the oceans are hardly barren, even at considerable depth. The abyssal plains that stretch endlessly across the floors of our oceans are populated by scavengers, decomposers and microbes that derive energy from geological processes, such as volcanic activity. An ocean may be dark, but life will occur if suitable nutrient and energy sources are available.

Surprises from Terrestrial Biodiversity

Investigations on terrestrial biodiversity produced a real surprise in 2018. If you examine the numbers of aquatic species there are then the unremarkable result is that there are far more species (that is to say, there is a higher species richness) in the tropical reef areas than in the mid-latitudes and the polar regions. Indeed, there is a fairly simple relationship between latitude and species richness. The greater the latitude the lower the species richness (the fewer *types* of species). This seems obvious, as the reef areas have abundant food, ample light and they are warm. So far so good. The simple expectation is then that new species will emerge (speciate) most frequently in these areas, as this is where the vast majority of species are found. Higher speciation (the formation of new species through natural selection) is then expected in tropical areas.

However, it turns out that this is untrue. Work by Daniel L. Rabosky (University of Michigan) and co-workers examined the relationship between different species using comparisons of their DNA. The phylogenetic analysis of ray-finned fish found that the fastest rate of speciation was nearest the poles, where there are relatively few species and where the conditions might be expected to be harshest. The question is why might this relationship hold?

There are a few possible answers. Firstly, the conditions may be very competitive with regular loss of species caused by small changes to their local environment. Many of the species found in polar waters are restricted to these areas. Known as species endemism, species found in Arctic and Antarctic waters tend to be localized to particular niches, while those in the tropics show far lower levels of endemism.

Interestingly, the Mediterranean Sea bucks the trend, with both low speciation and low endemism. However, this may reflect the relatively recent flooding of the basin 6 million years ago and/or the relative isolation of the Mediterranean basin from neighboring waters.

A simple idea is that species that are adapted to cold, dark conditions, such as those found on the abyssal plain, might naturally have higher levels of speciation than those living in more clement, tropical climes. However, this doesn't match the data, and there is no "depth-dependence" on the rate of speciation. Indeed, another basic tenant of evolutionary theory—that hotter conditions will drive faster metabolism and hence higher rates of evolutionary change—simply doesn't hold either. Interestingly, this might appear to contradict the commonly held idea discussed earlier that the Cambrian explosion was a metabolism-driven evolutionary adventure. However, this is not quite as clear cut as it seems. For one thing, cold water holds more oxygen than warm water, hence one possibility that despite metabolism generally scaling with temperature in poikilothermic ("cold-blooded") species, the greater availability of oxygen may facilitate higher levels of speciation in cold-dwelling species than warm. Hence, the relationship between metabolism and speciation may appear to be violated when it isn't.

Alternatively, one might conclude that the rapid, seasonal variation in climate may drive speciation, and with it seasonal changes to sea ice and coastal ice. As such, temporal variation may be critical to the development of new species.

Although these are all nice-sounding ideas, the likely solution to Rabosky's dilemma may come from another key study that was published in 2017. Hudson T. Pinheiro (California Academy of Sciences) and co-workers investigated marine speciation in tropical islands. The setting was the shallow waters off of the Trinidad island chain in the central Atlantic. These islands have repeatedly risen and sunk in the Atlantic, as the ice sheets waxed and waned over the last several tens of thousands of years. The team concluded that where species could island hop, the rate of speciation (the formation of new species) was low. Where the islands were spaced further apart, new species tended to emerge at a higher rate.

Presumably, where island hopping was possible, niches were filled with invasive species before new species could evolve to fill them. Indeed, one can see a simple relationship between these two studies. Where there was limited mobility of species (high endemism) in this study, speciation rates were higher. Therefore, we might conclude that endemism is critical to speciation, and the relationship seen in Rabosky's study is simply measuring the effect of invasiveness of species. Where waters are cold, immigration of species from one niche to another is limited, and the evolutionary rate is higher than it is in warm, tropical waters. In this scenario metabolism and other factors may be (largely) superfluous. That is not to say that they are not important for the survival and proliferation of species, merely that they are less important in speciation (the formation of new ones).

Remember, also, that the polar regions were glaciated until recently. Higher rates of speciation might reflect the opening up of new niches after the end of the last ice age.

Implications of Ecological Ideas

A fundamental shift has begun in considering what it is for a planet to be habitable. In part that has come from research into microbial life on Earth, which has revealed a plethora of extremophiles—organisms that are adapted to extreme environments that are often found away from the surface biosphere. This has allowed an expansion of "habitability" to include the sub-surface oceans of icy worlds or (perhaps) the cooler, higher-level, acidic clouds of Venus. It has also come from the gradual incorporation of evolutionary theories, which have begun to inform us of what kinds of life have been possible. In this context, we have brought together climate models and some ecological ideas to dissect the habitability and biological potential of planets. In this context the concept of niches is fundamental. Here, as described earlier, we can consider whether particular parts of a habitable planet's surface will be capable of supporting complex biospheres.

Life develops complexity in the context of the environment in which it develops (described above). However, with the use of the hypervolume concept of niches, one is then able to go further and compare terrestrial biology and its landscape with those that are likely on worlds we can only imagine. Astrophysical climate models of planets incorporate a combination of terrestrial observation of climate (and weather) and sophisticated computer models of atmospheres. Although these are for aquaplanets, or idealized swamp worlds with a simple, global moisture source, they can reproduce many of the features of climate and are directly applicable to aquaplanets.

The models presented in a 2018 paper (Evolutionary Exobiology II), could be tweaked to include some of the effects of mountains and other aspects of landmasses to provide reasonable first models of planetary habitability at the level of the niches that would be available for organisms. This is entirely feasible without a priori knowledge of a planet. Taking this further, Rodrick Wallace and this author are now looking at describing habitability for many worlds in a more rigorous but generally applicable mathematical context. This work is be published in 2019.

In a further, but connected, approach, we've considered the application of criteria that were first used to evaluate the tropical ecosystems in the 1940s. Here, Holdridge was able to mathematically define which kinds of vegetation would be found in which areas, in a manner that was dependent on the temperature and precipitation in this locale. The model has been successfully applied in a number of other terrestrial ecosystems and seemed a sensible approach to investigate biological diversity elsewhere. A key aspect of the model employed is the use of precipitation and temperature data. These are readily available in climate models for tidally locked planets. This allows one to bypass the requirement of humidity in calculations of niche amplitude, which we had previously considered. This has now been published in The International Journal of Astrobiology; and applies the Holdridge Life Zone Model (HLZM) to the Met Office climate model for Proxima b. The outcomes define the habitable region and clearly illustrate how limited life may be on tidally locked planets.

In a different but equally compelling light comes the mathematical tour de force of Claudio Maccone (SETI and Istituto Nazionale di Astrofisica, INAF, Italy). Claudio has taken the famous Drake equation and mathematically dissected it, yielding some surprising outcomes. Notably is his consideration of the lifetime of civilizations. Defining each variable of the Drake equation produces a lovely match for the lifetime of species—and for any complex, intelligent species that arises through natural selection.

Although Maccone's math is often fairly convoluted, the outcome appears to define both the famous "Red Queen" hypothesis of evolutionary theory and more prosaically for us, how long each human empire can exist. The key concept is the consideration of how quickly the species in question rises to its peak, in terms of numbers (or economics, when it comes to prosaic notions of imperial dominance). The rise time and peak then determine how long the organism (or empire) will take to decline to its extinction. The model accurately describes the British, Roman and Spanish empires. Interestingly, it also puts the end of the U. S. global superpower a generation away, in the year 2050. Most importantly, the peak of U. S. dominance was several decades ago, if Claudio Maccone's model is correct…

The Red Queen hypothesis is named after an explanation made by the Red Queen to Alice in Lewis Carroll's *Through the Looking-Glass* in her explanation of life in Looking Glass Land. Organisms have to keep moving to stay in one spot. The principle is straightforward enough; organisms compete with others and have to survive environmental change in order to stay put—i.e., survive. Why can't you just stay still, bathing in the glory of your success? The likeliest answer comes down to competition and parasitism. As organisms increase in number they will interact with an increasing number of other species. Moreover, new species can evolve in energy-rich environments to exploit them, either as a food source or a habitat for parasitism. When the first humans emerged from their common ancestors relatively few parasites and pathogens could have survived by reproducing within them.

However, increase the population size and suddenly a pathogen or parasite can get away (in its own evolutionary terms) with using this new organism to support its own replication. Take Ebola or HIV as recent examples of emerging pathogens. The first outbreak of Ebola was in 1976 and killed a limited number of people. The most recent outbreaks in West Africa killed hundreds and spread across geographical borders. HIV likely crossed the species barrier at some point in the 19th century. However, it wasn't until the middle of the 20th century that it really took off in terms of the numbers of people it infected.

Unfortunately, for humans (and some other species), we have two other Achilles heels: mass migration and (uniquely to humans) mass trade. These greatly facilitate the spread and further evolution of pathogens. Ebola was principally a pathogen of blood tears and occasionally sweat, killing almost all of those infected.

However, in the recent largest outbreak over 11,300 people were reported killed, and some of these were infected through sexual contact—a new mode of infection. The outbreak also saw the death rate fall (to about 70% of those infected, from nearly 100% of those infected in earlier outbreaks). However, chronic infection emerged in some of those infected. This is a natural consequence of evolutionary processes associated with the survival and propagation of the virus in large groups. Those viruses (or pathogens in general) that have very high mortality rates are killed with their hosts before they are transmitted to new hosts. Slower and less lethal strains can propagate successfully so mortality rates fall, even as the number of infected people increases. Typically, with many emergent pathogens lethality falls over time, but persistence increases. Ebola is now clearly not an exception to this general rule.

In this context, human civilizations are bound to fail in two generic ways. Firstly, it fails through the collapse of dominant powers, whether through recession, conflict, internal (intraspecific) competition or through its impact on its immediate environment. Secondly, the human race, with its 7.5 billion cohorts, is increasingly vulnerable to a burden of pathogens and parasitic life forms. We've pretty much done away with competing species, but we're also making a strong headway in eliminating other species that might be beneficial. There is a way around this: long term planning and innovation. Clearly this is something other species cannot do. Isolating oneself from novel ideas, new technologies or; conversely, welding oneself to older and outcompeted models of governance (at whatever societal level) leaves us (as a species, or as a national power) vulnerable to terminal competition. Bear that principle in mind at all times. Organizations, from high schools to national governments, can avoid the fate of the Red Queen by reorganizing and adopting new strategies and technologies rather than simply shutting oneself off from competition so that atrophy sets in. Although this is a uniquely human foible, isolationism allows internal machinery to become inefficient without competition. This results in economic costs to rise and complacency to become the norm. Meanwhile, your competitors move on around you. One returns to Maccone's global forecasts.

Taken more broadly, all species experience another source of change or conflict—their environment. Whether it's a changing climate or some sort of natural calamity, organisms must adapt at the species level to an environment that is constantly changing. Those that do not will also become extinct. Taken together the Red Queen's observations stand. Keep moving or lose the race.

Conclusions

This chapter is perhaps the most crucial in the book—and a worthy addition to the original chapters of *Under a Crimson Sun* (Springer, 2013). Here, we integrate the hard physics and chemistry of the first chapters to consider how life "fits in" to the worlds in which it inhabits. It is here, perhaps more than in any other part of the book, we consider the limitations of our knowledge. Applying ecological concepts tightens the rope and allows us to consider sensibly what kind of life may be found on worlds other than ours. However, as the SPH modelling inadvertently illustrates, biologists are still not talking clearly to astrophysicists and climate scientists (nor, it seems, the latter to one another, on occasion).

A planet is not an amorphous blob. Atmospheric circulation is determined by planet size, its spin and the amount of energy it receives and radiates. This dictates which parts of the surface are habitable for which kinds of species. Tidally locked planets are rather restrictive places, as this chapter and Chap. 5 illustrate. However, all is not lost, and one hopes that this chapter makes the concept of habitability altogether more sensible.

Now that we can see how life fits its worlds, we can begin to explore the kinds of red dwarf worlds we know exist. Chapter 8 reconsiders those that were the focus of the first edition of this book, while Chap. 9 takes us to the new worlds that have been discovered since 2015. One of these seems distinctly home-like, despite its red dwarf sun.

References

Barthlott, W., Mutke, J., Rafiqpoor, M. D., Kier, G., & Kreft, H. (2005). Global centres of vascular plant diversity. *Nova Acta Leopoldina, 92*, 61–83.

Blair Hedges, S., Blair, J. E., Venturi, M. L., & Shoe, J. L. (2004). A molecular timescale of eukaryote evolution and the rise of complex multicellular life. *BMC Evolutionary Biology, 4*, 1–9. http://www.biomedcentral.com/1471-2148/4/2.

Blonder, B. (2017). Hypervolume concepts in niche- and trait-based ecology. *Ecography, 40*, 001–013. https://doi.org/10.1111/ecog.03187.

Brin, G. D. (1983). The 'great silence': The controversy concerning extraterrestrial intelligent life. *Quarterly Journal of the Royal Astronomical Society, 24*, 283–309.

Buldyrev, S. V., Parshani, R., Paul, G., Stanley, H. E., & Havlin, S. (2010). Catastrophic cascade of failures in interdependent networks. *Nature, 464*, 1025–1028.

Calcott, B., & Sterelny, K. (2011). *The major transitions in evolution revisited*. Cambridge: Massachusetts Institute of Technology. ISBN: 978-0-262-01524-0.

Cardenas, R. (2014). Quantitative theory of habitability: From ecology to astrobiology. https://www.researchgate.net/publication/305220533.

Cardenas, R. (2017). On the quantification of habitability. In *Conference paper: II International Conference on Biogeosciences*.

Cardenas, R., Perez, N., Martinez-Frias, J., & Martin, O. (2015). On the habitability of aquaplanets. *Challenges, 5*, 284–293. https://doi.org/10.3390/challe5020284.

Carter, B. (1993). The anthropic selection principle and the ultra-darwinian synthesis. In F. Bertola & U. Curi (Eds.), *The anthropic principle* (pp. 33–63). Cambridge: Cambridge University Press.

Cirkovic, M. M. (2018). Woodpeckers and diamonds: Some aspects of evolutionary convergence in astrobiology. *Astrobiology, 18*(5), 491–502. https://doi.org/10.1089/ast.2017.1741.

Danovaro, R., Dell'Anno, A., Pusceddu, A., Gambi, C., Heiner, I., & Kristensen, R. M. (2010). The first metazoa living in permanently anoxic conditions. *BMC Biology, 8*, 30–40. http://www.biomedcentral.com/1741-7007/8/30.

Davison, J., Moora, M., Öpik, M., Ainsaar, L., Ducousso, M., Hiiesalu, I., Jairus, T., Johnson, N., Jourand, P., Kalamees, R., Koorem, K., Meyer, J.-Y., Püssa, K., Reier, Ü., Pärtel, M., Semchenko, M., Traveset, A., Vasar, M., & Zobel, M. (2018). Microbial island biogeography: isolation shapes the life history characteristics but not diversity of root-symbiotic fungal communities. *The ISME Journal, 12*, 2211–2224. https://doi.org/10.1038/s41396-018-0196-8.

de, L. C., Madec, G., Roquet, F., Holmes, R. M., & McDougall, T. J. (2017). Abyssal ocean overturning shaped by seafloor distribution. *Nature, 551*, 181–186. https://doi.org/10.1038/nature24472.

Gaston, K. J. (2000). Global patterns in biodiversity. *Nature, 405*, 220–227.

Gatti, R. C. (2016). The fractal nature of the latitudinal biodiversity gradient. *Biologia, 71/6*, 669–672. https://doi.org/10.1515/biolog-2016-0077.

Gillman, L. N., Wright, S. D., Cusens, J., McBride, P. D., Malhi, Y., & Whittaker, R. J. (2015). Latitude, productivity and species richness. *Global Ecology and Biogeography, 24*, 107–117. https://doi.org/10.1111/geb.12245.

Hanson, R. (1998) Must early life be easy? The rhythm of major evolutionary transitions. http://mason.gmu.edu/~rhanson/hardstep.pdf.

Holdridge, L. R. (1947). Determination of world plant formations from simple climatic data. *Science, 105*, 367–368.

Holdridge, L. R. (1967). *Life Zone Ecology. Revised Edition* (p. 206). San José: Tropical Science Center.

Ifo, S. A., Moutsambote, J.-M., Koubouana, F., Yoka, J., Ndzai, S. F., Nucia, L., Bouetou-Kadilamio, O., Mampouya, H., Jourdain, C., Bocko, M. Y. A. B., Mouanga-Sokath, M. M. D., Odende, R., Mondzali, L. R., Emmanue, Y., Wenina, M., Jenkins, C. N., Pimmb, S. L., & Joppac, L. N. (2013). Global patterns of terrestrial vertebrate diversity and conservation. *PNAS, 110*(28), E2602–E2610. https://doi.org/10.1073/pnas.1302251110.

Jenkins, C. N., Pimm, S. L., & Joppa, L. N. (2013). Global patterns of terrestrial vertebrate diversity and conservation. *PNAS*, 2602–2610. https://doi.org/10.1073/pnas.1302251110.

Kay, R. F., Madden, R. H., Van Schaik, C., & Higdon, D. (1997). Primate species richness is determined by plant productivity: implications for conservation. *Proceedings of the National Academy of Sciences of the United States of America, 94*(24), 13,023–13,027. https://doi.org/10.1073/pnas.94.24.13023.

Kiera, G., Krefta, H., Leeb, T. M., Jetzb, W., Ibischc, P. L., Nowickic, C., Mutkea, J., & Barthlotta, W. (2009). A global assessment of endemism and species richness across island and mainland regions. *PNAS, 106*(23), 9322–9327. https://doi.org/10.1073/pnas.0810306106.

Koonin, E. V. (2010). The origin and early evolution of eukaryotes in the light of phylogenomics. *Genome Biology, 11*, 209. http://genomebiology.biomedcentral.com/articles/10.1186/gb-2010-11-5-209.

Lebauer, D. S., & Treseder, K. K. (2008). Nitrogen limitation of net primary productivity in terrestrial ecosystems is globally distributed. *Ecology, 89*(2), 371–379.

Lenton, T. M., Boyle, R. A., Poulton, S. W., Shields-Zhou, G. A., & Butterfield, N. J. (2014). Co-evolution of eukaryotes and ocean oxygenation in the Neoproterozoic era. *Nature Geoscience, 7*(4), 257–265. ISSN: 1752-0894.

Lingam, M. & Loeb, A. (2018). Dependence of biological activity on the surface water fraction of planets. https://arxiv.org/pdf/1809.09118.pdf.

Lingam, M. & Loeb, A. (2018). Is extraterrestrial life suppressed on subsurface ocean worlds due to the paucity of bioessential elements? https://arxiv.org/pdf/1806.00018.pdf.

Loeb, A., Batista, R. A., & Sloan, D. (2016). Relative likelihood for life as a function of cosmic time. *Journal of Cosmology and Astroparticle Physics, 2016*, 040. https://doi.org/10.1088/1475-7516/2016/08/040. https://arxiv.org/pdf/1606.08448v2.pdf.

Maccone, C. (2010). The statistical Drake equation. *Acta Astronomica, 67*, 1366–1383. https://doi.org/10.1016/j.actaastro.2010.05.003.

Maccone, C. (2014). Evolution and mass extinctions as lognormal stochastic processes. *International Journal of Astrobiology, 13*(4), 290–309. https://doi.org/10.1017/S147355041400010X.

Mora, C., Tittensor, D. P., Adl, S., Simpson, A. G. B., & Worm, B. (2011). How many species are there on earth and in the ocean? *PLoS Biology, 9*(8), e1001127. https://doi.org/10.1371/journal.pbio.1001127.

Ouissika, B. C., & Joel, L. J. (2015). Tree species diversity, richness, and similarity in intact and degraded forest in the tropical rainforest of the Congo basin: Case of the forest of Likouala in the Republic of Congo. *International Journal of Forestry Research, 2016*, 7593681. https://doi.org/10.1155/2016/7593681.

Partin, C. A., Bekker, A., Planavsky, N. J., Scott, C. T., Gill, B. C., Li, C., Podkovyrov, V., Maslov, A., Konhauser, K. O., Lalonde, S. V., Love, G. D., Poulton, S. W., & Lyons, T. W. (2013). Large-scale fluctuations in Precambrian atmospheric and oceanic oxygen levels from the record of U in shales. *Earth and Planetary Science Letters, 369–370*, 284–293.

Penn, J., & Vallis, G. K. (2012). The thermal phase curve offset on tidally- and non-tidally-locked exoplanets: A shallow water model. *The Astrophysical Journal, 842*(2). https://arxiv.org/pdf/1704.06813.pdf.

Pinheiro, H. T., Bernardi, G., Simon, T., Joyeux, J.-C., Macieira, R. M., Gasparini, J. L., Rocha, C., & Rocha, L. A. (2017). Island biogeography of marine organisms. *Nature, 549*, 82–85. https://doi.org/10.1038/nature23680.

Price, T. D., Hooper, D. M., Buchanan, C. D., Johansson, U. S., Tietze, D. T., Alström, P., Olsson, U., Ghosh-Harihar, M., Ishtiaq, F., Gupta, S. K., Martens, J., Harr, B., Singh, P., & Mohan, D. (2014). Niche filling slows the diversification of Himalayan songbirds. *Nature, 509*, 222–225. https://doi.org/10.1038/nature13272.

Proença, V., Martin, L. J., Pereira, H. M., Fernandez, M., McRae, L., Belnap, J., Böhm, M., Brummitt, N., García-Moreno, J., Gregory, R. D., Honrado, J. P., Jürgens, N., Opige, M., Schmeller, D. S., Tiago, P., & van Swaay, C. A. M. (2017). Global biodiversity monitoring: From data sources to essential biodiversity variables. *Biological Conservation, 213*, 256–263. https://doi.org/10.1016/j.biocon.2016.07.014.

Rabosky, D. L., & Matute, D. R. (2013). Macroevolutionary speciation rates are decoupled from the evolution of intrinsic reproductive isolation in Drosophila and birds. *PNAS, 110*(38), 15,354–15,359. www.pnas.org/cgi/doi/10.1073/pnas.1305529110.

Rabosky, D. L., Chang, J., Title, P. O., Cowman, P. F., Sallan, L., Friedman, M., Kaschner, K., Garilao, C., Near, T. J., Coll, M., & Alfaro, M. E. (2018). An inverse latitudinal gradient in speciation rate for marine fishes. *Nature, 559*, 392–395. https://doi.org/10.1038/s41586-018-0273-1.

Raymond, S. N., Scalo, J., & Meadows, V. S. (2007). A decreased probability of habitable planet formation around low-mass stars. *The Astrophysical Journal, 669*, 606–614.

Ricklefs, R. E., & Heb, F. (2016). Region effects influence local tree species diversity. *PNAS, 113*(3), 674–679. https://doi.org/10.1073/pnas.1523683113.

Rosenzweig, M. L. (1968). Net primary productivity of terrestrial communities: prediction from climatological data. *The American Naturalist, 102*(923), 67–74. https://doi.org/10.1086/282523.

Schirrmeistera, B. E., de Vosb, J. M., Antonellic, A., & Bagheria, H. C. (2013). Evolution of multicellularity coincided with increased diversification of cyanobacteria and the Great Oxidation Event. *PNAS, 110*(5), 1791–1796.

Schopf, J. W. (1995). Disparate rates, differing fates: Tempo and mode of evolution changed from the precambrian to the phanerozoic. In W. M. Fitch & F. J. Ayala (Eds.), *Tempo and mode in evolution, genetics and paleontology 50 years after simpson* (pp. 41–61). Washington D.C.: National Academy Press.

Shannon, C. E. (1948). A mathematical theory of communication. *Bell System Technical Journal, 27*(3), 379–423. https://doi.org/10.1002/j.1538-7305.1948.tb01338.x. http://worrydream.com/refs/Shannon%20-%20A%20Mathematical%20Theory%20of%20Communication.pdf.

Shannon, C. E. (1948). A mathematical theory of communication. *Bell System Technical Journal, 27*(4), 623–666. https://doi.org/10.1002/j.1538-7305.1948.tb00917.x.

Šímová, I., Violle, C., Kraft, N. J. B., Storch, D., Svenning, J. C., Boyle, B., Donoghue, J. C., II, Jørgensen, P., McGill, B. J., Morueta-Holme, N., Piel, W. H., Peet, R. K., Regetz, J., Schildhauer, M., Spencer, N., Thiers, B., Wiser, S., & Enquist, B. J. (2015). Shifts in trait means and variances in North American tree assemblages: species richness patterns are loosely related to the functional space. *Ecography, 38*, 649–658. https://doi.org/10.1111/ecog.00867.

Sperling, E. A., Wolock, C. J., Morgan, A. S., Gill, B. C., Kunzmann, M., Halverson, G. P., Macdonald, F. A., Knoll, A. H., & Johnston, D. T. (2015). Statistical analysis of iron geochemical data suggests limited late Proterozoic oxygenation. *Nature, 523*, 451–454. https://doi.org/10.1038/nature14589.

Spiegel, D. S., & Edwin, L. T. (2011). Life might be rare despite its early emergence on Earth: A Bayesian analysis of the probability of abiogenesis. *PNAS*. https://doi.org/10.1073/pnas.0709640104. https://arxiv.org/pdf/1107.3835v1.pdf.

Stevenson, D. S. (2018). Evolutionary Exobiology II: investigating biological potential of synchronously-rotating worlds. *International Journal of Astrobiology*, 1–15. https://doi.org/10.1017/S1473550418000241.

Stevenson, D. S. (2018). Niche amplitude, tidal-locking and Fermi's Paradox. *International Journal of Astrobiology*, 1–7. https://doi.org/10.1017/S1473550418000253.

Stevenson, D. S. (2019). Planetary mass, vegetation height and climate. *International Journal of Astrobiology*, 1–6. https://doi.org/10.1017/1473550418000484.

Stevenson, D. S. (2019). Phytoclimatic mapping of exoplanets. *International Journal of Astrobiology*, 1–10. https://doi.org/10.1017/S1473550419000181.

Stevenson, D. S., & Large, S. (2017). Evolutionary exobiology: Towards the qualitative assessment of biological potential on exoplanets. *International Journal of Astrobiology, 16*, 1–5. https://doi.org/10.1017/S1473550417000349.

Stevenson, D. S. & Wallace, R. (2019). Biogeographical Modeling of Alien Worlds. (manuscript in preparation)

Szathmary, E., & Maynard Smith, J. (1995). The major transitions in evolution. *Nature, 374*, 227–232.

Wallace, R. & Stevenson, D. S. (2019). The Extraterrestrial Niche: Determining the Scope of Life on Other Worlds. (manuscript in preparation)

Wallace, R. (2009). Metabolic constraints on the eukaryotic transition. *Origins of Life and Evolution of the Biosphere, 39*, 165–176. https://doi.org/10.1007/s11084-009-9157-6.

Wallace, R. (2016). The metabolic economics of environmental adaptation. *Ecological Modelling, 322*, 48–53. https://doi.org/10.1016/j.ecolmodel.2015.11.017.

Wallace, R. (2017). The aerobic transition as an economic ratchet. *International Journal of Astrobiology*. https://doi.org/10.1017/S1473550417000325.

Walz, U. (2011). Landscape structure, landscape metrics and biodiversity. *Living Reviews in Landscape Research, 5*(3). http://www.livingreviews.org/lrlr-2011-3.

Watson, R., & Szathmary, E. (2016). How can evolution learn? *Trends in Ecology & Evolution, 31*(2), 147–157. https://doi.org/10.1016/j.tree.2015.11.009.

William Schopf, J. (1999). The oldest fossils and what they mean. In *Major events in the history of life* (pp. 29–63). Princeton: Princeton University Press. ISBN: 0867202688 9780867202687.

Xiong Jian-gang, X. J., Fan, Y., & Jun, L. (1994). The influence of topography on the nonlinear interaction of Rossby waves in the barotropic atmosphere. *Applied Mathematics and Mechanics, 15*(6), 585–594.

Yockey, H. P. (2005). *Information theory, evolution, and the origin of life*. Cambridge: Cambridge University Press. ISBN 0·521·80293-8.

Žliobaitė, I., Fortelius, M., & Stenseth, N. C. (2017). Reconciling taxon senescence with the Red Queen's hypothesis. *Nature, 552*, 92–95. https://doi.org/10.1038/nature24656.

Chapter 8
The Failure of Early Optimism

Introduction

When the first edition of this book was written there were only two plausible candidate systems that might host planets in their stars' habitable zone, Gliese 581 and Gliese 667C. Both systems hosted a number of super-terrans, whose numbers varied often by the month as radial velocity data was interpreted, reinterpreted, discarded, recovered—only for vexing to continue. Obviously, like many of the authors of the papers upon which the book was centered, hyperbole tended to replace considered thought as the possibility of fame and further grants beckoned. The Gliese 581 system was a source of particular ire as rival groups made claims for the existence of planet 581g, which was at the time the best candidate for an Earth-like world—albeit a tidally locked one. This chapter rounds up some of the acrimony and resolution that surrounded the dissection of these star systems and discusses, in the light of the earlier chapters, just how habitable these worlds really were.

Gliese 581: One Dwarf, Three (Likely) Worlds

The System

The red dwarf Gliese 581 has a mass of a little over three-tenths that of the Sun, with a radius somewhat less than one-third that of our star. The star lies 20.22 light years away and has a relatively low proportion of heavy elements compared to our Sun—roughly one-third. This implies that it is old—perhaps as much as 11 billion years. Whatever its true age is, it is likely to exceed 7 billion years, making it older than half the age of the Milky Way Galaxy.

With Gliese 581 weighing in at such a low mass, this M-dwarf releases only 0.2% of the visible radiation of our Sun. However, with a spectral peak in the infrared, the total radiation released is more like 1.3% that of the Sun. The total output

D. S. Stevenson, *Red Dwarfs*, https://doi.org/10.1007/978-3-030-25550-3_8

seems trivial beside our Sun's, but the distinction between visible output and total output is important. More than any other wavelength the abundance of infrared light will determine the temperature on the surface of any planet orbiting such a low mass star. Stellar output is grossly underestimated for both the least and most massive stars. The least massive stars have radiation dominated by the infrared, while the most massive have an output dominated by the ultraviolet. Visible output does clearly vary with mass, but because most telescopic surveys concentrate on visible light, they understate the energy from the stellar extremes. Determination of habitability must, therefore, take into account the full stellar energy budget, not just what we can see with our limited eyes.

What of the planets themselves? Initial determinations from the radial velocity data posited that there were six planets: b through to g. In the initial discoveries, planets b, c, d and e were found using HARPS, based at the European Southern Observatory in Chile. Planet e nuzzled in closest to its parent. At the time of its discovery this world held the title as the smallest identified world outside the Solar System. Orbiting less than 2 million km from the photosphere of Gliese 581, Planet e was determined to be a 1.94–3.10 Earth-mass sizzler (Fig. 8.1).

Moving further out, we encounter a planetary giant: Planet b. This 15–30 Earth-mass planet likely resembles Neptune—but will clearly be far hotter, making it a good match for Gliese 436. Temperatures likely exceed 400 °C, particularly if the atmosphere retains much of the incident radiation through the action of greenhouse

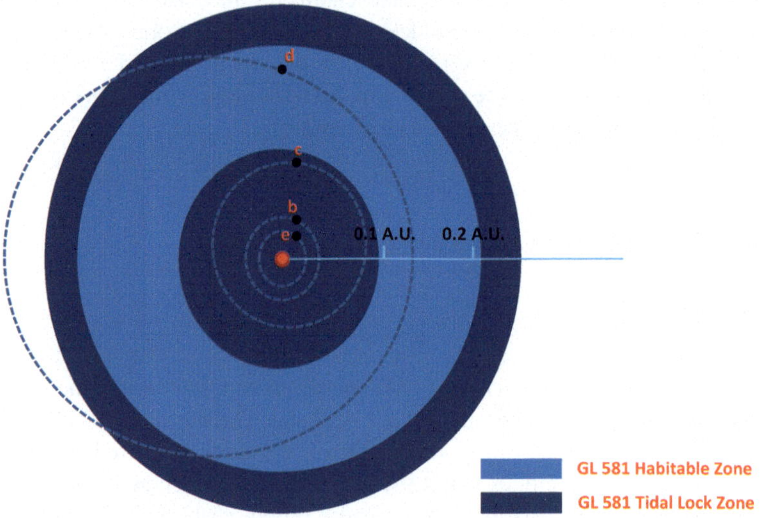

Fig. 8.1 The four-planet solution for the Gliese 581 planetary system, based on the HARPS data. Planets b, c and e are all tidally locked to their parent star, while d may be tidally locked or experience pseudosynchronous rotation, with days lasting almost a full planetary year. A debris disc lies between 25 and 60 AU from the star. Thus, despite the small scale of the planetary system compared to ours, Gliese 581 has a debris disc analogous to our Kuiper Belt

gases such as water, carbon dioxide and methane. Planet b is also tidally locked to its Sun, taking five days to swing around its host. Given the strong heating and probable thick atmosphere, the weather on such a world is likely to be wild, as planetary winds seek to even out the temperature differences from star-lit to night hemispheres. Linking this to our earlier discussions of atmospheric circulation, this world would experience strong super-rotation along its equatorial band and relatively slight differences in temperature across the terminator.

On the inner edge of Gliese 581's habitable zone lies Planet c, orbiting 11 million km from its star. Also confirmed in later analyses, Gliese 581c weighs in at 5–10 times the mass of Earth. Given its location and likely composition, this world might be a diminutive and very warm version of Neptune or something altogether different—a planet dominated by a hot, thick ocean of water and a deep, overwhelming body of cloud. It could be a massive Venus-like world. Planet c orbits its host star every 13 days and, like Planet b, will be tidally locked to the central star.

Ignoring the planet's mass, there remained the possibility that Planet c might be habitable. However, given its location in an analogous position to Venus, any atmosphere with a composition dominated by water vapor will have overheated long ago. If we were to envisage some sort of climate, superimpose the Rhines rotation model that we examined in Chap. 5. With an orbital period of 13 days, there should be relatively strong zonal (east-west) airflow along its equatorial and mid-latitude bands.

Moreover, remember that red dwarfs are far more luminous prior to birth than they are once they reach the main sequence. Therefore, Gliese 581c was likely super-heated at birth, rendering its surface uninhabitable. This contrasts with Venus, which may only have achieved its present greenhouse state in the last billion years or so, as the Sun slowly brightened. With such strong, early heating, only a planet with a very rich supply of volatile gases would have been able to form oceans at all. Perhaps, more likely Gliese 581c was born dry, with any of its original water split by EUV into hydrogen and oxygen. Although this process left Venus with a volcanic atmosphere of carbon dioxide and sulfur dioxide, a planet with a much greater supply of water might do something altogether different.

Rather than lose all of its water, photolysis of a terrestrial ocean or two of this would produce a dense atmosphere of hydrogen and oxygen, with a varying mass of residual water that was protected from photolysis by the other gases. Over time the hydrogen would largely escape to space, at a rate that was dependent on the mass of the planet and the nature of the radiation incident on it. If the rate of loss was high, such hydrodynamic escape might drag much of the oxygen with it, resulting in a strongly impoverished atmosphere. However, this depends on the supply of water from any underlying ocean, or hot ice layer. Any oxygen that didn't escape would react with surface materials rather than leave some oxygen-rich atmosphere.

Planet f and g have been discounted in most astronomical circles, leaving the final—and temperate—world open for discussion.

Gliese 581d

The final planet to be discovered was Planet d. This occupied a far more distant orbit around its star—and is not without its own controversy. The orbit was mapped by Steven Udry and co-workers in 2007, again using HARPS data. Planet d takes nearly 67 days (1,600 hours) to orbit its red dwarf. Estimates of the planet's mass range from 5 to 6 Earth masses, to as high as 10 times that of Earth (Fig. 8.2).

However, like the seemingly banished Gliese 581g, Planet d is not without its detractors. In 2012 a study by Roman Baluev concluded that the existence of Planet d was questionable. After filtering out red noise (so-called correlated noise that links its frequency with its magnitude) the significance of the radial velocity wiggles fell to a little over 2.2 standard deviations (or only marginally beyond what would be significant with a probability of 95%). This left the interpretation of the HARPS data problematic—but only, really, if you aren't comfortable with a planet that has an orbital period twice that of the rotation period of the star. However, the same analysis retained the stronger signals for Planets b and c, whose presence was confirmed independently at Keck. Planet e also stayed in the data, although the data from Keck was more problematic and suffered the same issue with red noise as the HARPS data for Planet d. Planet d, meanwhile, vanished from the Keck dataset.

Likewise, in 2014, Paul Robertson and colleagues carried out a further statistical analysis of the data and drew the same conclusion: Gliese 581d did not exist, but rather its detection was an artifact caused by variations in stellar activity. These have a period of 130 days—presumably the rotation period of the star—which is twice

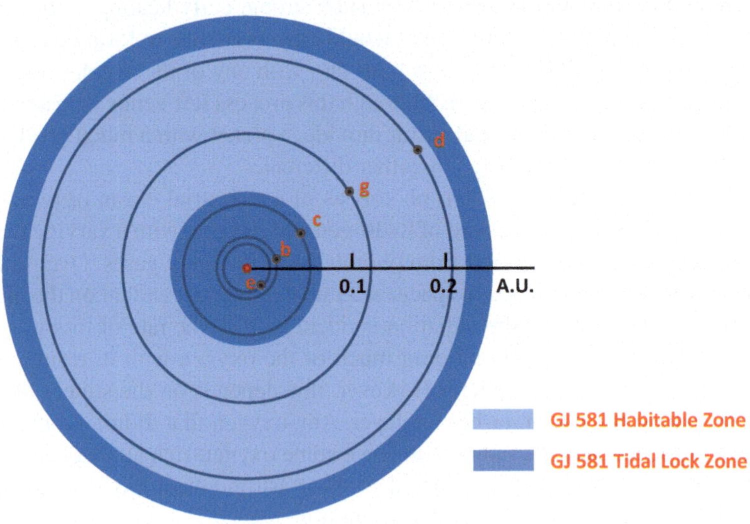

Fig. 8.2 Planet Gliese 581g? The revised 2012 orbital model of Steven Vogt et al, based on his extended analysis of HARP and HIRES data. If planet d is in a circular orbit, then the very habitable planet g *might* exist

the preferred orbital period of Planet d. If the HARPS RV measurements are clouded by these variations, then half the data needed to support the existence of the planet is spurious.

However, further analysis in 2015 by Guillem Anglada-Escudé (Queen Mary University, Belfast) and co-workers rejected the theory, pointing out flaws in Robertson and co-workers methodology—how they dissected the HARPS data using statistics. The rejection was fairly blunt and made reference to various tools that were normally used in planet detection, which were, themselves, validated. However, though Anglada-Escudé and co-workers concluded that the planet could exist, it was in itself not a confirmation. That will require further investigation and considerably more data. Until that occurs, Planet d's presence hangs somewhat in the balance. More importantly, Planet g—the former glory-child—can only exist if d does. This is because the determination of the former planet's orbital characteristics rely on the interpretation that some of the changes in the radial velocity of the star are caused by the gravitational pull of Planet d. That's currently a stretch.

One final, and rather interesting, analysis of Planet d's orbit was carried out by Valeri Makarov (U. S. Naval Observatory). Makarov investigated the effects of the various planets' orbital periods. Taking those of Planet b, c, d and e into account—and assuming that the mantle of Planet d was similar to that of Earth—led to Planet d slipping out of tidal locking. Instead, tidal forces between Planet d's mantle and the other three worlds might lead to Planet d rotating twice for every orbit of its star. Indeed, the spin-period could be even faster, with more axial spins per orbit. This would make the planet's day-year relationship rather similar to that of Mercury—and perhaps give it a day-night cycle and atmospheric kick more akin to that of Venus than a tidally locked world in which one hemisphere slumbers in perpetual darkness. Rather than the star occupying a fixed location in the sky, it would creep across it over several weeks. The days would certainly be peculiar and extended, but otherwise recognizable. Such dynamics were dependent on the orbit of Planet d being moderately eccentric (varying around the figure of 0.27, meaning that the orbit was roughly a third longer in one plane than the other).

Turning our attention to the characteristics of Planet d, its size of 5–7 Earth masses implies a fairly substantial amount of volatile material. Such compounds will strongly affect the habitability of the world, both in terms of providing the raw materials for life but also on the geological processes that are necessary for the sustenance of life. Bringing in the Kepler data suggests that this planet must have a deep ocean of water and likely a fairly massive atmosphere. Recall from earlier that planets with radii greater than 1.4 Earths will be waterworlds, and those above 1.6 Earths likely encased in a deep atmosphere. Given plausible compositions, at 5–7 Earth masses, Gliese 581d has a deep layer of volatiles that forms a carapace over its rocky interior. If the planet is old, there is likely more oxygen and less iron as well. This will favor a deeper layer of volatiles and a smaller core (Fig. 8.3).

Now, let's be optimistic about this (possible) world. The planet will have a convecting mantle—even after 7 billion years, as it has more than ample internal reserves of heat. However, the high mass has a number of less palatable implications, as we've seen in Chaps. 3 and 4. A likely wet mantle favors convection, but

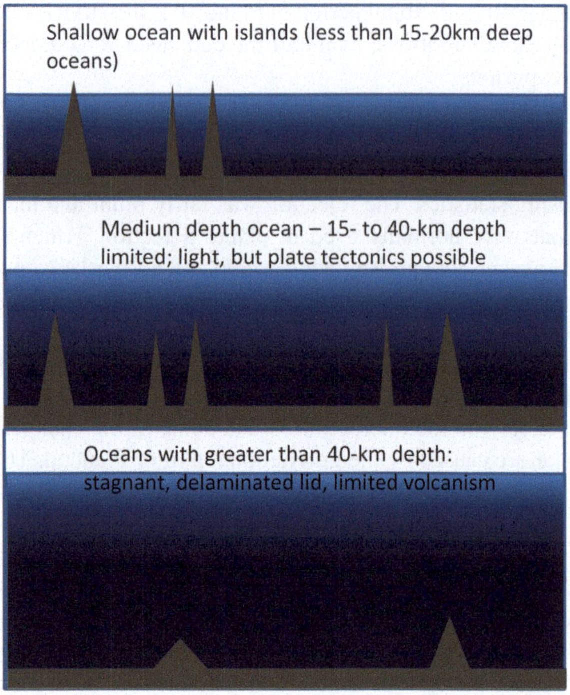

Fig. 8.3 The effect of ocean depth on the habitability of aquaplanets. If oceans are relatively shallow (top), light can reach the top of undersea mounts, and some islands will emerge. These can form abodes for life forms that depend on photosynthesis. As ocean depth increases, these niches vanish

the overbearing deep ocean cover will turn any basaltic crust into eclogite and cause it to delaminate in short measure. The planet will then have a granitoid crust overlying a dense mantle, with a deep ocean on top. Many of the measures of Laura Schaefer and Dimitar Sasselov (both of the Harvard-Smithsonian Center for Astrophysics) suggest that massive super-terrans will lack much evidence of volcanism from relatively early stages because of the presence of a thick granitoid lid—and because the depth of melting will be too great to allow magmas to reach the surface (Fig. 8.4). Therefore, even with more mass and more internal heat, Gliese 581d is likely as geologically dead as a doornail.

This world would then have deep oceans overlying a largely granitic crust, which in turn topped a still hot but largely inaccessible mantle. Some hydrothermal or volcanic activity might be possible, although the scope for this is low if the work that was discussed in Chap. 4 is applicable. The oceans will be saturated with carbon dioxide but be otherwise relatively impoverished in nutrients, in particular phosphate, which is normally delivered to them by erosion of exposed land. Clearly, this aquaplanet will not experience this kind of process. However, with an acidic ocean, phosphate ions can remain in solution. Yet, this is a small plus if sub-ocean tectonic processes are not occurring.

With a stagnant lid, the lack of plate tectonics might also limit the development and maintenance of a global magnetic field, particularly as the planet will rotate

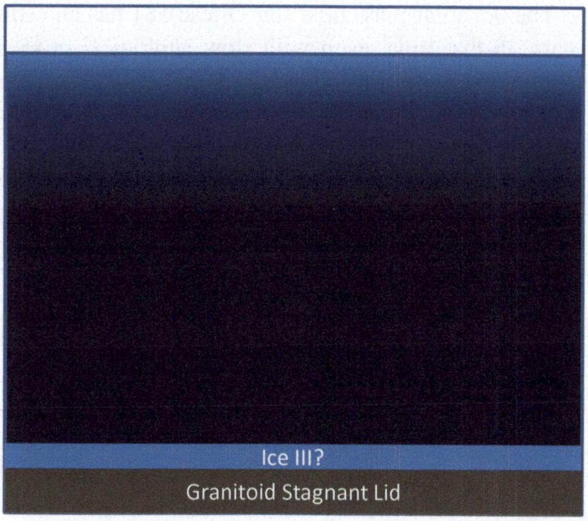

Fig. 8.4 Deep ocean-like worlds should be geologically inactive. Oceans greater than 50 km generate so much pressure on the underlying lithosphere that it is unstable. This forms an inert deep stagnant lid with little or no volcanism. A layer of Ice III nails the lid shut on this inert lithosphere, preventing living things in the ocean accessing nutrients from the underlying rock

slowly. This would have longer term impact on the sustainability of the planet as an abode for life. The atmosphere of Gliese 581d might be exposed to the full ravages of the stellar wind at very early times. Even if its atmosphere started out dense, atmospheric erosion could have whittled it down to something resembling that of Mars in the 7 billion years the planet has existed. This would clearly depend on the depth of any oceans, as well as the presence or sustenance of any initial magnetic field. However, the absence of a field—and erosion—might be a saving grace for Gliese 581. If the atmosphere is eroded, the depth of the ocean would decrease. Perhaps, then, as the global ocean shrivels, life—complex life—might find a foothold on any emerging land. The extent and longevity of any biosphere, will then depend on how much water is lost and how stagnant that crustal lid is.

If we return to the more likely (drowned planet) scenario, the availability of nutrients and energy will be far lower than it is on Earth. Here, there is good evidence that life started at or near hot springs (probably, but not necessarily, on the ocean floor) and then migrated to more temperate abodes. However, if the ocean is tens to hundreds of kilometers in depth, pressures might be so severe as to preclude the spread of even simple biology from the ocean depths. Consequently, unless circulation was very efficient, and the atmosphere peculiarly transparent to visible light and importantly, infrared radiation, photosynthesis would likely be impossible. Such a deep ocean world would not be a pleasant place to visit. If life did arise, it would likely not progress beyond microscopic forms (Chap. 9).

Yet, for anything that was able to survive these conditions, the environment would be unlikely to change over most of the life of the star system. The only measurable change would be the remorseless decline in geothermal activity as the planet cooled down. In the end, it would be the faltering internal clock that killed off life, rather than anything occurring in the atmosphere or the star above.

The 0.3 solar mass host star Gliese 581 has an estimated lifetime of 300 billion years. In that time, even with slow geological processes, almost all of the atmospheric carbon dioxide will end up on the ocean floor as carbonate—or dissolved in the ocean. The process is facilitated by low temperatures, leading to an interesting positive feedback. As carbon dioxide levels fall, the greenhouse effect will falter and temperatures will follow suit. As these decline two effects come into play. Carbon dioxide becomes more soluble in water as temperatures decline, and the growth of ice over these cooling waters leads to an increase in albedo that causes further cooling as the planet becomes more reflective. Gliese 581d is already near the outer limit of the habitable zone, making it particularly vulnerable to freezing over. Seven billion years might be more than enough time to turn this world into an ice ball, its vigorous atmosphere whipping snow over an icy shell.

Perhaps this is unnecessarily gloomy. With the radiation from Gliese 581 dominated by infrared, the buildup of snow and ice has a far less pronounced effect on global temperatures than visible light has on our world. On Earth the Sun's light is readily reflected by snow and ice and changes to the amount of land covered by these materials can cause a significant drop in global temperatures. During the period from 750 to 650 million years ago Earth appears to have largely frozen over several times. Only the steady exsanguination of carbon dioxide from volcanoes appears to have arrested the process.

For planets orbiting red dwarfs the story is a little different. Infrared radiation is less readily reflected by snow and ice than visible light is and so such an icy runaway can be avoided. Gliese 581d may thus be a tad slushy rather than completely frozen. A slush ball is a perfectly habitable place for a variety of organisms, including those that are complex and require oxygen. The limiting factor will be the productivity—the amount of usable food—the oceans can produce. If this is still high, complex webs of life will extend throughout its depths and the planet will be vibrant. However, a deficit of phosphate and other ions might preclude the spread of life in an otherwise water-rich world. Moreover, if the orbit is eccentric and the planet rotates non-synchronously, then the climate could be warmer, making the surface ice-free.

Either way, if Gliese 581 is 7 billion years old, life may at best be simply clinging on. It would seem unlikely that this planet will currently sustain anything more complex than organisms we might call bacteria. Alternatively, if Gliese 581d has very deep oceans, it was likely never habitable.

The Gliese 667 System: Two Super-Terrans, Three Stars

The System

Lying 22.1 light years from Earth is Gliese 667. Unlike Gliese 581, the Gliese 667 system is a far more interesting construct (Fig. 8.5). For one thing there are three stars, not one. Two of these stars, 667A and 667B, are low mass K-class orange

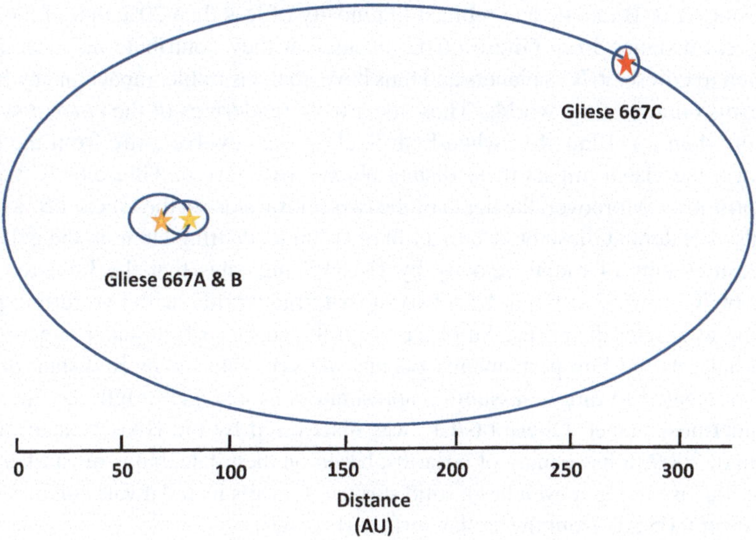

Fig. 8.5 The arrangement of the stellar component of Gliese 667. Stars A and B orbit one another within 20 AU. However, the planet hosting star Gliese 667C never approaches closer than 65 AU and at its most distant is 215 AU from the central K-dwarf pair

dwarfs orbiting their common center of gravity. The orbit of both K stars is very eccentric (with an eccentricity of 0.6), with their separation varying from 5 to 20 AU—roughly the Sun-Jupiter distance at the closest approach, to the Sun-Uranus distance when furthest apart. The stars weigh in at 0.73 and 0.69 solar masses, or roughly three-quarters that of our Sun. In terms of physical girth, each star more closely approaches that of the Sun rather than Jupiter.

The evolution of each K-dwarf will be completed in less than 50 billion years, leaving two white dwarfs—considerably less time than the red dwarf Gliese 581 will take to expire.

Lying on the fringes of the star system a third star orbits sluggishly, a red dwarf with a mass only slightly greater than that of Gliese 581. Separated by anything between 56 and 215 AU of space, this little star takes hundreds of years to complete one orbit of the central pair. For comparison Pluto orbits the Sun at an average distance of 39.4 AU, taking 248 years for each orbit. Thus, even if we were to swap the far more luminous Sol for Gliese 667A and B, they would still appear as two bright stars in the sky, rather than additional luminous suns, as many artistic renderings suggest. One particularly prominent rendering has Gliese 667C "setting" over its habitable and a very Utah-like landscape, with the two distant K-dwarfs forming a well-separated pair in the evening sky. Quite aside from the lack of sunsets on tidally locked worlds, the separation of the two orange K-class stars is likely too great in these images. When star C is at its maximum distance of 215 AU, the two distant suns, separated by less than 20, would simply appear as one bright, possibly peanut-shaped star.

As stars A & B radiate a combined luminosity of less than 20% that of the Sun, their great distance from Gliese 667C means that they contribute no meaningful radiation to Gliese 667C's planets and thus have no measurable impact on any living organisms on any of the worlds. Thus, the artistic renderings of the *current* system are more than a tad fanciful; when both K-class stars evolve away from the main sequence, the visual impact these distant objects will have on Gliese 667C will be more profound. Moreover, the death of the two central stars in the Gliese 667 system may also condemn Gliese 667Cc to a life of solitude, drifting alone in the galaxy.

Measurements of radial velocity by HARPS indicates that the little red star, Gliese 667C, carries with it at least two super-terran worlds, with two further planets lying at greater distances. An observer on the surface of the outer super-terran would have its red Sun permanently occupy the sky, with the more distant orange dwarfs relegated to rare, unassuming appearances as a single, bright daytime star. The innermost planet, Gliese 667Cb, was discovered by the HARPS team in the autumn of 2009. It has a mass of 6 Earths, based on their data, but is most likely too hot for life, assuming it even has a solid surface. It orbits its red dwarf sun in a week on an orbit 0.05 AU from the stellar surface (Fig. 8.6).

The outer world, Gliese 667Cc, was announced two years later and takes approximately four weeks to complete one orbit. The world receives approximately 90% of the energy of our planet, orbiting at 0.12 AU from its star. It is, therefore, clearly habitable, at least in terms of the amount of radiation it gets from Gliese 667C, even when one takes into account the increased proportion of infrared radiation in the star's spectrum.

Although the 90% figure implies that this planet would be cooler, the greater proportion of infrared radiation makes it marginally warmer than Earth, with an equilibrium temperature of 277.4 K depending on the reflectivity of the clouds and surface. With a mass of 3.9–5 Earths it is most likely a largely solid world, with

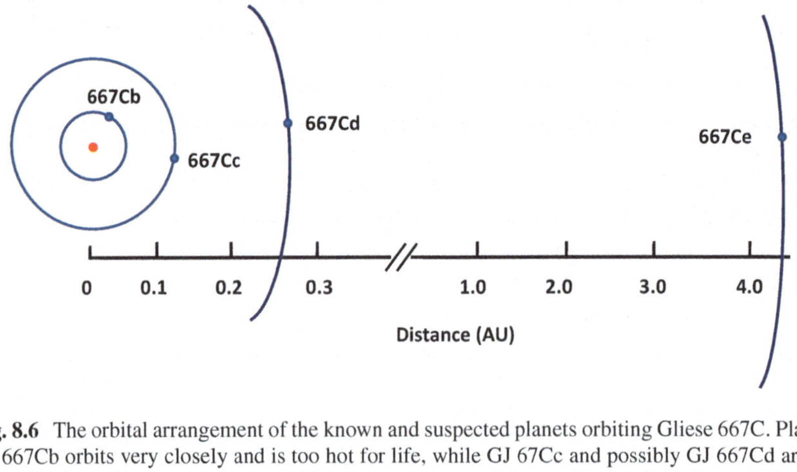

Fig. 8.6 The orbital arrangement of the known and suspected planets orbiting Gliese 667C. Planet GJ 667Cb orbits very closely and is too hot for life, while GJ 67Cc and possibly GJ 667Cd are in the stellar habitable zone. A suspected Saturn-mass planet orbits much more distantly at a little less than the Sun-Jupiter distance

perhaps a substantial fraction of more volatile materials, of which water is probably the most abundant. Thinking back to the Kepler data and analysis by Leslie Rogers we can be fairly certain that Gliese 667Cc is either an aquaplanet or one with a reasonably thick gaseous envelope—unless it has an unusual composition, or has been stripped by stellar radiation. However, with a lower mass than Gliese 581d, it is more likely to be habitable for complex life than the latter world.

If one were to stand (or float) on Gliese 667Cc, one would see a red sun, somewhat yellower than from the same vantage on Gliese 581d. Although carrying comparable masses, the lower metallicity of Gliese 667C renders the central star somewhat hotter and bluer than Gliese 581. From the planetary surface, the star would sit rigidly in one spot in the daytime sky, filling an area more than five times greater than that occupied by the Sun in ours. Standing at the sub-stellar point and staring upwards, the sky immediately surrounding the star might appear relatively blue. However, as you panned your gaze across the sky any blue would rapidly give way to yellower then red hues. Thus, we encounter the same issues associated with orbiting a red dwarf—the fixed location of the star in the sky and the dispersion of radiation with (angular) separation, or increasing distance, from the SSP. This effect is partly offset on Gliese 667Cc because there is a greater abundance of higher energy, and radiation to go around. This makes Gliese 667Cc a better candidate for supporting oxygenic photosynthetic life.

Thinking about the atmospheric circulation pattern, this would, again, be the fairly predictable anti-solar to sub-solar flow at the surface. With a long rotation period, convection would be confined to within 15°–30° or so of the sub-solar point, and there would be little or no capacity for frontal precipitation. Orographic precipitation on windward slopes would allow the development of complex life, should such slopes exist. Remember, the higher the planet's mass, the lower the relief on any landscapes. However, with a relatively high mass, we should remember that Gliese 667Cc is likely an aquaplanet, excepting the possibility that the planet was born oddly deficient in volatiles or has been stripped.

Modeling by Jack Lissauer (NASA Ames Research Center) implies this, but volatile poor is a relative term. A greater initial mass of material might still lead to a waterworld, even though proportionately less volatile material was initially present than was found on the young Earth. A 6 Earth-mass super-terran, with only 50% of the water content per unit mass found on Earth, will still have a deep global ocean and no dry land. Reduce the volatile content further to 15% that of Earth per unit mass, and some land will be visible, depending on how much water is locked into the planetary crust and mantle. All things are relative, and if we are to have super-terrans with dry land like Earth, they must be much drier than our world. This is evident in at least some of the discoveries that we outlined in Chap. 2. Most planets follow a rather regular pattern from small and dense to large and gaseous, but there are some small, ultra-dense worlds and a few rather mediocre "fluffy" ones. Therefore, Gliese 667Cc might be rather Earth-like, if fate dealt it a good hand.

There are other options for creating dry super-terrans. We already know that the process of planet formation is as catastrophic as it is frantic. Evidence for widespread orbital migration dominates our observations of the Solar System and many exoplanetary systems. In such environments there is plenty of scope for violent collisions between planetesimals. A late, violent collision might then drive off the bulk of a super-terran's volatile materials, leaving a largely dry planet. Although this seems contrived, it carries a reasonable probability and may well be an important source of super-terran planets that are habitable for complex land-dwelling life forms.

Finally, given suitable age, a 4–5 Earth-mass planet will be expected to begin drying out after a few billion years. Excepting the effect of the stellar wind, the mantle and lower crust will tend to remove surface volatile materials, until the majority of the water is returned to the mantle. Desiccation could be compounded by the steady surface cooling such planets might experience as the abundance of greenhouse gases declines. Growth of ice sheets will lower sea levels to the point at which dry land might emerge from the beneath the surface of any global ocean. Yet, it may turn out that *only* planets with masses comparable to Earth (or less) have suitable areas of dry land on which complex, multicellular life can develop. Further detailed observations will clearly be needed to confirm or refute this supposition.

Jupiter's Galilean satellites illustrate some of these effects. Io, the innermost Galilean satellite, is effectively dry, despite its location well outside the Sun's snow line. All of its water has been driven off by vigorous (tidal) heating within its interior—and possibly also by its proximity to a hot, young Jupiter. Such heating has been forced by the perpetual kneading action of its mass by Jupiter and the outlying satellites. A planet orbiting deep within the gravitational well of its red dwarf star may also experience bouts of violent heating, particularly early on, when its orbit may still be adjusting.

Rory Barnes (University of Washington), working with James Kasting (NASA Astrobiology Institute) and others proposed that such heating could be sufficiently strong as to drive much, if not all, of the water content from a developing planet. This will occur if the planet lies in an orbit close enough to its central star. The star itself doesn't directly heat the world through radiation. Gravity does the work, baking the interior until all of its water, carbon dioxide and other biologically useful gases have been driven into space. Although Barnes' work is theoretical, it does, once again, force us to re-examine our assumptions. To be habitable, one has to be more than a lump of rock at a set distance from your star.

The Prospects for ET Phoning from Gliese 667Cc

If we are to assume some sort of intelligent life can arise on a super-terran, it would seem likely that it will need dry land on which to do this. Not only does dry land provide a greater diversity of environments on which to develop and acquire sensory information, it also produces a more diverse set of challenges to the underlying genetics. Evolutionary processes are likely only dynamic where the environment is variable and can provide the necessary selective pressures. As Chap. 7 suggested, a waterworld might allow for the development of fish, but the fish are not likely to be intelligent, as the availability of oxygen is restricted by its solubility in water.

On Earth's surface at room temperature the air contains 21% oxygen, but only around 9% of the ocean's mass is dissolved oxygen. Oxygen is likely essential for intelligence. Brains are expensive, fuel-hungry organs to run. Limited oxygen supply will mean that respiration is restricted, body temperatures are likely lower and the processes needed for rapid cell-to-cell communication muted. Whales and dolphins may be smart, but they are air-breathing mammals and evolved first on land, later returning to the oceans. They do not provide an exception to our aquatic rule. Of course, intelligence could be created if organisms work in parallel, with each of their slower brains working in unison. A sort of aquatic hive might be envisaged if dispersion of its members in water is not an issue.

Of particular interest is the somewhat controversial discovery of complex, multicellular organisms living on the floor of the Mediterranean Sea. Roberto Danovaro and colleagues (Department of Marine Science, Polytechnic University of Marche) discovered a variety of simple multicellular organisms living in the sediments on the floor of the L'Atalante basin—a region filled with hypersaline (extremely dense and salty) water. These species, Locifera, were all new to science. Various techniques confirmed that these organisms were viable, metabolizing entities rather than strays that had simply fallen into the basin and died.

Unlike every other multicellular eukaryote thus far discovered, these Locifera did not operate using oxygen. Instead their cellular mitochondria have been replaced with another organelle (internal compartment) called a hydrogenosome. Rather than consuming oxygen, these organelles produce hydrogen, hence the name. These hydrogenosomes utilize a similar strategy to mitochondria to generate ATP, but it is less intricate and generates less useful energy per unit of fuel than standard mitochondria. However, like mitochondria, they are clearly derived from free-living bacteria through the process of endosymbiosis and at least in some cases contain their own chromosome that carries instructions for the manufacture of proteins needed to make the hydrogenosome work.

Intriguing though these life forms are, they suggest only that it is *possible* to evolve more complex organisms that operate without oxygen, not that you can evolve an energy-costly brain needed for intelligence. These Locifera, if confirmed, are hardly primates—or even as complex as fish. Oxygen appears essential, at least for the latter speciation.

Finally, even if we can envisage intelligent aquatic organism (perhaps something akin to an octopus), such life forms are clearly going to remain unable to communicate their status from the depths of an ocean. If organisms are to build devices capable of observing the universe and listening out for spammed Dorito ads from Earth—albeit 22 years later—they need dry land on which to construct them. An aquaplanet *may* be a hospitable abode for fish, but it is somewhat less likely a world on which to find the alien equivalent of Hedy Lamarr, Barbara McClintock or Richard Feynman. Indeed, for any aquaplanet with a deep ocean, the chances that life could evolve into anything more complex than bacteria seems unlikely. If the ocean has an ice layer overlying its stagnant, granitic lid, then the lack of exchange of ions between the crust, mantle and ocean would likely mean that such a world is barren. Where life does originate through chemical reactions in the ocean, the steady reduction in the supply of these as the planetary system evolves would make this world barren within a few hundred million years of its formation.

The only interesting scenarios are those that were outlined earlier. Here, in some situations, Gliese 581 and 667Cc may have been desiccated soon after birth (or were even born this way). In this case, water might return in more limited—and hence useful—volumes through outgassing of the mantle. If we can determine the radii of these worlds, we might be able to determine the likelihood that these are rocky worlds like Earth, rather than drowned equivalents; drowned being the operative word.

Red, Not Dead?

Since 2013 a number of planets have been found around other red dwarf stars. We now look at these under the spy glass of those factors used to largely condemn Gliese 581d and Gliese 667Cc. This final section looks at their potential for habitability as well as their capacity to host complex life.

Table 8.1 Summarizes the qualities of the other red dwarf worlds that are at least superficially habitable

Planet	Planet mass (Earth masses)	Radius (Earth radii)	Temperature (K)	Habitable for any life?	Waterworld or telluric world?
Gliese 581d	6.98	2.2?	181	Unlikely as thick lid	Waterworld but likely frozen
Gliese 667Cc	3.7	1.54?	277	Microbial seems likely, but desiccation may make it habitable for complex life	Most likely a waterworld
Kepler-1229b	2.7	1.4	213	Microbial but with opportunity for complex life depending on ocean depth	Transitional between telluric and waterworld
Kapteyn b	4.8	1.6	205	Microbial may be possible	Waterworld
Ross 128b	1.35	Unknown	213–301	Unlikely but a narrow strip near the terminator is possible.	Telluric—super-Venus or possible Saharan-like
Wolf 1061c	4.3	1.6	223	Microbial may be possible	Waterworld
Luyten b	2.89	1.35	259	Like Kepler-1229b this could have complex life	Transitional between telluric and waterworld

Most are a lot less "habitable" than initial appearances might suggest

Gliese 832c

This is an interesting world—but one that is unlikely to be habitable. Gliese 832c is a warm super-terran with a mass of 5.4 Earths—possibly more. As such at best it is an aquaplanet, unless it has been thoroughly stripped of volatiles by its star in its youth.

What makes it unusual is its eccentric orbit. This takes it into and out of the hot edge of its star's habitable zone. At periastron (the closest approach) the planet has an equilibrium temperature a good 25 °C higher than Earth, placing it in a similar position to Venus. However, at apastron the temperature could fall to –40° C—more than 20° colder than our world. One might imagine a world where temperatures oscillate wildly. However, if Gliese 832c has a reasonable atmosphere, water vapor and carbon dioxide would be expected to retain heat when the planet makes its closest approach. Therefore, the planet likely would retain enough heat on every passage of periastron so that it heats up. The end result would be a thermal runaway and the evolution of the planet to a super-Venus state.

In any circumstance, the climate is likely to be intriguing—with its eccentric orbit "unlocking" the planet from its star, in a pseudo-synchronous fashion.

Ross-128b

Currently, Ross-128b lies 10.89 light years away. However, in 79,000 years it will be the closest red dwarf to the Sun as it whisks its way through the galactic disc. Should we consider exploring it, if humanity still survives this long? In short, almost certainly no.

The 1.35 Earth-mass planet of Ross-128 (Ross-128b) is likely more a super-Venus than a super-Earth. This world has an orbital period of 9.9 days, placing its circulation in the Rhines-rotator regime. However, it receives 38% more energy from its star than Earth does. That alone should ring alarm bells, given that a 10% increase on our current insolation will evaporate our oceans. If we take into account the greater proportion of energy at longer wavelengths—and its enhanced absorption by water vapor—we can pretty much guarantee that Ross-128b underwent a runaway greenhouse effect a very long time ago. This leads to some interesting observations. Taking a lead from our Venusian world and Gliese 832c, Ross-128b would initially have had a moist greenhouse atmosphere that showed Rhines rotation. This may have been very transitory, as heat from the contracting infant Ross-128 inflated the troposphere, flooded the planet's stratosphere with moisture, then split it, causing the loss of hydrogen.

Within a few hundred million years—depending on the initial inventory of gases—Ross-128b would have lost the majority of its water vapor. Lacking water, like Venus, carbon dioxide would have steadily accumulated, and the temperatures would continue to rise. Within 1 billion years, Ross-128b would become a tidally locked Venus, with a thick, carbon dioxide-dominated atmosphere. Like Venus, volcanism would pump it full of sulfur dioxide. Photolysis of this compound and carbon dioxide would provide enough oxygen to convert this to sulfuric acid. Ross-128b would then be a faster-rotating Venus with a thick, gloopy sea of carbon dioxide swilling over its surface—a torrid aquaplanet with a difference—one with a sea of super-critical carbon dioxide rather than water.

In a kinder scenario, Ross-128b begins well within the hot zone of its contracting Sun's influence. Initially, Ross-128b is baked dry. However, as Ross-128 contracted and became less luminous, its companion world was left hot, dry, but possibly just beyond the inner edge of the habitable zone. With substantially less volatile material, the planet constituted a dry biosphere, with minimal water. Tidal locking in this scenario could leave an odd eyeball world that is baked dry on most of the sunlit hemisphere, temperate near the terminator and a frozen desert on the night hemisphere. In this case there might be a strip of biosphere on an otherwise dry world. This would give Ross-128b a Saharan SPH.

Kepler-1229b

Kepler-1229b is a far more interesting prospect for life. Its star sits on the M-K divide with 0.54 solar masses of material. This gives it a lifetime of 46 billion years or so. This relatively metal-poor star has a luminosity of 4% that of the Sun. Although small, this is sufficient to push the habitable zone to 86.5-day periods. Interestingly at this period, work by Jeremy Leconte suggests that tidal forces between the star and its planet may be overwhelmed by other processes in the planetary atmosphere. Here, depending on the mass of the atmosphere, heating of gas on the day-lit side may induce thermal tides—mass movements of air from the day to night sides. Although certainly controversial, this may cause sufficient acceleration to push the planet around faster than it would otherwise rotate around the star. Tidal locking would cease, and the planet might develop a slow, Venus-like rotation.

Although Kepler-1229b is a super-terran, its mass is relatively small, around 2.7 Earth masses. This is optimal in terms of the longevity of plate tectonics, should it operate here. It is massive enough to sustain a strong internal engine, but not so massive that the surface is likely to clog with granite in the first 1–2 billion years. This allows deep cycling of nutrients as well as water and carbon dioxide. Moreover, as the mantle can, in principle, permeate with water, the effect of lowering temperatures on its dynamism should be offset by a reduction in viscosity caused by an influx of water. Finally, the mantle should retain sufficient energy to avoid swallowing all of the surface water for several billion years. Thus, all things considered, Kepler-1229b should be habitable for some form of life.

What about the kind of life the planet could host? That depends critically on its volatile inventory. At 0.54 solar masses, the star is far enough removed from the planet to avoid blistering it with EUV and X-rays in quite the same manner as TRAPPIST-1 or Proxima Centauri (Chap. 9). Moreover, with more mass, descent onto the main sequence would be significantly shorter than either of the aforementioned stars. Therefore, there is far more opportunity for this world to retain enough volatiles. However, how much?

Remember there might be so much that the surface is drowned in a deep ocean. This, of course, is not a problem one can answer without further observation. If the planet has oceans with approximately the same mass as ours, there will be continents. Yet, recall, with a higher gravitational pull than Earth and a hotter mantle (at a given age) there is more opportunity for continental crust to spread. Hence, the oceans will need to be proportionately shallower for continents to ride high above them. Indeed, this is the problem besetting all super-terrans. Although they may have dynamic interiors and atmospheres, the extra burden of volatiles and the generally lower relief of any continents will mean sub-aerial (non-aquatic) life will likely be more restricted in its range. This must impact the evolutionary potential of these planets.

If there is exposed continental crust, we can envisage the formation of complex species. The precise potential depends on where those continents lie, with respect to the SSP. Close in, and things are good; more distant, and life becomes far more challenging, as rainfall will be limiting. In the oceans, if there are shallow regions—irrespective of whether there is any exposed land—a broad variety of species are possible, with complex food chains. Deeper water means less variety. Photosynthesis would be confined to the surface layer, with deeper water supplied by precipitation of organic matter and, potentially from below through chemosynthetic pathways. Here, undersea volcanism could support a dark biosphere.

Overall, Kepler-1229b represents a promising abode for life—and potentially complex life. If the planet is rotating relative to its star, the prospects improve still further, because the entire surface of the planet is available to support photosynthesis and with it complex and energetic food chains. Moreover, the climate, while distinct from that of Earth, would still be recognizable "terran" with a diurnal cycle, albeit a long one…

Luyten's World, Luyten b

Luyten b, aka Gliese 273b, is another super-terran, lying 12.2 light years from Earth. Its mass is likely somewhat greater than Kepler 1229b (at 2.89 Earth masses). However, though it has an orbital period far shorter than Kepler-1229b, its orbit appears to be somewhat more eccentric. Taken at face value, the 18.6-Earth day orbital period places it near the upper end of the Rhines-rotator regime.

However, the eccentric nature of the orbit might mean that it does rotate on its axis relative to its star. Such rotation may alter over relatively modest intervals of time, much like that of Venus. In this regard, Venus provides an interesting standard. Over the seven years *Venus Express* observed this world, the average wind speed at 50 km increased. When the probe arrived in 2006 the average wind speed was 300 km per hour. However, by the demise of the craft in 2014 wind speeds had increased to nearly 400 km per hour.

Igor Khatuntsev (Space Research Institute, Moscow) and, separately, a Japanese group, carried out a set of observations from Earth. Khatuntsev's team determined the changes in wind speed through observations of cloud features in images as they moved between frames. In all over 45,000 features were methodically tracked by hand and more than 350,000 further features automatically compared using a computer program. Independently, the Japanese team used a similar, but automated, process, to monitor the position of different cloud features over successive image frames.

Over shorter periods of time both teams also recorded rapid variation in cloud velocities, with some features circumnavigating the globe in less than 4 days. Yet a few orbits later, the speed of these features had decreased so that the same feature took over 5 days to cover the same distance.

Both teams conclude that such changes in Venusian wind speed might be caused by an exchange of momentum between the accelerating atmosphere and the decelerating planetary spin. Perhaps more surprising was the possibility that the atmosphere may be becoming spun up by an exchange of momentum between Earth and Venus as they passed one another in their orbits around the Sun. Similarly, an eccentric orbit of Gliese 273b (Luyten b) around its star will promote exchanges of momentum between the star's outer layers and the planet. As Luyten's star rotates over a period of 118 days, the planet will generate a small, yet significant, tidal bulge on the star as it orbits around it. Therefore, exchanges of momentum could alter wind velocities in a cyclic manner on this and other, similar, worlds.

Overall, the relatively low mass and a radius in the range of 1.4–1.6 Earths make it likely that this is an aquaplanet, with deeper oceans than Kepler-1229b. Again, this is supposition based purely on mass. One may argue that Luyten's world lies closer to its star and may have had lots more volatiles while the 4.4-billion-year-old, 0.29 solar mass star was dropping onto the main sequence. Therefore, although the planet is likely to have a more interesting climate than Kepler-1229b, it could also be substantially drier. This may be no bad thing if it meant this world shed enough water to expose continental crust. Temperature-wise, the planet is likely to be a lot warmer than Earth if it has an Earth-like atmosphere. The planet receives around 6% more energy from its Sun than Earth does. This, coupled to the higher proportion of infrared radiation, means Luyten b may be a better match for Earth in a few hundred million years' time. The equilibrium temperature is 4 °C higher than ours—warmer than our planet today, but still far cooler than Earth was in most of the Mesozoic or the Eocene.

Wolf-1061c

Wolf-1061c is a virtual twin of Gliese 667Cc. At 4.3 Earth masses and 1.6 Earth radii, this planet is likely drowned in deep oceans—assuming such water survived the star's formative years. Wolf-1061c is somewhat cooler than its twin, receiving only 60% of the energy that our planet does. This could make it a frozen aquaplanet, depending on the volume of water on its surface, the abundance of greenhouse gases and the presence (albeit an unlikely one) of continental crust.

Like its counterparts, Wolf-1061c is likely to be habitable for microbial life, but the evolution of more complex life is probably held in check. An interesting point worth considering is, could complex aquatic life, resembling fish or complex invertebrates, evolve in the oceans of an aquaplanet? The prosaic answer would be, "Yes," the more complex and likely answer is probably the contrary. Now, this is the author's opinion, but we will attempt to justify it with the following reasoning. Most of Earth's complex animal and plant lineages evolved in the Ediacaran and through the Cambrian. Such lineages were oxygen-utilizing, high energy species that survived either by eating aquatic plants or one another. Evolutionary development, as Rodrick Wallace proposes, was likely driven by the availability of copious amounts of energy. In essence without the energy made available by free-oxygen the Cambrian explosion would have been more of a gentle thud than the megaton detonation we observed. However, there is more to it. Although the energy supply is fundamental to supporting the development and diversification of large-bodied organisms, if there are no niches in which they can evolve, diversification may not happen or happen in a far more subdued form.

Think about the geology of this terrestrial period. In the Ediacaran and through the Cambrian, the supercontinent Pannotia disintegrated with the opening of the Iapetus Ocean. Growth of buoyant warm oceanic crust and erosion of the Grenville mountains meant that oceans inundated much of the continental surface in what is called a marine transgression. Consequently, large areas of land became shallow, warm seas. Such sunlit, temperate environments, with diverse chemistries and topologies, are just what evolution adores. While the oxygen allowed organisms to access vast quantities of energy in their food supplies, it seems the abundance of new aquatic landscapes allowed the development of a large variety of undersea forests and other niches for life to colonize. In this view, the Cambrian explosion is as much about the geophysics of the day as it is about the energetic environment that was emerging.

Thinking then about aquaplanets, one is confronted with a rather sorrowful truth. If there is only deep water, the presence of oxygen will be a necessary but insufficient driver of speciation. Life will remain microbial, but even if something more dynamic evolves, it will not proliferate to form the myriad forms of life we see on Earth. In an interesting discussion with Rodrick Wallace some time ago, Rodrick suggested that even on an aquaplanet one might have a "Cambrian explosion" if floating vegetation substituted for land. In this scenario, long-lived areas of vegetation would form platforms on which other species might colonize.

There is one situation where this might occur—and this is a situation that is wholly unique to tidally locked planets. In most oceans, a combination of wind and currents would likely disperse such floating "vegelands." Winds or currents might carry them to far cooler areas, or they would simply break up into small chunks, in a manner driven by wave action. However, on a tidally locked aquaplanet—particularly a planet with a long orbital period—winds tend to converge on the SSP. Under suitable circumstances, floating "vegelands" might develop and persist under the leaden skies, where precipitation falls unrelentingly. One presumes such conditions will facilitate the evolution of umbrellas…

Kapteyn b

Kapteyn b has an unusual origin. It is likely an outcast from the Milky Way's largest globular cluster, Omega Centauri. How it ended up wandering the galaxy is another matter (Table 8.1).

Globular clusters (or as Omega Centauri may be, the core of a stripped dwarf galaxy) are dynamic beasts that are discussed in detail by this author in *The Complex Lives of Star Clusters* (Springer, 2015). To precis this book, the stars within the cluster are bound to its core through gravity. This means that they have long looping orbits that take them into and out of the cluster's dense core. Here, gravitational interactions between lone stars and binaries lead to some interesting events. In essence, as the interloping lone star swings past any binary system, gravitational interactions transfer momentum from the denser and more massive binary to the interloping star. Over time, the orbit is expanded as the lone star gains momentum. Meanwhile, the binary stars lose momentum and their orbit shrinks. Ultimately, such interactions can lead to the merger of the binary stars and the ejection of the lone star. As the majority of stars are red dwarfs, with low mass, they tend to eject from the core of the cluster.

As they move further from the gravitational hub, they become susceptible to the gravitational pull of the galaxy and are gradually teased out from it. Indeed, if one looks at globular clusters they are flanked on their forward and leading edges by low-density tails of stars. Low mass clusters, such as Palomar 5, consist of a few tens of thousands of stars, topped and tailed by long streams of thousands of others that mark the path of the cluster around the Milky Way.

This kind of forced extraction may sound rather harsh. However, think about the long-term survival of the planetary system. The end-result is that when a red dwarf hosts a habitable (or potentially habitable) planet, encounters between the red dwarf and the other stars of the cluster can be the fortuitous ejection of the star and its planets from the cluster. Although this removes the red dwarf and its worlds from its beautiful siblings, it would also limit the chances of a subsequent catastrophic encounter with neighboring stars that might remove its planets. The Palomar clusters are in the process of melting away, leaving long trails of stars ahead of and behind the core of the cluster. Maybe, just maybe, within these evaporating trails of stars, there lurks a star with a habitable world, awaiting discovery.

Sun-like stars cast their worlds further afield, making them more prone to tidal stripping by close neighbors. Here, close encounters between stars will preferentially remove the planets. Indeed, it is the red dwarfs, with their moths circling close to their flames, which are the surest candidates for such a strong bond. As we look out into the halo and to these ancient star cities, it seems likely that their dimmest residents will be the best places to examine for the presence of habitable worlds. But moreover, as was mentioned earlier, with globulars evaporating red dwarfs like so much ancient blood, astronomers could well do worse than cast their eyes on the tidal tails that extend beyond the cluster walls when searching for planetary systems.

In one such thread of stars, ripped from Omega Centauri, lies Kapteyn's star. This star, also known as Gliese 191, is an ancient red dwarf that lies a little over 13 light years from Earth. Despite its current proximity to Earth, Kapteyn's star is an interstellar hobo, making a brief pass of our system of worlds on its long circuit through the halo of the Milky Way. In the summer of 2014 a team of astronomers, led by Guillem Anglada-Escudé from Queen Mary's University, announced the detection of two planets orbiting this star. The team had used the radial velocity method (HARPS, Chile and HiRES at Keck) to uncover these small, hidden worlds. The outer planet (Kapteyn c) has a mass roughly half that of Neptune but is certainly too cold for life. It may be a gaseous mini-Neptune or perhaps a frozen aquaplanet.

Meanwhile, the innermost planet (Kapteyn b) has 4.8 times the mass of Earth and orbits squarely within the habitable zone around the star. In this region temperatures are modest enough to allow the presence of liquid water on the planet's surface. Although such a massive world is certainly a poor prospect for the development of complex life, it may well be an aquaplanet with deep oceans and microbial life. The number of available niches is likely to be small; and most importantly, if oceans are deep, it becomes questionable whether anaerobic photosynthesis will evolve into forms that are oxygenic. From earlier, we know that photosynthesis almost certainly evolved from deep, hot anoxic respiration systems. If these are far below the surface of any ocean—and remain so—then the propensity for photosynthetic life to evolve into forms that use visible radiation becomes slight.

This conclusion may well be contentious, but it is reasonable given the development of oxygenic photosynthesis in shallow water, formally anaerobic communities, called stromatolites. Although oxygenic photosynthesis is certainly ubiquitous in shallow terrestrial waters, it could not have evolved here from other anaerobic forms, as the earliest of these anaerobic forms appear to use energy sources and chemical substrates (either hydrogen sulfide or small organic molecules) produced by volcanism. However, Kapteyn b might be a living world—and a very old one. Investigating its atmosphere and searching for evidence of such life should be a priority for James Webb or its successors.

Moreover, Kapteyn b opens the door to the possibility that there are millions of potentially habitable worlds orbiting, tightly, their red dwarf suns as they weave their way through the disc of the Milky Way. Were any living organisms detected on Kapteyn b, it would make Kapteyn b the oldest known habitable world in the galaxy, and likely one of the oldest in the universe. As a super-terran, a planet only slightly more massive than Earth, Kapteyn b opens up the probability that globular clusters could support stars with rocky planets—not simply gas giants. However, aside from debunking this "Rare Earth" idea, Kapteyn's star has a little more up its sleeve—and something that brings us back to globular clusters.

Kepler-186f

Here is a world for which we know the radius but not the mass. Its radius was determined by Kepler using the familiar transit method. At 1.17 Earth radii, it is almost certainly a low mass super-terran—as in has a mass between 1.5 and 3 Earths, depending on its composition. This makes it likely to be rocky and possibly a super-terran.

Table 8.2 The characteristics of the red dwarf planets that orbit inside or close to the habitable zone

Planet	Planet mass (Earth masses)	Radius (Earth radii)	Temperature (K)	Habitable for any life?	Waterworld or telluric world?
Kepler-186f	1.5–3.0	1.17	188	Probably too cold for anything other than extremophiles—but perhaps life in ice-covered oceans	Telluric
LHS-1140b	6.6	1.6–1.9?		Unlikely, but simple microbial if undersea volcanism	Sub-Neptune to super-terran aquaplanet
K2-155d	3–6?	1.64	289	Unlikely, as too hot	Super-Venus?
Gliese 832c	5.4	1.5?	253	The eccentric orbit might lead to thermal runaway, but microbial is possible	Waterworld; possibly too hot for life

Most are a lot less "habitable" than initial appearances might suggest. For reference Earth has an equilibrium temperature of −18 °C, in the absence of greenhouse gases. Therefore, always assume that a planet with an atmosphere will likely be warmer, on average, than one that does not have one

However, if its density is oddly low, say two-thirds that of Earth, then its mass will also be low. Such a low-density world will be an aquaplanet—or a world shrouded in a thick hydrogen-rich atmosphere. Yet, one has to say that such incantations seem frivolous. Sticking with what is reasonable, we should expect either a disproportionately dry super-terran or a less massive aquaplanet (Table 8.2).

Kepler-186f is a cold world. It receives only one-third (0.29) the amount of radiation that Earth does. Taking Mars as a reasonable comparison (although Mars has an insolation of 0.59-Earths), the Martian climate is on average considerably colder, with an equilibrium temperature a good 70° colder than Earth (−85 °C versus −18 °C for Earth). However, this is largely down to its thin and largely transparent atmosphere, which allows most of the re-radiated solar radiation to escape directly to space. If Kepler-186f has a thick atmosphere that is rich in poorly condensable greenhouse gases (so not water), then temperatures might be more clement. However, this is a push. At low temperatures carbon dioxide dissolves readily in any free water (assuming that it isn't frozen). Moreover, carbon dioxide clouds are poor insulators and tend to reflect more visible radiation than they absorb infrared radiation. Hence, they tend to cause cooling rather than warming. The end result is likely that Kepler-186f is a frozen ball—albeit one that might host life under its ice shell.

LHS 1140b

LHS1140b is unlikely to be a habitable world for complex life. With 6.6 times the mass of Earth, and a radius exceeding 1.4 Earths, this object is likely to be (at best) a massive aquaplanet. With roughly half Neptune's mass it is likely to have a thick atmosphere—if it has not suffered substantial loss of volatile materials when its 0.3 solar mass red dwarf was descending onto the main sequence.

The orbital period is 25 days, but given its radius this should still have an atmosphere that undergoes Rhines rotation. This gives this sub-Neptune or aquaplanet an interesting climate with strong westerly winds west of the SSP and easterlies east of the SSP. However, aside from an interesting climate, the same objections to habitability for complex life, at least, apply. However, as before, if the mantle is able to interact with the hydrosphere—the ocean and atmosphere—then something interesting in terms of biology may be possible. We can imagine that the world has been stripped of much of its volatile content, but this is theoretical and does nothing to solve the mass problem.

Let's not forget, with such a high mass, plate tectonics is likely impossible. Recall that this is nothing to do with the propensity for the hot mantle to convect, but rather the formation of a thick, stagnant, granitic lid. Moreover, with this mass, volcanism will be extremely muted, as the depth at which partial melting would be possible is far deeper than it is on Earth. Thus, once that thick lid forms, there will be very limited partial melting at depth beneath it. Lacking both fresh supplies of nutrients and energy, the prospects for life will be minimal at best.

K2-155d

We close the deal with a final world—and one that is not much more suited to habitability than Ross-128b. Indeed, all things considered this is likely worse than the aforementioned world. K2-155d is an even larger world than LHS1140. Discovered using the transit method, the mass is poorly constrained, but the radius (1.64 Earths) is well beyond the aquaplanet limit, and it likely has a substantial atmosphere. Although the orbital radius is relatively large, with a period of 40.7 Earth days, the host red dwarf is relatively luminous, and this places K2-155d and its inner sibling inside the hot edge of the habitable zone.

The equilibrium temperature for this planet is a good 34° Celsius higher than Earth. This is easily enough to vaporize its oceans, if it ever had them. Yes, there are issues with determining the luminosity of the star, but there would have to be substantial reductions in the estimates to consider K2-155d worthy of contention as a habitable planet. Therefore, like Ross-128b, this super-terran should probably be demoted to inhospitability.

All in all, the "haul" of planets has, to date, not given a significant number of ones we could comfortably say might be inhabited. Most of the planets that inhabit the circumstellar habitable zones are too massive to comfortably host conventional tectonism and likely have oceans that are too deep to allow land to pop above their surface. Indeed, in the case of Kapteyn b, LHS 1140b, Gliese 581d and Wolf-1061c the worlds may be so massive that there is not only a thick granitoid lid but also a mantle that is too pressurized to melt to any meaningful extent that would yield sub-(ocean) surface volcanism. Therefore, for these planets, life is likely not going to develop, or not develop very far in terms of its complexity and variation. We need to look further afield—and Chap. 9 offers two far more promising candidates, one of which is very close to home.

A Special Mention

Since the 1960s Barnard's star was considered home to a Jupiter-mass planet. Indeed, at one point, there was the suggestion the second closest star system to ours might have hosted two planets. However, all subsequent investigations turned up emptyhanded. However, as this book slipped towards submission, at the end of 2018, Ignasi Ribas (Institut de Ciències de l'Espai (ICE, CSIC)) and an international army of scientists announced that this small red dwarf appears to host a 2.3 Earth mass super-terran in a 233-day orbit. Although this frigid world is certainly not habitable for complex life, the 2.3-Earth mass object might make an interesting scaled-up version of Europa with a frozen ocean, overlying a tectonically active, rocky heart. In this regard, Barnard b might be habitable for some form of microbial life.

There remain many uncertainties about this planet—and whether it may be accompanied by a more distant sibling. However, it seems oddly gratifying that this 0.6 solar mass red dwarf has finally been paired with at least one fairly massive planet after decades of disappointment. Both its proximity to the Sun and its relatively wide orbit will make Barnard b an attractive target for future investigations of its atmosphere, as well as for direct imaging of the planet as a whole. In 2019 Teegarden's star, located 12.5 light years away, was found to host two super-terrans in the stellar habitable zone. Teegarden b has a likely surface temperature compatible with surface water, while the more distant Teegarden c is likely largely frozen, with a mean temperature of −47°C.

Conclusions

Gliese 581d and 667Cc are two close planetary worlds—lying within 25 light years of Earth. Gliese 581d takes the title as the first potentially habitable world found, while Gliese 667Cc is perhaps the most habitable world found at the time of writing. That said, we shouldn't jump in the nearest large spaceship and head off, just yet. Both worlds have potential problems when it comes to habitability for life forms like us. Gliese 581d appears fairly aged and most likely lacks a solid surface on which to land. It is also cold. Conversely, Gliese 667Cc is probably too young to have formed a breathable atmosphere and may, like Gliese 581d, have a surface dominated by deep oceans of water.

We may find that all planets with masses greater than 2 or 3 times that of Earth are waterworlds. We may need to focus our attentions to the smaller, Earth-mass planets, where water is less abundant and a solid surface more probable. However, if Jack Lissauer's research is correct, and the habitable worlds of red dwarfs are deficient in volatiles, then we may not need to worry so much. For the reasons previously discussed, *dry* planets are likely to be more habitable for complex life forms than waterworlds; they will certainly be the likeliest abodes for sophisticated, fashion-conscious aliens, capable of space travel and communication. It seems adversity might be the mother of invention.

These problems aside, the worlds found so far form the first of what will inevitably be countless potentially habitable planets. Should ESA's Darwin or NASA's Terrestrial Planetary Finder missions get off the ground, we should expect the discovery of a rich tapestry of habitable worlds within a decade. Sooner or later an Earth-sized world of suitable age will be found and a second home for future humanity might emerge. Importantly, there are many more variables to consider as to habitability than the mere location of the planet: mass, composition and age are all factors. Although tidal locking is unlikely to preclude the rise of life, the consequent failure of any geodynamo needs to be considered if the long-term habitability of the world is to be assessed.

These are early days of both discovery and understanding, and the brutality of hard data hasn't yet impacted on our imaginations. Although more hard work is due, there is an abundance of opportunity for daydreaming. Dream on.

References

Barnard b

Ribas, I., Tuomi, M., Reiners, A., Butler, R. P., Morales, J. C., Perger, M., Dreizler, S., Rodríguez-López, C., González Hernández, J. I., Rosich, A., Feng, F., Trifonov, T., Vogt, S. S., Caballero, J. A., Hatzes, A., Herrero, E., Jeffers, S. V., Lafarga, M., Murgas, F., Nelson, R. P., Rodríguez, E., Strachan, J. B. P., Tal-Or, L., Teske, J., Toledo-Padrón, B., Zechmeister, M., Quirrenbach, A., Amado, P. J., Azzaro, M., Béjar, V. J. S., Barnes, J. R., Berdiñas, Z. M., Burt, J., Coleman, G., Cortés-Contreras, M., Crane, J., Engle, S. G., Guinan, E. F., Haswell, C. A., Henning, T., Holden, B., Jenkins, J., Jones, H. R. A., Kaminski, A., Kiraga, M., Kürster, M., Lee, M. H., López-González, M. J., Montes, D., Morin, J., Ofir, A., Pallé, E., Rebolo, R., Reffert, S., Schweitzer, A., Seifert, W., Shectman, S. A., Staab, D., Street, R. A., Suárez Mascareño, A., Tsapras, Y., Wang, S. X., & Anglada-Escudé, G. (2018). A candidate super-Earth planet orbiting near the snow line of Barnard's star. *Nature, 563*, 365–368. https://doi.org/10.1038/s41586-018-0677-y.

Gliese 832c

Wittenmyer, R. A., Tuomi, M., Butler, R. P., Jones, H. R. A., Anglada-Escude, G., Horner, J., Tinney, C. G., Marshall, J. P., Carter, B. D., Bailey, J., Salter, G. S., O'Toole, S. J., Wright, D., Crane, J. D., Schectman, S. A., Arriagada, P., Thompson, I., Minniti, D., Jenkins, J. S., & Diaz, M. (2014). GJ 832c: A super-earth in the habitable zone. *The Astrophysical Journal., 791*(2), 114. https://doi.org/10.1088/0004-637X/791/2/114. https://arxiv.org/pdf/1406.5587.pdf.

K2-155d

Hirano, T., Dai, F., Livingston, J. H., Fujii, Y., Cochran, W. D., Endl, M., Gandolfi, D., Redfield, S., Winn, J. N., Guenther, E. W., Prieto-Arranz, J., Albrecht, S., Barragan, O., Cabrera, J., Cauley, P. W., Csizmadia, S., Deeg, H., Eigmüller, P., Erikson, A., Fridlund, M., Fukui, A., Grziwa, S., Hatzes, A. P., Korth, J., Narita, N., Nespral, D., Niraula, P., Nowak, G., Pätzold, M., Palle, E., Persson, C. M., Rauer, H., Ribas, I., Smith, A. M. S., & Van Eylen, V. (2018). K2-155: A bright metal-poor M dwarf with three transiting super-earths. *The Astronomical Journal, 155*, 124. https://arxiv.org/abs/1801.06957.

Luyten b (GJ 273b)

Astudillo-Defru, N., Forveille, T., Bonfils, X., Segransan, D., Bouchy, F., Delfosse, X., Lovis, C., Mayor, M., Murgas, F., Pepe, F., Santos, N. C., Udry, S., & Wunsche, A. (2017). The HARPS search for southern extra-solar planets XLI. A dozen planets around the M dwarfs GJ 3138, GJ 3323, GJ 273, GJ 628, and GJ 3293. *Astronomy & Astrophysics, 602*, 21. https://arxiv.org/pdf/1703.05386.pdf.

LHS 1140

Dittmann, J. A., Irwin, J. M., Charbonneau, D., Bonfils, X., Astudillo-Defru, N., Haywood, R. D., Berta-Thompson, Z. K., Newton, E. R., Rodriguez, J. E., Winters, J. G., Tan, T.-G., Almenara, J. M., Bouchy, F., Delfosse, X., Forveille, T., Lovis, C., Murgas, F., Pepe, F., Santos, N. C., Udry, S., Wünsche, A., Esquerdo, G. A., Latham, D. W., & Dressing, C. D. (2017). A temperate rocky super-Earth transiting a nearby cool star. *Nature, 544*, 333–336. https://arxiv.org/abs/1704.05556.

Proxima b

Boutle, I. A., Mayne, N. J., Drummond, B., Manners, J., Goyal, J., Lambert, F. H., Acreman, D. M., & Earnshaw, P. D. (2017). Exploring the climate of Proxima B with the met office unified model. *Astronomy & Astrophysics, 601*, 13. https://arxiv.org/abs/1702.08463.

Feng, F., & Jones, H. R. A. (2018). Was Proxima captured by alpha Centauri A and B? *Monthly Notices of the Royal Astronomical Society, 473*(3), 3185–3189. https://doi.org/10.1093/mnras/stx2576.

Ribas, I., Bolmont, E., Selsis, F., Reiners, A., Leconte, J., Raymond, S. N., Engle, S. G., Guinan, E. F., Morin, J., Turbet, M., Forget, F., Velasco, G., & Anglada-Escudé, G. (2016). The habitability of Proxima Centauri b I. Irradiation, rotation and volatile inventory from formation to the present. *Astronomy & Astrophysics, 596*, A111. https://arxiv.org/pdf/1608.06813.pdf.

Turbet, M., Leconte, J., Selsis, F., Bolmont, E., Forget, F., Ribas, I., Raymond, S. N., & Anglada-Escudé, G. (2016). The habitability of Proxima Centauri b. II. Possible climates and observability. *Astronomy & Astrophysics, 596*, A112. https://doi.org/10.1051/0004-6361/201629577. https://www.aanda.org/articles/aa/pdf/2016/12/aa29577-16.pdf.

Ross-128b

Bonfils, X., Astudillo-Defru, N., Díaz, R., Almenara, J.-M., Forveille, T., Bouchy, F., Delfosse, X., Lovis, C., Mayor, M., Murgas, F., Pepe, F., Santos, N. C., Ségransan, D., Udry, S., & Wünsche, A. (2016). A temperate exo-Earth around a quiet M dwarf at 3.4 parsecs. *Astronomy & Astrophysics, 613*, 9. https://arxiv.org/abs/1711.06177.

Teegarden b and c

Zechmeister, M., Dreizler, S., Ribas, I., Reiners, A., Caballero, J. A., Bauer, F. F., et al. (2019). The CARMENES search for exoplanets around M dwarfs—Two temperate Earth-mass planet candidates around Teegarden's Star. Astronomy & Astrophysics, 627, A49. https://doi.org/10.1051/0004-6361/201935460.

Wolf-1061c

LePage, A. (2017). I also draw the reader's attention to an article by Andrew LePage at: https://www.drewexmachina.com/2017/03/25/habitable-planet-reality-check-wolf-1061c-revisited/.

Wright, D. J., Wittenmyer, R. A., Bentley, C. G., Riedel, J. S., & Jinglin, Z. (2015). Three planets orbiting Wolf 1061. *The Astrophysical Journal., 817*, L20. https://doi.org/10.3847/2041-8205/817/2/L20. https://arxiv.org/abs/1512.05154v1.

Gliese 581d

Barnes, R., Mullins, K., Goldblatt, C., Meadows, V. S., Kasting, J. F., & Heller, R. (2013). Tidal Venuses: Triggering a climate catastrophe via tidal heating. *Astrobiology, 13*, 225–250. https://arxiv.org/pdf/1203.5104.pdf.

Barnes, R. (2017). Tidal locking of habitable exoplanets. *Celestial Mechanics and Dynamical Astronomy, 129*, 509–536. https://link.springer.com/article/10.1007/s10569-017-9783-7.

Gregory, P. C. (2011). Bayesian re-analysis of the Gliese 581 exoplanet system. *Monthly Notices of the Royal Astronomical Society, 415*, 2523–2545. https://arxiv.org/pdf/1101.0800.pdf.

Lestrade, J.-F., Matthews, B. C., Sibthorpe, B., Kennedy, G. M., Wyatt, M. C., Bryden, G., Greaves, J. S., Thilliez, E., Moro-Martín, A., Booth, M., Dent, W. R. F., Duchêne, G., Harvey, P. M., Horner, J., Kalas, P., Kavelaars, J. J., Phillips, N. M., Rodriguez, D. R., Su, K. Y. L., & Wilner, D. J. (2013). A debris disk around the planet hosting M-star GJ 581 spatially resolved with Herschel. *Astronomy & Astrophysics, 548*, 15. https://arxiv.org/pdf/1211.4898.pdf.

Lingam, M., & Loeb, A. (2018a). Dependence of biological activity on the surface water fraction of planets. *The Astronomical Journal, 157*, 25. https://arxiv.org/pdf/1809.09118.pdf.

Lingam, M., & Loeb, A. (2018b). Is extraterrestrial life suppressed on subsurface ocean worlds due to the paucity of bioessential elements? *The Astronomical Journal, 156*, 151. https://arxiv.org/pdf/1806.00018.pdf.

Loeb, A., Batista, R. A., & Sloan, D. (2016). Relative likelihood for life as a function of cosmic time. *Journal of Cosmology and Astroparticle Physics, 2016*, 10. http://arxiv.org/pdf/1606.08448v2.pdf.

Rogers, L. A. (2015). Most 1.6 Earth-radius planets are not rocky. *The Astrophysical Journal, 801*(41), 1–13. https://doi.org/10.1088/0004-637X/801/1/41.

Tuomi, M. (2011). Bayesian re-analysis of the radial velocities of Gliese 581: Evidence in favor of only four planetary companions. *Astronomy & Astrophysics, 528*, L5. https://arxiv.org/pdf/1102.3314.pdf.

Tuomi, M., Pinfield, D., & Jones, H. R. A. (2011). Application of Bayesian model inadequacy criterion for multiple data sets to radial velocity models of exoplanet systems. *Astronomy & Astrophysics, 532*, A116. https://arxiv.org/pdf/1106.5981.pdf.

Wheatley, P. J., Louden, T., Bourrier, V., Ehrenreich, D., & Gillon, M. (2016). Strong XUV irradiation of the Earth-sized exoplanets orbiting the ultracool dwarf TRAPPIST-1. *Monthly Notices of the Royal Astronomical Society: Letters, 465*, L74–L78. https://arxiv.org/pdf/1605.01564.pdf.

Wordsworth, R. D., Forget, F., Selsis, F., Millour, E., Charnay, B., & Madeleine, J.-B. (2011). Gliese 581d is the first discovered terrestrial-mass exoplanet in the habitable zone. *The Astrophysical Journal Letters, 733*, L48. https://arxiv.org/pdf/1105.1031.pdf.

Zendejas, J., Segura, A., & Raga, A. C. (2010). Atmospheric mass loss by stellar wind from planets around main sequence M stars. *Icarus, 210*, 539–544. Preprint available at http://arXiv.org/pdf/1006.0021v1.pdf.

Chapter 9
A New Hope

Introduction

In the last decade two new planetary systems have emerged from radial velocity data, those of Proxima Centauri and TRAPPIST-1. Proxima is a 0.12 solar mass red dwarf and the closest star to our Sun, making it a particularly attractive target for further investigation. Meanwhile, the older TRAPPIST-1 system has seven planets, several of which could be habitable—at least in the crudest sense. Although each system has its detractors and flaws, both could host sophisticated life and are easily the best systems we know of that could be habitable.

This final chapter examines both of these systems and considers their habitability in the light of all of the factors we considered in this book. It's crucial that all of these considerations are taken into account, rather than just the position of the planet in the circumstellar zone—and its bulk size. To this end, the key (and simple) concept of niche amplitude can be used to dissect bulk habitability, while age, mass and composition can be used to consider the nature of life on these worlds and their potential to survive. Moreover, we can now apply some twentieth century phytoclimatic (coupled vegetation-climate) modeling to investigate the true biological potential of planets.

Above all, within the dream world-like models, which are discussed here, there is a critical assessment of the relevant geophysics. TRAPPIST-1e is old, and the age of the planet is taken into account in the consideration of its atmosphere. Proxima b may be very dry by terrestrial standards. This is also modeled, with reasonable impacts on the geophysics of the planet considered, alongside the impact on (potential) habitability. Although the planets envisaged here are hypothetical, perhaps even fanciful, we underpin the speculation with sufficient hard science to determine just what kinds of life *could* be possible given these circumstances. The outcomes may be surprising.

D. S. Stevenson, *Red Dwarfs*, https://doi.org/10.1007/978-3-030-25550-3_9

Proxima b

The System

Lying a paltry 4.22 light years away is the 4.85-billion-year-old Proxima Centauri. As Chaps. 1 and 5 explored, this star has a mass roughly one-eighth that of the Sun and is in the stellar class M5.5Ve. The "e" at the end is significant, as it indicates that this red dwarf releases periodic flares. Although these are not particularly powerful as such stars go, they do result in significant increases in the intensity of UV and X-rays that Proxima b must experience. Proxima Centauri is the outlier of the Alpha Centauri double system and, therefore, is a reasonable, albeit less massive, analog of Gliese 667C.

The planet Proxima b has a mass of 1.3 times that of Earth, a figure that was clearly defined by the radial velocity method. This mass places it clearly in the telluric class—but potentially with sufficient volatiles to be an aquaplanet. The radius is clearly not well-defined, and estimates vary from 0.8 to 1.4 Earths, depending on which model composition you decide to choose. In terms of its insolation, Proxima b receives roughly two-thirds the amount of radiation as Earth, likely making it cooler than Earth, despite the relatively higher proportion of infrared radiation emitted by the star. Best fit models give it an equilibrium temperature of around 234 K (compared to 255 K for Earth). Proxima b lies relatively close to the inner edge of its star's habitable zone, just like Earth. In the case of Proxima Centauri, this region stretches from 0.0423 to 0.0816 AU, giving planets within this range orbital periods that last from 9.1 to 24.5 days. Table 9.1 summarizes this information.

At the moment, we know of no other planet in this system, making Proxima b one of the few red dwarf systems with only one world. (Barnard b is another.) Guillem Anglada-Escudé (Queen Mary University) and colleagues did detect a weak secondary signal in the range of 60 to 500 days, but its nature is still unclear due to stellar activity and incomplete sampling of data. Although Proxima b appears

Table 9.1 The physical properties of Proxima b

Property	Proxima b
Mass (Earths)	1.3
Radius (Earths)	0.88–1.4
Density (Earths)	0.9–1.4
Orbital radius (AU)	0.05
Length of year (Earth-days)	11.2
Equilibrium temp. (°C)	−39.1 (234 K)
Surface water?	Possible
Habitable (possible life)?	Yes

to be an only child, like the other red dwarfs that have potentially habitable planets, it is a super-terran—albeit one of relatively low mass.

Finally, the location of the planet *suggests* that it didn't form in its current orbital location. More likely the planet either migrated into its current orbit or perhaps has had its orbit maneuvered in whatever encounter brought Proxima into the Alpha Centauri system (Chap. 1). Thinking firstly of migration, if the planet underwent Type I migration, then interactions between the gas disc and the planet would bring a relatively volatile-rich world inwards. However, if migration occurred at the pebble interaction stage, impacts would deplete the planet of volatile materials.

If we consider the more dramatic scenario, illustrated in Fabo Feng and Hugh Jone's simulations, then, conceivably, Proxima b may have formed and initially remained in a far more distant orbit. Here, it would accrete volatile-rich material beyond the snow-line. Then, as Proxima was captured, gravitational interactions between the three stars may have bodily shoved into its new orbit. Finally, interactions between Proxima b and the plausible more distant planet may have maneuvered Proxima b into its current short-period orbit. Any of these scenarios would have significant impacts on the planet's inventory of volatile substances and hence its long-term habitability.

These kinds of issues may be resolved when the James Webb telescope finally becomes operational. Construction of it is now running a decade over its original operational planned date, with the latest slated time for orbital insertion being 2022. Ignas Snellen (Leiden Observatory) and co-workers have already decided what observations they will carry out with this orbiting observatory. At superior conjunction, where the planet lies roughly between Earth and Proxima b, there should be sufficient irradiation of the atmosphere to allow the spectral detection of carbon dioxide at 15 microns with good resolution. Water vapor should also be detectable in the near-infrared. These measurements will be critical in determining both the broad composition of the atmosphere and the likely habitability within this sea of gas. The composition will set the temperature and also modulate the motion of gases.

The Planetary Geosphere

Let's turn our attention to the lump of rock that forms the bulk of Proxima b. Although we don't know an awful lot about it, aside from its mass (1.3 Earths), we can make some reasonable assumptions. Firstly, it is likely to be roughly chondritic in composition—that is, it will be mostly rock, which means that it likely has around 0.2% of its mass as volatile substances. Some models of red dwarfs suggest planets may be born with more water than Earth was. However, this is balanced by greater loss of water when the infant star is still coalescing (Chap. 5). Assuming that this bulk composition is broadly correct, then all things being equal, the surface *could* have oceans and an atmosphere that is largely transparent to visible radiation. So far, so good.

Given its mass and likely density, the mantle should be hot enough to convect and the lithosphere thick enough to conduct force, without becoming so rigid that it won't subduct (Chap. 3). Thus, while we cannot be certain that the lithosphere has plate tectonics, if the planet has a hydrosphere and the mantle remains suitably hot, the lithosphere should be of the right consistency to subduct and recycle, like Earth. Of course, the planet *is* more massive than Earth, which makes any continental lithosphere more prone to spreading and hence more likely to impede plate tectonics (Chap. 3).

Proxima Centauri is estimated to be marginally older than the Sun, at 4.83 billion years. If the planet has plate tectonics, something that is utterly reasonable with a modest inventory of volatiles and a broadly chondritic composition, then it should have synthesized at least as much continental crust as Earth. However, beware. Adrian Lenardic (Rice University) has suggested plate tectonics might shut off when continents cover 50% of the surface. Earth has 40% coverage at 4.6 billion years of age. Therefore, it is possible that Proxima b has reached a threshold where plate tectonics gums up. And with the continental landmasses more prone to spreading on more massive planets, this could be a problem. If we are kind enough to Proxima b, there may be a greater volume of granitoid crust, but not so much that the possibility of plate tectonics is excluded.

Next, what about the presence of a magnetic field? The stellar field strength at Proxima b is many times that experienced by Earth. Proxima's source field is 600 times stronger than the Sun (600 versus 1G), and such intense magnetic fields give rise to strong X-ray (and extreme ultraviolet, EUV) fluxes. Indeed, at Proxima b's location, this super-terran receives approximately 400 times the atmospheric bombardment that Earth does. However, if Proxima b began life volatile-rich, even this seemingly catastrophic irradiation may not have stripped the atmosphere and decimated the planet's potential for life.

If plate tectonics is operational, then the planet is likely to have a circulating iron-nickel core and a magnetic field. The low rotation rate (11.2 days) certainly reduces the likely strength of any dipole field, and suitable mantle convection could drive sufficient circulation to produce a more complex field than is seen on Earth. Thinking more deeply about this, Mars and the Moon have (had, in the case of the Moon) a singular plume—degree-1 convection. Earth has a degree-2 circulation with two broad superplumes underlying Africa and the Pacific. Given its likely diameter, Proxima b is likely to have mantle convection like Earth does, with two hemispheric plumes flanked by zones of subduction (or more general zones of mantle-downwelling). This broad circulation pattern should facilitate circulation in the iron-nickel core and hence make the formation of a planetary field more likely.

Yes, there are many uses of the words "likely" and "probably" here, but these are reasonable, albeit undetermined quantities so far. Observations with the long overdue James Webb telescope should resolve these issues. One should also bear in mind that changes to the composition of the planet are worth considering. Lower the abundance of iron and nickel and increase the volume of the mantle (or increase the abundance of non-metals in the core), and we could increase the rate of convection in the core. This could offset the effect of limited (or episodic) circulation.

To sum up, Proxima b has a mass and an orbital separation that are highly amenable to the origin and development of habitable conditions for life. These conditions are offset by a number of factors; these include the probability of atmospheric erosion by Proxima Centauri and the possibility that plate tectonics has ceased because the surface has clogged with granite. That said, the probability that the surface is habitable is not eliminated by these constraints.

Possible Climates

As Colin Goldblatt (University of Victoria) illustrates in his models, the climate of Proxima b is likely to be temperate if we assume the planet has an atmosphere with a density at least 10% as dense as Earth's—and with a similar nitrogen-dominated composition.

Taking this somewhat further are the Met Office models by Ian A. Boutle and colleagues. Their models of Proxima b show that the temperature is above freezing within 50°–60° longitude of the SSP and in a somewhat greater breadth of latitude on either side of the SSP. Altering the atmosphere from an Earth-like composition to one that has nitrogen only reduces the temperature distribution and the extent of super-rotation, as we might expect from results presented in Chap. 5. The somewhat elongated shape of the zone of precipitation is partly due to the Rhines-circulation pattern, which tends to lengthen the area of convergence in a north-south direction.

The Boutle models also show modest super-rotation around the planetary equator that extends to the mid-latitudes, though air super-rotates aloft. This model is similar to, but not quite the same as, Eric Wolf's, Ravi Koparappu's and Jacob Haqq-Misra's model, which would be applicable here. The Met Office model has a modest crab claw-like pattern that is focused on the SSP. Maximum precipitation extends to around 30° either side of the SSP, but with a Rhines-like circulation pattern, this is extended quite far to the north and south of the SSP (indeed, almost as far as 60° north and south). The temperature map broadly matches the distribution of precipitation. Temperatures are nowhere extremely high on the day-side, peaking in the mid-high twenties Celsius. However, once you move beyond 50° on either side of the SSP, temperatures fall away quickly to levels that are not habitable for complex life that requires sunlight. A deep biosphere is possible, but nothing more, despite access to visible radiation.

A rather interesting—and unique—addition to these models is the manner in which precipitation occurs—and something that is ignored in all of the other models this author has seen. An obvious question is whether the precipitation is in the form of rain or snow. The Met Office model shows that around the edge of the modest crab-claw rain zone, precipitation is always of snow. Thus, in a distribution that is approximately 20° wide, precipitation falls as snow in a region 50°–70° east and west of the SSP; and 50°–75° north and south of the SSP. This is illustrated in Fig. 9.1.

The upshot is that in these regions there should be a net growth of ice sheets on any land. In turn, this may lower global temperatures and reduce evaporation and precipitation around the SSP until the climate is stabilized. It also means that any land in these regions, despite potentially viable temperatures for complex life, will

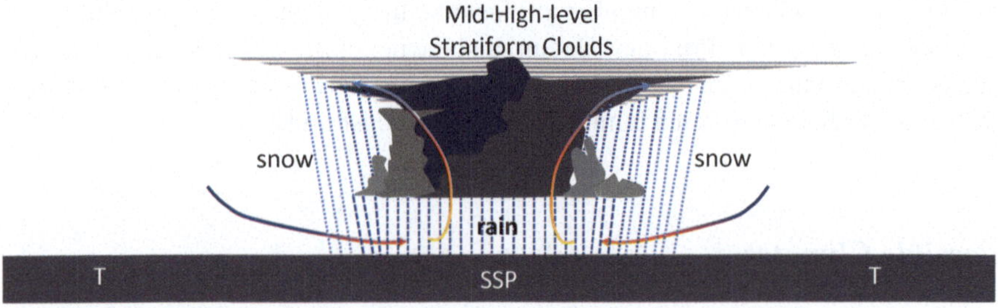

Fig. 9.1 Snow on Proxima b. As air ascends above the sub-stellar point, the SSP, it brings in colder air from all sides. As this air is day-side cold, precipitation falling into the inflow, turns to snow. Snow fall will be a permanent feature in this region and this will limit the zone of temperate climate by promoting the formation of ice sheets. *T* Terminator, *SSP* Sub-stellar point

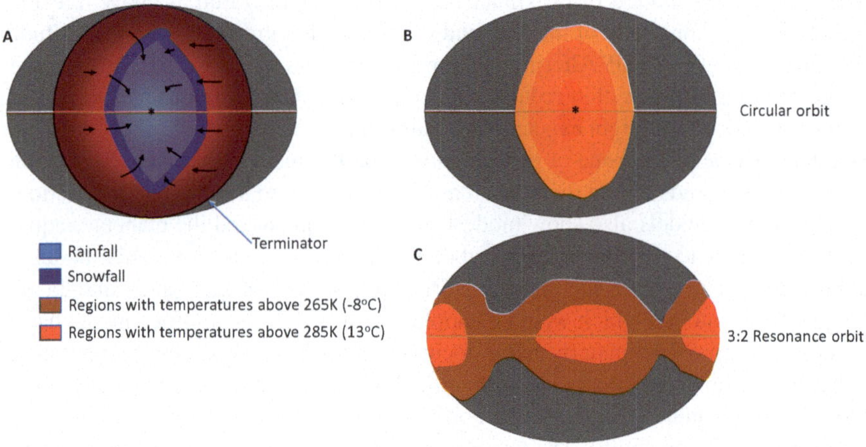

Fig. 9.2 The Met Office simulation of climate on Proxima b. (**a**) The distribution of rainfall and snowfall; (**b**) the temperature if the planet is tidally-locked; and (**c**) if the planet is rotating as Mercury does in a 3:2 resonance—meaning three rotations of the planet for every two orbits of the star. Note the presence of snowfall in the region surrounding the SSP but well inside the terminator. The asterisk marks the SSP in (**a**) and (**b**). Modified from Bouttle et al. (2017)

only experience simple life in a sub-surface biosphere. Life that is more complex could adapt and migrate onto this region, but it could not evolve from simpler forms that are confined to these terrains. This is mostly because any food chains that can persist there must obtain nutrients from elsewhere. Sub-ice life—in an ocean or any liquid but ice-capped bodies of water—is the exception, should these exist. (Figure 9.2 summarizes the Met Office Proxima b climate model.)

What about the effect of land distribution? As the terrestrial Pleistocene ice ages showed, ice did not grow evenly across the northern hemisphere. Ice sheets grew where winds delivered moisture to the growing ice mass. Thus Scandinavia, Scotland and eastern and central North America saw the greatest growth of ice, along with the Pacific Northwest. The colder bulk of Siberia saw virtually no glaciation simply because the air that flowed over it was dry. A similar phenomenon occurred over North America, with the Rockies intercepting moisture and leaving a dry, cold corridor between the Laramide ice sheet and the main Laurentian cap. This ice-free corridor had important ramifications for the dispersal of species, including modern humans. Therefore, we should not expect evenly glaciated land regions flanking the SSP; rather, areas where precipitation is high will see a build-up of ice, while areas shielded from moist inflow (by land or mountains) will remain ice free.

The Met Office Proxima b models illustrate a couple of other key points, such as the effect of clouds and the effect of the overall circulation. In most models, the atmosphere of planets remains relatively moist and cloudy. However, in the Met Office models, strong subsidence away from the SSP effectively renders the lower atmosphere cloud-free, or with such thin cloud that heat readily escapes to space. Overall, this renders Proxima b colder than in other, cloudier models. Subsidence-driven heating is minimal above the dark hemisphere, with the lowermost kilometer or so of air able to shed heat effectively. Moreover, as suggested in an article published in 2018, the depth of convection varies as you approach the SSP and with it the kinds of clouds that will be present.

Most of the zone of inflow to the SSP will have only very shallow convection, and in the author's empirical model, virtually no rainfall in areas close to the SSP that have long stretches of land over which the air is flowing. Think of the Sahel. Despite strong, albeit shallow, convection over Mali, Chad and neighboring regions, precipitation is limited because the air flowing into this zone is bone dry (and is known as the Saharan air layer). Convective precipitation only occurs where there is an influx of moist air from the equatorial Atlantic.

As you cross the Atlantic from the Saharan coast you can see the effect of moisture and land distribution. The Saharan air becomes mixed with moist Atlantic air. There is less subsidence and ultimately the overlying inversion breaks by the time one reaches the Caribbean. This allows moist orographic and convective precipitation in abundance over the latter region and neighboring Florida in the summer. Ditto but in a similar vein, Hawaii. This lies some considerable distance from the equator, yet the windward slopes of Hawaii, Maui and Oahu all receive copious precipitation as moist air is forced up over the volcanic peaks of these islands. Context, therefore, is crucial.

Another interesting outcome of the Met Office models is the manner in which convection operates near to the SSP. Unlike Earth, which has a Sun that migrates across the equatorial sky in a diurnal manner, heating is always from above and is focused on one area. This has the effect of heating and stabilizing the upper atmosphere. Consequently, only around 50% of the convection that occurs produces deep clouds—and this is only likely to consist of transient, strongly convecting plumes. The remaining convection—even at the SSP—is low level. The net effect is that only 50% of the precipitation in the areas flanking the SSP is convective. The remainder falls from a higher level, but persistent stratiform clouds populate the sky (principally altostratus and altocumulus). It is these clouds that precipitate rainfall and snowfall away from the SSP out towards the terminator. As suggested in the author's second 2018 paper, additional precipitation may fall along the equatorial band where the air tends to converge in the short-period models of Jacob Haqq-Misra and colleagues. Again, this comes from higher-level clouds that tend to persist over broader areas than are seen on the tropical parts of Earth. This is uncertain at present and would almost certainly require orographic forcing.

A key observation is that the planetary hydrological cycle is limited to 60° on either side of the SSP. With limited or non-existent precipitation outside this region, the sole job of the atmosphere is to transport heat in effectively dry air. Again, this strongly constrains the habitability of tidally locked planets, at least when one considers what could exist on land.

Finally, the Met Office group modified the period of the planet so that it orbited in a 3:2 resonance—that is, completed three spins for every two orbits. This configuration gave a model Earth-like climate, but the surface temperature was only above freezing in a band extending 35° north and south of the equator. This model did not include a terrestrial atmospheric composition, only a nitrogen-dominated one. Therefore, the effect of greenhouse warming on the overall temperature and precipitation cannot be assumed. More importantly, the group found that if you include the effects of precipitation as snow, the likeliest outcome is that Proxima b would descend into a snowball state if its spin-orbit cycle was in a 3:2 resonance. Such a planet has a day and night that each last approximately 12 Earth days (288 h). Here, the eccentricity is set to 0.3, giving an orbit similar to that of Mercury. This frigid state never becomes a hard snowball, but it is still rather chilly by modern terrestrial standards. In the most eccentric 3:2 resonance models, only the sides of the planet that face the star at periastron (the closest approach) remain ice free. These worlds have a slushy snowball state, rather than a hard snowball, because of the cooling at apastron (the furthest point in the orbit). Overall, the areas that are either ice free or have temperatures that are above the freezing point of water are greater in the 3:2 resonance model than the tidally locked model.

Using the tidally-locked model for Proxima b, Fig. 9.3a represents a hypothetical land configuration on this map of this world and indicates mantle upwelling and subduction. Figure 9.3b assumes that the surface is partly clogged with granite, so that any surface horizontal tectonics is regional, rather than global—something unfamiliar on Earth. For the sake of simplicity, this is not considered further here, but must be considered when one examines habitability on super-terrans.

Figure 9.4 adds the appropriate atmospheric circulation pattern, which in this case is most likely a Rhines-rotator pattern if sufficient atmosphere remains. This has once again been tweaked to reflect the effects of the mountains and basins superimposed on top. The effect of these is to add extra "swirls" to the overall flow towards the SSP. The interesting and relatively unique impact of the overall circulation is that the flow at the surface is relatively shallow. On Earth, the equivalent trade winds extend through the entire depth of the troposphere (the atmosphere's lowest layer). However, on tidally locked worlds this is capped by gently downwelling airflow at a few thousand meters—and probably less than 3000—furthest from the SSP (Chap. 5). This means that the air flowing towards the SSP at the surface is trapped into a layer that will readily be blocked by mountains or even a

b

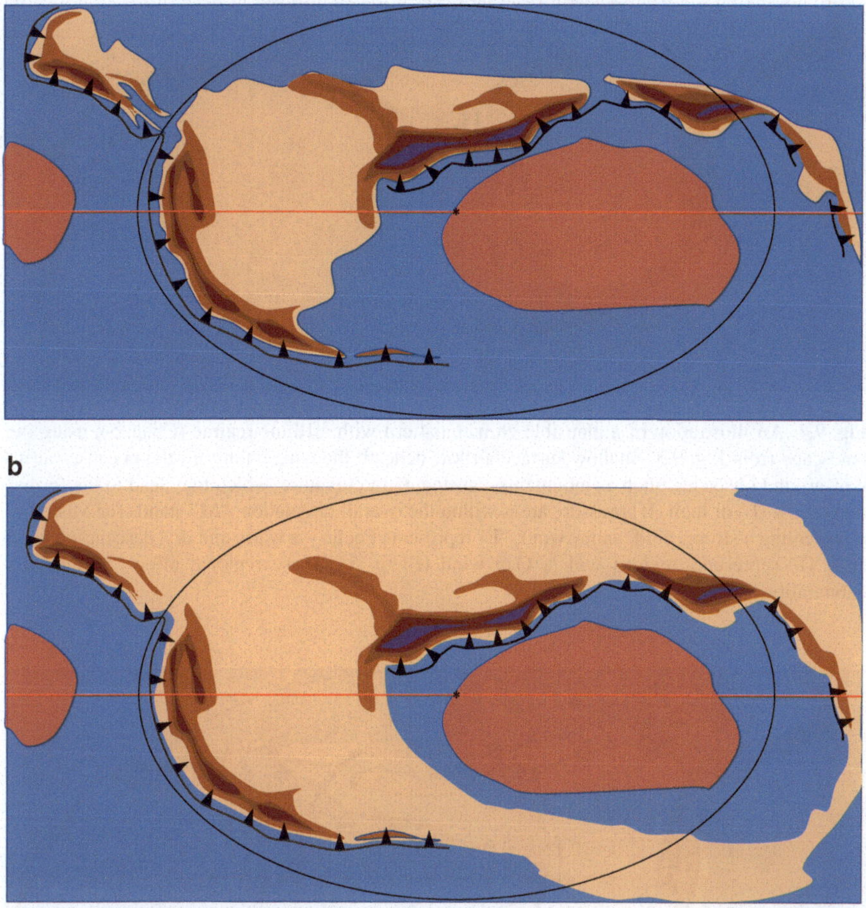

Fig. 9.3 (**a**) A hypothetical Proxima b—assuming volatiles are present. Continents are brown; oceans are blue and mountainous regions are color-coded, from mid-brown to purple. Mantle super-plumes (LLSVPs) are indicated in translucent orange-red; subduction by black lines and triangles. The SSP is marked with a black asterisk. The terminator is marked by the dark oval line in this and the two subsequent figures. (**b**) A hypothetical Proxima b—assuming volatiles are present but with continental spread. Continents are brown; oceans are blue and mountainous regions are color-coded, from mid-brown to purple. Mantle super-plumes (LLSVPs) are indicated in translucent orange-red; subduction by black lines and triangles. The SSP is marked with a black asterisk. The terminator is marked by the dark oval line in this and the two subsequent figures. Here, the continental crust is so extensive that while subduction is possible in some areas, much of the surface is static

Hawaiian-sized islands. Any moisture is then forced out onto windward slopes rather than squeezing easily over their summits. This makes orographic rainfall more probable—or where the air is dry, forces the flow around the sides of the mountains in a strong flanking-flow. Figure 9.5 illustrates the effect of topography on precipitation on Proxima b if it has retained volatiles.

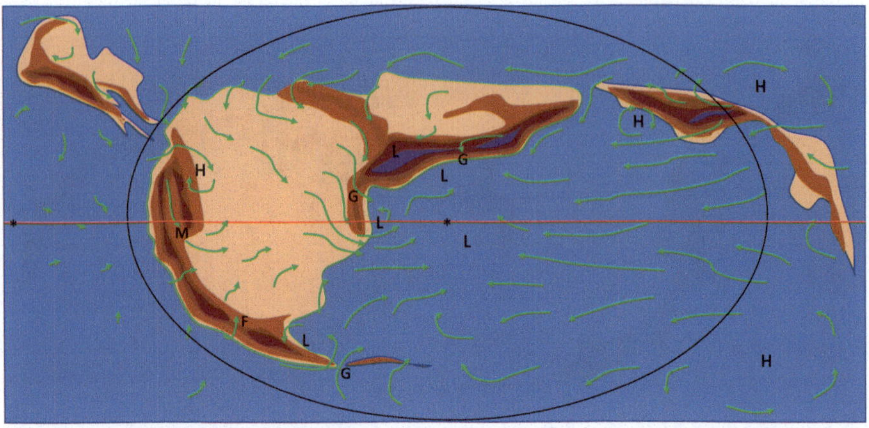

Fig. 9.4 An illustration of a plausible climate pattern with a Rhine-regime (Chap. 5), using the landscape from Fig. 9.3. Shallow surface airflow beneath the temperature inversion causes air to flow around obstacles, such as mountains, causing extra curvature, which may lead to the formation of low (L) or high (H) pressure areas within the overall circulation. "**M**" stands for Mistral—representing a strong, cold, valley wind; "**F**" represents Foehn—a warm and dry, descending wind and "**G**" represents regions with a Gap wind (Chap. 5). These regional effects will impact habitability

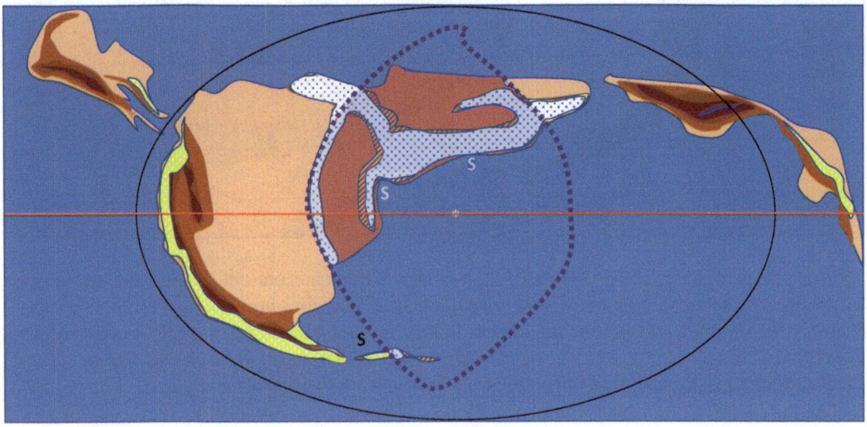

Fig. 9.5 Precipitation on model Proxima b with volatiles. As with the model in Chap. 5, this model incorporates the precipitation pattern from Bouttle et al., with snow falling around the periphery of the main zone of convection. Color-coding is the same as in Chap. 5. Red is convective, yellow orographic, striped is mixed convective and orographic and stippled, snow. No frontal precipitation is indicated in this model as these regions lie outside areas where there is likely to be any significant precipitation in this model. Additional (likely light) snow fall is indicated in yellow stippled regions on windward slopes

Proxima b's Prospects for Habitability

Overall, Proxima b falls into the "not bad" group—if it has not been stripped by Proxima's flares. The overall low temperature of the planet will make it less likely that all of the planet's volatiles will have been lost to space, and estimates from Ignasi Ribas (Institut de Ciències de l'Espai (IEEC-CSIC) and colleagues suggest that Proxima b would have "only" lost an Earth's ocean-worth of water and other volatiles during the collapse phase of its star. As Earth has twice this in total (an equal volume of volatiles in the oceans and the mantle) and Proxima b is 1.3 times as massive, the planet should have had more than enough to support an early biosphere after its formation. The questions that remain are: Did it have the same proportions of volatile materials at birth; and did mantle outgassing replenish anything that was lost to space?

The relatively low mass of the planet also makes it less likely that the surface will be in a stagnant-lid regime, driven by the build-up of continental crust. However, the final word on habitability will come down to the retention of volatiles. This will constrain both the abundance of life-giving water, the overall temperature of the planet and the sustainability of plate tectonics—or other surface tectonism. We must wait until decent measurements of the atmosphere arrive to determine these characteristics, for it will be a measurement of the surface temperature and carbon dioxide abundance that tells us how dense the atmosphere is likely to be and, by inference, how much water (if any) is retained in liquid form. The status of Proxima b currently, thus, hovers between that of an exo-Earth and an exo-Mars—albeit one which is likely to be far more geologically active. Nevertheless, of the planets that we've discussed so far, Proxima b is by far the best candidate—and it's close enough to realistically consider traveling to, in one form or another, in a little over a generation.

To sum up what is possible—and hopefully probable—Proxima b is a good candidate for life, if it has retained enough water to form oceans or smaller seas. Figure 9.6 celebrates what might be possible. Bear in mind "a good candidate for life" does not equate to a good candidate for complex life. That will depend on the number and variety of available niches.

Now, all of this is a best-case scenario. What about if Proxima b has been baked by repeated flares and coronal mass ejections, as seems rather likely? We now enter the realm of utter speculation and have some fun with it. Figure 9.7 suggests a dry Proxima b that still retains at least a Mars-like atmosphere. Unlike Mars, Proxima b is very likely to be volcanically active—perhaps as much as Earth. In this scenario the planet has some volatiles, but these are largely trapped in the coldest parts of the planet. This is predominantly in the dark hemisphere, but also on mountain sides, shielded from the Sun. Ice sublimates on the day side into the thin atmosphere.

Figure 9.7 illustrates a world with hemispheric tectonics. Here, the sun-lit, dry hemisphere is underlain by an LLSVP. New lithosphere is produced by diffuse spreading above this, with it extending slowly outwards towards the terminator, where it is cold enough and dense enough to subduct. In this model we assume that

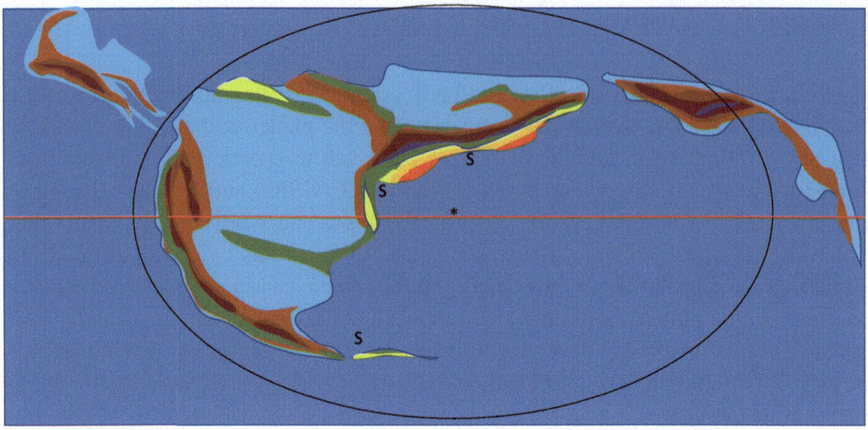

Fig. 9.6 Niche volume on a model Proxima b, with volatiles. As with the outcome in Chap. 7, the likelihood of Proxima b hosting complex life likely comes down the availability of liquid water. If we assume Figs. 9.3, 9.5 and 9.6 are reasonable, then the niche amplitude follows the water as we approach the SSP, where precipitation is greatest. In this model, the SSP is over the ocean, so much of the land remains dry and a poor, desert-like, abode for life. Red areas have the highest niche volume, grading through orange, yellow and green to blue, with the lowest niche volume, meaning that red areas will likely host the greatest numbers of species (Chap. 7). Brown represents the outline of mountains as in previous figures. Areas colored blue are likely to have only sub-surface biospheres, where geothermal energy or sources of nutrients exist.

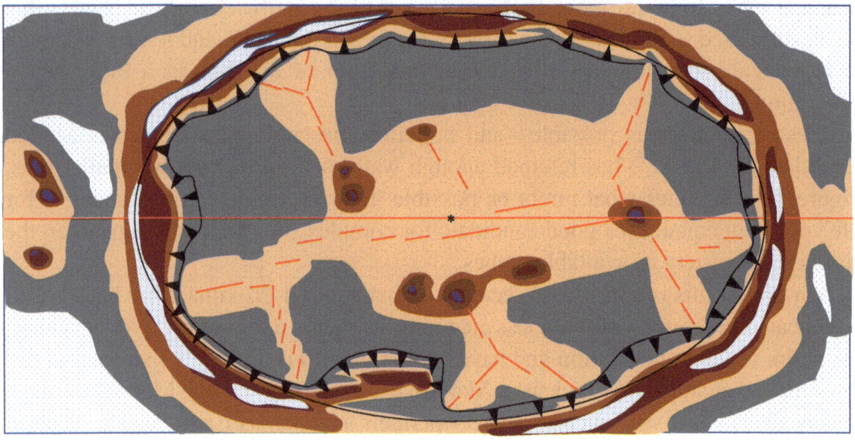

Fig. 9.7 The first of two, related, "what-if" volatile poor-world, exo-Mars-like worlds. The sun-lit hemisphere is dry and experiences hot-spot volcanism focused around the edges of an LLSVP, which is centered on the SSP. This is analogous to the Procellarum Basin facing the Earth. The dark hemisphere has volatiles (principally water and carbon dioxide) as ice (stippled regions). Hemispheric tectonism subducts thinner lithosphere in an outward-facing ring near the terminator. Lithosphere is produced diffusely near the SSP (red lines). Subduction produces highlands and volcanism along the terminator. SSP is a black asterisk. Upland areas graded in tan, through to purple. In this model there is no plate tectonics beyond the terminator as the lithosphere is cold and thick. The terminator is marked by the dark oval line

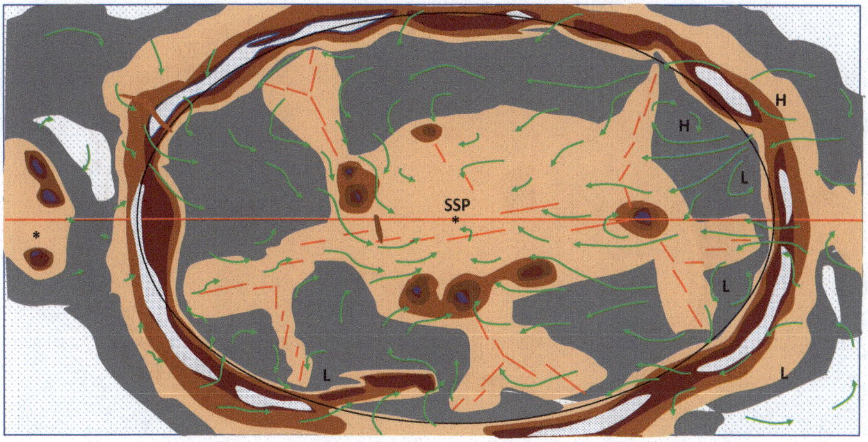

Fig. 9.8 Circulation model for this dry world. While it is superficially similar to the wet model, the increased surface friction and the presence of substantial obstacles to airflow result in far more turbulent, eddy-ridden flow. As the air is intrinsically dry, precipitation is barely affected by these

the dark hemisphere does not have conventional plate tectonics, as the lithosphere is cold and thick. However, this may not be reasonable, and either plate tectonics or a form of vertical squishy-lid tectonics may operate in a manner analogous to Venus.

This world is unlikely to be completely sterile. The presence of volcanism and some surface water (as ice on the surface) allows this Proxima b to have a deep, or sub-ice, biosphere. Even the illuminated side may have extremophilic life that is adapted to living in hot springs wherever there is volcanic activity, which vents water from the mantle. Figure 9.8 illustrates the airflow patter for this planet, and Fig. 9.9 the corresponding niche volume in this version of a dry Proxima b.

Alternatively, Fig. 9.10 shows broadly the same figure as Fig. 9.7, but here, subduction is occurring on inward dipping zones that extend towards the SSP and the dark hemisphere subducts under the warmer lithosphere of the illuminated hemisphere. In this scenario, the relatively young lithosphere of the lit hemisphere is warmer and has less water and other volatile substances than shown in Figs. 9.7 and 9.8. Again, the reasoning can be overturned to make the lithosphere as viable as in the earlier figures. However, the regions where new lithosphere is produced are viable biospheres, where volatile substances are released from the mantle. Most will freeze out in the dark, cold atmosphere. However, here, microscopic life is very possible. This limited but extensive biosphere is illustrated by a scattered and low niche volume in Fig. 9.11.

These figures are, of course, based on pure (albeit qualified) speculation. However, one hopes that they serve to illustrate the possibilities even when things appear bleak. They also illustrate the range of outcomes that can be achieved simply by tweaking the conditions one considers. To (poorly) paraphrase Mark Twain, "Reports of the demise of the viability of red dwarf planets are greatly exaggerated." News from the front has been very encouraging.

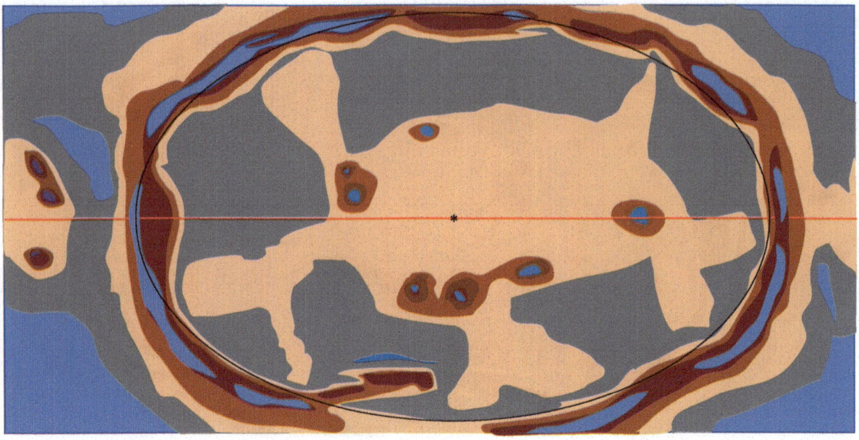

Fig. 9.9 Niche volume for the first dry model. The planet's biosphere is constrained by the lack of water. However, where there is ice, life could exist under its surface if there are supplies of geothermal energy and nutrients (Chap. 6). These are abundant where there is subduction-driven magmatism. The ice and dust-covered ocean basins on the dark hemisphere could also support a deep biosphere, albeit one that is more limited in extent. The lit hemisphere is effectively sterile except where geothermal activity delivers water and nutrients in sub-surface vents or aquifers

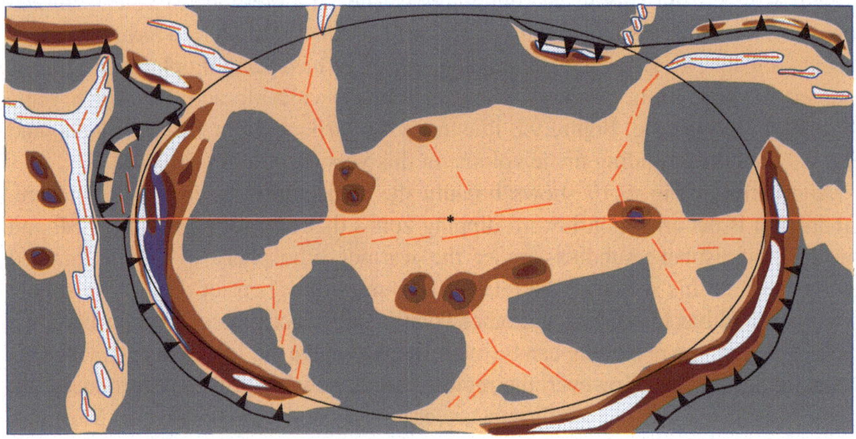

Fig. 9.10 An alternative, "what-if" volatile poor-world, Super-Mars-like world. This has the same general set-up as the last figure, with hemispheric tectonics. However, on this one the lithosphere of the hotter, dry hemisphere is "plumbed into" the dark hemisphere through regions of new lithosphere creation (red lines). Subduction is directed *towards* the SSP and LLSVP. Here, subduction retreats outwards until a ridge is reached and compressive forces are relieved. This should fill-up a large part of the dark hemisphere with buoyant subduction-resistant lithosphere over time, while the lit side remains as a domed volcanically-active region with thinner lithosphere. The dark hemisphere's ocean floor is younger, warmer and may be poorer in ice and volatiles. This makes it, a poorer abode for life than in the other scenario. Ice may form along volcanic fissures on the ridges in the dark hemisphere

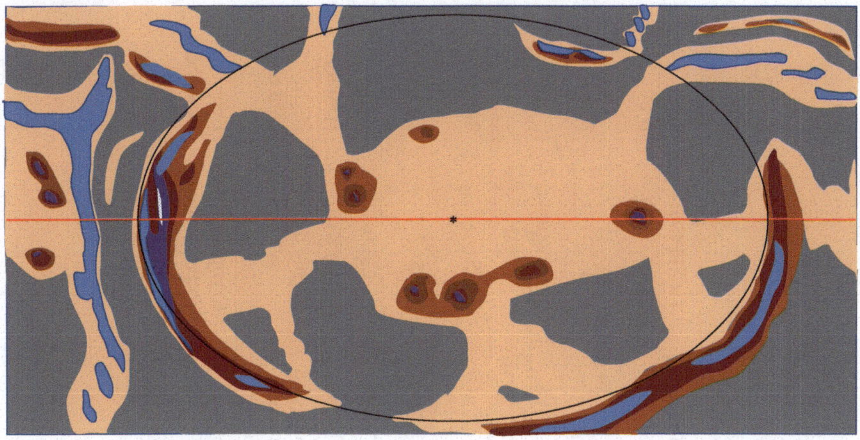

Fig. 9.11 Niche-volume for the alternative model with inward-facing subduction. Overall, the result is similar, except that if the ocean crust is young, then it will be warmer and less likely to retain volatiles. However, this depends on the overall abundance of water; how quickly the crust cools and how much water is supplied and sublimates. Volcanic ridges and the undersides of ice-caps on volcanic peaks are the likeliest abodes for microbial life (blue). Like the previous hypothetical world, this Proxima b is a poor abode for anything other than microbial life. SSP is marked with an asterisk and the terminator with a black line

TRAPPIST-1

The System

Twenty-six parsecs from Earth lies the 0.08 solar mass red dwarf TRAPPIST-1. This Jupiter-sized star is orbited by a family of Mars-sized to modest super-terran-sized planets, three of which appear to lie in the star's circumstellar habitable zone. Table 9.2 lists the properties of these seven worlds. Although the transit method was employed in their initial detection, transit-timing variation was subsequently employed to determine their respective masses. From this and the initial transit data, the densities of each of the planets was secured (Chap. 2).

Overall, their radii are relatively similar to one another, with values somewhere between 13% larger than Earth down to around that of Mars. However, though the densities of the terrestrial planets Mercury, Venus and Earth are all fairly similar, those of TRAPPIST-1b-h vary significantly; only TRAPPIST-1e has a density similar to Earth's. The others are more Mars-like, generally rocky, but clearly with substantial veneers of volatile materials and/or potentially smaller cores. TRAPPIST-1d, which has the greatest similarity to Earth in terms of the energy it receives from its star, is the least dense, the smallest and something of an anomaly.

Planet d aside, the others form something of a trend in terms of density, beginning relatively low in the case of Planet b (with a density, if not radius, similar to Mars), peaking at TRAPPIST-1e, before declining once more to TRAPPIST-1h.

Table 9.2 TRAPPIST-1 planets and their characteristics

Property	TRAPPIST 1-b	TRAPPIST 1-c	TRAPPIST 1-d	TRAPPIST 1-e	TRAPPIST 1-f	TRAPPIST 1-g	TRAPPIST 1-h
Mass (Earths)	0.85	1.38	0.41	0.62	0.68	1.34	0.33
Radius (Earths)	1.09	1.06	0.77	0.92	1.04	1.13	0.76
Density (Earths)	0.73	0.88	0.62	1.02	0.82	0.76	0.72
Orbital radius (AU)	0.011	0.015	0.021	0.028	0.037	0.045	0.063
Length of year (Earth-days)	1.51	2.42	4.04	6.10	9.21	12.35	20
Equilibrium temp. (°C)	123 (391.8 K)	61.7 (334.8 K)	9 (282.1 K)	−27 (246 K)	−54 (219 K)	−74.5 (198.6 K)	−104 (169 K)
Surface water?	Only under day-side high pressure	Marginal—if under pressure	Yes (moist greenhouse), perhaps along terminator	Yes, with green-house gases	Yes, on day-side abundant green-house gases	Highly unlikely	Highly unlikely
Habitable (any life)?	Bio-molecules unstable: no	Possible extremophiles	Thermophiles	Yes	Yes	Cryophiles and deep biosphere only	Deep bio-sphere only

Planet e is by far the most habitable

TRRAPPIST-1e is somewhat denser than Earth but a little smaller, suggesting that it has lost volatile material. The radii and hence volumes follow no particular pattern, making the entire system somewhat enigmatic. However, the misplaced, small and relatively low-density TRAPPIST-1d seems to indicate that this seemingly rather orderly system has undergone some fairly extraordinary events.

When one compares this to our system of eight, dare one say nine, worlds, the rather simple trends fall apart. In the Solar System, the density of planets generally declines from the Sun out to Saturn, before rising somewhat again at Uranus and Neptune. This reflects the abundance of material and the ability of planets to accrete hydrogen and helium from the circumstellar medium. This is clearly not the case with TRAPPIST-1, where the planets follow the opposite trend (TRAPPIST-1d excepted).

Initial Assessments of Habitability

The sheer number of potentially habitable planets in the TRAPPIST-1 system has focused the minds of most astrobiologists. Planets d through to g could be, with various tweaks, amenable to the development of life. Planet d is likely to have begun its life with a moist greenhouse, if it has retained sufficient water since inception. With an approximate 4-day-long year, the atmosphere should be strongly super-rotating, resulting in a broad region along its equatorial belt over which precipitation is likely if water has been retained.

The current equilibrium temperature is around 30° hotter than the present Earth. Given a likely age of 8 billion years, TRAPPIST1-d should have evolved into a moist greenhouse some time ago—and, indeed, may never have been cool enough for extensive oceans. Instead, the moist greenhouse that would have existed early on should, at least in principle, have evolved into a far drier state, governed by the photolysis (light-driven splitting) of water vapor in its stratosphere. The end result depends on how much water the planet had to begin with. However, if we compare its density with Earth and Mars something unexpected emerges. Mars is less massive than TRAPPIST-1d, but its density is greater. This means that a substantial portion of TRAPPIST-1d consists of volatile materials—or that it lacks a significant metallic core. Indeed, the best match in terms of density of the terrestrial worlds is the Moon.

Likewise, as we've seen, TRAPPIST-1d also bucks the trend in density across the TRAPPIST-1 system. These observations imply that TRAPPIST-1d didn't form where it now lies and that it still retains a substantial mass of low-density material. Alternatively, like the Moon, TRAPPIST-1d may be the product of a collision between larger protoplanets early in the system's history—leaving TRAPPIST-1d the equivalent of the Moon—only rather larger. Such a world would be as volatile poor as it would have been thoroughly desiccated during its violent birth. Consequently, TRAPPIST-1d would hardly be a fruitful abode for life.

In the near future, one hopes that the system will be observed in enough detail to probe the composition of its planets' atmospheres. To this end Julien deWit (MIT) and colleagues have made the first measurements of the atmospheres of Planets d, e and f. None of these worlds appear to have cloud-free atmospheres that are dominated by hydrogen. Planets g and h may well be more Neptune-like, but the results were far from conclusive. However, as far as the potentially habitable worlds go, this is good news. Hydrogen-rich atmospheres will, by definition, be poor in oxygen. The two gases are poor bedfellows, as any chemist would observe. If oxygen is necessary for the evolution of complex multicellular life, then the last thing we want to see is a Jupiter or Neptune-like planet.

Moreover, an absence of clouds would also preclude complex biology for reasons that one hopes are apparent from earlier. No clouds means no precipitation and, therefore, limited surface water. Therefore, all three worlds (TRAPPIST-1d, e and f) appear to have atmospheres of some sort, and these are likely to be cloudy. Unfortunately, the resolution of these first observations was insufficient to conclude much else. However, the best fit for the inner planets is that they have atmospheres dominated by one or more of the following: nitrogen, carbon dioxide or water. Further, higher resolution observations with the James Webb should clarify matters. In a very positive sense, the presence of atmospheres on the innermost planets suggests that atmospheric stripping by red dwarfs is either less fearsome than some expect—or that it is limited to a small number of red dwarf planetary systems. In this regard, the planets of TRAPPIST-1 may just be lucky.

Climate Models

Turning our attention to the climate, things are a little more interesting. As TRAPPIST-1 is such a small star, all of the potentially habitable worlds must have very short orbital periods. Planet d has a "year" of a little over 4 Earth-days; Planet e, 6.1 Earth-days; and Planet f has a period of 9.2 days. This puts Planet d in the super-rotating regime of Jacob Haqq-Misra and colleagues. In this situation (if the planet rotates in the same sense as Earth does, cold air will penetrate westwards towards the SSP, while warm air will be driven in a strong, turbulent flow towards the northeast of the SSP.

The upper atmosphere of Planet d will have strong super-rotation from west to east, with a pronounced pair of Rossby waves flanking the updraft above the SSP. Air flows towards the SSP over the poles along the approximate line of longitude of the SSP. Counterflow, away from the SSP, is principally at height, but near to the ASP air also flows poleward to return across the pole as well. This means that on these worlds some *relatively* moist air flows over the poles away from the ASP. However, since the air coming from the dark hemisphere is very cold, the relative humidity is going to remain low, assuming that there is any moisture available (below).

Planets e and f, with their longer periods, are likely in the Rhines-rotator regime, much like Proxima b. Here, we can expect a fairly broad area of precipitation cen-

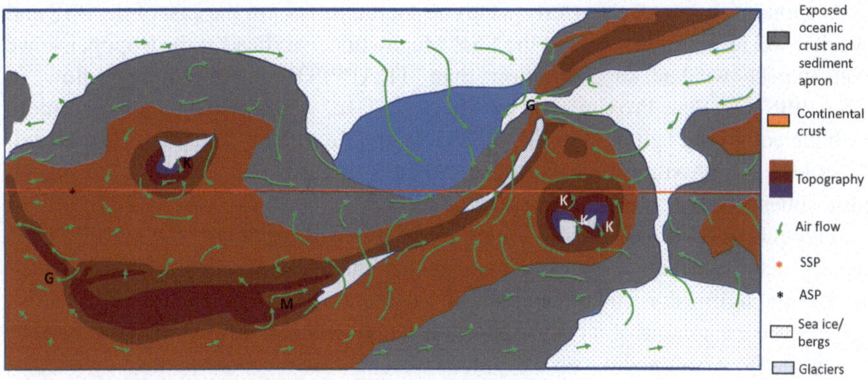

Fig. 9.12 A plausible TRAPPIST-1e climate for a drier, but still viable telluric world. This world has a transitional climate between the superrotating and Rhines regimes. TRAPPIST-1e, in this model, is assumed to be relatively dry, with more exposed land, of which much is sedimentary cover overlying the parched ocean basins (grey). The planet is cold. Oceans are assumed to be shallower than on Earth and could be frozen, except near the SSP. There is no plate tectonics. Airflow is diverted around the edges of the continental slopes, as well as mountainous regions. The highest topography is associated with hot-spot volcanism on an otherwise relatively inert surface. M is a cold valley wind; **K**, katabatic wind and **G** a gap wind

tered on the SSP, but adopting a rather elongated configuration in a north-south plane. As such, depending on the position of the planet, precipitation is likely as snow around the edges of the principal area of rainfall, as it is in Proxima b. This wet or icy region will cover a good fraction of the illuminated hemisphere, if there is no land (an aquaplanet). If, however, there is a considerable land-fraction, then the availability of moisture will be reduced, and mountainous regions could intercept much of the available moisture that might otherwise fall in the vicinity of the SSP. Planet e has the shortest period, and the circulation should be transitional between the super-rotation and Rhines regimes. This is the configuration used in Fig. 9.7.

Figure 9.12 superimposes the rotation pattern on a hypothetical planet, to take into account the distribution of land and mountainous areas that will intercept moisture. If one assumes that the atmosphere still retains moisture—and there is at least some evidence to suggest each of these worlds has an atmosphere—then these assumptions (or models) become reasonable expectations. Unlike our earlier model of Proxima b, this planet has been left with a drier surface that is far quieter, geologically, than Proxima b. Remember, this is a model. The geological state might be accurate, but the abundance of surface water is an inference at best (from the planet's age). Like Proxima b, much of the planet is likely to be frozen, and this reduces the amount of moisture available in the atmosphere, which further restricts the areas in which precipitation will fall. Much of the precipitation is likely to be snow rather than rain.

Next, we should consider the equilibrium temperature. How warm are the worlds and how much precipitation will be rain or snow, when compared to Proxima b? Firstly, both planets have shorter orbital periods than Proxima b. This will produce correspondingly stronger easterly winds along the equator to the east of the SSP and weaker westerlies to the west of the SSP. Meanwhile, overhead at the tropopause we can expect weaker super-rotation than TRAPPIST-1d but stronger flow in TRAPPIST-1e than Proxima b. For TRAPPIST-1e, winds will be deflected more into the equatorial belt than in TRAPPIST-1f and Proxima b. The polar regions will be colder and possibly drier on Planet e than if this was equally warm but had an atmosphere that circulated differently.

Precipitation will still occupy much of the sunlit hemisphere, if the planet is an aquaplanet. Place mountains and continental landmasses on the sunlit hemisphere, but near to the terminator, and the amount of precipitation is likely to be limited, as this will reduce the amount of moisture available to the air that is flowing towards the SSP.

Similarly, when landmasses are placed under the SSP but are flanked by oceans, the land will be provided with an abundance of precipitation from these water sources. In the region extending 65°–70° or more to the east or west of the SSP, temperatures will be low enough for precipitation to fall as snow. This is also shown in Fig. 9.12.

All other things being equal, of the two planets, TRAPPIST-1e will be colder than Planet d, both because it is further from its star but also because changes in the circulation regime favor more limited transport of heat around the planet. This should expand the region in which precipitation will fall as snow around the SSP.

In the Rhines-rotator planets TRAPPIST1-e and f, convergence is shifted from the equatorial band to the longitudinal region 60° on either side of the SSP. Beginning approximately 60° N, 60° W, arcing through the SSP at 0° N, 0° E, then southwestwards to 60° S, 60° W, air strongly converges from the east and west. The upper atmosphere is still moving broadly in a west to east flow above this region. Here, Mai-Yu front-like convergence seems possible. Perhaps disturbances could develop along the northern and southern extremities of this zone of convergence before propagating towards the SSP.

Airflow also propagates to the north and south of this region towards the polar regions. The models of Jacob Haqq-Misra and colleagues suggest enhanced cloud cover and elevated temperatures, which could favor the formation of extended regions where orographic rainfall (snowfall) or frontal rainfall (snowfall) is possible. In this case, as above, the *front* would be a semi-permanent feature resembling a stalled warm front. Colder air would be flowing southwards along its western side, with warmer, subtropical air moving northwestward (in the northern hemisphere) and southwestward (in the southern hemisphere) over this colder, equator-ward moving air masses.

In the super-rotator planets, air flows towards the SSP along the equator, with the strongest inflow from the east. This movement drives air north of the equator, where it encounters westerly flow around the top of the SSP. Here, strong sustained convergence should lead to a substantial uplift of air from the south over the westerly flow. A semi-permanent and potentially undulating frontal boundary, similar to the intertropical convergence zone, should exist if this flow is reliable. Precipitation will be enhanced and sustained along this entire region. However, as TRAPPIST-1e is likely colder than Earth, much of the precipitation will be as snow on windward slopes. This will lead to formation of glaciers and ice sheets over windward slopes (shown as white, stippled regions in Fig. 9.12).

Similarly, to the east and west of the SSP in the super-rotator planets, converging air from the northeast and southeast should give rise to uplift and precipitation. The extent of such precipitation will depend on the amount of energy available. If the air is generally cool, as would be expected in TRAPPIST-1e, then cloud cover may be in the form of small convective towers that are capped by an inversion. Precipitation will then be limited. Similarly, if the air is intrinsically dry, then even strong convergence and uplift may fail to produce much precipitation.

Planet e would likely be cold by our standards. Ice would cover perhaps 80% of the surface, and any oceans would be frozen, except near the SSP and downwind of this region. The amount of ice declines with carbon dioxide levels, until at a few percent CO_2 (or hundreds of times our atmospheric level) most of the surface is ice free. Although this concentration of carbon dioxide is possible with tectonism and age, this level seems largely unsustainable, given the propensity for carbon dioxide to react with water and various metal ions. Therefore, carbon dioxide levels are likely low and, at 8 billion years and with limited volcanism, lower than they are on Earth. This leaves the TRAPPIST-1e pretty icy. To offset this, if the air is dry, much of the icy ocean basins may be covered in eroded dust. This would decrease surface albedo, which may make the planet warmer overall.

Planet d, which is likely too hot for most life, has a small get-out clause if the planet is largely dry. Jeremy Leconte's work suggests that if a planet has limited water vapor in its atmosphere (something that is likely for planet d), then a thermal runaway can be avoided and the planet remain cool enough for residual water in a narrow strip around the terminator (Chap. 5). This leaves it analogous to Gliese 832c. Most of the sunlit hemisphere would be hot desert. There would be no oceans, the dark hemisphere would be cold and dry—and perhaps ice free—and the terminator region might support a limited biosphere consisting of microbial life that huddles around geothermal springs and other transient sources of water.

Eric Wolf's work with the TRAPPIST system clearly suggests that TRAPPIST-1f is too cold for liquid water on the surface. Indeed, the temperatures are likely so cold that both water vapor and carbon dioxide have precipitated out from the atmosphere, leaving this planet very cold and dry—too cold and dry for any life on the surface but not beneath it. Any precipitation will be likely be transient carbon dioxide and water snow, following volcanic activity that belches it into the frigid atmosphere. Likewise, TRAPPIST-1e may be largely ice-covered but have some ice-free areas. This is illustrated in Fig. 9.12, where a hypothetical, relatively dry but largely frozen world is shown. Planet e should be habitable with only modest amounts of greenhouse gases, if we assume that it has circulating oceans and geological activity—both of which are reasonable given the modest age of this system. This leaves the TRAPPIST-1 system the potential to have a range of habitable planets, depending on the nature of the life forms that might exist.

There is a further consideration—one that is irrelevant on Earth but a significant factor on Mars. If you were to drain Earth's oceans you would reveal vast plateaus rising 3–4 km over the ocean floor. These are our continents. For us, these are the

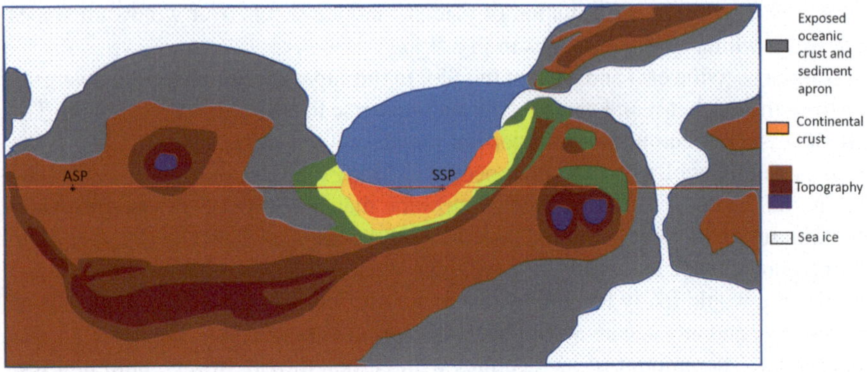

Fig. 9.13 Habitability of an icy TRAPPIST-1e. The niche volume is illustrated in different colors as shown before in Chap. 7 and Fig. 9.7. Red areas have the greatest number of species, grading through orange and yellow to green, with the latter the least. This ignores microbial life, which could be abundant under the surface—or in or under areas of ice, where there is liquid water. Land ice has been removed for simplicity. The exposed oceanic crust serves as new land for colonization of life. Green areas on upland regions are assumed to be affected by orographic precipitation—although this is likely to be snow

base level on which the biosphere of the land is constructed. However, if TRAPPIST-1e is like Mars—or had plate tectonics or other processes that built continents—then the base level (sea level) would lie 2–4 km below the level of the mean continental surface. This would place most of the land at considerable altitude and likely render it too cold for habitability of complex species. Meanwhile, the partially dried ocean basins, replete with debris aprons from the upland continental areas, would form the new land areas for colonization of various species.

Figure 9.13 develops this idea and illustrates qualitatively where life could be abundant around the SSP, even on what are really ocean floors. These dry areas will be low enough to be warm, will experience greater geothermal flow than continental crust (as the lithosphere is thinner) and will receive water draining from the continental slopes, where orographic rainfall may be abundant. Imagine a gently sloping plain, dotted with "islands" and flanked by enormous rock walls that rise kilometers upwards on all sides. Mars forms something of a useful analogy, with its large northern basin and tall southern highlands.

However, remember that this is a model world. TRAPPIST-1e might have oceans as deep as those on Earth, or none at all. It may also never have had plate tectonics, and therefore the idea of large continental uplands would be meaningless. In this model, we presume that its mass, availability of volatiles and a telluric composition make plate tectonics likely in the past. At 8 billion years old, this may have ceased, but this is (of course) speculation. However, it's always nice—and useful—to play with ideas!

Thinking further, with due consideration to deep cycling and loss of volatiles to the mantle or to space, Planet e remains a solid candidate for life, while f and g are likely to have lost most of their carbon dioxide to the surface and the mantle. Under these circumstances, although planets TRAPPIST-1f and g may have been habitable initially—albeit rather like early Mars—they are almost certainly frozen ice balls now, irrespective of the amount of energy these received from their star. Remember that this is not the last word in habitability for life in general. In both cases, each could support a deep biosphere under a layer of ice or in any remaining frozen oceans. This leaves a system that could still have up to four habitable worlds, habitable being a relative term.

To summarize: TRAPPIST-1d might have a narrow ring of microbial (and possibly green) habitability; TRAPPIST-1e might be analogous to Neoproterozoic snowball Earth; while TRAPPIST-1f and g might be either Mars-like or Europa-like, depending on how much water each has. In either instance, both will only have deep biospheres, or cryospheres, depending on the availability of nutrients, energy sources (likely geothermal) and liquid water. Complex surface biology, in terms of multicellular life, seems unlikely.

Overall, our model of Planet e is a poor candidate for life over the majority of its surface. This comes down to two factors: the overall chill, and the low sea levels that leave the continental surfaces at high altitudes. This makes them colder and drier than they would otherwise be, were the ocean basins full of water. However, near the SSP, temperatures are high enough for liquid water, and a reasonable area of dry land is available with sufficient water, warmth and light to sustain complex life. Remember that this world has had its airflow patterns tweaked, its carbon dioxide levels assumed and the amount of water inferred. That's quite a lot of assumptions—even if they are reasonable. You could, of course, consider a whole battery of worlds with the same starting conditions that were assumed by Eric Wolf and colleagues. Although the outcome would be similar, alterations in the distribution of the land and the amount there is of it will dramatically alter the climate models and hence habitability. However, that's a lot of additional modeling.

Taken together, and in the light of SETI's SPH classification system, TRAPPIST-1e has a low score; most likely it would fall into the "Saharan" class of worlds—or, if it was lucky, Pampian. You, the reader, may construct your own worlds and play with them. That's the beauty of this area of research. The only limitation—beyond the underpinning biological, geological and climatological constraints—is that imposed by your imagination.

Note, for example, the work of Tommaso Alberti (Università della Calabria). He has taken habitability a little further and included the effects of any surface vegetation on albedo (surface reflectivity). In these models, Planets d and e are habitable, but f, g and h soon freeze over. However, unlike other models (e.g., those of Eric Wolf and colleagues), Planet e becomes uninhabitable. As Alberti suggests, differences between the outcomes of his model and those of other researchers likely depends on how the models are constructed. The upshot is simply models are just that: models. The best that can be said at the moment is that the TRAPPIST-1 system is likely to host habitable worlds—at least in terms of temperature. Habitability means that microbial life will be the dominant or sole form. None of the worlds are likely to host complex life, if the climate models are correct. However, this depends on a range of factors, which are not tightly constrained at present.

Although planets with high levels of carbon dioxide (or methane) will be warm enough, the question is, will these planets have these abundances of greenhouse gases? For example, as Turbet and colleagues make clear, even if the planets hold onto sufficient amounts of volatile substances, methane will never be one of these. The reason is simple and comes back to observations of Titan and the planets of the outer Solar System. Here, ultraviolet radiation chemically activates the bonds in methane, displacing one or more hydrogen atoms and causing the remaining structures to react with one another. The end-product is a reddish slurry of compounds known as tholins, which precipitate onto the surface. Charon is covered in them, as is the surface of Titan. These reactions will rapidly deplete the atmosphere of its inventory of methane, unless there is some mechanism to replace it. Carbon dioxide is unlikely to persist on the surface of cold TRAPPIST-1 planets any more than it is on Mars. Therefore, without some means to trap energy from the star, the outer TRAPPIST-1 planets are unlikely to remain warm enough to support extensive oceans and with them life.

Aquaplanets can retain higher abundances of greenhouse gases over the 8 billion years the system has existed and thus will be proportionately warmer. However, here, too, relatively limited insolation from the star likely condemned each of the outer planets to an icy and permanent oblivion. The only relief will come once TRAPPIST-1 approaches the end of the main sequence in around 6 trillion years and becomes sufficiently luminous to melt the ice and turn each world into a Venusian hell. As one of the last of the universe's conventional stars to pass away, this is a somewhat depressing end to life in the universe.

In this regard, one should remember that the principle of habitability in Alberti's models depends on their having vegetation in the first instance. Of course, if the planet overheats early on (as TRAPPIST-1d likely has), then there won't be any vegetation to affect albedo and hence climate. Likewise, if TRAPPIST-1e cooled down too quickly, then vegetation would never evolve. Therefore, in this sense the models are abstract constructions that only work if all the planets are habitable to the point at which complex, multicellular life evolves—unless we imagine a cyano-bacterial sludge over the surface…

This brings us back to the importance of realism. Eric Wolf's and colleagues' models remain viable, as they only depend on the position of the planet in the habitable zone and the orbital period. These can apply to any planet, no matter its actual habitability. This is also true of competing atmospheric models. It is better to have the circulation models first, rather than envisage the effect of vegetation—which seems somewhat akin to placing the cart before the horse. Assumptions of atmospheres that retain several bars (or tens of bars) of carbon dioxide simply aren't feasible on planets this old. Therefore, Planets f, g and h have deep cryospheric biospheres at best. Nonetheless, Alberti's models retain credibility *once* habitability is established. Remember, TRAPPIST-1 is around 8 billion years old. There has been plenty of time for the star to settle, the planets to evolve through any moment of tempestuosity (if that's even a word) and settle into a near steady state that will prevail until their parent star dies. Only the slowly declining sound of the internal engine casts a thin shadow. Here, we can assume that their internal engines will die long before their star does. Then the winds, Rhines rotating or otherwise, will continue to blow over lifeless deserts until the universe fades.

The elephant in the room is the amount of water the planets lose, compared to their initial inventory. Since many models predict substantial amounts of water are likely present in red dwarf systems, the end result depends on the billion years or so the smallest red dwarfs (TRAPPIST-1 included) spend drifting onto the main sequence. Accurate models of planet formation and water loss are needed, which also include degassing (in-gassing) to (from) the atmosphere. Models that are specifically designed to cope with tidal locking would also be an obvious addition to the shopping basket. At present many models are based on terrestrial parameters and then tweaked to account for slow rotation and tidal locking.

The Wildcard: Induction Heating and Planetary Meltdowns

We leave the TRAPPIST-1 system with an odd, yet rather fascinating, idea that's surely worthy of a B-movie: planetary meltdowns instigated by electromagnetic induction. The proximity of a planet to its star, particularly to very low mass red dwarfs, allows the dipole field to directly plumb into the field of the planet and direct energetic flares and coronal mass ejections straight into the planet's atmosphere. As we've already discussed, this energizes and can strip it, given sufficient time. But Kristina Kislyakova (University of Vienna) and colleagues suggest that if the planet has a relatively conducting mantle and a metallic core, then changes in the orientation of the field could lead to electromagnetic heating of the planet's interior.

To make this work, the star's field needs to be inclined with respect to its rotation pole, so that as the star rotates on its axis the orientation of its field also alters, effectively egg-beating the interior of the planet and melting the upper mantle. Heating is initially mantle-wide, but as the temperatures rise, the upper mantle, which is at the lowest pressures, melts first. This increases its conductivity to the point that almost all of the subsequent heating then occurs in this layer. The upper mantle then forms a magma ocean. The lower mantle may solidify once more, depending on how heat is transferred from the upper layer.

However, the very fluid state of the upper mantle enables it to rapidly degas. If there were plate tectonics, this likely ceases, and tectonism would likely become an Io-like heat pipe. In this situation the lithosphere of the planet is degraded so that the bulk of it is simply thick crust that overlies the hot, molten asthenosphere below. Contrast this with Earth, where the bulk of the lithosphere is cool upper mantle, with a crust forming the uppermost layer of it. Remember that the hotter the upper mantle, the greater the degree of partial melting and the thicker the crust is, relative to the lithospheric mantle.

When applied to the TRAPPIST-1 system, the effects are rather impressive. The innermost planet TRAPPIST-1b is heated strongly enough to thoroughly melt its interior, generating an Io-like world. Meanwhile, further out, TRAPPIST-1c and d are heated enough to accelerate outgassing. If this process was active in the

TRAPPIST-1 system, then TRAPPIST-1c and d would be thoroughly desiccated at an early stage. This would guarantee that these worlds would have little or no reserves of volatiles now. We await further data on this system to determine whether such a process has (or still is) active.

Fortunately, the slow-rotation period of Proxima Centauri (83.5 days) ensures that induction heating has no measurable effect in our nearest planetary neighbor.

Problems, Considerations and the Way Forward

In some regards this chapter is a little depressing. It's not that red dwarfs can't host habitable worlds. It's just that despite their vast numbers, they make proportionately less suitable ones. The somewhat dismal return on searches so far is a reflection of the method of detection that favors larger planets and ones that are close to their stars. One hopes that, if nothing else, when the myriad factors are taken into account, you can see that they can host both life and, likely, at least in their oceans, complex life. Increasing numbers of finds of planets in the 0.7–2 Earth-mass range will likely bring in much more promising abodes.

The issue is more complex than a consideration of tidal locking or the theoretical loss of volatiles. At a prosaic level the "potentially habitable" red dwarf worlds that we know of are simply not as good as many scientists would like to believe. Yes, Proxima b and TRAPPIST-1e are genuinely worthy candidates, but after that we are very much off the green and into the rough. The problem in terms of doing new research is partly grant-driven. The reality for the research community is that these types of discoveries neither (or rarely) contribute directly to the advancement of society nor deliver Earth 2.0. The end result is merely a better understanding of our place in the universe and with it, one hopes, a respect for nature and the fragility of our world. These are not important considerations for many funders of research.

We cannot simply take off and colonize Proxima b, Ross-128b and their ilk when we make a mess of this world. The universe is a pretty unforgiving place when it comes to biology—as the author's 8-year old daughter found out when all but one of the 20-odd Cabbage-White caterpillars we recovered from the garden were parasitized. We (both the planet and the species) are that one caterpillar that made it through the blizzard of events that raged since the origin of our world. That is not to say we are a "rare Earth," merely that, all things considered, Earth has a variety of advantages that make it suitable for the evolution of complex—and at least vaguely intelligent—species. Moreover, the serendipitous evolution of humanity on Earth does not preclude the development of other intelligent species elsewhere. Simply spoken, the stochastic nature of reality means that evolution has many hands with which to work and only one of these (or a very few) make humans. The rest might make other intelligent life forms, but most do not.

Perhaps the best way of looking at this is to consider the biological entropy of life on Earth and relate it to the geological entropy over time. Planets with the lowest landscape diversities will always give rise to the lowest entropy of living organisms. If we want intelligent species, we need planets that have long-lasting biospheres and complex and active geologies. Are red dwarfs out of the picture? Let's look again at the possibilities.

When we consider complex and possibly intelligent life, what we are looking for is a planet that is tectonically active—and, therefore, has a mass somewhere between 0.5 and 3 Earths. It will not be an aquaplanet, and it is unlikely to be tidally locked to its star. Let's look at this more closely. The kind of planet we are looking for has a high number of available niches. These are both spatial and preferably temporal, i.e., there is a day-night cycle. A best-fit world is geologically active, as this maximizes the exchange of gases between the interior and the surface. Geological activity has its own "best-fit"—plate tectonics—or an equivalent, because this allows rapid exchange between the interior and exterior of the planet. However, other possibilities exist that, while less efficient, might still suffice to replenish carbon dioxide, nitrogen, phosphorous and water vapor. Plate tectonics, or extensive delamination, *may* also be required to stir the mantle, and with it any metallic core that generates a magnetic field. However, this is conjecture, albeit conjecture with some theoretical underpinning.

Next, if the planet is tidally locked we have to accept two issues that will limit the planet's potential to host complex life: the effect of plate tectonics on niche availability and the lack of temporal niches. If there is plate tectonics, then there is a finite limit on how long species have to evolve before surface tectonism delivers the region across the terminator to the dark hemisphere. Here, the biosphere on any continental land or aquatic margin will be terminated if it is dependent on light as an energy supply. This limits light-driven biological productivity to half the planet. Therefore, the maximum number of geographical niches may be half those on an asynchronously rotating planet (one that is not tidally locked to its star).

Next, for each climatological zone, the area available is less than the equivalent region on Earth. Consider Earth's tropics—approximate 60° of latitude and 360° of longitude. On a tidally locked planet, there is a roughly spherical region, which subtends 30° on either side of the SSP: this is a far smaller area. If there is plate tectonics, organisms may spend relatively little time in the area in which particular adaptations can develop.

What about super-plumes? In Chap. 3 we saw how these organize the general circulation of the terrestrial mantle. Small worlds (with Mars-like diameters or smaller) have a single mantle super-plume (a degree-1 plume). Earth has two such plumes, with a third likely under western Antarctica. These plumes are flanked by subduction zones that return mantle material to the core-mantle boundary.

In general, it appears that Africa (underlain by its plume) has barely moved in the last 300 million years. Meanwhile, the Pacific basin has remained open in one form or another for at least 1 billion years. If Earth-sized planets typically show two plumes, one needs to consider the impact of their stability on the distribution of continents. If there was a large super-plume in the mantle, on the SSP hemisphere (the lit hemisphere), then this might have the effect of shuffling land over the terminator and into darkness. The impact of this may last over a billion years if the gravitational force on this mantle upwelling was similar to Earth's on the Moon's.

Although certainly conjecture, there must be a consideration of mass distribution and orbital stability for tidally locked planets. There seems to be an assumption that tidal forces won't influence the distribution of continents because the mass of the continents (even supercontinents) is insufficient to gather the forces needed to lock them into one orientation. However, if we consider the underlying mantle superplume (LLSVP) the mass is considerably different, and the effect could be profound if tidal forces act to orientate the large, underlying mantle mass (Chap. 3). To date, this is not a published consideration in terms of habitability.

Next, think about the temporal niches. On Earth, there are at least three: day, night and crepuscular (dawn and dusk). There is only one on a tidally locked planet, and this is dependent on your geographical location. Close to the terminator, a dawn/dusk-like niche exists, while closer to the SSP (let's say at 60°–70°) something more day-like is the enduring feature. Aside from the temporal fixture enforced by the tidal locking another temporal absence will be seasons. Tidal locking does away with these as well. The only means of escape is the possibility that the planet will have an eccentric orbit. In this case, tidal locking may be avoided, and the kinds of temporal niches we are familiar with may emerge.

Comparisons with K-Class Stars and Their Worlds

Table 9.3 lists the candidate habitable worlds for orange dwarf stars.

The K-class dwarfs effectively avoid all of these limitations on niche availability. In most calculations K-class stars, with masses in excess of 0.6 solar masses, will have planets that are unable to tidally lock unless they have very thick atmospheres. As such, if they did, they would be poor candidates for the development of all but the most primitive life. Therefore, what do we think about the possibilities for life—complex or otherwise—on the potentially habitable worlds of K-class stars?

Kepler 62e and 62f

Kepler-62e and f are super-terrans that orbit a K2 orange dwarf 1200 light years from Earth. Neither world should be tidally locked to its star, meaning that aside from their orbital period being less than Earth's, the experience one might have would not be particularly different to that experienced here.

However, aside from the possibility that these planets might be Earth-like (which they aren't), the main thing to consider would be the effect of shortening the year. We'll return to this shortly.

The radii of both worlds suggest that they are either aquaplanets or mini-Neptunes. As both worlds formed outside the tidal lock radius—and far enough from their star to avoid desiccation during its approach to the main sequence—one

Table 9.3 Orange dwarf candidate planets and their characteristics

Planet	Stellar class	Radius (Earths)	Mass (Earths)	Equilibrium temperature (K)	Semi-major axis (AU)	Likely planet type	Habitable?
HD 40307g	K2.5V	2.39	7± 3	277	0.60	Sub-Neptune/ aquaplanet	Unlikely
HD 219134g	K3V	2.40	10.8	298	0.38	Neptune	Unlikely
HD 40307g	K2.5V	2.39	7.0	278	0.60	Sub-Neptune/ aquaplanet	Unlikely
Kepler-442b	K?V	1.34	2.3	233	0.41	Telluric/ aquaplanet	Possible, particularly microbial
Kepler-61b	K7V	2.15	6.65	273	0.26	Aquaplanet/ sub-Neptune	Possible, particularly microbial
Kepler-62e	K2V	1.61	4.5	270	0.42	Aquaplanet	Microbial; more complex life possible
Kepler-62f	K2V	1.41	2.8	208	0.72	Telluric or frozen aquaplanet	Potentially habitable for many life forms; but only if enough green-house gases

Most are clearly too massive to be considered “habitable”. Kepler-62f is an appropriate size, but clearly day-side cold

must assume both planets are volatile-rich. Importantly, as K-class stars take less than 100 million years to hit the main sequence, the opportunity for desiccation is limited—unless there is a violent collision early in the planet’s history. Therefore, we must conclude that the principle that a planet is very wet if its radius exceeds 1.4 Earths, and low density (i.e. gaseous) if it exceeds 1.6 Earths, applies. Consequently, an optimist has two aquaplanets in orbit around Kepler-62, with masses around 4.8 and 2.8 Earths, if we assume reasonable densities. The planets are around 7 billion years old, likely giving them stable climates.

Inside the 2.8-Earth mass Kepler-62f, the mantle likely convects, but the base of the ocean likely overlies a largely granitic crust. Meanwhile, the more massive Kepler-62e probably suffers the same fate as LHS 1140b. Here, the granitic crust (and lithosphere as a whole) is likely so thick that the depth at which the mantle is hot enough to partially melt is now so deep that little or no partial melting can now occur (Chap. 3). This makes Kepler-62e geologically dead, despite its considerable reserve of internal heat.

Kepler-62e, with its 122.4-day orbital period, lies near the inner edge of its star's habitable zone, while Kepler-62f lies near the outer edge, taking slightly more than 267 days to complete one orbit. With an equilibrium temperature of 244 K (−29 °C), Kepler-62f is only slightly colder than Earth; and with a reasonable inventory of greenhouse gases, it could be pleasantly warm. Conversely, Kepler-62e may have undergone a runaway, moist greenhouse many billions of years ago. This would leave an odd, hot, rocky super-terran. Therefore, though both worlds are touted as potentially habitable, only the outer Kepler-62f may retain the possibility of habitability, deep in its oceans. It seems altogether more probable that Kepler-62f emulates other super-terrans. This world may simply be a ball of water, encased by a thick atmosphere lying atop a geologically inert ball of rock.

Kepler-442

Kepler 442b is an altogether more likely abode for life. With 0.61 solar masses of material, the star is only just K-class, and the planet is not far from the tidal lock radius, where it would present one face to its star. The mass is likely to be 2–3 Earths, but could be as great as 8.2 or as small as 1 Earth mass. However, the median values are more reasonable, given reasonable densities. As we've already seen, the relatively low mass increases the chances that there is dry land and sufficient light reaches the ocean floor in places to encourage both the development of oxygenic photosynthesis and biological diversity. It also makes it less likely that the surface has formed a stagnant lid, as continental crust will spread less, and it is less likely that the base of any basaltic crust will have transformed into eclogite.

The host K-class star is around 2.9 billion years old, which would imply that there is a lower probability that complex life could have arisen. However, remember that this assumes that life arises and evolves at a similar pace on all worlds, which is clearly a big assumption. However, if there is liquid water (i.e., oceans), then 2.9 billion years should be ample time to develop oxygenic photosynthesis, given suitable evolutionary fodder. The rise of oxygenic photosynthesis may be detectable in the spectra of the planet, given instrumentation with suitable resolution. That this world transits its host star makes its atmosphere amenable to the detection of an oxygen signal in the background spectrum.

The orbital period is 112.3 Earth days, making its year as long as one of our seasons; and despite its proximity to its star, this world is cold by our standards. Indeed, this may be its greatest flaw. With an equilibrium temperature of −40 °C, this world is a good 22° colder than Earth. There is every chance that this planet is a frozen Hoth. However, perhaps, with a dark ocean surface and an atmosphere relatively rich in carbon dioxide, Kepler-442b might be a slush world—or water-band world. Here, the equatorial region remains largely ice free. Whether it is or not depends on the amount of land, which affects the weathering and draw-down of carbon dioxide and with it the effective temperature.

The Problems of Being an Orange World

K-dwarf planets have one major problem all of their own—the length of their year. Most worlds will not be tidally locked to their star, but they will have a year that is a terrestrial season-long—if the planet has a tilt, like ours. Why is this a problem? Think about terrestrial seasons. Plant growth, predominantly, occurs in spring and summer. This occurs when the tilt of Earth maximizes insolation in one hemisphere and then the other. If Kepler-442 or any other planet is tilted so that the planet has seasons, then the growth season will be very short indeed—approximately one month long per hemisphere. Yes, plants might evolve that reproduce in this time (think of desert-flowering plants, for example); however, this will place quite a burden on evolution, with growth seasons far shorter than most complex organisms can thrive and reproduce in. It's worth asking if all life will be limited by a short growing season?

Moreover, the atmospheric circulation patterns would be rather interesting on tilted planets, with rapid changes from one seasonal pattern to another, as the planet completes each orbit of its star. This could make them rather stormy worlds to inhabit. Imagine that the planet has a 20° tilt. If it has an orbital period of 180 days then every 90 days the sun migrates 40° across the sky, approximately double the rate of the comparable migration on Earth. Each hemisphere would be subjected to alternate periods of rapid heating and cooling, particularly over continental areas, where a lack of circulation of the surface and a lower specific heat capacity (than neighboring oceans) would lead to the greatest changes.

Climate flip-flops would be least on aquaplanets, where the oceans would likely buffer temperature changes that would be manifest over such short time intervals. Indeed, here, the rapid changes in insolation may go largely "unnoticed" as the oceans cushion the temperature of the air above. However, the atmosphere above the oceans would be losing heat rapidly in the winter months and gaining it rapidly in the summer. Would this lead to particularly stormy conditions in the temperate and polar regions for much or all of the year?

Terraforming a K-dwarf planet that was tilted might be a challenging experience, with the brevity of the main growing seasons. However, on such tilted planets tropical areas (the regions lying between the maximum northward and southward positions of the overhead sun) would experience the least variation over the year; and hence these areas would be habitable for the greatest numbers of species, as they are on Earth. Temperature environments, lying distant to oceans, might well be harsh, with particularly strong diurnal variation in the winter season. Although this is speculative, one might imagine we would find this environment habitable but likely harsh.

Taken more generally, the number of available niches for tilted planets that orbit K-dwarfs might be similar to Earth (given a planet of comparable size). Indeed, if organisms adapt by reproducing more rapidly, there might be something of a "niche explosion," with a large number of seasonally temporal niches.

If the planet of interest is not tilted, then (aside from variations in insolation between periastron and apastron) climatic variation would be nil. On this planet, the effect of the shorter orbital period would also be nil, and life would evolve under steadier conditions than it does on Earth. Here, of course, a lack of seasonal variation might remove some temporal niches—seasons where some organisms reproduce, for example, the Bluebell. On this world, the closest analogy would be the equatorial region on Earth, where there is little variation in sunlight and temperature over the year. There would be fixed latitudinal bands that experienced relatively constant climatic conditions. Indeed, the only variation on a day-to-day or year-to-year basis would be the passage of frontal boundaries associated with Rossby waves. At the latitudes of Vancouver, London or Christchurch imagine a constant spring-like climate, with modest temperatures and the passage of frequent frontal systems. There would be no summer (or winter) monsoons over the continents, and the polar regions would be permanently sub-zero and covered in snow. At least the weather from season to season would be predictable to the point of being banal.

Imagining the Grand Scheme of Things

We conclude by presenting a means of considering habitability in its broadest sense. The final tables and figures consider what kinds of niches are possible and how probable it would be to encounter these on an exoplanet. For example, mid-ocean ridge systems may be present on any planet with plate tectonics. Deep ocean environments might be expected to be similarly common if the planet is to have a suitable source of water. We can consider life temporally (its occurrence in time) and spatially (it's prevalence in a given space).

Table 9.4 links the evolution of life in the context of the evolution of our planet. Here, what were regarded as evolutionary transitions occurring randomly in time are linked to geological events. It is these, rather than the underlying biology, that determines when these are possible. Spatially, Table 9.5 then shows us one means of subdividing life on Earth. Geographical constraints limit what some planets can accommodate. This may be obvious—there is no savannah on an aquaplanet—but the impact may be less obvious. With a low variety of niches, life is constrained.

The take home message from this book is two-fold. Life is inevitable (given simple circumstances). However, the route life takes after inception is locked into the evolution of the planet on which it finds itself. Biological complexity requires time and circumstance, which while following trends that are set by the aging of their host planet, are not rigid. If plate tectonics were universal (which it likely isn't) there is no guarantee where the continental platforms, on which complex species evolve, will end up. Their location and the distribution of topography affects rainfall and precipitation and with it what range and abundance of species can exist. Evolution, through natural selection, is a complex beast that responds sensitively to its environment. Chapters 8 and 9 put this concept in hypothetical but realistic contexts, but there is far more work to be done.

Table 9.4 A timeline for the evolution of complex life

Evolutionary transition	Driving geological/astronomical event(s)	Impact
Advent of IR-driven photosynthesis (Late Hadean-early Archaean)	Cessation of widespread cometary/asteroidal impacts	Organic molecules become limiting factors for living organisms
Advent of light-driven photosynthesis (Mid-Archaean)	Emergence of land (volcanic islands/shallow submarine volcanism)	Shallow, marine environments with abundant light
Advent of oxygenic photosynthesis (Latest Archaean)	Reduction in rate of volcanism and supply of hydrogen sulphide; emergence of more land and shallow marine environments with more light; more luminous Sun	New electron donors needed to support photosynthesis; more energy available from sunlight in shallower water
Rise of eukaryotes (Mid-Proterozoic)	Reduction in volcanism; oxidation of ferrous iron and sulphides in ocean causes their precipitation and loss	Oxygen concentrations rise and favor the formation of aerobic cells; oxygen supports complex, internal cellular machinery
Rise of multicellular life (Mid-late Proterozoic)	Re/In-gassing of cooling mantle and reduction in ocean depth; possible acceleration in rate of plate tectonics as mantle becomes less viscous	Emergence of continents produces niches for life that are (later) protected by ozone; phosphate levels rise in oceans because of increased weathering
Diversification of living organisms (Ediacaran-Cambrian onwards)	Reduction in ocean depth; continued formation of continental crust; oxygenation of oceans extensive marine transgression and formation of many shallow water niches	Further elaboration of biological niches to match increase in availability of habitable land

The so-called major transitions of life must be considered in a geological framework. Here is a suggested model where life transitions from simpler to more complex forms based on critical geological events. If, for whatever reason, such events are absent on planets that are otherwise habitable, the chances of the evolutionary transition occurring will be vastly reduced. This implies that for any planet with deep oceans, complex life is impossible

Conclusions

At present, when one looks beyond the hyperbole, the number of truly interesting "habitable" planets is rather limited. Although all of those red dwarf-associated planets, which are considered here, might be likely habitable for something we would recognize as living, only Proxima b makes a good candidate. TRAPPIST-1e comes in a likely runner-up. However, this conclusion is only valid if each world has a suitable atmosphere.

Proxima b is a good size to maintain an internal engine for billions of years, and it is not so massive that any oceanic (basaltic) crust is likely to destabilize. This

Table 9.5 Terrestrial biomes

Terrestrial ecosystem	Deep aquaplanet		Shallow aquaplanet		Telluric planet	
	Asynchronous	Synchronous	Asynchronous	Synchronous	Asynchronous	Synchronous
Wet Coastal	No	No	Yes (Island)	Yes (Day-side Island)	Yes	Yes (Day-side)
Dry Coastal	No	No	Yes (Island)	Yes (Day-side Island)	Yes	Yes (Day-side)
Polar and Alpine Tundra	No	No	Yes (Island)	Yes (Day-side Island)	Yes	Yes (Day-side)
Mires: Swamp, Bog, Fen, and Moor	No	No	Yes (Island)	Yes (Day-side Island)	Yes	Yes (Day-side)
Temperate Deserts and Semi-Deserts	No	No	Yes (Island)	Yes (Day-side Island)	Yes	Yes (Day-side)
Coniferous Forests	No	No	Yes (Island)	Yes (Day-side Island)	Yes	Yes (Day-side)
Temperate Deciduous Forests	No	No	Yes (Island)	Yes (Day-side Island)	Yes	Yes (Day-side)
Natural Grasslands	No	No	Yes (Island)	Yes (Day-side Island)	Yes	Yes (Day-side)
Heathlands and Shrublands	No	No	Yes (Island)	Yes (Day-side Island)	Yes	Yes (Day-side)
Temperate Broad-Leaved	No	No	Yes (Island)	Yes (Day-side Island)	Yes	Yes (Day-side)
Evergreen Forests	No	No	Yes (Island)	Yes (Day-side Island)	Yes	Yes (Day-side)
Mediterranean-Type Shrublands	No	No	Yes (Island)	Yes (Day-side Island)	Yes	Yes (Day-side)
Hot Deserts and Arid Shrublands	No	No	Unlikely	Unlikely	Yes	Yes
Tropical Savannas			Yes (Island)	Yes (Day-side Island)	Yes	Yes (Day-side)
Tropical Rain Forest	No	No	Yes (Island)	Yes (Day-side Island)	Yes	Yes (Day-side)
Wetland Forests	No	No	Yes (Island)	Yes (Day-side Island)	Yes	Yes (Day-side)
Coastal Aquatic	No	No	Yes (Island)	Yes (Day-side Island)	Yes	Yes (Day-side)
Shallow aquatic	No	No	Yes (Island)	Yes (Day-side Island)	Yes	Yes (Day-side)
Deep aquatic	Yes	Yes	Yes	Yes	Yes	Yes (Day-side)
Hydrothermal vents (ocean)	Early	Early	Yes	Yes	Yes	Yes
Inland Aquatic	No	No	Yes (Island)	Yes (Day-side Island)	Yes	Yes (Day-side)
Rivers and Streams	No	No	Yes(Island)	Yes (Day-side Island)	Yes	Yes (Day-side)
Lakes	No	No	Yes (Island)	Yes (Day-side Island)	Yes	Yes (Day-side)
Caves	No	No	Yes (Island)	Yes (Island)	Yes	Yes

A comparison of the kinds of biomes found on Earth and the possibility of finding them on other planets. Aquaplanets are niche-poor, to say the least. Early Earth may have resembled a shallow aquaplanet with limited land

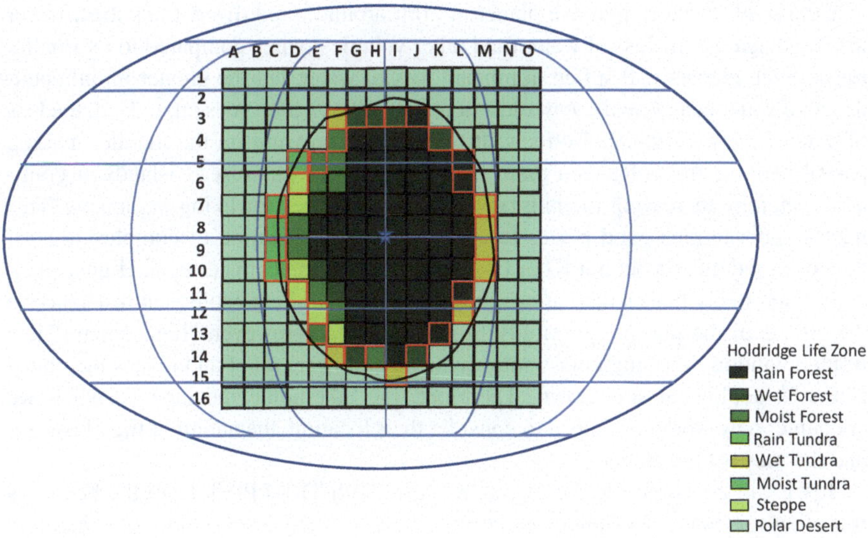

Fig. 9.14 Potential for life on Proxima b. Holdridge Life Zones when applied to Proxima b with on a hypothetical land surface covering the region where temperatures are high enough for life at the surface. This is an application of terrestrial ecological models to the climate model for Proxima b. We assume only that vegetation has evolved, not that it can evolve

makes plate tectonics all the more probable if there is liquid water on the planet's surface. Although plate tectonics is neither necessary nor sufficient for habitability, the effect on several parameters that are needed for habitability make its presence a likely boost for habitability in general. Figure 9.14 illustrates an application of a terrestrial ecological model to Proxima b. The Holdridge Life Zone model allows us to predict what could be possible with terrestrial vegetation.

The K-dwarf candidates are limited in number and also relatively massive. This is likely a simple effect of the methods in which the planets were found. Smaller planets are proportionally harder to detect, either through the radial velocity method or through transits. General modeling makes planets with masses greater than 3 Earths likely poor candidates for plate tectonics because they will likely form a stagnant lid early on. However, they could be volcanically active and have oceans of suitable depth. In that regard, the depths of oceans are critical. For super-terrans orbiting within the circumstellar habitable zones of K-class and G-class stars, the greater distance from their host star makes it likely that there will be relatively little loss of volatiles, compared to planets with the same mass orbiting within the habitable region of a red dwarf.

Although this seems good, remember the problem. The deeper the ocean, the greater the likelihood that the oceanic basaltic crust will become unstable. Higher masses make the topography softer, and this lowers the possibility of active surface tectonism and volcanism (despite the hotter mantle). Thus, these worlds may be habitable for some life, but this will be restricted by limitations in the number of niches and (potentially) the availability of energy.

Conversely, a more massive planet in orbit around a red dwarf finds itself better insulated against the loss of water (and other volatiles) when compared to its smaller peers. Such planets will suffer significant loss of water, but the greater initial abundance of water can leave these worlds more habitable as time goes on. Indeed, the loss of water has two beneficial effects. It reduces the pressure on the mantle, making partial melting and volcanism more likely and it exposes crust as islands or continents, thereby increasing the number of available niches for living organisms. This makes super-terrans good places to consider for the evolution of complex life—if they orbit red dwarfs but not K- or G-class stars. The context is critical. Higher planetary mass likely makes their atmospheres denser and offers geological resistance to the effects of the planet's proximity to its parent star. However, "higher mass" is a limited window. Very high mass softens the crust, lowers relief and makes the planet prone to geological shut down. Therefore, the take-home message is, when we encounter new worlds, we need to consider their location, the nature of their host star and the mass of the planet.

If we can constrain the density, as we have with TRAPPIST-1, all the better, as this allows us to narrow down whether the planet is likely a solid lump or something else that is drowned in a deep ocean or oppressive atmosphere. In the context of atmospheric stripping, the nature of the star is critical—as are the transit observations of planets. In particular, these constrain many of the loose mathematical theories and often (unfortunately) show them wanting. It is likely that there is an overestimation of how much material can be stripped by various stellar processes—and, more fundamentally, a serious underestimation of how much volatile material ends up in the crust and mantle, which can then be returned to the surface. Higher resolution of the TRAPPIST-1 planets should help resolve this issue.

Moreover, the processes whereby planets form are also poorly constrained. Are all telluric planets low in volatiles or is there a range that is independent of their distance from the star? Although there are plenty of models, there needs to be an "empirical reckoning," where reality steps in and waves its hand at us.

Finally, if we take 100 Earth-like planets and place them in the habitable zones of various low or very low-mass stars, what is the biological outcome? Will an Earth always produce complex life—or life at all? Will the atmosphere always become oxidizing, through the action of photosynthesis? How long can the biosphere persist? Consider whether Earth would remain habitable if it were placed at comparable insolation around TRAPPIST-1. At 8 billion years of age, would Earth retain an atmosphere that would support photosynthesis, or would we see a planet where life had retreated into the dark spaces from where it emerged billions of years earlier? Would it have an atmosphere at all—or something pallid and wispy, like that of Mars?

The emergence and persistence of life is a complex beast in its own right. Although biologists are now at the cusp of defining how life likely emerged and its key facets evolved, the timescales for these processes are solely constrained by what we have recorded on Earth. This doesn't even tell us how easy it was to evolve it, merely what survived and evolved here. This often appears to cause confusion, even among biologists.

The genetic record, though better than the fossil record, is a record of survival. There is little, if any, evidence of failed branches. Was all life on Earth based around the replication of DNA (or RNA); or did something else emerge only to fail? We don't know. A rather well-known physicist once commented on the evolution of oxygenic photosynthesis, stating that its emergence only once implied that it was biologically difficult to do. However, the emergence and preservation problem rears its head again. In this regard, if one examines the setup of the photosynthetic machinery, it doesn't look too difficult to see how the final, oxygen-liberating system emerged. The process might even have been easy.

There is still a long way to go at nearly every step in the chain from nebula to astronomer. Fundamental physical and chemical problems remain to be solved. However, one thing is gratifying—the involvement of evolutionary biologists and ecologists in developing a deeper understanding of what is biologically possible with exoplanets. Perhaps the first edition of this book got ball rolling on that one—and that more recent work by the author and others has driven that forward. In the coming years we should see fully integrated ecological (or geographical) models of planets that can incorporate the increasingly detailed climate models as well as realistic geological models of planets.

Therefore, we will conclude by addressing, once more, a reviewer's comment regarding the content of the first edition of this book. That book, and more recent projects, have sought to show how important it is to integrate all physical aspects of planetary existence. We can only address habitability if we first consider the geography of planets from rock, through soil and climate, through to the evolutionary potential in terms of the variety of available niches. Therefore, we must always first consider (and draw the reader's attention to) the multitude of factors that control evolution. That is how we will better understand our world and the universe.

References

Proxima b

Boutle, I. A., Mayne, N. J., Drummond, B., Manners, J., Goyal, J., Lambert, F. H., Acreman, D. M., & Earnshaw, P. D. (2017). Exploring the climate of Proxima B with the Met Office Unified Model. *Astronomy & Astrophysics, 601*, A120. https://arxiv.org/abs/1702.08463.

Feng, F., & Jones, H. R. A. (2018). Was Proxima captured by alpha Centauri A and B? *Monthly Notices of the Royal Astronomical Society, 473*(3), 3185–3189. https://doi.org/10.1093/mnras/stx2576.

Ribas, I., Bolmont, E., Selsis, F., Reiners, A., Leconte, J., Raymond, S. N., Engle, S. G., Guinan, E. F., Morin, J., Turbet, M., Forget, F., Velasco, G., & Anglada-Escudé, G. (2016). The habitability of Proxima Centauri b I. Irradiation, rotation and volatile inventory from formation to the present. *Astronomy & Astrophysics, 596*, A111. https://arxiv.org/pdf/1608.06813.pdf.

Stevenson, D. S. (2019). Phytoclimatic mapping of exoplanets. *International Journal of Astrobiology*, 1–10. https://doi.org/10.1017/S1473550419000181.

Turbet, M., Leconte, J., Selsis, F., Bolmont, E., Forget, F., Ribas, I., Raymond, S. N., & Anglada-Escudé, G. (2016). The habitability of Proxima Centauri b. II. Possible climates and observability. *Astronomy & Astrophysics, 596*, A112. https://doi.org/10.1051/0004-6361/201629577. https://www.aanda.org/articles/aa/pdf/2016/12/aa29577-16.pdf.

TRAPPIST-1

Barnes, J. R., Jenkins, J. S., Jones, H. R. A., et al. (2014). Precision radial velocities of 15 M5-M9 dwarfs. *Monthly Notices of the Royal Astronomical Society, 439*(3), 3094–3113. https://doi.org/10.1093/mnras/stu172. https://arxiv.org/abs/1401.5350.

Bourrier, V., de Wit, J., Bolmont, E., Stamenkovi, V., Wheatley, P. J., Burgasser, A. J., Delrez, L., Demory, B.-O., Ehrenreich, D., Gillon, M., Jehin, E., Leconte, J., Lederer, S. M., Lewis, N., Triaud, A. H. M. J., & Van Grootel, V. (2017). Temporal evolution of the high-energy irradiation and water content of TRAPPIST-1 exoplanets. *The Astronomical Journal, 154*, 121. https://arxiv.org/pdf/1708.09484.pdf.

Burgasser, A. J., & Mamajek, E. E. (2017). On the Age of the TRAPPIST-1 System. *The Astrophysical Journal, 845*(2), 110. https://arxiv.org/abs/1706.02018.

de Wit, J., Wakeford, H. R., Lewis, N. K., Delrez, L., Gillon, M., Selsis, F., Leconte, J., Demory, B.-O., Bolmont, E., Bourrier, V., Burgasser, A. J., Grimm, S., Jehin, E., Lederer, S. M., Owen, J. E., Stamenković, V., & Triaud, A. H. M. J. (2018). Atmospheric reconnaissance of habitable-zone earthsized exoplanets. *Nature Astronomy, 2*, 214–219. https://arxiv.org/abs/1802.02250.

Delrez, L., Gillon, M., Amaury, H. M. J., et al. (2018). Early 2017 observations of TRAPPIST-1 with Spitzer. *Monthly Notices of the Royal Astronomical Society, 475*(3), 3577–3597. https://doi.org/10.1093/mnras/sty051. https://arxiv.org/abs/1801.02554.

Gillon, M., Triaud, A. H. M. J., Demory, B.-O., Jehin, E., Agol, E., Deck, K. M., Lederer, S. M., de Wit, J., Burdanov, A., Ingalls, J. G., Bolmont, E., Leconte, J., Raymond, S. N., Selsis, F., Turbet, M., Barkaoui, K., Burgasser, A., Burleigh, M. R., Carey, S. J., Chaushev, A., Copperwheat, C. M., Delrez, L., Fernandes, C. S., Holdsworth, D. L., Kotze, E. J., Van Grootel, V., Almleaky, Y., Benkhaldoun, Z., Magain, P., & Queloz, D. (2017). Seven temperate terrestrial planets around the nearby ultracool dwarf star TRAPPIST-1. *Nature, 542*, 456–460. https://doi.org/10.1038/nature21360.

Grimm, S. L., Demory, B.-O., Gillon, M., Dorn, C., Agol, E., Burdanov, A., Delrez, L., Sestovic, M., Triaud, A. H. M. J., Turbet, M., Bolmont, É., Caldas, A., de Wit, J., Jehin, E., Leconte, J., Raymond, S. N., Van Grootel, V., Burgasser, A. J., Carey, S., Fabrycky, D., Heng, K., Hernandez, D. M., Ingalls, J. G., Lederer, S., Selsis, F., & Queloz, D. (2018). The nature of the TRAPPIST-1 exoplanets. *Astronomy & Astrophysics, 613*, A68. https://www.eso.org/public/archives/releases/sciencepapers/eso1805/eso1805a.pdf.

Kislyakova, K. G., Noack, L., Johnstone, C. P., Zaitsev, V. V., Fossati, L., Lammer, H., Khodachenko, M. L., Odert, P., & Güdel, M. (2017). Magma oceans and enhanced volcanism on TRAPPIST-1 planets due to induction heating. *Nature Astronomy, 1*, 878–885.

Krisztián, V., Kővári, Z., Pál, A., et al. (2017). Frequent flaring in the TRAPPIST-1 system—Unsuited for life. *The Astrophysical Journal, 841*(2), 124. https://doi.org/10.3847/1538-4357/aa6f05. https://arxiv.org/abs/1703.10130.

Shields, A. L., Barnes, R., Agol, E., Charnay, B., Bitz, C., & Meadows, V. S. (2016). The effect of orbital configuration on the possible climates and habitability of Kepler-62f. *Astrobiology, 16*. https://arxiv.org/pdf/1603.01272.pdf.

Tamayo, D., Rein, H., Petrovich, C., & Murray, N. (2017). Convergent migration renders TRAPPIST-1 long-lived. *The Astrophysical Journal Letters, 840*(2), L19. https://doi.org/10.3847/2041-8213/aa70ea..Available. https://arxiv.org/abs/1704.02957.

Van Grootel, V., Fernandes, C. S., Gillon, M., Jehin, E., Manfroid, J., Scuflaire, R., Burgasser, A. J., Burdanov, A., Delrez, L., Demory, B.-O., de Wit, J., Queloz, D., & Triaud, A. H. M. J. (2018). Stellar parameters for TRAPPIST-1. *The Astrophysical Journal, 853*(1), 30. arXiv:1712.01911.

Wheatley, P. J., Louden, T., Bourrier, V., Ehrenreich, D., & Gillon, M. (2017). Strong XUV irradiation of the Earth-sized exoplanets orbiting the ultracool dwarf TRAPPIST-1. *Monthly Notices of the Royal Astronomical Society: Letters, 465*(1), L74–L78. Preprint available: https://arxiv.org/pdf/1605.01564v1.pdf.

Wolf, E. (2017). Assessing the habitability of the TRAPPIST-1 system using a 3D climate model. *The Astrophysical Journal Letters, 839*, 6. http://iopscience.iop.org/article/10.3847/2041-8213/aa693a/pdf.

K-Dwarfs

Borucki, W. J., Agol, E., Fressin, F., Kaltenegger, L., Rowe, J., Isaacson, H., Fischer, D., Batalha, N., Lissauer, J. J., Marcy, G. W., Fabrycky, D., Désert, J.-M., Bryson, S. T., Barclay, T., Bastien, F., Boss, A., Brugamyer, E., Buchhave, L. A., Burke, C., Caldwell, D. A., Carter, J., Charbonneau, D., Crepp, J. R., Christensen-Dalsgaard, J., Christiansen, J. L., Ciardi, D., Cochran, W. D., DeVore, E., Doyle, L., Dupree, A. K., Endl, M., Everett, M. E., Ford, E. B., Fortney, J., Gautier, T. N., III, Geary, J. C., Gould, A., Haas, M., Henze, C., Howard, A. W., Howell, S. B., Huber, D., Jenkins, J. M., Kjeldsen, H., et al. (2013). Kepler-62: A five-planet system with planets of 1.4 and 1.6 Earth radii in the habitable zone. *Science, 340*, 587–590. https://arxiv.org/ftp/arxiv/papers/1304/1304.7387.pdf.

Torres, G., Kipping, D. M., Fressin, F., Caldwell, D. A., Twicken, J. D., Ballard, S., Batalha, N. M., Bryson, S. T., Ciardi, D. R., Henze, C. E., Howell, S. B., Isaacson, H. T., Jenkins, J. M., Muirhead, P. S., Newton, E. R., Petigura, E. A., Barclay, T., Borucki, W. J., Crepp, J. R., Everett, M. E., Horch, E. P., Howard, A. W., Kolbl, R., Marcy, G. W., McCauliff, S., & Quintana, E. V. (2015). Validation of twelve small Kepler transiting planets in the habitable zone. *The Astrophysical Journal, 800*, 99. https://arxiv.org/pdf/1501.01101.pdf.

Models and Problems

Armstrong, J. C., Barnes, R., Domagal-Goldman, S., Breiner, J., Quinn, T. R., & Meadows, V. S. (2014). Effects of extreme obliquity variations on the habitability of exoplanets. *Astrobiology, 14*(4), 277–291. https://doi.org/10.1089/ast.2013.1129. https://arxiv.org/pdf/1404.3686.pdf.

Boutle, I. A., Mayne, N. J., Drummond, B., Manners, J., Goyal, J., Lambert, F. H., Acreman, D. M., & Earnshaw, P. D. (2017). Exploring the climate of Proxima B with the Met Office Unified Model. https://arxiv.org/pdf/1702.08463.pdf.

Gale, J., & Wandel, A. (2015). The potential of planets orbiting red dwarf stars to support oxygenic photosynthesis and complex life. *International Journal of Astrobiology, 16*, 1–9. https://arxiv.org/abs/1510.03484.

Haqq-Misra, J., Kopparapu, R. K., & Wolf, E. T. (2018). Why do we find ourselves around a yellow star instead of a red star? *International Journal of Astrobiology, 17*(1), 77–86. https://doi.org/10.1017/S1473550417000118.

Kiang, N. Y., Segura, A., Tinetti, G., Govindjee, R. E. B., Cohen, M., Siefert, J., Crisp, D., & Meadows, V. S. (2007). Spectral signatures of photosynthesis. II. Coevolution with other stars and the atmosphere on extrasolar worlds. *Astrobiology, 7*(1), 252–274.

Kiang, N. Y., Siefert, J., Govindjee, & Blankenship, R. E. (2007). Spectral signatures of photosynthesis. I. Review of Earth organisms. *Astrobiology, 7*(1), 222–251.

Kislyakova, K. G., Noackz, L., Johnstoney, C. P., Zaitsevx, V. V., Fossati, L., Lammer, H., Khodachenko, M. L., Odert, P., & Güdely, M. (2017). Magma oceans and enhanced volcanism on TRAPPIST-1 planets due to induction heating. *Nature Astronomy, 1*, 878–885. https://arxiv.org/pdf/1710.08761.pdf.

Parke Loyd, R. O., Shkolnik, E. L., Schneider, A. C., Barman, T. S., Meadows, V. S., Pagano, I., & Peacock, S. (2018). HAZMAT. IV. Flares and superflares on young M stars in the far ultraviolet. *The Astrophysical Journal, 867*, 70. (13pp). http://imgsrc.hubblesite.org/hvi/uploads/science_paper/file_attachment/359/Lloyd_published_ApJ_paper_100118.pdf.

Shields, A. L., Ballard, S., & Johnson, J. A. (2016). The habitability of planets orbiting M-dwarf stars. *Physics Reports, 663*, 1–38. https://arxiv.org/abs/1610.05765.

Spiegel, D. S. & Turner, E. L. (2011). Life might be rare despite its early emergence on Earth: A Bayesian analysis of the probability of Abiogenesis. https://www.researchgate.net/profile/Dave_Spiegel/publication/51924468_Bayesian_analysis_of_the_astrobiological_implica-

tions_of_life%27s_early_emergence_on_Earth/links/00b7d520312842b7d6000000/Bayesian-analysis-of-the-astrobiological-implications-of-lifes-early-emergence-on-Earth.pdf.
Stern, R. J. (2016). Is plate tectonics needed to evolve technological species on exoplanets? *Geoscience Frontiers, 7*, 573–580. https://doi.org/10.1016/j.gsf.2015.12.002.
Stevenson, D. S. (2018). Evolutionary Exobiology II: Investigating biological potential of synchronously-rotating worlds. *International Journal of Astrobiology*, 1–15. https://doi.org/10.1017/S1473550418000241.
Stevenson, D. S. (2018). Niche amplitude, tidal-locking and Fermi's Paradox. *International Journal of Astrobiology*, 1–7. https://doi.org/10.1017/S1473550418000253.
Stevenson, D. S. (2019). Biogeographical Modeling of Alien Worlds. Manuscript in preparation.
Stevenson, D. S. (2019). Planetary mass, vegetation height and climate. *International Journal of Astrobiology*, 1–6. https://doi.org/10.1017/S1473550418000484.
Stevenson, D. S., & Large, S. (2017). Evolutionary exobiology: towards the qualitative assessment of biological potential on exoplanets. *International Journal of Astrobiology, 16*(4), 1–5. https://doi.org/10.1017/S1473550417000349.
Tarter, J. C., Backus, P. R., Mancinelli, R. L., Aurnou, J. M., Backman, D. E., Basri, G. S., Boss, A. P., Clarke, A., Deming, D., Doyle, L. R., Feigelson, E. D., Freund, F., Grinspoon, D. H., Haberle, R. M., Hauck, S. A., II, Heath, M. J., Henry, T. J., Hollingsworth, J. L., Joshi, M. M., Kilston, S., Liu, M. C., Meikle, E., Neil Reid, I., Rothschild, L. J., Scalo, J., Segura, A., Tang, C. M., Tiedje, J. M., Turnbull, M. C., Walkowicz, L. M., Weber, A. L., & Young, R. E. (2007). A Reappraisal of the Habitability of Planets Around M Dwarf Stars. *Astrobiology, 7*(1), 30–65.
Walz, U. (2011). Landscape structure, landscape metrics and biodiversity. *Living Reviews in Landscape Research, 5*(3). http://www.livingreviews.org/lrlr-2011-3.
Zeng, L., Sasselov, D. D., & Jacobsen, S. B. (2016). Mass–radius relation for rocky planets based on PREM. *The Astrophysical Journal, 819*(127), 5. https://doi.org/10.3847/0004-637X/819/2/127.

Glossary

Abiotic (factor) An environmental factor, such as light intensity or temperature, that is non-living

Adiabatic lapse rate The rate at which dry air cools or warms with height

Aerobic respiration Process of energy conversion that uses oxygen

AGB A type of pulsating red giant star that alternates energy production between hydrogen and helium fusion

AGB-Manqué A relatively low-mass AGB star with a thin hydrogen envelope

Albedo The reflectivity of a surface. High albedo means the surface is highly reflective

Alpha Centauri Double-star system that is itself part of a triple star system, which includes Proxima Centauri. The closest star system to Earth

Amazonian planet A planet that has highly abundant vegetation

Ammonia A noxious nitrogen-hydrogen chemical that is critical for the formation of biomolecules

Anaerobic respiration Less efficient forms of respiration that do not involve oxygen

Anthropocene The latest addition to the geological calendar, where human activities dominate alterations to the surface geology and atmosphere of Earth

Anti-stellar point (ASP) The point on a tidally locked planet that is diametrically opposite the sub-stellar point, where the star is directly overhead

Aquaplanet A planet that is wholly or mostly (greater than 90% of the surface) covered in water

Aquatic primary habitability index A mathematical measure of habitability of oceans or other bodies of water

Archaeal cell A type of organism that has many of the features of a bacterial cell, but has the internal machinery of eukaryote cells (and is likely their evolutionary source)

Archaean Geological era extending from 4.0 to 2.45 billion years ago

Asgard The name given to the collection of Archaea that appear to be the modern derivative of the direct ancestors of eukaryotes

Asthenosphere The hot, plastic, uppermost layer of the "fluid" mantle and which underlies the frozen lithosphere

D. S. Stevenson, *Red Dwarfs*, https://doi.org/10.1007/978-3-030-25550-3

Astrometry The technique used to identify planets that involves sensitive measurements of stellar wobbles in images that partly presages the radial velocity method

Asynchronous rotation Where a planet orbits a star (or satellite a planet), but the period of the orbit is different from the rotation period of the planet

Atmospheric collapse Where one or more gases condense from an atmosphere, causing its pressure to rapidly drop

Atmospheric stripping Any process that actively removes gases from a planet's atmosphere, taking them into interplanetary space

ATP synthase The dynamo-like enzyme that manufactures ATP

Back-arc The region in front of a subducting slab of lithosphere but behind the volcanic front; often a site of crustal tension and magmatism

Bacteriorhodopsin A molecule containing a chemical related to vitamin A that absorbs light in the middle of the visual range of humans

Barnard's Star A relatively massive red dwarf that is in the second-closest star system to the Sun

Basalt Iron and magnesium-rich, silica-poor, dense volcanic rock that forms the bulk of the ocean crust and is common in all terrestrial planets

Basaltic Any rock with a composition similar to basalt

BEER An acronym that describes a combination of techniques used to identify planets

Bicarbonate A compound ion consisting of one carbon atom, one hydrogen atom and three oxygen atoms, bearing an overall negative charge. Produced by the dissolution of carbon dioxide in water

Biodiversity A blanket terms describing the ecological, genetic and environmental richness of an environment

Blue Hook A location on the Hertzsprung–Russell diagram where helium burning stars of low mass curve onto the white dwarf cooling track

Boring billion A somewhat disingenuous term given to the bulk of the Proterozoic era, where life seems to have become "stuck," at least in evolutionary terms, roughly lasting from 2 to 1 billion years ago

Boron A simple non-metallic element that colors diamonds blue and can be used to trace subduction of volatile substances into the mantle

Boundary-layer convection (mantle) A term used to describe model-mantle convection where there is no internal heating but heating from below and cooling from above

Cambrian explosion A period lasting at most a few million years where life greatly diversified in richness of species

Carbonate A compound ion produced by the removal of a hydrogen ion from bicarbonate. Metal carbonates are the principle reservoir of carbon dioxide on Earth and possibly Mars.

Carbonate-silicate cycle A series of chemical reactions that remove then return carbon dioxide from the atmosphere by weathering, then return it to the surface by volcanism, the principal means through which planetary temperatures are maintained in the presence of liquid water

Carbon cycle The collection of biological processes that remove and produce carbon dioxide and interact with the carbonate-silicate cycle

Carbon dioxide A compound of carbon and two oxygen atoms and the principle greenhouse gas after water vapor in the atmospheres of terrestrial planets

Carbon monoxide A noxious compound of one carbon atom and one oxygen atom—also a greenhouse gas, but fortunately rarer

Chlorophyll A complex compound containing carbon, nitrogen, hydrogen, oxygen and critically (usually) one magnesium ion. This absorbs sunlight (and derivative, infrared radiation) and drives photosynthesis

Chinook A warm mountain wind produced by dry adiabatically warmed air (see Fohn)

Columbia Proterozoic supercontinent

Compensation point The point at which the amount of carbon dioxide absorbed by photosynthesis matches the amount released by respiration

Convection The mass movement of liquids and gases driven by changes in their density

Core-mantle boundary The boundary between the core and the base of the mantle

Coriolis effect An inertial effect associated with gases or liquids moving towards or away from the center of a spinning mass

Craton A stable piece of continental crust, usually Archaean or Proterozoic in age

Cyanobacteria A type of bacterium that carries out photosynthesis and has a complex internal structure; the ancestor of the chloroplast

Cyanide A compound of hydrogen, carbon and nitrogen that is the cornerstone of the synthesis of nucleobases in interstellar space

Degree-1 Mantle convection focused on one large upwelling

Degassing Any process that removes gases/volatile substances from the hot, rocky mantle of a planet or satellite

Delamination A process through which a dense (usually eclogite) root of the crust falls off into the underlying plastic mantle

Delta Scuti variables A type of F-A-class main sequence star that pulsates

Diamonds A dense form of carbon, where every atom of carbon is bound to its neighbors through four covalent bonds

Dinoflagellate A type of eukaryote (protist) that has characteristics of animal and plant cells

DNA (deoxyribonucleic acid) The principal information storage molecule of life on Earth, consisting of two complimentary strands of sugar (deoxyribose) and phosphate, bonded through four nucleobases, adenine, thymine, cytosine and guanine

Eclipse timing variation A method of planet detection involving sensitive measurements of eclipses of one star by another, which are altered by the motion of a planet around their common center of mass

Eclogite A dense, metamorphic rock produced by heat and pressure acting on basaltic material (usually, first altered to amphibolite)

Electron transport (chain) A series of proteins and chemicals arranged in the membrane of a cell (or mitochondrial and chloroplast inner membranes), which conduct electrons from "food molecules" to another substance, such as oxygen or nitrate

Ellipsoidal variation Variations in the shape of a star caused by the gravitational pull of a planet

El Niño A season or two long reversal of winds in the Pacific, driven by momentum transfer across the Pacific Ocean

Endemism A description of living organisms that are found in a particular geographical location

Endosymbiosis The process through which mitochondria and chloroplasts arose in modern eukaryote cells

Entropy A measure of the number of particle microstates, often confused with disorder and linked to their energy and number

Episodic tectonics A hypothetical state where plate tectonics stops and starts on a telluric world. Venus may exhibit this form of tectonism

Equatorial Rossby wave A particular form of atmospheric wave that moves predominantly from east to west along the equator and consists of pairs of alternating low- and high-pressure regions

E-type diamonds Diamonds that bear the chemical signature of eclogite

Eukaryote Any complex cell that contains a variety of internal structures

Evolution A generic terms given to the progressive increase in complexity of life on Earth. This can be measured in terms of species richness (how many species there are) or in terms of their individual structural complexity

EUV Extreme ultraviolet—ultraviolet radiation of very short wavelength and overlapping the longest wavelengths of X-rays; emitted by very hot material

Extreme horizontal branch (EHB) A region on the HR diagram where hot, low mass stars are located, are burning helium and have a very thin canopy of hydrogen

Faint Sun paradox The problem of keeping the infant Earth from freezing when the Sun was much dimmer than it is now

Far UV Ultraviolet wavelengths close to those of visible light

FEBs (falling evaporating bodies) Comet-like bodies observed in the protoplanetary discs of young stars, likely consisting of protoplanetary mass icy materials that are being irradiated and evaporated by their star

Flare An energetic eruption on the surface (or just above the photosphere) of a star, where magnetic fields have "shorted-out" and released their energy

Foehn (Föhn) A warm wind analogous to the Chinook, where dry air warms at the dry adiabatic lapse rate as it descends a mountain slope

Front A distinct boundary between different air masses with different temperatures and/or relative humidity

Gaia James Lovelock's idea that planetary climate is altered and stabilized by feedback between the atmosphere and living organisms. A "soft Gaia," whereby living organisms contribute to climate is preferable to one where it has a dominant role.

Gap wind Any wind associated with airflow across mountain chains, through gaps in the range(s): e.g., Mistral

Gondwana (Gondwanaland) A supercontinent that persisted from 550 to 200 million years ago

Grand Tack hypothesis, the The idea that Jupiter migrated close to the Sun as it was forming, until an encounter with the infant Saturn drew it back outwards again

Granite A light-colored igneous rock with high concentrations of silicon dioxide (silica); the intrusive equivalent of rhyolite

Granitoid Any silica-rich rock that is chemically similar to granite

Gravitational lensing The effect of the curvature of space-time on the path of light through space

Great Oxidation Event (GOE) The period beginning approximately 2.45 billion years ago, where the atmosphere and oceans began to fill with oxygen, and much of the ferrous iron in the oceans precipitated as ferric iron onto the ocean floor

Greenstone A metamorphosed igneous ultrabasic (highly magnesium-rich) rock found in Archaean and early Proterozoic continental regions. Somewhat analogous to (and often confused with) modern-day ophiolites

Hadley (circulation) A simple circulation model where warm air rises over the regions of strongest heating, spreads out and cools at height, then descends and flows back towards its source

Heat pipe (tectonics) A means whereby a hot planet or satellite loses heat through volcanism. Io is the prototype heat-pipe world.

Helium fusion Three helium nuclei fuse to form carbon nuclei; or four fuse to form oxygen nuclei

Hemispheric tectonics A hypothetical form of plate tectonics that affects one hemisphere of a planet only. Perhaps common on hot telluric planets that are tidally locked to their star

Hertzsprung–Russell diagram The eponymous diagram relating stellar temperature to luminosity. Often abbreviated HR diagram.

Hoggar Part of the central Sahara, where isolated volcanism occurred recently

Hollandite A compound of calcium, manganese and oxygen that forms the core of the oxygen-evolving complex in photosynthesis

Horizontal branch A region on the HR diagram where stars are burning helium

Hot Jupiter A massive, gaseous world with a short period orbit around its star, leaving it tidally locked and very hot

Hydrogen fusion A number of processes that combine hydrogen nuclei to make helium nuclei

Hydronium (ion) An ionic form of hydrogen, consisting of three hydrogen nuclei and two electrons

Hypervolume A term used to describe a biological niche in terms of three or more variables that control its size

Ignimbrite A rock produced by pyroclastic flows

Induction heating A process through which planetary interiors are heated by the effects of stellar magnetic fields that conduct through them

Ingassing/regassing The return of volatiles by a variety of processes to the interior of a planet; plate tectonics should be the most efficient of these

Insolation A measure of the amount of energy received by a planet's surface

Intertropical Convergence Zone (ITCZ) The region of convergence of the trade winds, where convection is strongest

Inversion (Temperature) A region of stable or increasing atmospheric temperature with increasing altitude

Iron sulfide protein A protein that contains a compound of iron and sulfur and is used to conduct electrons

Jet stream A relatively narrow band of high velocity winds in the middle-upper troposphere of a planet caused by rapid variation in temperature and the Coriolis effect

Katabatic wind A cold wind that falls under gravity from high altitude, frigid areas

K-dwarf An orange main sequence star with an intermediate mass between class M and G (approximately 0.6–0.8 solar masses)

Keel tectonics A hypothetical form of plate tectonics where movement of plates is driven by the force of convecting mantle on the lithospheric roots of continental crust

Kelvin wave An atmospheric or ocean wave that moves from west to east. Atmospheric waves are likely driven by convection and the effects of rain-driven cooling

Komatiite An ultrabasic, magnesium-rich extrusive volcanic rock

Kozai-Lidov mechanism A mechanism of planetary migration, driven by an exchange of orbital eccentricity

KREEP An acronym for potassium (K), REE (rare-earth elements) and P (for phosphorus)

Lake Baikal Earth's deepest lake, which lies in an area of continental rifting that is driven by the collision of India and Asia

Late Heavy Bombardment The term given to an era where Earth, the Moon and other terrestrial planets may have been pummeled by cometary bodies roughly 4.2–3.9 billion years ago

L-dwarf An object that is cooler than an M-class dwarf. Some are true, very low mass stars (0.08–0.075 solar masses) while others are young brown dwarfs that will cool and fade

Lee cyclogenesis The formation of low pressure areas (cyclones) in the lee of mountains

Lee cyclone Any low pressure area that has formed in the wake of a mountain

Lithosphere The upper, solid portion of a planet, consisting of the crust and frozen upper mantle

Large Low Shear Velocity Provinces (LLSVP) Extensive regions with lower than average seismic velocities and which correspond to regions of upwelling in the mantle

Lokiarchaeota A new sub-class of archaeal cell that is likely the derived form of the direct ancestors of eukaryote cells (see Asgard)

MACHOs Massive, compact halo objects—a proposed massive component of dark matter

M-dwarf A red dwarf; a star with a mass between 0.08 and 0.6 solar masses

Main sequence The region on the HR diagram where stars spend most of their lives and fuse hydrogen into helium

Mantle transition zone The region lying between the upper and lower mantle, which is approximately 440–660 km beneath Earth's surface

Meiosis A form of cell division that produces gametes (sex cells)

Mei-Yu front A near-stationary boundary between northward moving tropical air and eastward moving temperate air, which lies over east Asia during the summer months

Meriodonal The movement of air along roughly longitudinal lines (see zonal)

Metagenomics The use of computing and DNA sequencing on collections of material in an unsorted sample, e.g., sediment containing hundreds of different species of microbe. Computing is used to determine which DNA came from which species and assemble the information correctly.

Microlensing The gravitational effect of a planet, or other small body, on the intensity of radiation received on Earth

Microstates A term given to the determined or probabilistic properties of particles (see entropy)

Mistral A cold valley wind found in the winter and spring months in southern France (a form of gap wind)

Moist greenhouse A term used to describe a planet warmed to relatively high temperatures by the heat-trapping effects of water vapor. A dry greenhouse is one driven largely or wholly by the effect of carbon dioxide

Monsoon A seasonal change in airflow on a planet caused by uneven heating of the surface

Mountain wind Any wind generated by airflow between, down or up mountains

Natural selection The processes whereby the environment selects for the survival of particular species or individuals on the basis of their characteristics

Neoproterozoic The latest Proterozoic era, extending from 1 billion to 541 million years ago and including the Ediacaran

Net Primary Productivity (NPP) A measure of the increase in biomass that takes into account the effects of photosynthesis and respiration

Niche A term referring the conditions in which an organism survives

Niche volume (See hypervolume)

Nitrate An ionic compound of nitrogen and oxygen

Nitrogen A relatively common element found in many biomolecules

Nitrogen cycle The biological and non-biological processes that remove and return nitrogen gas to the atmosphere of a planet

Nucleobase A compound of nitrogen, carbon, hydrogen and often oxygen that is used to store information in organisms

Oldoinyo Lengai A fascinating volcano in eastern Africa that erupts sodium carbonate ash and lava

Olivine A common iron-magnesium silicate, which is critical to the formation of biomolecules in oceanic hydrothermal systems

Ophiolite A "fossil" of oceanic crust that is swept up and preserved in continental collisions

Orbital period The time it takes a planet to orbit its star

Orographic rainfall Rainfall caused by the forced ascent and cooling of moist air masses over mountains or upland areas

Oxygen-evolving complex The enzyme that contains hollandite and releases oxygen from the oxidation of water in some photosynthetic organisms

Paleo-Tethys Ocean An ocean that existed between Gondwanaland and Laurasia in the Permian; this followed the Rheic Ocean and preceded the Tethys

Pampian planet A planet that is relatively dry overall; Earth is Pampian

Pangaea The most recent supercontinent on Earth

PAR (Photosynthetically Active Radiation) The wavelengths of radiation that are used by photosynthesis

Pebble accretion The process through which planets may grow by the slow accretion of small, pebble-sized rocks, perhaps in a dense gaseous disc

Peridotite The intrusive equivalent of Komatiite and the most common dense rock in the upper mantle

Phosphate An ionic compound of phosphorous and oxygen, which is integral to energy storage and signaling in cells; and a component of RNA and DNA

Phosphate cycle A series of biological and non-biological processes that cycle phosphorous on Earth

Photometry The process of planetary detection involving direct imaging and measurements of the brightness of a star

Photosynthesis A series of related chemical reactions that use light (or radiant heat) to power the capture and reduction of carbon dioxide gas

Plate tectonics The processes through which a planet's surface may be divided into varying semi-rigid sections that move independently of one another. Terrestrial planet tectonics is driven by subduction of dense, oceanic lithosphere

Plume (Mantle) A narrow upwelling of hot mantle, most likely from the core-mantle boundary

Porcellarum Basin A large interconnected series of basins on the Moon's surface that are underlain by a fossilized super-plume

Porphyrin A ring molecule that forms the core of chlorophyll and various other molecules

Proterozoic (era) The geological era that spans 2.45–0.54 billion years ago

Protoplanet An object, measuring up to a few hundred kilometers across that are the nuclei of terrestrial planets. Ceres is a battered analogous body

Protoplanetary disc The disc of gas and dust surrounding a protostar

Protostar A collapsing and heating sub-stellar object that is yet to begin sustained nuclear reactions

Proxima Centauri The nearest star to ours and part of the Alpha Centauri triple-star system

Pseudo-synchronous rotation A type of planetary rotation, where the orbital period is not quite aligned with the planet's rotation period, often in a 3:2 ratio. Mercury exhibits this style of rotation.

P-type diamonds Diamonds bearing the chemical signature of peridotite

Purple/yellow edge A hypothetical absorption pattern that would be visible on a planet where bacteriorhodopsin was the principal photosynthetic material instead of chlorophyll

Pyrite A compound of iron and sulfur and the functional core of iron-sulfur proteins

Pyroxene A silicate that contains calcium or magnesium

Quantitative Habitability Theory (QHT) A quantitative (mathematical) description of a planet's propensity to host life

Radial velocity Measurements of changes in the determined motion of a star around the center of gravity in a planetary system

Rain shadow The region lying in the lee of mountains where air has dried and now descends onto neighboring plains. Little or no precipitation falls.

Red edge The dividing wavelength around 700 nm on Earth, where absorption from chlorophyll ceases

Red giant A large, evolved cool star

Re-gassing (See in-gassing)

Red Queen hypothesis The concept that an organism must continue to adapt and change or it will become extinct

Respiration A series of related reactions that release chemical energy from food molecules

Rhines rotation A form of atmospheric circulation that is determined by the radius of a planet, its spin, and the ability of air to form eddies

Rhyolite The extrusive equivalent of granite or granitoid rock

Ringwoodite A dense silicate found near or in the mantle transition zone that could hold mantle water

Roll-back (of subduction) The process of the retreat of a subduction zone from the original margin, driven by subduction that is occurring faster than the overriding plate can move

RNA (ribonucleic acid) A molecule consisting of ribose, phosphate and four nucleobases (adenine, cytosine, guanine and uracil)

Rodina The Neoproterozoic supercontinent

Rossby wave A packet of waves that moves predominantly from east to west in a planet's atmosphere

Saharan metacraton The continental lithosphere underlying the eastern Sahara

Sequestration The removal of volatile substances from an atmosphere by the surface of a planet

Serpentine A water-altered form of olivine that is abundant in oceanic lithosphere

Serpentinite A rock that consists mostly of serpentine

SETI The organization dedicated to the search for extra-terrestrial intelligence

Shannon entropy (of information) A measure of the information content and its propensity to change upon transmission from one area to another; the effective equivalent of entropy of particle microstates

Shannon Diversity Index A means of determining the species complexity in an area

Shannon Landscape Diversity Index The analogous measure of landscape richness and distribution to the Species Diversity Index (above)

Shield The generic term given to a stable piece of continental crust (analogous to craton)

Single-layer convection (mantle) A model pattern of convection where the mantle is only heated from within and cools by conduction at its top

Speciation The formation of new species through natural selection

Species An organism that is able to form fertile offspring upon reproduction

Species richness The numbers or types of species in an area

Squishy lid A type of stagnant lid tectonics where the lithosphere is relatively warm and soft and intruded by copious amounts of magma. Likely on the early Earth and possibly also present on Venus

Stagnant lid Any planetary lithosphere that is stable and exhibits no regional (lateral) movement
Standard Primary Habitability (SPH) A measure of planetary habitability (see Quantitative Habitability Theory)
Star spot A cooler region on the surface of a star that has energy transport limited by magnetic fields
Stationary front A boundary between air masses that exhibits only limited north-south or east-west movement
Subduction The process through which colder, denser lithosphere returns to the mantle
Subdwarf B (sdB) A hot, usually helium-burning, star on the horizontal branch of the HR diagram
Subdwarf M (sdM) A low metallicity (and ancient) M-class star that is slightly hotter than corresponding stars of solar composition
Subdwarf O (sdO) A hotter, bluer version of an sdB star
Sub-stellar point (SSP) The point on a planet's surface where its star is directly overhead
Super-deep diamonds Diamonds that originate in deeply subducted material
Super-plume A generic term for an LLSVP and where hotter and lower density mantle is ascending
Super-rotation The movement of airflow in the same sense of direction as a planet rotates but faster than the velocity at the equator
Super-rotator A planet in which the dominant means of air flow is super-rotation
Super-terran A telluric planet with a mass a few times that of Earth
Synchronous rotation Where one body rotates on its axis at the same rate it orbits a larger (more massive) body
Tectonic Activity Index (TAI) A measure of a planet's surface tectonism
Tectonism The movement of areas of rock in the lithosphere
Teegarden's Star A nearby low mass red dwarf thought to host at least two planets, one of which may be habitable
Tehuantepecer A gap wind originating in Mexico and extending into the Pacific
Telluric A term for a rocky Earth/super-Earth/super-terran sized planet
Temperature inversion (See inversion)
Tharsis A large, volcanic domed region on Mars
Tidal heating The process through which the gravitational pull of a larger body on a smaller one kneads and heats its interior as it orbits
Tidal locking The process through which a planet (or its satellite) alters its rotation period to match its orbital one
Trade wind The northeasterly (northern hemisphere) and southeasterly (southern hemisphere) winds that blow between the Horse Latitudes and the Doldrums and form the return leg of our planet's Hadley cells
Transit The movement of a planet across the face of its star, temporarily obscuring part of its surface
Transit Duration Variation (TDV) The length of transits will vary as a transiting

planet either slows down or accelerates in its orbital motion, with respect to the center of gravity of the system.

Transit Timing Variation (TTV) The time at which a planet transits its star will vary if another planet pulls on the star and the transiting planet. Importantly, this effect can be used to determine the mass of both the seen and unseen planets.

True Polar Wander (TPW) The bulk movement of the lithosphere over the mantle or core of the planet over a relatively short period of time

TTG (Tonalite-Trondhjemite-Granodiorite) A type of granitoid rock produced by the melting of eclogite

Type I migration Migration of a planet driven by the gravitational pull of gas and dust interior to the embryo planet

Type II migration Planetary migration caused by gravitational interactions of a massive planet with its gas disc and the steady accretion of gas and dust onto the protostar

Type III migration Planetary migration caused by uneven accretion of material onto an embryo planet, with material moving in a horseshoe-shaped path along the orbit of the planet

Ultra-short period planets Planets with orbital periods measured in hours rather than days

UV Ceti stars A type of variable red dwarf star that emits powerful flares

Wadsleyite A dense silicate mineral that may be found in the mantle transition zone and that may contain water

Water An abundant, universal solvent, likely critical for all life

Water-belt (planet) A planet analogous to a "slush-ball" Earth, where most of the planet is frozen but a thawed, ice-free equatorial belt exists. TRAPPIST-1e may be such a world (but with a spot, rather than a belt, which remains ice-free)

Weathering Any process through which rocks are degraded by the action of chemicals, or meteorological conditions

WIMPs Another 1980s acronym for dark matter, this time consisting of small and as yet hypothetical particles

W UMa variables A type of variable star where both stars are in contact with one another

X-class flare A powerful stellar flare

Zonal flow The movement of air in a broadly west–east (prograde) or east–west (retrograde) direction

Index

D. S. Stevenson, *Red Dwarfs*, https://doi.org/10.1007/978-3-030-25550-3

The manufacturer's authorised representative in the EU is Springer Nature Customer Service Centre GmbH, Europaplatz 3, 69115 Heidelberg, Germany. If you have any concerns regarding our products, please contact ProductSafety@springernature.com

Printed and bound by CPI Group (UK) Ltd, Croydon, CR0 4YY
05/10/2026
02228130-0013